Structural Biology

Quincy Teng

Structural Biology

Practical NMR Applications

Second Edition

Quincy Teng
Department of Pharmaceutical
and Biomedical Sciences
College of Pharmacy
University of Georgia
Athens, GA, USA

ISBN 978-1-4614-3963-9   ISBN 978-1-4614-3964-6 (eBook)
DOI 10.1007/978-1-4614-3964-6
Springer New York Heidelberg Dordrecht London

Library of Congress Control Number: 2012942118

© Springer Science+Business Media New York 2013
This work is subject to copyright. All rights are reserved by the Publisher, whether the whole or part of the material is concerned, specifically the rights of translation, reprinting, reuse of illustrations, recitation, broadcasting, reproduction on microfilms or in any other physical way, and transmission or information storage and retrieval, electronic adaptation, computer software, or by similar or dissimilar methodology now known or hereafter developed. Exempted from this legal reservation are brief excerpts in connection with reviews or scholarly analysis or material supplied specifically for the purpose of being entered and executed on a computer system, for exclusive use by the purchaser of the work. Duplication of this publication or parts thereof is permitted only under the provisions of the Copyright Law of the Publisher's location, in its current version, and permission for use must always be obtained from Springer. Permissions for use may be obtained through RightsLink at the Copyright Clearance Center. Violations are liable to prosecution under the respective Copyright Law.
The use of general descriptive names, registered names, trademarks, service marks, etc. in this publication does not imply, even in the absence of a specific statement, that such names are exempt from the relevant protective laws and regulations and therefore free for general use.
While the advice and information in this book are believed to be true and accurate at the date of publication, neither the authors nor the editors nor the publisher can accept any legal responsibility for any errors or omissions that may be made. The publisher makes no warranty, express or implied, with respect to the material contained herein.

Printed on acid-free paper

Springer is part of Springer Science+Business Media (www.springer.com)

Preface

The second edition of "Structural Biology: Practical NMR Applications" retains the focus of the previous edition, which is to provide readers with a systematic understanding of fundamental principles and practical aspects of NMR spectroscopy. At the beginning of each section, questions and objectives highlight key points to be learned, and homework problems are provided at the end of each chapter, except Chap. 9. One hundred multiple-choice questions with answers are provided to further aid in understanding the topics. In response to comments and suggestions from readers, and based on my own research and teaching experiences, I have made improvements and added a new chapter (Chap. 9) on metabolomics, which is an important application of NMR spectroscopy.

Over the years since NMR was first applied to solve problems in structural biology, it has undergone dramatic developments in both NMR instrument hardware and methodology. While it is established that NMR is one of the most powerful tools for understanding biological processes at the atomic level, it has become increasingly difficult for authors and instructors to make valid decisions concerning the content and level for a graduate course of NMR spectroscopy in structural biology. Because many of the details in practical NMR are not documented systematically, students entering into the field have to learn the experiments and methods through communication with other experienced students or experts. Often such a learning process is incomplete and unsystematic. This book is meant to be not only a textbook but also a handbook for those who routinely use NMR to study various biological systems. Thus, the book is organized with experimentalists in mind, whether they are instructors or students. For those who have a little or no background in NMR structural biology, it is hoped that this book will provide sufficient perspective and insight. Those who already have NMR research experience may find new information or different methods that are useful to their research.

Because understanding fundamental principles and concepts of NMR spectroscopy is essential for the application of NMR methods to research projects, the book begins with an introduction to basic NMR principles. While detailed mathematics and quantum mechanics dealing with NMR theory have been addressed in several well-known NMR books, Chap. 1 illustrates some of the fundamental principles and concepts of NMR spectroscopy in a more descriptive and straightforward manner. Such questions as, "How is the NMR signal generated? How do nuclear spins behave during and after different radio-frequency pulses? What is the rotating frame, and why do we need it?" are addressed in Chap. 1. Next, NMR instrumentation is discussed starting with hardware components. Topics include magnetic field homogeneity and stability, signal generation and detection, probe circuits, cryogenic probes, analog-to-digital conversion, and test equipment. A typical specification for an NMR spectrometer is also included in the chapter. There is also a chapter covering NMR sample preparation, a process that is often the bottleneck for the success of the NMR project. Several routine strategies for preparing samples for macromolecules as well as complexes are dealt with in detail.

Chapter 4 discusses the practical aspects of NMR, including probe tuning, magnet shimming and locking, instrument calibrations, pulse field gradients, solvent suppression, data acquisition and processing, and homonuclear two-dimensional experiments. In Chap. 5, experiments that are routinely used in studying biological molecules are discussed. Questions to be addressed include how the experiments are setup and what kind of information we can obtain from the experiments.

The next chapter focuses on the application of NMR techniques to the study of biological molecules. The use of NMR in studying small biological molecules such as ligands, drugs, and amino acids involved in different biological pathways is covered. Then, applications in studies of macromolecules such as proteins, protein–peptide, and protein–protein complexes are discussed in Chap. 7. Chapter 8 deals with dynamics of macromolecules, important information that can be obtained uniquely by NMR methods.

Chapter 9 discusses essential principles and applications of NMR-based metabolomics. First, fundamentals of multivariate analysis are addressed in a simple and easily understood manner. The next section focuses on sample preparation, which includes detailed procedures and protocols on collecting and preparing biofluid samples, quenching cells and tissues, and extracting metabolites from cells and tissues. Practical aspects of NMR experiments routinely used in metabolomics are also discussed in detail, including experimental setup, data processing and interpretation. In the next section, a number of examples are worked out in detail to illustrate statistical analyses of NMR data and interpretation of the statistical models. Several protocols for using software packages for multivariate analysis are also provided in this section. The last four sections focus on applications of NMR-based metabolomics, including metabolomics of biofluids, cellular metabolomics, live cells, and applications to cancer research.

I would like to thank many colleagues who have used the previous edition in their teaching, and those who have contributed directly or indirectly to this book. I am particularly grateful to Dr. Jun Qin for writing sections of Chaps. 3 and 7, and for numerous discussions, and Drs. Kristen Mayer, Weidong Hu, Steve Unger, Fang Tian, John Glushka, Chalet Tan, Drew Ekman, and Timothy Collette for reviewing all or part of the text and providing corrections, valuable comments, and encouragement. I am appreciative of the investigators and publishers who have allowed me to use their figures in this text. I am also indebted to the editors and staff at Springer Science+Business Media, especially to Editor Portia Formento, Senior Editor Andrea Macaluso, and Senior Publication Editors Patrick Carr and Felix Portnoy, who made this book happen.

Athens, GA, USA   Quincy Teng

Contents

1 Basic Principles of NMR ... 1
  1.1 Introduction ... 1
  1.2 Nuclear Spin in a Static Magnetic Field ... 2
    1.2.1 Precession of Nuclear Spins in a Magnetic Field ... 2
    1.2.2 Energy States and Population ... 4
    1.2.3 Bulk Magnetization ... 5
  1.3 Rotating Frame ... 6
  1.4 Bloch Equations ... 11
  1.5 Fourier Transformation and Its Applications in NMR ... 13
    1.5.1 Fourier Transformation and Its Properties Useful for NMR ... 13
    1.5.2 Excitation Bandwidth ... 15
    1.5.3 Quadrature Detection ... 17
  1.6 Nyquist Theorem and Digital Filters ... 18
  1.7 Chemical Shift ... 20
  1.8 Nuclear Coupling ... 26
    1.8.1 Scalar Coupling ... 26
    1.8.2 Spin Systems ... 29
    1.8.3 Dipolar Interaction ... 30
    1.8.4 Residual Dipolar Coupling ... 31
  1.9 Nuclear Overhauser Effect ... 36
  1.10 Relaxation ... 39
    1.10.1 Correlation Time and Spectral Density Function ... 40
    1.10.2 Spin–Lattice Relaxation ... 41
    1.10.3 T_2 Relaxation ... 44
  1.11 Selection of Coherence Transfer Pathways ... 47

	1.12	Approaches to Understanding NMR Experiments	48
		1.12.1 Vector Model	48
		1.12.2 Product Operator Description of Building Blocks in a Pulse Sequence	49
		1.12.3 Introduction to Density Matrix	54
	Appendix A: Product Operators		59
	References		61
2	**Instrumentation**		65
	2.1	System Overview	65
	2.2	Magnet	65
	2.3	Transmitter	69
	2.4	Receiver	74
	2.5	Probe	77
	2.6	Quarter-Wavelength Cable	85
	2.7	Analog/Digital Converters	87
	2.8	Instrument Specifications	90
	2.9	Test or Measurement Equipment	92
		2.9.1 Reflection Bridge	92
		2.9.2 Oscilloscope	93
		2.9.3 Spectrum Analyzer	96
		2.9.4 System Noise Measurement	98
	References		101
3	**NMR Sample Preparation**		103
	3.1	Introduction	103
	3.2	Expression Systems	104
		3.2.1 Escherichia coli Expression Systems	104
		3.2.2 Fusion Proteins in the Expression Vectors	105
		3.2.3 Optimization of Protein Expression	106
	3.3	Overexpression of Isotope-Labeled Proteins	107
	3.4	Purification of Isotope-Labeled Proteins	108
	3.5	NMR Sample Preparation	109
		3.5.1 General Considerations	109
		3.5.2 Preparation of Protein–Peptide Complexes	109
		3.5.3 Preparation of Protein–Protein Complexes	110
		3.5.4 Preparation of Alignment Media for Residual Dipolar Coupling Measurement	111
	3.6	Examples of Protocols for Preparing ^{15}N/^{13}C Labeled Proteins	113
		3.6.1 Example 1: Sample Preparation of an LIM Domain Using Protease Cleavage	113
		3.6.2 Example 2: Sample Preparation Using a Denaturation–Renaturation Method	114
	References		115

4 Practical Aspects ... 117

- 4.1 Tuning the Probe ... 118
- 4.2 Shimming and Locking ... 120
- 4.3 Instrument Calibrations ... 123
  - 4.3.1 Calibration of Variable Temperature ... 123
  - 4.3.2 Calibration of Chemical Shift References ... 124
  - 4.3.3 Calibration of Transmitter Pulse Length ... 125
  - 4.3.4 Calibration of Offset Frequencies ... 127
  - 4.3.5 Calibration of Decoupler Pulse Length ... 130
  - 4.3.6 Calibration of Decoupler Pulse Length with Off-Resonance Null ... 132
- 4.4 Selective Excitation with Narrow Band and Off-Resonance Shaped Pulses ... 133
- 4.5 Composite Pulses ... 136
  - 4.5.1 Composite Excitation Pulses ... 136
  - 4.5.2 Composite Pulses for Isotropic Mixing ... 136
  - 4.5.3 Composite Pulses for Spin Decoupling ... 137
- 4.6 Adiabatic Pulses ... 139
- 4.7 Pulsed Field Gradients ... 141
- 4.8 Solvent Suppression ... 145
  - 4.8.1 Presaturation ... 146
  - 4.8.2 Watergate ... 147
  - 4.8.3 Water Flip-Back ... 148
  - 4.8.4 Jump-Return ... 149
- 4.9 NMR Data Processing ... 150
  - 4.9.1 DC Drift Correction ... 150
  - 4.9.2 Solvent Suppression Filter ... 150
  - 4.9.3 Linear Prediction ... 151
  - 4.9.4 Apodization ... 151
  - 4.9.5 Zero Filling ... 154
  - 4.9.6 Phase Correction ... 154
- 4.10 Two-Dimensional Experiments ... 158
  - 4.10.1 The Second Dimension ... 158
  - 4.10.2 Quadrature Detection in the Indirect Dimension ... 159
  - 4.10.3 Selection of Coherence Transfer Pathways ... 161
  - 4.10.4 COSY ... 163
  - 4.10.5 DQF COSY ... 164
  - 4.10.6 TOCSY ... 166
  - 4.10.7 NOESY and ROESY ... 167
- References ... 171

5 Multidimensional Heteronuclear NMR Experiments ... 173
5.1 Two-Dimensional Heteronuclear Experiments ... 173
5.1.1 HSQC and HMQC ... 174
5.1.2 HSQC Experiment Setup ... 176
5.1.3 Sensitivity-Enhanced HSQC by PEP ... 177
5.1.4 Setup of seHSQC Experiment ... 179
5.1.5 HMQC ... 180
5.1.6 IPAP HSQC ... 182
5.1.7 SQ–TROSY ... 184
5.2 Overview of Triple-Resonance Experiments ... 187
5.3 General Procedure of Setup and Data Processing for 3D Experiments ... 189
5.4 Experiments for Backbone Assignments ... 190
5.4.1 HNCO and HNCA ... 191
5.4.2 HN(CO)CA ... 198
5.4.3 HN(CA)CO ... 201
5.4.4 CBCANH ... 203
5.4.5 CBCA(CO)NH ... 206
5.5 Experiments for Side-Chain Assignment ... 209
5.5.1 HCCH–TOCSY ... 209
5.6 3D Isotope-Edited Experiments ... 214
5.6.1 ^{15}N-HSQC–NOESY ... 214
5.7 Sequence-Specific Resonance Assignments of Proteins ... 216
5.7.1 Assignments Using ^{15}N-Labeled Proteins ... 216
5.7.2 Sequence-Specific Assignment Using Doubly Labeled Proteins ... 217
5.8 Assignment of NOE Cross Peaks ... 218
References ... 219

6 Studies of Small Biological Molecules ... 223
6.1 Ligand–Protein Complexes ... 223
6.1.1 SAR-by-NMR Method ... 224
6.1.2 Diffusion Method ... 227
6.1.3 Transferred NOE ... 229
6.1.4 Saturation Transfer Difference ... 232
6.1.5 Isotope-Editing Spectroscopy ... 233
6.1.6 Isotope-Filtering Spectroscopy ... 235
6.2 Study of Metabolic Pathways by NMR ... 238
References ... 244

7 Protein Structure Determination from NMR Data ... 247
7.1 Introduction and Historical Overview ... 247
7.2 NMR Structure Calculation Methods ... 249
7.2.1 Distance Geometry ... 250
7.2.2 Restrained Molecular Dynamics ... 251

7.3	NMR Parameters for Structure Calculation		253
	7.3.1	Chemical Shifts	253
	7.3.2	J Coupling Constants	254
	7.3.3	Nuclear Overhauser Effect	255
	7.3.4	Residual Dipolar Couplings	256
7.4	Preliminary Secondary Structural Analysis		258
7.5	Tertiary Structure Determination		259
	7.5.1	Computational Strategies	259
	7.5.2	Illustration of Step-by-Step Structure Calculations Using a Typical XPLOR Protocol	260
	7.5.3	Criteria of Structural Quality	264
	7.5.4	Second-Round Structure Calculation: Structure Refinement	265
	7.5.5	Presentation of the NMR Structure	265
	7.5.6	Precision of NMR Structures	266
	7.5.7	Accuracy of NMR Structures	267
7.6	Protein Complexes		267
	7.6.1	Protein–Protein Complexes	267
	7.6.2	Protein–Peptide Complexes	268

Appendix B: sa.inp—Xplor Protocol for Protein Structure Calculation ... 269
Appendix C: Example of NOE Table ... 276
Appendix D: Example of Dihedral Angle Restraint Table ... 278
Appendix E: Example of Chemical Shift Table for Talos ... 281
Appendix F: Example of Hydrogen Bond Table ... 283
Appendix G: Example of Input File To Generate A Random-Coil Coordinates ... 284
Appendix H: Example of Input File to Generate a Geometric PSF File ... 285
References ... 286

8 Protein Dynamics ... 289
8.1 Theory of Spin Relaxation in Proteins ... 290
8.2 Experiments for Measurement of Relaxation Parameters ... 298
    8.2.1 T_1 Measurement ... 298
    8.2.2 T_2 and $T_{1\rho}$ Measurements ... 302
    8.2.3 Heteronuclear NOE Measurement ... 304
8.3 Relaxation Data Analysis ... 306
References ... 308

9 NMR-Based Metabolomics ... 311
9.1 Introduction ... 311
9.2 Fundamentals of Multivariate Statistical Analysis for Metabolomics ... 313

9.3	Sample Preparation		321
	9.3.1	Phosphate Buffer for NMR Sample Preparation	321
	9.3.2	Urine Samples	322
	9.3.3	Blood Plasma and Serum	323
	9.3.4	Tissue Sample Quench, Storage, and Extraction	324
	9.3.5	Culture Adherent Cells	326
	9.3.6	Quench and Extract Cells	328
9.4	Practical Aspects of NMR Experiments		332
	9.4.1	Calibration	332
	9.4.2	Automation	332
	9.4.3	NMR Experiments	336
9.5	Dada Analysis and Model Interpretation		343
	9.5.1	NMR Dada Processing and Normalization	344
	9.5.2	Analysis of Metabolomic Data	351
9.6	Metabolomics of Biofluids		361
9.7	Cellular Metabolomics		370
	9.7.1	Experiments	370
	9.7.2	NMR Data Processing	373
	9.7.3	Principal Component Analysis	373
	9.7.4	Partial Least Squares for Discriminant Analysis	376
9.8	Metabolomics of Live Cell		377
9.9	Metabolomics Applied to Cancer Research		382
	9.9.1	Silibinin Anticancer Efficacy	383
	9.9.2	Metabolomic Profiling of Colon Cancer	385
References			388

Appendix I: Multiple Choice Questions 393

Appendix J: Nomenclature and Symbols 413

Index 417

Chapter 1
Basic Principles of NMR

1.1 Introduction

Energy states and population distribution are the fundamental subjects of any spectroscopic technique. The energy difference between energy states gives raise to the frequency of the spectra, whereas intensities of the spectral peaks are proportional to the population difference of the states. Relaxation is another fundamental phenomenon in nuclear magnetic resonance spectroscopy (NMR), which influences both line shapes and intensities of NMR signals. It provides information about structure and dynamics of molecules. Hence, understanding these aspects lays the foundation to understanding basic principles of NMR spectroscopy.

In principle, an NMR spectrometer is more or less like a radio. In a radio, audio signals in the frequency range of kilohertz are the signals of interest, which one can hear. However, the signals sent by broadcast stations are in the range of 100 MHz for FM and of up to 1 GHz for AM broadcasting. The kilohertz audio signals must be separated from the megahertz transmission frequencies before they are sent to speakers. In NMR spectroscopy, nuclei have an intrinsic megahertz frequency which is known as the Larmor frequency. For instance, in a molecule, all protons have the same Larmor frequency. However, the signals of interest are the chemical shifts generated by the electron density surrounding an individual proton, which are in the kilohertz frequency range. Many of the protons in the molecule have different chemical environments which give different signals in the kilohertz range. One must find a way to eliminate the megahertz Larmor frequency in order to observe the kilohertz chemical shifts (more details to follow).

1.2 Nuclear Spin in a Static Magnetic Field

1.2.1 Precession of Nuclear Spins in a Magnetic Field

As mentioned above, energy and population associated with energy states are the bases of the frequency position and the intensity of spectral signals. In order to understand the principles of NMR spectroscopy, it is necessary to know how the energy states of nuclei are generated and what are the energy and population associated with the energy states.

Key questions to be addressed in this section include the following:

1. What causes nuclei to precess in the presence of magnetic field?
2. What kind of nuclei will give NMR signals?
3. How do nuclear spins orient in the magnetic field?

Not any kind of nucleus will give NMR signals. Nuclei with an even number of both charge and mass have a spin quantum number of zero, e.g., ^{12}C. These kinds of nuclei do not have nuclear angular momentum and will not give raise to NMR signal; these are called NMR inactive nuclei. For nuclei with nonzero spin quantum number, energy states are produced by the nuclear angular moment interacting with the applied magnetic field. Nuclei with nonzero spin quantum number possess nuclear angular momentum whose magnitude is determined by:

$$P = \hbar\sqrt{I(I+1)} \tag{1.1}$$

in which I is the nuclear spin quantum number and $\hbar$ is the Plank constant divided by 2π. The value of I is dependent on the mass and charge of nucleus, and it can be either an integral or half integral number. The z component of the angular momentum is given by:

$$P_z = \hbar m \tag{1.2}$$

in which the magnetic quantum number m has possible values of $I, I-1, \ldots, -I+1, -I$, and a total of $2I+1$. This equation tells us that the projection of nuclear angular momentum on the z axis is quantized in space and has a total of $2I+1$ possible values. The orientations of nuclear angular momentum are defined by the allowed m values. For example, for spin ½ nuclei, the allowed m are ½ and −½. Thus, the angular momentum of spin ½ ($I=$½) has two orientations, one is pointing up (pointing to z axis) and the other pointing down (pointing to $-z$ axis) with an angle of 54.7° relative to the magnetic field (Fig. 1.1).

The nuclei with a nonzero spin quantum number will rotate about the magnetic field B_0 due to the torque generated by the interaction of the nuclear angular momentum with the magnetic field. The magnetic moment (or nuclear moment), μ, is either parallel or antiparallel to their angular momentum:

$$\mu = \gamma P = \gamma \hbar \sqrt{I(I+1)} \tag{1.3}$$

1.2 Nuclear Spin in a Static Magnetic Field

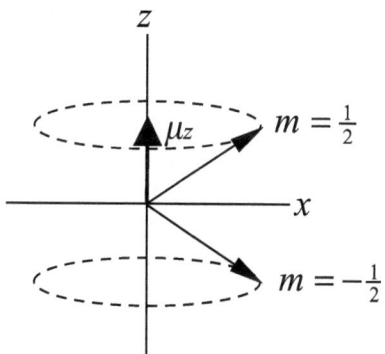

Fig. 1.1 Orientation of nuclear angular moment μ with spin ½ and its z component, μ_z. The vectors represent the angular moment μ rotating about the magnetic field whose direction is along the z axis of the laboratory frame

in which γ is the nuclear gyromagnetic ratio, which has a specific value for a given isotope. Thus, γ is a characteristic constant for a specific nucleus. The angular momentum P is the same for all nuclei with the same magnetic quantum number, whereas the angular moment μ is different for different nuclei. For instance, ^{13}C and ^{1}H have same angular momentum P because they have same spin quantum number of ½, but have different angular moments μ because they are different isotopes with different γ. Therefore, nuclear angular moment μ is used to characterize nuclear spins. The moment μ is parallel to angular momentum if γ is positive or antiparallel if γ is negative (e.g., ^{15}N). Similar to the z component of angular momentum, P_z, the z component of angular moment μ_z is given by:

$$\mu_z = \gamma P_z = \gamma \hbar m \tag{1.4}$$

The equation indicates that μ_z has a different value for different nuclei even if they may have same magnetic quantum number m. When nuclei with a nonzero spin quantum number are placed in a magnetic field, they will precess about the magnetic field due to the torque generated by the interaction of the magnetic field B_0 with the nuclear moment μ. The angle of μ relative to B_0 is dependent on m. Nuclei with nonzero spin quantum numbers are also called nuclear spins because their angular moments make them spin in the magnetic field.

In summary, the nuclear angular momentum is what causes the nucleus to rotate relative to the magnetic field. Different nuclei have a characteristic nuclear moment because the moment is dependent on the gyromagnetic ratio γ, whereas nuclei with the same spin quantum number possess the same nuclear angular momentum. Nuclear moments have quantized orientations defined by the value of the magnetic quantum number, m. The interaction of nuclei with the magnetic field is utilized to generate an NMR signal. Because the energy and population of nuclei are proportional to the magnetic field strength (more details discussed below), the frequency and intensity of the NMR spectral signals are dependent on the field strength.

1.2.2 Energy States and Population

It has been illustrated in the previous section that nuclei with nonzero spin quantum numbers orient along specific directions with respect to the magnetic field. They are rotating continuously about the field direction due to the nuclear moment μ possessed by nuclei. For each orientation state, also known as the Zeeman state or spin state, there is energy associated with it, which is characterized by the frequency of the precession.

Key questions to be addressed in this section include the following:

1. What is the energy and population distribution of the Zeeman states?
2. What are the nuclear precession frequencies of the Zeeman states and the frequency of the transition between the states, and how are they different?
3. How are energy and population related to the measurable spectral quantities?

The intrinsic frequency of the precession is the Larmor frequency ω_0. The energy of the Zeeman state with magnetic quantum number m can be described in terms of the Larmor frequency:

$$E = -\mu_z B_0 = -m\hbar\gamma B_0 = m\hbar\omega_0 \tag{1.5}$$

in which B_0 is the magnetic field strength in the unit of tesla, T, and $\omega_0 = -\gamma B_0$ is the Larmor frequency. Therefore, the energy difference in the allowed transition (the selection rule is that only single-quantum transition, i.e., $\Delta m = \pm 1$, is allowed), for instance, between the $m = -\frac{1}{2}$ and $m = \frac{1}{2}$ Zeeman states is given by:

$$\Delta E = \hbar\gamma B_0 \tag{1.6}$$

Because $\Delta E = \hbar\omega$, the frequency of the required electromagnetic radiation for the transition has the form of:

$$\omega_0 = \gamma B_0 \tag{1.7}$$

which has a linear dependence on the magnetic field strength. Commonly, the magnetic field strength is described by the proton Larmor frequency at the specific field strength. A proton resonance frequency of 100 MHz is corresponding to the field strength of 2.35 T. For example, a 600 MHz magnet has a field strength of 14.1 T. While the angular frequency ω has a unit of radian per second, the frequency can also be represented in hertz with the relationship of:

$$\nu = \frac{\omega}{2\pi} \tag{1.8}$$

As the magnetic field strength increases, the energy difference between two transition states becomes larger, as does the frequency associated with the Zeeman transition. The intensity of the NMR signal comes from the population difference of

1.2 Nuclear Spin in a Static Magnetic Field

two Zeeman states of the transition. The population of the energy state is governed by the Boltzmann distribution. For a spin ½ nucleus with a positive γ such as ^{1}H, or ^{13}C, the lower energy state (ground state) is defined as the α state for $m = ½$, whereas the higher energy state (excited state) is labeled as the β state for $m = -½$. For ^{15}N, $m = -½$ is the lower energy α state because of its negative γ. The ratio of the populations in the states is quantitatively described by the Boltzmann equation:

$$\frac{N_\beta}{N_\alpha} = e^{-\Delta E/kT} = e^{-\hbar\gamma B_0/kT} = \frac{1}{e^{\hbar\gamma B_0/kT}} \tag{1.9}$$

in which N_α and N_β are the population of α and β states, respectively, T is the temperature in Kelvin and k is the Boltzmann constant. The equation states that both the energy difference of the transition states and the population difference of the states increases with the magnetic field strength. Furthermore, the population difference has a temperature dependence. If the sample temperature reaches the absolute zero, there is no population at β state and all spins will lie in α state, whereas both states will have equal population if the temperature is infinitely high. At T near room temperature, ~300 K, $\hbar\gamma B_0 \ll kT$. As a consequence, a first-order Taylor expansion can be used to describe the population difference:

$$\frac{N_\beta}{N_\alpha} \approx 1 - \frac{\hbar\gamma B_0}{kT} \tag{1.10}$$

At room temperature, the population of the β state is slightly lower than that of the α state. For instance, the population ratio for protons at 800 MHz is 0.99987. This indicates that only a small fraction of the spins will contribute to the signal intensity due to the low energy difference and hence NMR spectroscopy intrinsically is a very insensitive spectroscopic technique. Therefore, a higher magnetic field is necessary to obtain better sensitivity, in addition to other advantages such as resolution and the TROSY effect (transverse relaxation optimized spectroscopy).

1.2.3 Bulk Magnetization

Questions to be addressed in this section include the following:

1. What is the bulk magnetization and where is it located?
2. Why do no transverse components of the bulk magnetization exist at the equilibrium?

The observable NMR signals come from the assembly of the nuclear spins in the presence of the magnetic field. It is the bulk magnetization of a sample (or macroscopic magnetization) that gives the observable magnetization, which is the vector sum of all spin moments (nuclear angular moments). Because nuclear

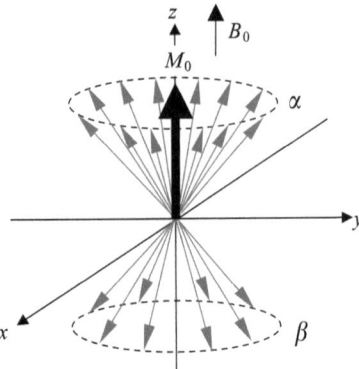

Fig. 1.2 Bulk magnetization of spin ½ nuclei with positive γ. x, y, and z are the axes of the laboratory frame. The *thin arrows* represent individual nuclear moments. The vector sum of the nuclear moments on the xy plane is zero because an individual nuclear moment has equal probability of being in any direction of the xy plane. The bulk magnetization M_0, labeled as a *thick arrow*, is generated by the small population difference between the α and β states, and is parallel to the direction of the static magnetic field B_0

spins precess about the magnetic field along the z axis of the laboratory frame, an individual nuclear moment has equal probability of being in any direction of the xy plane. Accordingly, the transverse component of the bulk magnetization at the equilibrium state is averaged to zero and hence is not observable (Fig. 1.2). The bulk magnetization M_0 results from the small population difference between the α and β states. At equilibrium, this vector lies along the z axis and is parallel to the magnetic field direction for nuclei with positive γ because the spin population in the α state is larger than that in the β state. Although the bulk magnetization is stationary along the z axis, the individual spin moments rotate about the axis.

1.3 Rotating Frame

Question to be addressed in this section include the following:

1. What is the rotating frame and why is it needed?
2. What is the B_1 field and why must it be an oscillating electromagnetic field?
3. How does the bulk magnetization M_0 react when a B_1 field is applied to it?
4. What is the relationship between radio frequency (RF) pulse power and pulse length?

The Larmor frequency of a nuclear isotope is the resonance frequency of the isotope in the magnetic field. For example, ^{1}H Larmor frequency will be 600 MHz for all protons of a sample in the magnetic field of 14.1 T. If the Larmor frequency were the only observed NMR signal, NMR spectroscopy would not be useful because there would be only one resonance signal for all ^{1}H. In fact, chemical shifts are the NMR signals of interest (details in Sect. 1.7), which have a frequency range

1.3 Rotating Frame

of kilohertz, whereas the Larmor frequency of all nuclei is in the range of megahertz. For instance, the observed signals of protons are normally in the range of several kilohertz with a Larmor frequency of 600 MHz in the magnetic field of 14.1 T. How the Larmor frequency is removed before NMR data are acquired, what the rotating frame is, why we need it and how the bulk magnetization changes upon applying an additional electromagnetic field are the topics of this section.

Since the Larmor frequency will not be present in any NMR spectrum, it is necessary to remove its effect when dealing with signals in frequency range of kilohertz. This can be done by applying an electromagnetic field B_1 along an axis on the xy plane of the laboratory frame, which rotates at the Larmor frequency with respect to the z axis of the laboratory frame. This magnetic field is used for the purposes of (a) removing the effect of the Larmor frequency and hence simplifying the theoretical and practical consideration of the spin precession in NMR experiments, and (b) inducing the nuclear transition between two energy states by its interaction with the nuclei in the sample according to the resonance condition that the transition occurs when the frequency of the field equals the resonance frequencies of the nuclei. This magnetic field is turned on only when it is needed. Because the Larmor frequency is not observed in NMR experiments, a new coordinate frame is introduced to eliminate the Larmor frequency from consideration, called the rotating frame. In the rotating frame, the xy plane of the laboratory frame is rotating at or near the Larmor frequency ω_0 with respect to the z axis of the laboratory frame. The transformation of the laboratory frame to the rotating frame can be illustrated by taking a carousel (also known as merry-go-round) as an example. The carousel observed by one standing on the ground is rotating at a given speed. When one is riding on it, he is also rotating at the same speed. However, he is stationary relative to others on the carousel. If the ground is considered the laboratory frame, the carousel is the rotating frame. When the person on the ground steps onto the carousel, it is the transformation from the laboratory frame to the rotating frame. The sole difference between the laboratory frame and the rotating frame is that the rotating frame is rotating in the xy plane about the z axis relative to the laboratory frame.

By transforming from the laboratory frame to the rotating frame, the nuclear moments are no longer spinning about the z axis, i.e., they are stationary in the rotating frame. The term "transforming" here means that everything in the laboratory frame will rotate at a frequency of $-\omega_0$ about the z axis in the rotating frame. As a result, the bulk magnetization does not have its Larmor frequency in the rotating frame. Since the applied B_1 field is rotating at the Larmor frequency in the laboratory frame, the transformation of this magnetic field to the rotating frame results in a stationary B_1 field along an axis on the xy plane in the rotating frame, for example the x axis. Therefore, when this B_1 magnetic field is applied, its net effect on the bulk magnetization is to rotate the bulk magnetization away from the z axis clockwise about the axis of the applied field by the left-hand rule in the vector representation.

In practice, the rotation of the B_1 field with respect to the z axis of the laboratory frame is achieved by generating a linear oscillating electromagnetic field with the magnitude of $2B_1$ because it is easily produced by applying electric current through

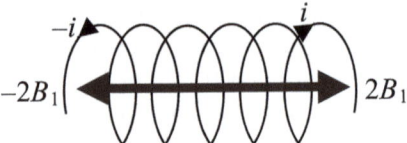

Fig. 1.3 The electromagnetic field generated by the current passing through the probe coil. The magnitude of the field is modulated by changing the current between $-i$ and $+i$. The electromagnetic field is called the oscillating B_1 field

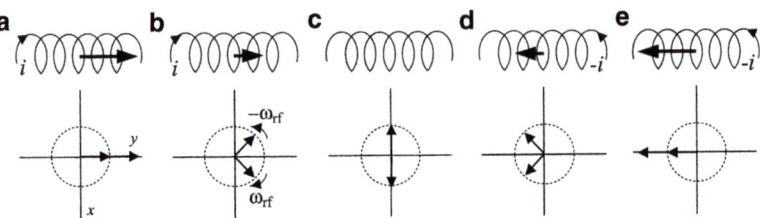

Fig. 1.4 Vector sum of the oscillating B_1 field generated by passing current through a probe coil. The magnitude of the field can be represented by two equal amplitude vectors rotating in opposite directions. The angular frequency of the two vectors is same as the oscillating frequency ω_{rf} of the B_1 field. When $\omega_{rf} = \omega_0$, B_1 is said to be on resonance. (**a**) When the current reaches the maximum, the two vectors align on y axis. The sum of the two vectors is the same as the field produced in the coil. (**b**) As the B_1 field reduces, its magnitude equals the sum of the projections of two vectors on the y axis. (**c**) When the two vectors are oppositely aligned on the x axis, the current in the coil is zero. (**d**) As the current reduces, both components rotate into the $-y$ region and the sum produces a negative magnitude. (**e**) Finally, the two components meet at the $-y$ axis, which represents a field magnitude of $-2B_1$. At any given time the two decomposed components have the same magnitude of B_1, the same frequency of ω_{rf} and are mirror image to each other

the probe coil (Fig. 1.3). The oscillating magnetic field has a frequency equal to the Larmor frequency of the nuclei. As the current increases from zero to maximum, the field proportionally increases from zero to the maximum field along the coil axis ($2B_1$ in Fig. 1.3). Reducing the current from the maximum to zero and then to the minimum (negative maximum, $-i$) decreases the field from $2B_1$ to $-2B_1$. Finally, the field is back to zero from $-2B_1$ as the current is increased from the minimum to zero to finish one cycle. If the frequency of changing the current is v_{rf}, we can describe the oscillating frequency as ω_{rf} ($\omega = 2\pi v$). Mathematically, this linear oscillating field (thick arrow in Fig. 1.4) can be represented by two equal fields with half of the magnitude, B_1, rotating in the xy plane at the same angular frequency in opposite directions to each other (thin arrows). When the field has the maximum strength at $2B_1$, each component aligns on the y axis with a magnitude of B_1. The vector sum of the two is $2B_1$ (Fig. 1.4a). When the current is zero, which gives zero in the field magnitude, each component still has the same magnitude of B_1 but aligns on the x and $-x$ axes, respectively, which gives rise to a vector sum of zero (Fig. 1.4c). As the current reduces, both components rotate into

1.3 Rotating Frame

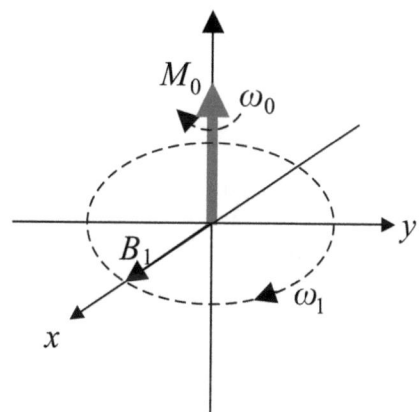

Fig. 1.5 B_1 field in the laboratory frame. The bulk magnetization M_0 is the vector sum of individual nuclear moments which are precessing about the static magnetic field B_0 at the Larmor frequency ω_0. When the angular frequency ω_{rf} of the B_1 field is equal to the Larmor frequency, that is, $\omega_{rf} = \omega_0$, the B_1 field is on resonance

the $-y$ region and the sum produces a negative magnitude (Fig. 1.4d). Finally, the two components meet at the $-y$ axis, which represents a field magnitude of $-2B_1$ (Fig. 1.4e). At any given time the two decomposed components have the same magnitude of B_1, the same frequency of ω_{rf} and are mirror image to each other.

If the frequency of the rotating frame is set to ω_{rf} which is close to the Larmor frequency ω_0, the component of the B_1 field which has ω_{rf} in the laboratory frame has null frequency in the rotating frame because of the transformation by the $-\omega_{rf}$. The other with $-\omega_{rf}$ in the laboratory frame now has an angular frequency of $-2\omega_{rf}$ after the transformation. Since the latter has a frequency far away from the Larmor frequency it will not interfere with the NMR signals which are in the range of kilohertz. Therefore, this component is ignored throughout the discussion unless specifically mentioned. The former component with null frequency in the rotating frame is used to represent the B_1 field. If we regulate the frequency of the current oscillating into the coil as ω_0, then setting ω_{rf} to equal the Larmor frequency ω_0, the B_1 field is said to be on resonance (Fig. 1.5). Since in the rotating frame the Larmor frequency is not present in the nuclei, the effect of B_0 on nuclear spins is eliminated. The only field under consideration is the B_1 field. From the earlier discussion we know that nuclear magnetization will rotate about the applied field direction upon its interaction with a magnetic field. Hence, whenever B_1 is turned on, the bulk magnetization will be rotated about the axis where B_1 is applied in the rotating frame. The frequency of the rotation is determined by:

$$\omega_1 = \gamma B_1 \tag{1.11}$$

This should not be misunderstood as ω_{rf} of the B_1 field since ω_{rf} is the field oscillating frequency determined by changing the direction of the current passing through the coil, which is set to be the same as or near the Larmor frequency. The frequency ω_{rf} is often called the carrier frequency or the transmitter frequency. The frequency ω_1 is determined by the amplitude of the B_1 field, i.e., the maximum strength of the B_1 field. By modulating the amplitude and time during which B_1

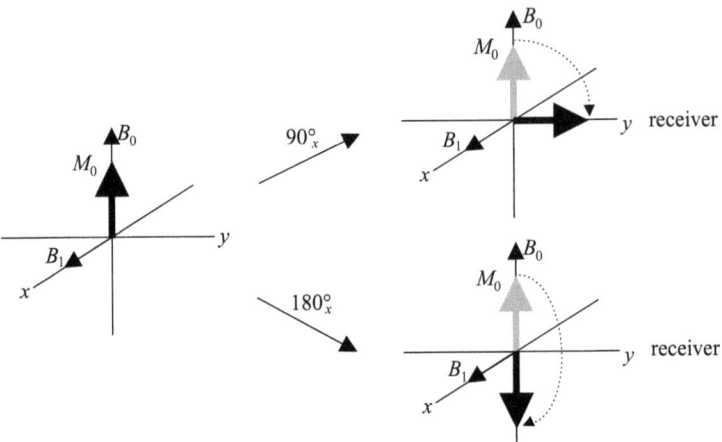

Fig. 1.6 Vector representation of the bulk magnetization upon applying a 90° pulse and a 180° pulse by the B_1 along the x axis in the rotating frame. The maximum signal is obtained when a 90° pulse is applied, which rotates the bulk magnetization (M_0) onto the xy plane. No signal is observed when a 180° pulse is applied, which rotates M_0 to $-z$ axis

is turned on, the bulk magnetization can be rotated to anywhere in the plane perpendicular to the axis of the applied B_1 field in the rotating frame. If B_1 is turned on, and then turned off when M_0 moves from the z axis to the xy plane, this is called a 90° pulse. The corresponding time during which B_1 is applied is called the 90° pulse length (or the 90° pulse width), and the field amplitude is called the pulse power. A 90° pulse length can be as short as a few microseconds and as long as a fraction of a second. The pulse power for a hard (short) 90° pulse is usually as high as half of a hundred watts for proton and several hundred watts for heteronuclei (all nuclei except ^{1}H). Because heteronuclei have lower gyromagnetic ratios than proton, they have longer 90° pulse lengths at a given B_1 field strength.

The 90° pulse length (pw$_{90}$) is proportional to the B_1 field strength:

$$v_1 = \frac{\gamma B_1}{2\pi} = \frac{1}{4\text{pw}_{90}} \tag{1.12}$$

$$\text{pw}_{90} = \frac{\pi}{2\gamma B_1} = \frac{1}{4v_1} \tag{1.13}$$

in which v_1 is the field strength in the frequency unit of hertz. A higher B_1 field produces a shorter 90° pulse. A 90° pulse of 10 μs is corresponding to a 25 kHz B_1 field. Nuclei with smaller gyromagnetic ratios will require a higher B_1 to generate the same pw$_{90}$ as that with larger γ. When a receiver is placed on the transverse plane of the rotating frame, NMR signals are observed from the transverse magnetization. The maximum signal is obtained when the bulk magnetization is in the xy plane of the rotating frame, which is done by applying a 90° pulse. No signal is observed when a 180° pulse is applied (Fig. 1.6).

1.4 Bloch Equations

As we now know, the nuclei inside the magnet produce nuclear moments which cause them to spin about the magnetic field. In addition, the interaction of the nuclei with the magnetic field will rotate the magnetization towards the transverse plane when the electromagnetic B_1 field is applied along a transverse axis in the rotating frame. After the pulse is turned off, the magnetization is solely under the effect of the B_0 field. How the magnetization changes with time can be described by the Bloch equations, which are based on a simple vector model.

Questions to be addressed in the current section include the following:

1. What phenomena do the Bloch equations describe?
2. What is free induction decay?
3. What are limitations of the Bloch equations?

In the presence of the magnetic field B_0, the torque produced by B_0 on spins with the angular moment μ causes precession of the nuclear spins. Felix Bloch derived simple semiclassical equations to describe the time-dependent phenomena of nuclear spins in the static magnetic field (Bloch 1946). The torque on the bulk magnetization, described by the change of the angular momentum as a function of time, is given by:

$$T = \frac{dP}{dt} = M \times B \qquad (1.14)$$

in which $M \times B$ is the vector product of the bulk magnetization M (the sum of μ) with the magnetic field B. Because $M = \gamma P$ (or $P = M/\gamma$) according to (1.3), the change of magnetization with time is described by:

$$\frac{dM}{dt} = \gamma(M \times B) \qquad (1.15)$$

When B is the static magnetic field B_0 which is along the z axis of the laboratory frame, the change of magnetization along the x, y, and z axes with time can be obtained from the determinant of the vector product:

$$\frac{dM}{dt} = i\frac{dM_x}{dt} + j\frac{dM_y}{dt} + k\frac{dM_z}{dt} = \gamma \begin{vmatrix} i & j & k \\ M_x & M_y & M_z \\ 0 & 0 & B_0 \end{vmatrix} = i\gamma M_y B_0 - j\gamma M_x B_0 \qquad (1.16)$$

in which i, j, k are the unit vectors along the x, y, and z axes, respectively. Therefore,

$$\frac{dM_x}{dt} = \gamma M_y B_0 \qquad (1.17)$$

$$\frac{dM_y}{dt} = -\gamma M_x B_0 \tag{1.18}$$

$$\frac{dM_z}{dt} = 0 \tag{1.19}$$

The above Bloch equations describe the time dependence of the magnetization components under the effect of the static magnetic field B_0 produced by the magnet of an NMR spectrometer without considering any relaxation effects. The z component of the bulk magnetization M_z is independent of time, whereas the x and y components are decaying as a function of time and the rate of the decay is dependent on the field strength and nuclear gyromagnetic ratio. The Bloch equations can be represented in the rotating frame, which is related to their form in the laboratory frame according to the following relationship:

$$\left(\frac{dM}{dt}\right)_{rot} = \left(\frac{dM}{dt}\right)_{lab} + M \times \omega = M \times (\gamma B + \omega) = M \times \gamma B_{eff} \tag{1.20}$$

in which $B_{eff} = B + \omega/\gamma$ and ω is the angular frequency of the rotating frame and is the same as ω_{rf}. The motion of magnetization in the rotating frame is the same as in the laboratory frame, provided the field B is replaced by the effective field B_{eff}. When $\omega = -\gamma B = \omega_0$, the effective field disappears, resulting in time-independent magnetization in the rotating frame. It is worth noting that $-\gamma B$ is the Larmor frequency of the magnetization according to (1.5), whereas γB is the transformation from the laboratory frame to the rotating frame, $-\omega_{rf}$ (Sect. 1.3).

Since the bulk magnetization at equilibrium is independent of time, based on the Bloch equations it is not an observable NMR signal. The observable NMR signals are the time-dependent transverse magnetization. However, at equilibrium the net xy projections of the magnetization (nuclear angular moments) are zero due to the precession of the nuclear spins. The simple solution to this is to bring the bulk magnetization to the xy plane by applying the B_1 electromagnetic field. The transverse magnetization generated by the B_1 field (90° pulse) will not stay in the transverse plane indefinitely; instead it decays under the interaction of the static magnetic field B_0 while precessing about the z axis and realigns along the magnetic field direction or the z axis of the laboratory frame (Fig. 1.7). The decay of the transverse magnetization forms the observable NMR signals detected by the receiver in the xy plane in the rotating frame, which is called the free induction decay, or FID.

The Bloch theory has its limitations in describing spin systems with nuclear interactions other than chemical shift interaction, such as strong scalar coupling. In general, the Bloch equations are applied to systems of noninteracting spin ½ nuclei. Nevertheless, it remains a very useful tool to illustrate simple NMR experiments.

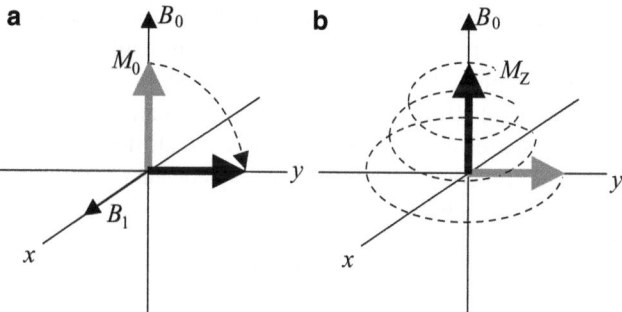

Fig. 1.7 Vector model representation of a one-pulse experiment. (**a**) The equilibrium bulk magnetization shown in the *shaded arrow* is brought to y axis by a 90° pulse along the x axis. (**b**) After the 90° pulse, the transverse magnetization decays back to the initial state while precessing about the z axis. An FID is observed by quadrature detection on the transverse plane

1.5 Fourier Transformation and Its Applications in NMR

The free induction decay is the sum of many time domain signals with different frequencies, amplitudes and phases. These time domain signals are detected and digitized during the signal acquisition period. In order to separate the individual signals and display them in terms of their frequencies (spectrum), the time domain data (FID) are converted to a frequency spectrum by applying the Fourier transformation, named after the discovery by French mathematician Joseph Fourier.

Question to be addressed in this section include the following:

1. What are the properties of the Fourier transformation useful for NMR?
2. What is the relationship between excitation bandwidth and pulse length in terms of the Fourier transformation?
3. What is quadrature detection and why is it necessary?

1.5.1 Fourier Transformation and Its Properties Useful for NMR

The Fourier transformation describes the connection between two functions with dependent variables such as time and frequency ($\omega = 2\pi/t$), called a Fourier pair, by the relationship (Bracewell 1986):

$$F(\omega) = \text{Ft}\{f(t)\} = \int_{-\infty}^{\infty} f(t)e^{-i\omega t}dt \qquad (1.21)$$

$$f(t) = \text{Ft}\{F(\omega)\} = \frac{1}{2\pi}\int_{-\infty}^{\infty} F(\omega)e^{i\omega t}d\omega \qquad (1.22)$$

Although there are many methods to perform the Fourier transformation for the NMR data, the Cooley–Tukey fast FT algorithm is commonly used to obtain NMR spectra from FIDs combined with techniques such as maximum entropy (Sibisi et al. 1984; Mazzeo et al. 1989; Stern and Hoch 1992) and linear prediction (Zhu and Bax 1990; Barkhuusen et al. 1985). For NMR signals described as an exponential function (or sine and cosine pair) with a decay constant $1/T$, the Fourier pair is the FID and spectrum with the forms of:

$$f(t) = e^{(i\omega_0 - (1/T))t} \tag{1.23}$$

$$F(\omega) = \text{Ft}\{f(t)\} = \int_{-\infty}^{\infty} e^{(i\omega_0 - (1/T))t} e^{-i\omega t} dt = \frac{i(\omega_0 - \omega) + (1/T)}{(\omega_0 - \omega)^2 + (1/T^2)} \tag{1.24}$$

which indicate that the spectrum is obtained by the Fourier transformation of the FID and the frequency signal has a Lorentzian line shape. Some important properties of the Fourier transformation useful in NMR spectroscopy are discussed below (Harris and Stocker 1998):

1. *Linearity theorem.* The Fourier transform of the sum of functions is the same as the sum of Fourier transforms of the functions:

$$\text{Ft}\{f(t) + g(t)\} = \text{Ft}\{f(t)\} + \text{Ft}\{g(t)\} \tag{1.25}$$

This tells us that the sum of time domain data such as an FID will yield individual frequency signals after the Fourier transformation.

2. *Translation theorem.* The Fourier transform of a function shifted by time τ is equal to the product of the Fourier transform of the unshifted function by a factor of $e^{i\omega\tau}$:

$$\text{Ft}\{f(t+\tau)\} = e^{i\omega\tau}\text{Ft}\{f(t)\} = e^{i2\pi\phi}\text{Ft}\{f(t)\} \tag{1.26}$$

This states that a delay in the time function introduces a frequency-dependent phase shift in the frequency function. A delay in the acquisition of the FID will cause a first-order phase shift (frequency-dependent phase shift) in the corresponding spectrum. This also allows the phase of a spectrum to be adjusted after acquisition without altering the signal information contained in the time domain data $f(t)$ (FID). The magnitude representation of the spectrum is unchanged by spectral phasing because the integration of $|\exp(i\omega t)|$ over all possible ω yields unity. Similarly,

$$\text{Ft}\{f(\omega - \omega_0)\} = e^{i\omega_0 \tau} f(t) \tag{1.27}$$

A frequency shift in a spectrum is equivalent to an oscillation in the time domain with the same frequency. This allows the spectral frequency to be calibrated after acquisition.

3. *Convolution theorem.* The Fourier transform of the convolution of functions f_1 and f_2 is equal to the product of the Fourier transforms of f_1 and f_2:

$$\text{Ft}\{f_1(t) * f_2(t)\} = \text{Ft}[f_1(t)] * \text{Ft}[f_2(t)] \qquad (1.28)$$

in which the convolution of two functions is defined as the time integral over the product of one function and the other shifted function:

$$f_1(t) * f_2(t) = \int_{-\infty}^{\infty} f_1(\tau) f_2(t-\tau) \, d\tau \qquad (1.29)$$

Based on this theorem, desirable line shapes of frequency signals can be obtained simply by applying a time function to the acquired FID prior to the Fourier transformation to change the line shape of the spectral peaks, known as apodization of the FID (see Sect. 4.9.4).

4. *Scaling theorem.* The Fourier transform of a function with which a scaling transformation is carried out ($t \rightarrow t/c$) is equal to the Fourier transform of the original function with the transformation $\omega \rightarrow c\omega$ multiplied by the absolute value of factor c:

$$\text{Ft}\{f(t/c)\} = |c|F(c\omega) \qquad (1.30)$$

According to this theorem, the narrowing of the time domain function by a factor of c causes the broadening of its Fourier transformed function in frequency domain by the same factor, and vice versa. This theorem is also known as similarity theorem.

5. *Parseval's theorem.*

$$\int_{-\infty}^{\infty} |f(t)|^2 dt = \int_{-\infty}^{\infty} |F(v)|^2 dv = \int_{-\infty}^{\infty} |F(\omega)|^2 d\omega \qquad (1.31)$$

This theorem indicates that the information possessed by the signals in both time domain and frequency domain is identical.

1.5.2 Excitation Bandwidth

In order to excite the transitions covering all possible frequencies, the excitation bandwidth is required to be sufficiently large. This requirement is achieved by applying short RF pulses. In certain other situations, the excitation bandwidth is required to be considerably narrow to excite a narrow range of resonance frequency such as in selective excitation. The following relationships of Fourier transformation pairs are helpful in understanding the process.

A Dirac delta function $\delta(t-\tau)$ in the time domain at $t=\tau$ gives rise to a spectrum with an infinitely wide frequency range and uniform intensity:

$$\text{Ft}\{\delta(t-\tau)\} = \int_{-\infty}^{\infty} \delta(t-\tau)e^{-i\omega t}dt = e^{-i\omega\tau} \quad (1.32)$$

which produces a frequency domain function with a perfectly flat magnitude at all frequencies because $|e^{-i\omega\tau}| = 1$. The δ function can be considered as an infinitely short pulse centered at τ. This infinitely short pulse excites an infinitely wide frequency range. When τ equals zero, each frequency has the same phase. Equation (1.32) means that in order to excite a wide frequency range, the RF pulse must be sufficiently short. Alternatively, for selective excitation, a narrow range of frequency is excited when a long RF pulse is used. A δ function in the frequency domain representing a resonance at frequency ω_0 with a unit magnitude has a flat constant magnitude in the time domain lasting infinitely long in time:

$$\text{Ft}\{e^{i\omega_0 t}\} = \int_{-\infty}^{\infty} e^{i\omega_0 t}e^{-i\omega t}dt = 2\pi\delta(\omega - \omega_0) \quad (1.33)$$

For a single resonance excitation, the RF pulse is required to be infinitely long. In practice, the short pulses are a few microseconds, which are usually called hard pulses, whereas the long pulses may last a few seconds, and are called selective pulses. Shown in Fig. 1.8 are the Fourier transforms of the short and long pulses. The bandwidth of the short pulse may cover several kilohertz and the selectivity of a long pulse can be as narrow as several hertz. A Gaussian function is the only function whose Fourier transformation gives another same-type (Gaussian) function (Fig. 1.8d, h):

$$\text{Ft}\{e^{-(t^2/\sigma^2)}\} = \int_{-\infty}^{\infty} e^{-(t^2/\sigma^2)}e^{-i\omega t}dt = -\sigma\sqrt{\pi}e^{-(\omega^2\sigma^2/4)} \quad (1.34)$$

A Gaussian shaped pulse will selectively excite a narrow range of frequency. The value of σ determines the selectivity of the pulse. The broadening of a Gaussian pulse results in narrowing in the frequency domain. More details on selective shaped pulses are discussed in Chap. 4.

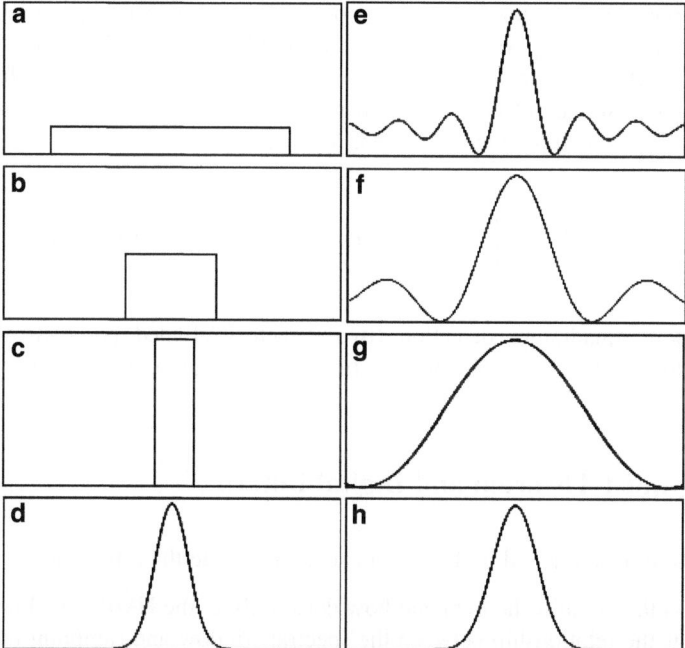

Fig. 1.8 RF pulses and their Fourier transforms. Long and short rectangular pulses (**a–c**) and corresponding Fourier transforms (**e–g**), Gaussian shaped pulse (**d**) and its Fourier transform (**h**)

1.5.3 Quadrature Detection

Two important time domain functions in NMR are cosine and sine functions. The Fourier transformations of the two functions are as follows:

$$\text{Ft}\{\cos(\omega_0 t)\} = \int_{-\infty}^{\infty} \frac{1}{2}(e^{i\omega t} + e^{-i\omega t})e^{-i\omega t}dt = \frac{1}{2}[\delta(v - v_0) + \delta(v + v_0)] \quad (1.35)$$

$$\text{Ft}\{\sin(\omega_0 t)\} = \int_{-\infty}^{\infty} \frac{1}{2i}(e^{i\omega t} - e^{-i\omega t})e^{-i\omega t}dt = \frac{1}{2i}[\delta(v - v_0) - \delta(v + v_0)] \quad (1.36)$$

The time domain signal may be considered as a cosine function. If an FID is detected by a single detector in the xy plane in the rotating frame during acquisition after a 90° pulse, the Fourier transformation of the cosine function gives rise to two frequency resonances located at v_0 and $-v_0$, as described by the δ functions in (1.35). This indicates that time domain signal detected by a single detector does not have information on the sign of the signal. As a result, each resonance will have a pair of peaks in frequency domain. In order to preserve the information on the sign of the resonance, a second detector must be used, which is placed perpendicular to the first one in the *xy* plane. The signal detected by the second detector is a sine function (a time function which is 90° out of phase relative to the cosine function). The combined signal detected by the quadrature detector is a complex sinusoidal function, $e^{i\omega_0 t}$, which produces a resonance at ω_0 or v_0:

$$\text{Ft}\{e^{i\omega t}\} = \int_{-\infty}^{\infty} e^{i\omega_0 t} e^{-i\omega_0 t} dt = 2\pi\delta(\omega - \omega_0) = \delta(v - v_0) \quad (1.37)$$

Therefore, quadrature detection (two detectors aligned perpendicularly) is required to detect NMR signals with distinguishable sign in the frequency domain.

1.6 Nyquist Theorem and Digital Filters

Questions to be addressed in the current section include the following:

1. What is the Nyquist theorem and how does it affect the NMR signal detection?
2. What is the relationship between the spectral window and sampling rate of the detection?
3. What is digital filtering and it application to NMR?

In the analog-to-digital conversion (ADC) of the NMR signal to the digital form, the analog signal is sampled with certain sampling rate by the ADC (Sect. 2.7, Chap. 2). The Nyquist theorem (Bracewell 1986) states that for the frequency to be accurately represented, the sampling rate is required to be at least twice the frequency. In other words, the highest frequency which can be accurately sampled is half of the sampling rate. The frequency is called the Nyquist frequency (f_n), which is defined as the spectral width, or the spectral window (SW). As a result, two points must be recorded per period of a sinusoidal signal. The time interval of sampling is called the dwell time (DW) and 1/DW is the sampling rate, which based on the Nyquist theorem has the relationship with SW:

$$f_n = \text{SW} = \frac{1}{2\text{DW}} \quad (1.38)$$

If a signal frequency, v, is higher than the Nyquist frequency, it will appear at a different frequency in the spectrum, called an aliasing frequency:

$$v_a = v - 2kf_n \quad (1.39)$$

in which v is the frequency of the signal, f_n is the Nyquist frequency, v_a is the aliased frequency, and k is an integer. For example, if the spectral width is set to 8 kHz, a signal with frequency of 9 kHz will appear at 1 kHz. A signal with a frequency

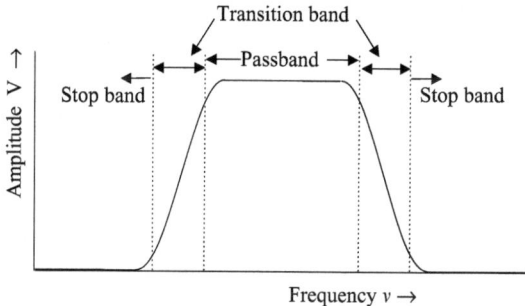

Fig. 1.9 Characterization of a bandpass filter. The signals with the frequencies inside the pass band of the filter pass through the filter, while those with the frequencies in the stop band are filtered out. The amplitudes of the signals with frequencies inside the transition band are attenuated (Winder 1997)

outside the SW can also be folded into the spectrum, called the folded frequency. For an f_n of 8 kHz, for instance, a signal with frequency 9 kHz will be folded at 7 kHz. Usually, a folded peak in an NMR spectrum shows a different phase than the unfolded peaks. In order to remove the aliased or folded signals, analog filters can be utilized before the signal is digitized. However, frequencies beyond the pass band (including noise) can still appear in the spectrum as the aliased or folded frequencies because the transition band between the pass band and stop band of analog filters is rather large (see Fig. 1.9 for the definition of transition band, pass band and stop band of a filter). The real solution to avoid folding-in of signals and noise is to utilize digital filters combined with oversampling (Winder 1997; Moskau 2001).

Oversampling denotes that the time domain signal is acquired with a larger spectral width and a larger number of data points than necessary. Because the spectral width is determined by the sampling rate (Nyquist theorem), tenfold oversampling increases both the spectral width and the number of data points by ten times (whereas the acquisition time is not changed). The role of the oversampling is to reduce the "quantization noise" produced by the ADC in the case that the receiver gain has to be set to a high value, by spreading the noise over the larger spectral width. As a result, the dynamic range as well as the signal-to-noise ratio can be increased. Additionally, application of oversampling leads a flatter baseline of the spectrum. A significantly larger number of data points is generated by the oversampling, which requires much disk storage space. For example, 20-fold oversampling of 32k data points will generate 640k data points (k is equivalent to multiplying the number by 1,024). To avoid the unnecessary larger data sets, a real-time digital filter is used to reduce the spectral width to that of interest by removing the frequency range outside the desired spectral width before the data are stored on disk. The digital filtering is achieved by digital signal processor integrated with the ADC circuits (real time) before time averaging, or by software (postacquisition) after the acquired FID is transferred from the console to the host computer prior to the storage on disk.

Similar to analog filters, a digital filter is characterized by the passband (spectral window of interest) and the shape of the stop band, which describes the steepness of the cutoffs of the filters (stop band is the region to be filtered out, Fig. 1.9). For postacquisition filters, the steepness of the cutoff is determined by the number of coefficients. A larger number of coefficients defines a filter with sharper cutoffs and flatter passband (brick-wall type with narrower transition band), whereas a smaller number of coefficients characterizes filters with slower cutoffs (wider transition band). The real time digital filters can be characterized as two types: brick-wall type with sharpest cutoffs and analog-alike with gradual cutoffs.

1.7 Chemical Shift

From the previous sections, we know how a detectable magnetization is generated and the FID is acquired with quadrature detection. Now, we would like to know what kind of signals we are going to observe and how the information can be used.

In the current section, the following questions are addressed:

1. What is chemical shift?
2. Where does chemical shift originate?
3. What are references and units of chemical shift?

All nuclear spins of the same isotope would have the same resonance frequency if there were no other kinds of interaction in addition to the Zeeman interaction. In fact, for a given isotope, dispersion of the NMR signals of nuclei is caused by the difference in the environment surrounding the nuclei. One of the factors causing the difference in frequency is the electronic shielding. The torque generated by the magnetic field also causes a precession of electrons around the magnetic field direction. The directional electronic precession produces a local magnetic field with a magnitude proportional to B_0, which shields some portion of the static field from the nuclei. This electronic precession is different from the random motion of electrons. The net effect can be described using a quality called shielding constant σ by:

$$v = \frac{\gamma}{2\pi} B_0 (1 - \sigma) \tag{1.40}$$

The shielding constant is always less than 1 because the induced local magnetic field will not be larger than the applied magnetic field.

The absolute zero of chemical shift is the one obtained for a bare nucleus without electrons. Although the absolute value of chemical shift may be obtained for bare nuclei such as protons, it is convenient to use a specific compound as a reference, whose resonance frequency is set to the chemical shift value of zero. The chemical shifts of other resonances are expressed as the difference in electron shielding to the reference nucleus:

$$\delta = \frac{v - v_{\text{ref}}}{v_{\text{ref}}} 10^6 \tag{1.41}$$

in which v and v_{ref} are the resonance frequencies of nucleus under study and the reference nucleus in units of megahertz, respectively, and δ is the chemical shift in unit of ppm (parts per million) of the nucleus with frequency v. Chemical shift δ is independent of the magnetic field strength, that is, the resonances in ppm present in a spectrum remain the same when obtained at different magnets with different field strengths. The reference compound is required to have the following properties: (a) stability in a variety of solvents, (b) an unchanged chemical shift value over a wide range of temperature and pH values, (c) easy to handle. Two compounds are commonly used for ^{1}H NMR reference: tetramethylsilane (TMS), which is the standard reference adopted by IUPAC (International Union of Pure and Applied Chemistry) and 2,2-dimethyl-2-silapentane-5-sulfonic acid (DSS), which is a secondary reference by IUPAC (Harris et al. 2001). Either of the reference compounds can be added into an NMR sample as an internal reference or used alone as an external reference. For internal referencing, the reference compound is dissolved with the sample, which clearly has limitations such as solubility, miscibility or reaction with the sample. For external referencing, a reference compound is dissolved alone in a specific solvent and the chemical shift is measured for the reference either in its own NMR tube or in a capillary insert tube inside the sample NMR tube. The zero frequency is set to the resonance frequency of the reference nucleus, which is used for all other experiments with the same isotope. Because of the high stability and homogeneity of NMR spectrometers, the external reference is of practical use for biological samples. In fact, once the chemical shift reference is calibrated on a spectrometer, the reference frequency will not change unless the ^{2}H lock frequency is adjusted. The graduate drift of the magnetic field is corrected by the z_0 current of the shimming assembly (see Sect. 4.2 in Chap. 4).

TMS is commonly used as the reference in ^{1}H and ^{13}C spectra for samples in organic solvents. Chemical shifts for all isotopes should include two decimal digits. DSS is commonly chosen as ^{1}H and ^{13}C external references for biological samples, which is dissolved in water in a pH range of 2–11 and used at 25 °C. The chemical shift of water or HDO has a temperature dependence that can be expressed as referenced to DSS over a wide range of temperature:

$$\delta_{H_2O}(\text{ppm}) = 4.76 - (T - 25)0.01 \tag{1.42}$$

For ^{31}P NMR, 85 % H_3PO_4 is the IUPAC standard reference which can be used externally (with a capillary insert). The chemical shift reference for ^{15}N is complicated and sometimes very confusing. Because there is no similar compound to DSS or TMS available for ^{15}N referencing as in the case of ^{1}H or ^{13}C, a variety of reference systems has been used to define 0.00 ppm for ^{15}N. Although CH_3NO_2 is the IUPAC ^{15}N reference, liquid NH_3 is the most popular ^{15}N reference for biological NMR. The disadvantage is the difficult handling of the sample. Indirect reference compounds are usually used such as ^{15}N urea in dimethyl sulfoxide (DMSO), saturated ammonium chloride in water. ^{15}N urea in DMSO is

Table 1.1 Frequency ratio Ξ of indirect heteronuclear references

Nucleus	Reference compound	Sample condition	$\Xi/\%$	References
^{1}H	DSS	Internal	100.0000000	By definition
^{2}H	DSS	Internal	15.3506088	Markley et al. (1998)
^{13}C	DSS	Internal	25.1449530	Wishart et al. (1995)
^{15}N	Liquid NH$_3$	External	10.1329118	Wishart et al. (1995)
^{31}P	(CH$_3$O)$_3$PO	Internal	40.4808636	Markley et al. (1998)
^{31}P	H$_3$PO$_4$	External	40.4807420	Harris et al. (2001)

a convenient sample as an indirect reference and has a ^{15}N chemical shift of 77.6 ppm relative to liquid ammonium (Sibi and Lichter 1979). It should be noted that an ^{15}N urea reference sample must be locked at the frequency of ^{2}H$_2$O. A simple method to achieve this is to place a capillary tube with ^{2}H$_2$O inside the NMR tube of the urea sample. An alternative way to obtain the correct reference frequency using the ^{15}N urea sample is to acquire the spectrum without ^{2}H locking. After the lock frequency is set to be on resonance of ^{2}H$_2$O using a ^{2}H$_2$O sample, the ^{15}N urea sample is placed into the probe without altering the lock frequency. The shimming can be done with ^{2}H gradient shimming. After setting the resonance frequency of ^{15}N urea to 77.6 ppm, the frequency at 0.00 ppm is the reference frequency for ^{15}N experiments in aqueous solutions.

A more convenient referencing system has been introduced that uses the ^{1}H reference for heteronuclei through the frequency ratio Ξ of the standard reference sample to DSS (or TMS):

$$\frac{\Xi}{\%} = 100 \frac{\nu_X}{\nu_{DSS}} \qquad (1.43)$$

in which ν_{DSS} and ν_X are the observed ^{1}H frequency of DSS and the observed frequency of X nucleus of the reference sample. The values of Ξ for different isotope reference samples are listed in Table 1.1. The reference frequency for X nuclei can be calculated from the ^{1}H reference frequency on the spectrometer according to:

$$\nu_{ref}^X = \Xi_{ref} \nu_{DSS} = \frac{(\Xi_{ref}/\%)\nu_{DSS}}{100} \qquad (1.44)$$

For example, if liquid NH$_3$ is used as the ^{15}N reference sample, the ^{15}N reference frequency is given by:

$$\nu_{ref}^{^{15}N} = \frac{10.1329118 \nu_{DSS}}{100} \qquad (1.45)$$

1.7 Chemical Shift

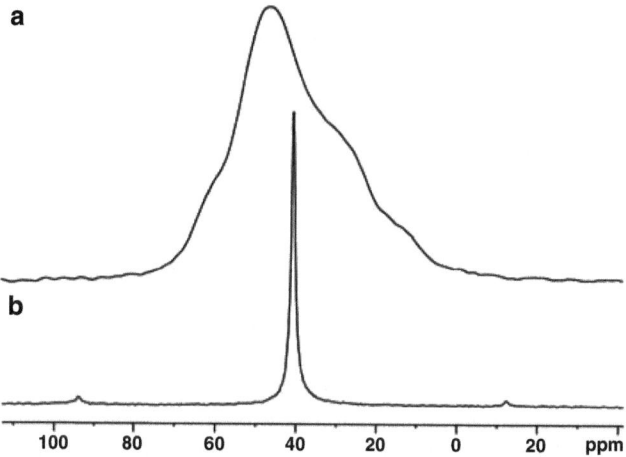

Fig. 1.10 (a) ^{1}H decoupled ^{13}C powder pattern and (b) magic-angle-spinning spectra of ^{13}C$_2$-Glycine

To set the reference frequency for heteronuclei, the ^{1}H reference frequency ν_{DSS} is first obtained for an aqueous sample containing DSS (1 mM). The chemical shift reference for heteronuclei is defined through the ratio Ξ according to (1.44).

The reference frequency does not change if the spectrometer frequency is unchanged. Often the spectrometer frequency is defined based on the deuterium lock frequency. Therefore, the lock frequency is kept unchanged during normal operation of the spectrometer. The calibrated reference is recorded corresponding to a lock frequency. It is a good practice to keep a record of the spectrometer deuterium lock frequency when calibrating chemical shift references.

The shielding constant, σ, is dependent on the distribution of electron density surrounding the nucleus. Because the distribution generally is spherically asymmetrical, σ usually has an anisotropic value (chemical shift anisotropy, or CSA) with an orientational dependence. An NMR spectrum of a solid sample looks like the one shown in Fig. 1.10 with a broad line width due to the orientation distribution of molecules in the sample. The broad chemical shift powder pattern can be averaged to its isotropic value, σ_{iso}, which is the trace of the three components of its chemical shift tensor, by mechanically rotating the sample about the direction along the magic angle, 54.7° at a rate larger than the anisotropic line width, for example, 10 kHz.

$$\sigma_{iso} = \frac{\sigma_{xx} + \sigma_{yy} + \sigma_{zz}}{3} \quad (1.46)$$

The technique is well known as magic angle spinning, or in short, MAS (Andrew et al. 1959; Lowe 1959). In solution, however, rapid molecular tumbling averages out the chemical shift anisotropy that the nuclei possess, resulting in sharp peaks at the isotropic resonances. If the molecular tumbling motion is slowed down, the

Table 1.2 Chemical shift range in proteins and peptides

Nucleus	NH^{backbone}	NH^{sidechain}	CH^{aromatic}	C$^\alpha$H	C'	C$^\beta$H
^{1}H	8–10	6.5–8	6.5–8	3.5–5		1–4
^{13}C			110–140	40–65	170–185	20–75
^{15}N	110–140					

Note: ^{1}H and ^{13}C chemical shifts in ppm are referenced to DSS, ^{15}N in ppm is referenced to liquid NH$_3$ through its frequency ratio Ξ

anisotropic property can be restored in a solution sample, which contains rich information on molecular structures. Chemical shift anisotropy has been used to determine structures of proteins in both solid and solution samples.

Several factors can contribute to the shielding constant, one of which originates from a spherical electronic distribution (s orbital electrons) called diamagnetic shielding, σ_{dia} (Friebolin 1993). The term diamagnetic indicates that the induced field has an opposite direction to the external static field, B_0. The shielding effect from a nonspherical electronic distribution (electron orbitals other than the s orbital), in which the induced local field has the same direction as B_0, is called paramagnetic shielding, σ_{para}.

$$\sigma = \sigma_{dia} + \sigma_{para} \qquad (1.47)$$

It is important to note that the term "paramagnetic shielding" has nothing to do with the effect of unpaired electrons referred to as paramagnetic NMR spectroscopy. It is named for the opposite sign to the diamagnetic shielding. Because σ_{dia} and σ_{para} have opposite contributions to the shielding constant, some of the effects are canceled out. The contribution of σ_{para} is proportional to $(m^2 \Delta E)^{-1}$ (the mass of the nucleus, m and the excitation energy to the lowest excited molecular orbital, ΔE) and the asymmetry of electronic distribution, whereas σ_{dia} is proportional to m^{-1} and the symmetry of electronic distribution. For protons, because the energy gap is large, the paramagnetic shielding is very small even when bonding causes distortion of the spherical distribution, resulting in a small shift range, normally 10 ppm. For ^{13}C, the σ_{para} becomes an important contribution to the shielding because ΔE is small. The distortion of the spherical electronic distribution induced by the bonding environment near the nuclei can significantly affect the value of the nuclear chemical shift. Hence, ^{13}C has a wider range of chemical shifts (approximately 300 ppm) compared to ^{1}H (Tables 1.2 and 1.3). The net effect of electronic precession produces a local magnetic field in the opposite direction to the magnetic field, B_0. Paramagnetic contributions usually have a dominant effect over the diamagnetic term in heteronuclei, which is responsible for the fact that the chemical shift range for heteronuclei is usually much larger than that of ^{1}H (Table 1.2).

The other factor that contributes to the shielding constant is called ring current effect (Lazzeretti 2000), which arises from the delocalized electrons of the

1.7 Chemical Shift

Table 1.3 Average chemical shifts in proteins and peptides

Residue	$^1H^N$	^{15}N	$^{13}C'$	$^{13}C^\alpha$	$^1H^\alpha$	$^1H^\beta$	$^1H^{other}$
Ala	8.15	122.5	177.6	52.2	4.33	1.39	
Arg	8.27	120.8	176.6	56.0	4.35	1.89, 1.79	γCH_2 1.70, δCH_2 3.32, NH 7.17, 6.62
Asn	8.38	119.5	175.6	52.7	4.74	2.83, 2.75	γNH_2 7.59, 6.91
Asp	8.37	120.6	176.8	53.9	4.71	2.84, 2.75	
Cys	8.23	118.0	174.6	56.8	4.54	3.28, 2.96	
Gln	8.27	120.3	175.6	56.0	4.33	2.13, 2.01	γCH_2 2.38, δNH_2 6.87, 7.59
Glu	8.36	121.3	176.6	56.3	4.33	2.09, 1.97	γCH_2 2.31, 2.28
Gly	8.29	108.9	173.6	45.0	3.96		
His	8.28	119.1	174.9	55.5	4.60	3.26, 3.20	2H 8.12, 4H 7.14
Ile	8.21	123.2	176.5	61.2	4.17	1.90	γCH_2 1.48, 1.19, γCH_3 0.95, δCH_3 0.89
Leu	8.23	121.8	176.9	55.0	4.32	1.65	γH 1.64, δCH_3 0.94, 0.90
Lys	8.25	121.5	176.5	56.4	4.33	1.85, 1.76	γCH_2 1.45, δCH_2 1.70, ϵCH_2 3.02, ϵNH_3^+ 7.52
Met	8.29	120.5	176.3	55.2	4.48	2.15, 2.01	γCH_2 2.64, ϵCH_3 2.13
Phe	8.30	120.9	175.9	57.9	4.63	3.22, 2.99	2,6 H 7.30, 3,5 H 7.39, 4 H 7.34
Pro	–	128.1	176.0	63.0	4.42	2.28, 2.02	γCH_2 2.03, δCH_2 3.68, 3.65
Ser	8.31	116.7	174.4	58.1	4.47	3.88	
Thr	8.24	114.2	174.8	62.0	4.35	4.22	γCH_3 1.23
Trp	8.18	120.5	173.6	57.6	4.66	3.32, 3.19	2H 7.24, 4H 7.65, 5H 7.17, 6H 7.24, 7H 7.50, NH 10.22
Tyr	8.28	122.0	175.9	58.0	4.55	3,13, 2.92	2,6H 7.15, 3,5H 6.86
Val	8.19	121.1	176.0	62.2	4.12	2.13	γCH_3 0.97, 0.94

Note: The chemical shifts of ^1H, ^{13}C and ^{15}N are in ppm, and referenced to DSS, DSS and liquid NH_3, respectively. The $^1H^N$, $^1H^\alpha$, ^{15}N, and $^{13}C'$, are from Wishart et al. (1991), $^{13}C^\alpha$ are from Spera and Bax (1991), $^1H^\beta$ and $^1H^{other}$ are from Wüthrich (1986)

p orbital moving between bonded atoms in an aromatic ring. A classical example of the ring current is that the ^1H chemical shift of benzene has a higher frequency (downfield shift) at 7.27 ppm compared to the resonance frequency of ethylene at 5.28 ppm. The π electrons of benzene circulating above and below the aromatic ring, when placed in a magnetic field, produce an additional magnetic field whose direction is opposite to the external static magnetic field at the center of the aromatic ring and along the external field at the outside edge of the ring. As a result, the field at the center of the ring has been reduced (more shielding), whereas the protons directly attached to the ring experience a field larger than the external field due to the addition of the induced field (deshielding). This phenomenon is called the ring current effect. The ring current has less effect on the ^{13}C chemical shifts of aromatic compounds. This has been explained by considering the fact that carbon nuclei are located approximately where the induced field changes direction between shielding and deshielding, that is, the induced field is close to zero.

1.8 Nuclear Coupling

Electronic shielding is one of the nuclear interactions contributing to the resonance frequency of nuclei. Interactions other than chemical shift are entirely independent of the magnetic field strength. These interactions provide information on the structures and dynamics of biological molecules.

Questions to be addressed in current section include the following:

1. What is scalar coupling and where does it originate?
2. What is its magnitude range and how can it be measured?
3. How does scalar coupling provide information on molecular structure?
4. What is the nuclear dipolar interaction and where does it originate?
5. How is its magnitude characterized?
6. Where does the nuclear Overhauser effect (NOE) come from and how is it generated?
7. How does NOE provide information on molecular structure?
8. What is the residual dipolar interaction and why does it exist in solution sample?

1.8.1 Scalar Coupling

Scalar (J, indirect, or spin–spin) coupling is the effect on nuclear spin A caused by the local magnetic field of its neighbor spin B. The orientation of spin B in the magnetic field produces a small polarization of the electrons mostly in the s orbital surrounding spin B. This polarization affects the electron density distribution of spin A directly bonded to spin B. Because the interaction depends on the s orbital electron density at the pair of nuclei, the electron density of the nuclei must be correlated, that is, in Fermi contact. Consequently, J coupling propagates only along chemical bonds.

Although the J coupling is anisotropic due to the asymmetric environment surrounding the nuclear spin, the interaction is averaged to an isotropic value in solution by the rapid molecular tumbling motion. The magnitude of the J coupling reduces significantly as the number of bonds separating the nuclei increases. Two-bond and three-bond couplings are at least one order of magnitude smaller than one-bond J couplings. Couplings longer than three-bond are close to zero, with an exception that the long range coupling beyond three bonds is observable in double-bonded compounds. Listed in Table 1.4 is the range of J couplings in different molecular bonds. When the frequency difference Δv in chemical shifts of spin I and S is much larger than the J coupling (weak coupling approximation, $\Delta v \gg J$), all peaks have equal intensity. This gives rise to a first-order spectrum. When the frequencies of two coupled nuclei are closer in magnitude to the J coupling ($\Delta v/J \leq 10$), second-order character will appear in the spectrum, which complicates the spectrum and produces uneven spectral intensities.

1.8 Nuclear Coupling

Table 1.4 Range of J coupling constants

	J_{HH}	J_{CH}	J_{CC}	J_{NC}	J_{NH}
^1J	276	120–250	30–80	<20	60–95
^2J	−30 to 0	−10 to 30	<20	<10	
^3J	<20	<10	<5	<1	

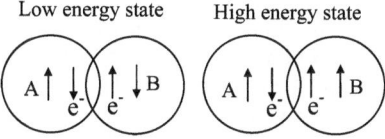

Fig. 1.11 Dirac model for one-bond J coupling (two-spin system). When both nuclei A and B have the same sign of gyromagnetic ratio γ, the effect of the coupling in which nuclei A and B are in antiparallel configuration stabilizes the low energy state, resulting in positive J

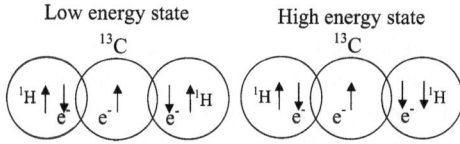

Fig. 1.12 Dirac model for J coupling of geminal protons. *Arrows* represent nuclei or electrons as labeled. For homonuclear J coupling, the effect of the coupling in which two protons in the low energy state are in parallel configuration destabilizes the low energy state, resulting in negative J

Homonuclear J coupling comes from the interaction between nuclei with the same gyromagnetic ratio, γ, whereas heteronuclear J coupling comes from those with different γ. According to the Dirac vector model, the low energy state is that in which the magnetic moments of nuclei A and B are in antiparallel configuration to the magnetic moments of their bonding electrons (Fig. 1.11). Since A and B are bonded, the electron of spin B in the bonding pair is also in antiparallel configuration with the electron of spin A. As a result, the nuclear spins A and B are antiparallel in the low energy state. If A and B both have positive gyromagnetic ratios such as in a CH bond, the effect of the coupling in which nuclei A and B are in an antiparallel configuration stabilizes the low energy state, resulting in positive J. Another example is the J coupling between geminal protons as shown in Fig. 1.12. The two geminal protons bond to a common carbon. In order to form covalent bond, the electrons of two protons must be in antiparallel configuration with the electron of the carbon. The energy state in which both protons have antiparallel configuration with their own electron has low energy. Since the two protons in the low energy state have a parallel configuration to each other, the coupling of the two protons destabilizes the low energy state, leading to a negative J coupling.

Since only s orbital electrons of the pair of nuclei contribute to the J coupling according to the Fermi contact mechanism, one-bond scalar couplings depend on the fraction of s orbital electrons involved in the bonding. For instance, the one-bond

Table 1.5 Correlation of J coupling with the contribution of s orbital electrons

Bond	s Fraction in CH bond (%)	$^1J_{CH}$	s Fraction in CC bond (%)	$^1J_{CC}$
CH3–CH3	25 × 100	125	25 × 25	35
CH2=CH2	33 × 100	156	33 × 33	67
CH≡CH	50 × 100	249	50 × 50	171

J coupling constants between ^{1}H and ^{13}C or between ^{13}C carbons are different for carbons with sp^3, sp^2, and sp orbitals. Because a sp carbon has a 50 % s character and a sp^3 carbon has a 25 % s character, ethyne has $^1J_{CH}$ and $^1J_{CC}$ couplings twice and fourfold as large as those of methane, respectively (Table 1.5).

Heteronuclear coupling constants between two nuclear isotopes I and S will be indicated by subscripts as J_{IS}, for example, J_{CH} for the J coupling constant between ^{13}C and ^{1}H. For homonuclei, subscripts are used to indicate the position of coupled protons. $^3J_{23}$ denotes the vicinal coupling constant between protons at positions 2 and 3. The number of bonds between the coupled nuclei is indicated by a superscript number before J, for example, $^2J_{HH}$ for the J coupling constant between two geminal protons and $^3J_{HH}$ for a vicinal H–H coupling (three-bond). Sometimes the number is ignored in one-bond coupling constants such as J_{CH} or J_{NH}. Since there is no one-bond coupling between protons in proteins and polypeptides, the coupling between protons is much weaker than the one-bond coupling of proton with carbon, nitrogen, or other nuclear isotopes.

Three-bond H–H coupling constants $^3J_{HH}$ contain information on the relative orientation of the coupled protons. Numerous studies have been performed to understand the relationship of coupling constants with dihedral angles. Karplus (1959) has theoretically described the dependence of the vicinal coupling constant $^3J_{HH}$ on the dihedral angle formed by the vicinal protons:

$$^3J_{HH} = 8.5\cos^2\phi - 0.28 \quad \text{(when } 0° \leq \phi \leq 90°\text{)} \tag{1.48}$$

$$^3J_{HH} = 9.5\cos^2\phi - 0.28 \quad \text{(when } 90° \leq \phi \leq 180°\text{)} \tag{1.49}$$

The general Karplus equation can be written as:

$$^3J = A \cos^2\theta + B \cos\theta + C \tag{1.50}$$

in which A, B, and C are the constants that depend on the specific coupled nuclei and θ is the dihedral angle (Karplus 1963). Semiempirical methods have been used to obtain values of constants $A, B,$ and C by studying the correlations of observed ^3J values to the dihedral angles in known protein structures. The relationship of $^3J_{H^N H^\alpha}$ to the dihedral angle ϕ has been derived from the structure of ubiquitin (Wang and Bax 1996):

$$^3J = 6.98 \cos^2\theta - 1.38 \cos\theta + 1.72 \tag{1.51}$$

1.8 Nuclear Coupling

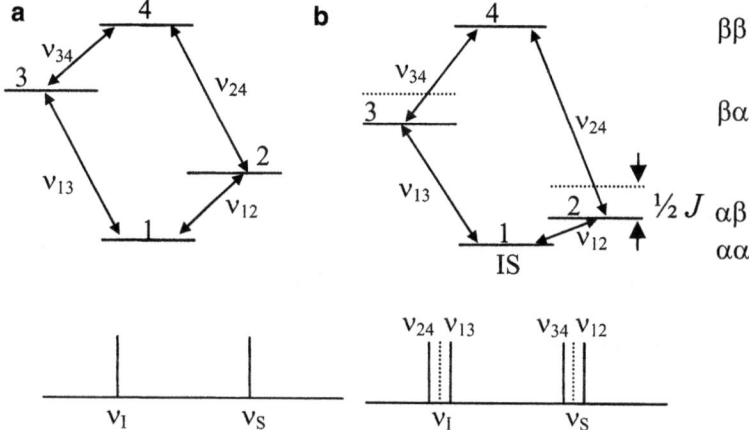

Fig. 1.13 Standard energy diagrams and spectra of AX spin systems (**a**) without and (**b**) with J coupling. Frequencies v_{13} and v_{24} are the transitions of spin I, whereas v_{12} and v_{34} are those of spin S. The *dashed lines* in (**b**) represent the energy levels or frequency positions without J coupling

In which $\theta = \phi - 60$. It should be noted that as many as four possible dihedral values can be derived from the above equation, which is clearly visible in the corresponding Karplus curve in Fig. 7.2 (Chap. 7). The ambiguity needs to be taken into account in the application of dihedral angle restraints in structure determination.

1.8.2 Spin Systems

The spin system refers to a group of nuclei that are coupled through J coupling. If a spin couples to one neighboring nucleus with a large chemical shift separation (weak coupling approximation, $\Delta v \gg J$), the two-spin system is said to be an AX system. In contract, when the frequency difference of the two nuclei are in the same order of magnitude as J coupling ($\Delta v \approx J$), the spin system is called AB spin system, which has close chemical shifts. In the absence of J coupling, the frequencies of the transitions for an AX spin system formed by spin I and spin S as shown in Fig. 1.13a are:

$$v_{12} = v_S; \quad v_{34} = v_S$$
$$v_{13} = v_I; \quad v_{24} = v_I$$

In the presence of J coupling (Fig. 1.13b), the frequencies become:

$$v_{12} = v_S - \tfrac{1}{2}J_{IS}; \quad v_{34} = v_S + \tfrac{1}{2}J_{IS}$$
$$v_{13} = v_I - \tfrac{1}{2}J_{IS}; \quad v_{24} = v_I + \tfrac{1}{2}J_{IS}$$

The frequency of J coupling is

$$v_J = \tfrac{1}{2} J_{IS}$$
$$\omega_J = 2\pi v_J = \pi J_{IS}$$

Each of the two chemical shift resonances now is split into two lines with the same intensity separated by J Hz, producing four lines in the spectrum. In molecules, there will often be protons coupling to two or more equivalent vicinal protons through three bonds. For a proton weakly coupled to n equivalent protons, the spin system is denoted as AX_n. In AX_n spin system, the number of split peaks and their relative intensities follow Pascal's triangle rule:

				1					n=0
			1		1				n=1
		1		2		1			n=2
	1		3		3		1		n=3
1		4		6		4		1	n=4

in which the number indicates the relative intensity of the signal peak.

1.8.3 Dipolar Interaction

Dipolar interaction plays an important role in structural and dynamic studies by NMR spectroscopy because of its dependence on the orientation and distance between dipole-coupled nuclei. In this section, the following questions are addressed:

1. What is the nuclear dipolar interaction?
2. How is the interaction described in equation form?
3. Where does NOE originate?
4. Under what conditions can the dipolar interaction be observed by solution NMR spectroscopy?

For weakly coupled spins, the dipolar contribution to the observed resonance in the high field limit is given by a simple orientation and distance dependence:

$$v_D = v_{\parallel} \frac{(3\cos^2\theta - 1)}{2} \tag{1.52}$$

in which θ is the angle between the dipolar vector and the magnetic field (Fig. 1.14) and $v_{\parallel}$ is the magnitude of the dipolar vector, or dipolar coupling constant, which is given by:

$$v_{\parallel} = -\frac{\gamma_1 \gamma_2 \hbar}{4\pi^2 r^3} \tag{1.53}$$

1.8 Nuclear Coupling

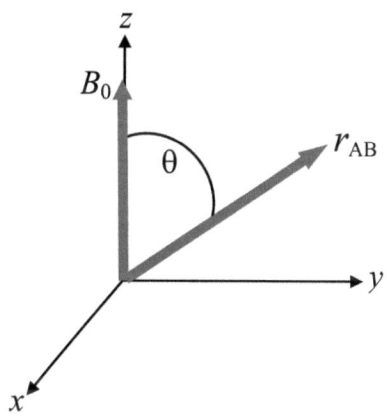

Fig. 1.14 Relative orientation of the internuclear vector r_{AB} and the magnetic field B_0 in the laboratory frame. B_0 is parallel to the z axis of the laboratory frame

$\hbar$ is reduced Plank's constant, and r is the distance between spins 1 and 2 which have the gyromagnetic ratios of γ_1 and γ_2, respectively. In the spectrum, ν_D is the frequency shift caused by dipolar coupling, whereas $\nu_\parallel$ is the dipolar coupling when the dipolar vector is parallel to the magnetic field. The dipolar interaction between two nuclear spins occurs through space rather than through molecular bond as in J coupling. The magnitude of the dipolar coupling for two protons separated by 3 Å is about 4.5 kHz, which is much larger than a J coupling. The dipolar coupling constant decreases rapidly with an increase of the dipole–dipole (DD) distance as a function of r^{-3}. In solution, the dipolar splitting is not observable because the orientational term of the dipolar interaction with respect to the magnetic field direction is averaged to zero by rapid molecular tumbling. However, the effect of dipolar interaction on molecular relaxation still exists at any instant, which is the origin of the nuclear Overhauser effect (NOE). The DD interaction is the most important contribution to spin relaxation of molecules in solution.

1.8.4 Residual Dipolar Coupling

Questions to be answered in this section include the following:

1. What is residual dipolar coupling (RDC) and how is it generated?
2. Why can RDC exist and how is it characterized?
3. How is the order of an alignment medium transferred to the macromolecules dissolved in the medium?
4. What are the requirements of the alignment media used to generate RDCs?

In a typical solution, the nuclear dipolar coupling is averaged to zero owing to the rapid molecular tumbling. Approaches have been developed to align macromolecules in solution with anisotropic media to regain the dipolar coupling (Tjandra and Bax 1997; Prestegard et al. 2000; de Alba and Tjandra 2002; Lipstitz and Tjandra 2004). The partial alignment of the macromolecules in media such as liquid crystals

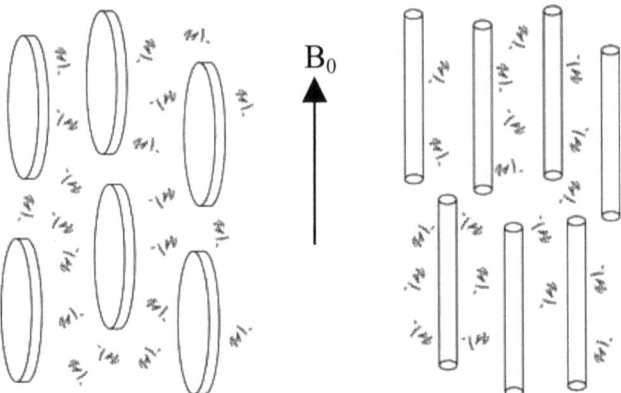

Fig. 1.15 Alignment of biomolecules in two liquid crystal media (reproduced with permission from Tjandra (1999), Copyright © 1999 Elsevier). (**a**) Bicelles are believed to be disk-shaped pieces of lipid bilayers aligning with their bilayer normal perpendicular to the applied magnetic field B_0. (**b**) Rod-like particles represent filamentous phage aligning with their long axis parallel to B_0

(Fig. 1.15) leads to the incomplete cancelation of the dipolar coupling, called RDC, which is the time or ensemble average of the dipolar coupling:

$$\nu_D = \nu_{\parallel} \frac{\langle 3\cos^2\Theta - 1 \rangle}{2} = \nu_{\parallel} \langle P_2(\cos\Theta) \rangle \qquad (1.54)$$

in which angular brackets refer to the averaging due to the molecular reorientations and internal motions, $P_2(x) = (3x^2 - 1)/2$ is the second Legendre polynomial, and other parameters are defined as in (1.52) and (1.53).

Since the dipolar interaction between heteronuclei is along the molecular bond, the sole unknown variable is the averaged molecular orientation, Θ, of the dipolar interaction with respect to the laboratory frame, or the magnetic field direction B_0. When the molecules are partially aligned in the magnetic field, the relative orientation of the dipolar interaction in the laboratory frame can be obtained by transforming the internuclear vector to an arbitrary molecular frame by angles α_x, α_y and α_z between the x, y, and z axes of the molecular frame and the dipolar vector. Then, the vector is further transformed from this molecular frame into the laboratory frame by a set of angles β_x, β_y and β_z between the x, y, and z axes of the molecular frame and the magnetic field direction (the z axis of the laboratory frame, Fig. 1.16). The angular dependence of the RDC can be represented by the two sets of angles as (Bax et al. 2001):

$$\begin{aligned}
\langle P_2(\cos\Theta) \rangle &= \frac{3}{2} \langle (\cos\alpha_x \cos\beta_x + \cos\alpha_y \cos\beta_y + \cos\alpha_z \cos\beta_z)^2 \rangle - \frac{1}{2} \\
&= \frac{3}{2} \langle (\cos^2\alpha_x \cos^2\beta_x + \cos^2\alpha_y \cos^2\beta_y + \cos^2\alpha_z \cos^2\beta_z \\
&\quad + 2\cos\alpha_x \cos\alpha_y \cos\beta_x \cos\beta_y + 2\cos\alpha_x \cos\alpha_z \cos\beta_x \cos\beta_z \\
&\quad + 2\cos\alpha_y \cos\alpha_z \cos\beta_y \cos\beta_z) \rangle - \frac{1}{2}
\end{aligned} \qquad (1.55)$$

1.8 Nuclear Coupling

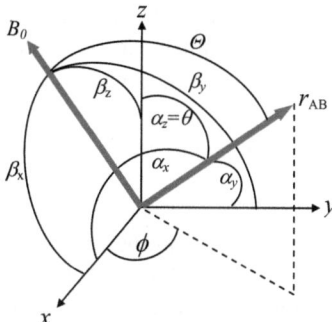

Fig. 1.16 The relative orientations of the internuclear vector r_{AB} and B_0 with respect to the x, y, and z axes of the molecular frame. The orientation of the internuclear vector r_{AB} with respect to the molecular frame is defined by the α angles, and the orientation of B_0 with respect to the molecular frame is defined by the β angles. The orientation between B_0 and r_{AB} is represented by angle Θ

If $c_{ij} = \cos\alpha_i \cos\alpha_j$ and $C_{ij} = \cos\beta_i \cos\beta_j$ with $i, j = (x, y, z)$, (1.55) can be rewritten as:

$$\langle P_2(\cos\Theta)\rangle = \frac{3}{2} \sum_{i,j=(x,y,z)} \langle C_{ij}c_{ij}\rangle - \frac{1}{2} \tag{1.56}$$

The variable C_{ij} is affected by the molecular tumbling, whereas c_{ij} is influenced by the internal motions. When the molecule is rigid, c_{ij} is not changed by the internal motions and can be treated as constant, then $\langle C_{ij}c_{ij}\rangle = \langle C_{ij}\rangle c_{ij}$. The equation of $\langle P_2(\cos\Theta)\rangle$ becomes:

$$\langle P_2(\cos\Theta)\rangle = \sum_{i,j=(x,y,z)} S_{ij}c_{ij} \tag{1.57}$$

The 3 × 3 matrix S is referred to as the Saupe order matrix, Saupe order tensor, or order tensor which is defined as:

$$S_{ij} = \frac{3}{2}\langle C_{ij}\rangle - \frac{1}{2}\delta_{ij} \tag{1.58}$$

in which δ_{ij} is the Kronecker delta function. The Saupe order matrix is symmetric since $\langle C_{ij}\rangle = \langle C_{ji}\rangle$, and traceless ($S_{xx} + S_{yy} + S_{zz} = 0$) because $\sum\langle C_{ii}\rangle = 1$. The symmetric condition eliminates three variables and the traceless condition reduces the final number of independent variables to five. In principle, if the molecular structure is known and hence c_{ij} is known, the five independent elements in the matrix can be obtained using the RDCs for at least five internuclear vectors in the molecule, provided they are not parallel to the magnetic field. In practice, the number of the measured dipolar couplings is much more than five.

The orientational information contained in the equation can also be solved in a molecular axis system in which the order matrix is diagonal. The molecular axis system is called the principal axis system, in which the RDC is given by:

$$v_D(\alpha_x, \alpha_y, \alpha_z) = v_\| \sum_{i=(x,y,z)} S_{ii} c_{ii} \qquad (1.59)$$

Now matrix S is a measurable quantity in the principal axis system because it is diagonal. S_{ii} also represents the probability of finding the ith axis along the magnetic field direction, which has the maximum value of 1 when the axis aligns with the magnetic field. The above equation tells us that the order information is contained in the order matrix and once the three elements of the order matrix in the principal axis system are known for a rigid molecule, the molecular structure may be determined by the RDCs of the macromolecule. Equation (1.59) can also be represented in polar coordinates by using the relationships of $c_{xx} = \sin^2\theta \cos^2\phi$, $c_{yy} = \sin^2\theta \sin^2\phi$ and $c_{zz} = \cos^2\theta$ (Fig. 1.16):

$$v_D(\theta, \phi) = v_\|(S_{xx}\sin^2\theta \cos^2\phi + S_{yy}\sin^2\theta \sin^2\phi + S_{zz}\cos^2\theta) \qquad (1.60)$$

in which θ is the angle between the internuclear vector and the z axis of the molecular frame or the principal axis frame (Fig. 1.16), which should not be confused with θ in (1.52) that defines the orientation of the nuclear vector with respect to B_0 (Fig. 1.14). In the principal axis system, only the differences in the principal values S_{ii} contribute to the RDC and the order tensor remains traceless $(-S_{zz} = S_{xx} + S_{yy})$ with the most ordered axis $|S_{zz}| > |S_{yy}| > |S_{xx}|$. After rearranging (1.60) using the relationship of $\cos^2\phi = 1/2(1 + \cos 2\phi)$, $\sin^2\phi = 1/2(1 - \cos 2\phi)$, and $\sin^2\theta = 1 - \cos^2\theta$, the RDC is described as:

$$v_D(\theta, \phi) = v_\|[S_{zz}P_2(\cos\theta) + \frac{1}{2}(S_{xx} - S_{yy})\sin^2\theta \cos 2\phi] \qquad (1.61)$$

By defining the principal alignment tensor A with an axial component $A_a = S_{zz}$ and a rhombic component $A_r = \frac{2}{3}(S_{xx} - S_{yy})$, the RDC is rewritten as:

$$v_D(\theta, \phi) = v_\|[A_a P_2(\cos\theta) + \frac{3}{4} A_r \sin^2\theta \cos 2\phi] \qquad (1.62)$$

In dilute liquid crystal media, the observed RDC is in the range of several hertz to several tens of hertz (0.1 % of dipolar coupling constant), which means that the values of A_a are on the order of 10^{-3}. The observed residual dipolar splitting is the difference of the two dipolar shifts, which is therefore given by:

$$\Delta v_D(\theta, \phi) = D_a[(3\cos^2\theta - 1) + \frac{3}{2} R \sin^2\theta \cos 2\phi] \qquad (1.63)$$

1.8 Nuclear Coupling

in which $D_a = A_a \nu_\parallel$ is the magnitude of the dipolar coupling normalized to the N–H dipolar coupling, and $R = A_r/A_a$ is the rhombicity of the dipolar coupling. The residual dipolar splitting is sometimes written as:

$$\Delta \nu_D(\theta, \phi) = D_a[(3\cos^2\theta - 1) + \eta \sin^2\theta \cos 2\phi] \quad (1.64)$$

in which η is the asymmetric parameter defined as $\eta = (S_{xx} - S_{yy})/S_{zz}$ or $\eta = \frac{3}{2} R$. The residual dipolar splitting can be described according to (1.60):

$$\Delta \nu_D(\theta, \phi) = D_{xx}\sin^2\theta \cos^2\phi + D_{yy}\sin^2\theta \sin^2\phi + D_{zz}\cos^2\theta \quad (1.65)$$

in which $D_{ii} = 2\nu_\parallel S_{ii}$ are the principal components of the RDC. The above equation resembles a powder pattern with three principal axes. It should be noted that the tensor D is also traceless, i.e., $D_{xx} + D_{yy} + D_{zz} = 0$. They are related to D_a and R by the relationship:

$$D_{zz} = 2D_a \quad (\text{obtained when } \theta = 0°, \phi = 0°) \quad (1.66a)$$

$$D_{xx} = -D_a\left(1 - \frac{3}{2}R\right) \quad (\text{when } \theta = 90°, \phi = 0°) \quad (1.66b)$$

$$D_{yy} = -D_a\left(1 + \frac{3}{2}R\right) \quad (\text{when } \theta = 0°, \phi = 0°) \quad (1.66c)$$

When the molecular structure is known, the principal components A_a and A_r of the alignment tensor can be determined from the measured RDCs. However, when the molecular structure is not available, these components are estimated from the distribution of the observed RDCs (Fig. 1.17). Because the magnitudes of the dipolar couplings are different for different types of internuclear bonds, it is necessary to normalize the observed residual couplings to the N–H dipolar coupling by scaling the observed dipolar coupling between nuclei i and j by a factor defined as $\gamma_N \gamma_H r_{ij}^3 / \gamma_i \gamma_j r_{NH}^3$ in order to take into account all dipolar couplings together. r_{ij} is the distance between nuclei i and j, and r_{NH} is the NH bond length. The distribution of the observed dipolar couplings can be obtained by plotting the histograms of all normalized dipolar couplings as shown in Fig. 1.17. For macromolecules, the histogram displays the powder pattern corresponding to that described by (1.65). Therefore, the alignment tensor elements A_a and R can be estimated from the singularities of residual dipolar splittings shown in the histogram. The estimated values are further optimized during the structural calculation by grid-searching the dipolar energy force as a function of the alignment tensor elements A_a and R.

From (1.63), it is clear that each observed RDC produces two or more oppositely oriented cones of possible internuclear vector orientations. For an asymmetric alignment tensor, possible bond orientations on as many as eight sets of cones can be obtained from each dipolar coupling. This ambiguity makes it very difficult

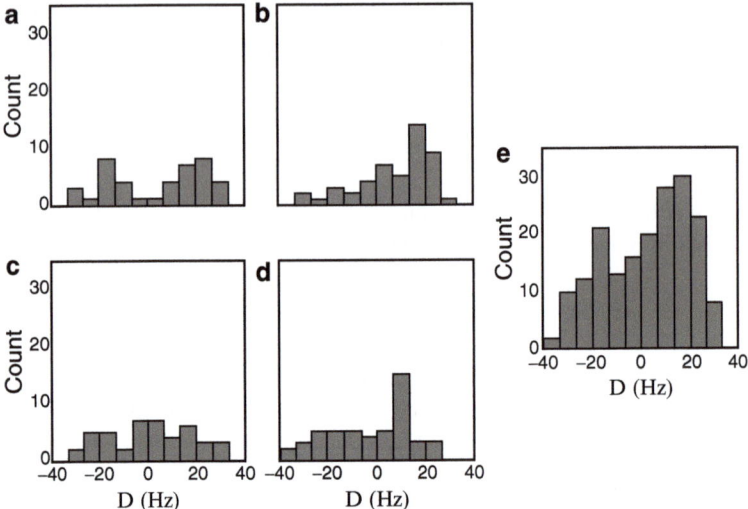

Fig. 1.17 Histograms of normalized residual dipolar couplings (RDCs) of (**a**) $^1D_{NH}$, (**b**) $^1D_{C^\alpha H^\alpha}$, (**c**) $^1D_{C'N}$, (**d**) $^1D_{NH}$. Because of the insufficient number of dipolar couplings, none of the individual type of dipolar couplings provides a powder pattern. (**e**) The addition of all the normalized dipolar couplings resembles a good powder pattern distribution (reproduced with permission from Baber et al. (1999), Copyright © 1999 Elsevier)

to determine the unique internuclear vector orientation without additional structural information. The degeneracy of the vector orientations can be reduced when two or more alignment tensors are obtained in different alignment media. A different alignment medium provides a different principal axis system, resulting in a different alignment tensor and thus resulting in a different set of orientation cones for a given dipolar coupling. The angle at the interception between two cones defined by the two alignment tensors yields the orientation of the vector.

1.9 Nuclear Overhauser Effect

When the resonance of a spin in an NMR spectrum is perturbed by saturation or inversion of the magnetization, it may cause the spectral intensities of other resonances in the spectrum to change. This phenomenon is called the nuclear Overhauser effect or NOE. The intensity change caused by NOE originates from the population changes of Zeeman states of coupled spins after the perturbation through the dipolar interaction. The origin of a steady-state NOE can clearly be illustrated for a two-spin-½ system, in which the two spins are coupled by dipolar interaction but there is no J coupling between the spins (Solomon 1955). Because the populations of spin states are changed by NOE, it is necessary to look at the process of the change in population in order to understand the effect.

1.9 Nuclear Overhauser Effect

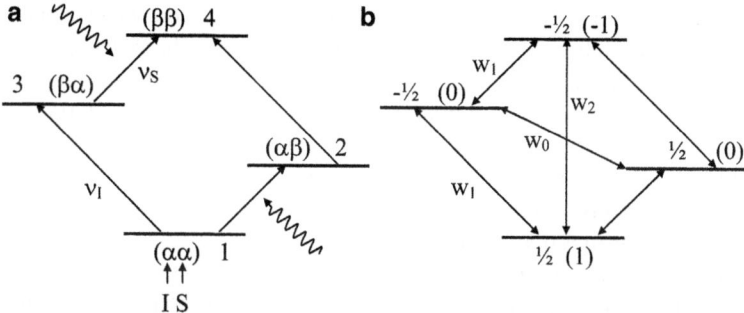

Fig. 1.18 Standard energy diagram for a two-spin system with $J_{IS} = 0$. (**a**) The four states are labeled as 1–4 corresponding to $\alpha\alpha$, $\alpha\beta$, $\beta\alpha$, and $\beta\beta$, respectively. The two S transitions of $1 \to 2$ and $3 \to 4$ have equal frequencies, as do the I transitions of $1 \to 3$ and $2 \to 4$. The S transitions are saturated by continuous RF irradiation at the resonance frequency of spin S. (**b**) The probabilities for zero-, single- and double-quantum transitions are represented by w_0, w_1, and w_2, respectively. The *numbers in parentheses* indicate the relative initial populations of energy levels, whereas the fractions are the populations after the saturation

The energy diagram for the two-spin system contains four energy states as shown in Fig. 1.18. The two spins are coupled via a dipolar interaction. For simplicity, the J coupling of the two spins is assumed to be zero. The resonance frequency of spin S is saturated by continuous radiation, which is denoted by transitions S of energy states $\alpha\alpha \to \alpha\beta$ and $\beta\alpha \to \beta\beta$. The energy states are labeled in such a way that the first state represents that for spin I and the second for spin S. For example, $\alpha\beta$ indicates that spin I is in state α and spin S in state β. To simplify the notation, the four energy levels are denoted as 1–4 with the lowest energy level being named as 1. Therefore, there are two S ($1 \to 2$, $3 \to 4$) and two I transitions ($1 \to 3$, $2 \to 4$). Because J coupling is absent between the two spins, the two transitions for spin I have the same resonance frequency, yielding a singlet in the NMR spectrum, as do the two transitions of spin S. Since in general the chemical shifts are different for the two spins in either homonuclear or heteronuclear system, resonances of the spins I and S do not overlap.

Upon saturating transitions of spin S, the population of level 1 is equal to that of 2 and levels 3 and 4 have equal populations. As a result, levels 1 and 3 are less populated compared to the equilibrium whereas the populations at levels 2 and 4 are increased by the saturation of the spin S resonance. After the irradiation, the system will try to restore the equilibrium through all allowable relaxation processes. The normal spin–lattice relaxation (see Sect. 1.10.2) of spin S, labeled as w_1, does not alter the population difference for spin I because the two transitions have the same relaxation rate, resulting in no change in population difference between the spin states. Therefore, the relaxation w_1 cannot change the intensity of spin I. However, in addition to spin–lattice relaxation via the above single-quantum transition there exist two other relaxation processes: w_0 via a zero-quantum transition with $\Delta m = 0$ and w_2 via a double-quantum transition with $\Delta m = 2$. Although these transitions are not directly observable in the NMR spectrum since both are forbidden according to the selection rule (see Sect. 1.2.2), they are allowed pathways for spin relaxations, known as cross relaxation, which is the relaxation caused by an exchange of magnetization between spins. The system perturbed by saturation will try to relax via these relaxation pathways to get back to equilibrium. These relaxations are governed primarily by nuclear DD interaction. In order words, they exist only when the two spins are close to each other and coupled by the dipolar interaction. Either of the two relaxation processes changes the intensity of spin I. Whether the intensity is enhanced or reduced after saturation depends on which relaxation mechanism is dominant.

Assuming that the equilibrium populations are $1, 0, 0$, and -1 at the energy levels 1–4, respectively, the populations are changed by the saturation of spin S transitions to $½, ½, -½$, and $-½$ (Fig. 1.18b). The fractions represent the relative populations to those at level 2 or 3 at equilibrium state, which are set to zero for simplicity. When the perturbed system relaxes back to equilibrium via w_0 alone, the populations at levels 2 and 3 eventually reach their equilibrium number, yielding the population distribution of $½, 0, 0, -½$. If it is observed and measured now, the intensity of spin I is reduced because the population difference for transitions $1 \rightarrow 3$ and $2 \rightarrow 4$ is decreased to $½$ from 1 at the initial equilibrium state. As a consequence, the intensity of spin I corresponding to the transitions is reduced by w_0 relaxation. On the other hand, if the system relaxes via the w_2 relaxation pathway the final populations at level 1 and 4 reach their equilibrium value but those of levels 2 and 3 do not relax, resulting in populations of $1, ½, -½, -1$ for levels 1–4, respectively. Now the population difference for transitions $1 \rightarrow 3$ and $2 \rightarrow 4$ is increased to $1½$ from the equilibrium value of 1. Hence, the enhanced intensity by NOE can be observed.

Which of the cross relaxation pathways is favorable in a system depends on the rate of molecular motion in solution, or the correlation time of molecules. For the dipolar interaction to cause cross relaxations, the local field produced by the dipolar interaction must fluctuate at a rate in the same scale as the frequency corresponding to the transition to be relaxed. It is easy to see from the energy diagram in Fig. 1.18b that the frequency corresponding to w_2 relaxation is in the megahertz range of the Larmor frequency, whereas that corresponding to w_0 is in the range of hertz to kilohertz because of the small energy gap for the transition. For small molecules that are tumbling fast at the frequency range of megahertz, w_2 is the dominant relaxation, resulting in observed intensity enhancement—positive NOE. On the other hand, large biological molecules tumbling slowly produce a local field fluctuating at the frequency range of hertz to kilohertz and hence favor w_0 relaxation, causing the intensity reduction, that is, negative NOE. For medium sized molecules, the two relaxation pathways are competing in the system. When w_0 and w_2 are compatible, causing the crossover between regimes such as for molecules with molecular weight of 1,000–3,000 Da, the NOE is very weak, if not zero.

The NOE enhancement factor is limited by the ratio of the cross-relaxation rate to the total relaxation rate of spin I:

$$\eta = \frac{\gamma_S}{\gamma_I} \frac{w_2 - w_0}{2w_1 + w_2 + w_0} \qquad (1.67)$$

in which γ_S and γ_I are the gyromagnetic ratios of spin S and spin I, respectively, the term $(2w_1 + w_2 + w_0)$ describes the total relaxation rate or dipolar spin–lattice relaxation, and $(w_2 - w_0)$ is the cross-relaxation rate. The enhancement factor can also be described by the intensities of spin I in the absence of the saturation, I_0, and in the presence of the irradiation, I:

$$\eta = \frac{I - I_0}{I_0} \qquad (1.68)$$

and the enhanced intensity is given by:

$$I = I_0(1 + \eta) \qquad (1.69)$$

For small molecules with short correlation times (extremely narrowing limit, $w_0 < w_2$, see Sect. 1.10.1), $w_0:w_1:w_2 = 2:3:12$. The maximum enhancement factor, η_{max} can be expressed in terms of gyromagnetic ratios of the two spins (Howarth 1987; Neuhaus and Williamson 1989):

$$\eta_{max} = \frac{\gamma_S}{2\gamma_I} \qquad (1.70)$$

Based on the above equation the maximum NOE enhancement obtained for small molecules in a homonuclear system is a factor of 0.5. For heteronuclear NOEs, the maximum factor is 1.99 for $^{13}C-^1H$ (I=^{13}C, S=^1H) and 2.24 for ^{31}P. For ^{15}N observation, the sign of the NOE enhancement is reversed compared to ^{13}C and ^{31}P, due to the negative gyromagnetic ratio, which generates an enhancement factor of -4.94. For large molecules with long correlation time (spin diffusion limit, $w_2 = w_1 = 0$), η_{max} is limited by:

$$\eta_{max} = -\frac{\gamma_S}{\gamma_I} \qquad (1.71)$$

Therefore, in large molecules, nuclei with positive gyromagnetic ratios give rise to negative NOEs. For instance, homonuclear NOE enhancement in large molecules is -1.0, and heteronuclear $^{13}C-^1H$ and $^{31}P-^1H$ have twice the magnitude relative to small molecules, -3.98 and -4.48, respectively. The $^{15}N-^1H$ NOE enhancement in biomolecules is nearly tenfold, at 9.88.

1.10 Relaxation

The macroscopic magnetization M_0 along the z axis of the rotating frame is rotated onto the y axis after a $90°_x$ pulse. The transverse magnetization will find ways to return to the z axis in the presence of the static magnetic field, B_0, which is its equilibrium state. In this section, questions to be addressed are regarding such issues in spin relaxations as the following:

1. What is the correlation time and how is it related to spin relaxation?
2. What are the autocorrelation function, the spectral density function, and their relationship?
3. What are T_1 and T_2 relaxations and where do they originate?
4. How can they be characterized?
5. How can they be determined experimentally?

The nuclear relaxation from excited states back to the ground states in the presence of a magnetic field undergoes different mechanisms than emission relaxation in optical spectroscopy because the spontaneous and stimulated emissions for the nuclear system are much less efficient relaxations, in that the resonance frequencies are several orders of magnitude smaller compared to those in optical spectroscopy. The magnetization perturbed by RF pulses relaxes back to the thermal equilibrium magnetization via two types of processes: T_1 relaxation along the static magnetic field direction and T_2 relaxation in the transverse plane perpendicular to the field direction (Abragam 1961). The former is a result of the nuclear coupling to the surroundings, which is characterized by spin–lattice or longitudinal relaxation time, whereas the latter is due to coupling between nuclei, which is characterized by the spin–spin or transverse relaxation time. During T_1 relaxation, the nuclei exchange energy with their surrounding or lattice, whereas there is no energy exchange with the lattice during T_2 relaxation. Compared to the electrons, the nuclear relaxation is very slow, which is in the order of milliseconds to minutes.

1.10.1 Correlation Time and Spectral Density Function

Although the DD interaction is averaged to zero in solution, the field of the DD interaction is not zero at any given instant. The nuclear relaxations are essentially caused by fluctuating interactions. The strength of the fluctuating field is measured by the autocorrelation function $G(\tau)$, which is the time average of the correlation between a field measured at time t and the same field measured at $(t + \tau)$ (Neuhaus and Williamson 1989):

$$G(\tau) = \overline{f(t)f(t+\tau)} \tag{1.72}$$

In which the bar represents the time average. $G(\tau)$ rapidly decays to zero as τ increases. For isotropic rotational diffusion of a rigid rod with a spherical top, the decay is frequently assumed to be exponential with a time constant τ_c. With this assumption, the autocorrelation function is reduced to:

$$G(\tau) = \frac{e^{-(\tau/\tau_c)}}{5} \tag{1.73}$$

1.10 Relaxation

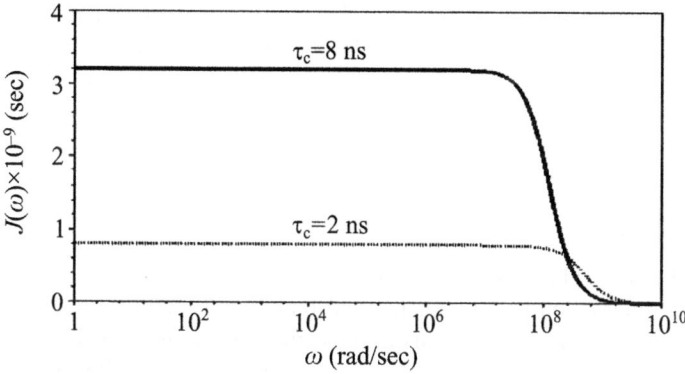

Fig. 1.19 Spectral density functions for an isotropic rotor with τ_c values of 2 ns (*dotted line*) and 8 ns (*solid line*) using (1.74)

in which τ_c is the decay time constant of the autocorrelation function, called the correlation time, which is defined as the mean time between reorientations or repositioning of a molecule. The correlation time is used to describe the rate of random motions and is expressed as the average time between collisions for translational motions or the time for a molecule to rotate one radian in rotational motions. Often, expressing the fluctuation of the field as a function of frequency is of interest. Then, the Fourier transformation of the autocorrelation function gives the correlation as function of frequency known as the spectral density function:

$$J(\omega) = \int_{-\infty}^{\infty} G(\tau)\, d\tau = \int_{-\infty}^{\infty} \frac{e^{(-\tau/\tau_c)}}{5} e^{-i\omega\tau} d\tau = \frac{2}{5} \frac{\tau_c}{(1+\omega^2\tau_c^2)} \quad (1.74)$$

which is a Lorentzian function. As shown in Fig. 1.19, $J(\omega)$ is unchanged when $\tau_c\omega \ll 1$ and decreases rapidly when $\tau_c\omega \approx 1$. The relationship between the autocorrelation function and the spectral density function is similar to that between a FID consisting of a single exponential decay and its Fourier transformed spectrum. Similar to autocorrelation function $G(\tau)$, $J(\omega)$ measures the strength of the fluctuating field in the frequency domain.

1.10.2 Spin–Lattice Relaxation

The nuclear relaxations are essentially caused by fluctuating interactions. The spin–lattice relaxation time T_1 describes the recovery of the *z* magnetization to its thermal equilibrium at which populations of the energy states reach the Boltzmann distribution. During T_1 relaxation, exchange of energy with the environment ("lattice") occurs due to various intra- and intermolecular interactions, including DD relaxation

(T_1^{DD}), chemical shift anisotropy (T_1^{CSA}), spin-rotation relaxation (T_1^{SR}), scalar coupling (T_1^{SC}), electric quadrupolar relaxation (T_1^{EQ}), interactions with unpaired electrons in paramagnetic compounds (T_1^{UE}), etc., which are summarized in the term:

$$\frac{1}{T_1} = \frac{1}{T_1^{DD}} + \frac{1}{T_1^{CSA}} + \frac{1}{T_1^{SR}} + \frac{1}{T_1^{SC}} + \frac{1}{T_1^{EQ}} + \frac{1}{T_1^{UE}} + \cdots \tag{1.75a}$$

$$R_1 = R_1^{DD} + R_1^{CSA} + R_1^{SR} + R_1^{SC} + R_1^{EQ} + R_1^{UE} + \cdots \tag{1.75b}$$

in which R_1 is T_1 relaxation rate constant.

The DD interaction is a dominant contribution to T_1 relaxation and causes the most efficient relaxation of protons in molecules in solution. Nuclear spins at the excited state can transit to ground states via energy exchange with surroundings or between nuclei. The energy exchanged to the lattice may be transformed into motions of translations, rotations and vibrations. The process of energy exchange is caused by the time-dependent fluctuation of magnetic (or electric) fields at or near nuclear the Larmor frequency. The fluctuating fields may be produced by vibrational, rotational, or translational motions of other surrounding nuclei, changes in chemical shielding, or unpaired electrons. In order for these time-dependent fluctuating fields to have significant effects on the nuclear relaxation, the random molecular motions or chemical shielding must have the same time scale as that of NMR, that is, close to the Larmor frequency. Consequently, molecular rotation and diffusion are the most efficient causes of nuclear relaxation in solution.

For an aqueous solution of normal viscosity, T_1 relaxation is inversely proportional to correlation time τ_c:

$$\frac{1}{T_1} \propto \tau_c \tag{1.76}$$

It indicates that slower random motions are responsible for a shorter T_1 relaxation time. Because the molecular random motions are influenced by the size of molecules, the magnitude of correlation time is significantly dependent on the molecular weight. For small molecules with molecular weight less than 100 Da, τ_c is in the range of 10^{-12} to 10^{-13} s, whereas macromolecules may have a τ_c as large as 10^{-8} s. Calculation of correlation time is a very complicated procedure in which many factors such as the shape of the molecule and different kinds of molecular motions must be taken into account. In general, τ_c is best estimated experimentally. One of the methods to estimate τ_c from experimental data is to make use of the spectral density functions (Sect. 1.10.1 and Chap. 8).

The time dependence of the process at which the initial macroscopic magnetization recovers through the spin–lattice relaxation is characterized by the T_1 relaxation time, which is described by the Bloch equation:

$$\frac{dM_z}{dt} = \frac{M_z - M_0}{T_1} \tag{1.77}$$

1.10 Relaxation

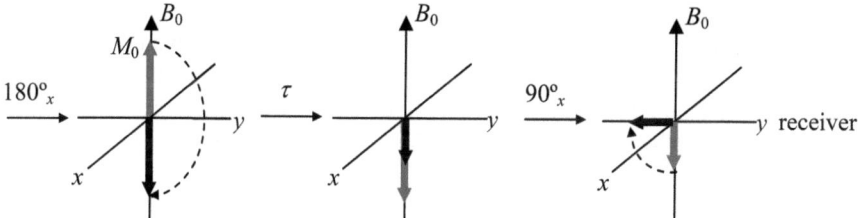

Fig. 1.20 Vector representation of T_1 relaxation time measurement. The 180° pulse rotates the magnetization M_0 to the $-z$ axis. The M_z recovers via the T_1 relaxation (along the z axis). After a delay τ, a 90° pulse along the x axis rotates the magnetization remaining on the $-z$ axis onto the transverse plane, where it is detected

The equation describes the T_1 relaxation for a system without spin coupling between the nuclei. The solution for the above Bloch equation is readily obtained by integrating the equation:

$$\frac{dM_z}{M_z - M_0} = \frac{dt}{T_1} \tag{1.78}$$

$$\int_0^{M_z} \frac{d(M_z - M_0)}{M_z - M_0} = \int_0^t \frac{dt}{T_1} \tag{1.79}$$

$$\ln M_0 - \ln(M_z - M_0) = \frac{t}{T_1} \tag{1.80}$$

$$M_z = M_0(1 - e^{-t/T_1}) \tag{1.81}$$

Because M_z is not observable directly, T_1 is experimentally determined by the following pulse sequence:

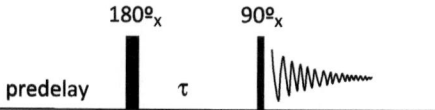

A pulse sequence is a train of pulses used to manipulate the magnetization of nuclei in a static magnetic field to produce time domain signals (FID). In a pulse sequence, the x axis represents the time events and the y axis describes the amplitude of the pulses. The 180° pulse of the T_1 experiment rotates the magnetization M_0 to the $-z$ axis. Since there exists no transverse magnetization after the 180° pulse, the recovery of the M_z comes only via the T_1 relaxation (along the z axis). As a result, there is no contribution from T_2 transverse relaxation (Fig. 1.20). After a delay τ, a 90° pulse along the x axis rotates the magnetization remaining on the $-z$ axis into the transverse plane, where it is detected. By arraying parameter τ, the signal intensities of the resonances change from negative to positive. When τ is set to be much longer than T_1, the intensities reach the maximum, which corresponds to full recovery of M_0. The T_1 constant can be calculated using the intensities at different τ values according to the following equations.

By integrating the Bloch equation for M_z, the time dependence of M_z is given by:

$$M_0 - M_z = A e^{-t/T_1} \qquad (1.82)$$

Time parameter t is the same as the delay τ in the inverse experiment and M_z is the transverse magnetization (signal intensity) created by the last 90° pulse. When $t=0$, $M_z = -M_0$. Therefore, $A = 2M_0$. By substituting A with $2M_0$ and taking logarithms on both sides of (1.82), we obtain:

$$\ln(M_0 - M_z) = \ln 2M_0 - \frac{t}{T_1} \qquad (1.83a)$$

$$\ln(I_0 - I) = \ln 2I_0 - \frac{t}{T_1} \qquad (1.83b)$$

in which I and I_0 are the peak volumes or intensities at $t=\tau$ and at $t=\infty$, respectively. In practice, I_0 is obtained with a τ value long enough (30–60 s) to allow the magnetization to fully relax back to the equilibrium state. When $t \equiv \tau_{null} = T_1 \ln 2$, $M_z = 0$ and thus $I_0 = 0$ (τ_{null} is the value of the delay τ at which M_z is zero). Therefore, T_1 can be determined by:

$$T_1 = \frac{\tau_{null}}{\ln 2} = 1.443 \tau_{null} \qquad (1.84)$$

T_1 can be calculated using (1.84) for all peaks in the spectrum.

1.10.3 T_2 Relaxation

T_2 relaxation (also known as spin–spin or transverse relaxation) describes the decay of transverse magnetization characterized the by the Bloch equations:

$$\frac{dM_x}{dt} = -\frac{M_x}{T_2} \qquad (1.85)$$

or

$$\frac{dM_y}{dt} = -\frac{M_y}{T_2} \qquad (1.86)$$

1.10 Relaxation

in which the time constant T_2 is called spin–spin or transverse relaxation time, which describes how fast the transverse magnetization M_x or M_y decays to zero. Because the transverse magnetization M_x and M_y are observable signals, T_2 relaxation time determines the decay rate of an FID, e^{-t/T_2}, which corresponds to the signal line width in frequency domain, $\Delta v_{1/2} = 1/(\pi T_2)$. $\Delta v_{1/2}$ is defined as the line width at half height of the signal amplitude.

T_2 relaxation does not cause population changes in the energy states and the energy of the system is not affected by the relaxation. The process is adiabatic. In the presence of fluctuating spin–spin interactions, energy is exchanged between nuclei. During spin–spin relaxation, the transition of one nucleus from a high energy state to a lower one causes another nucleus to move simultaneously from the lower state to the higher one. There is no energy exchange with the environment and hence no gain or loss in energy of the nuclear system. As a result, the phase coherence of the spins generated by the B_1 field is lost.

In practice, magnetic field inhomogeneity is the dominant contribution to the transverse relaxation. Each nucleus across the sample volume experiences a slightly different B_0 field caused by the inhomogeneity. Conversely, some of the chemically equivalent nuclei precess faster and some slower. This results in the fanning-out of individual magnetization vectors. The net effect is a loss in phase coherence similar to that caused by fluctuating spin interactions. By taking into account the effect of B_0 inhomogeneity, the transverse relaxation is described by the effective transverse relaxation time T_2^*:

$$M_y = M_0 e^{-(t/T_1^*)} \tag{1.87}$$

and

$$\Delta v_{1/2} = \frac{1}{\pi T_2^*} = \frac{1}{\pi T_2} + \gamma \Delta B_0 \tag{1.88}$$

The equation for M_y tells us that T_2^* is the time when the amplitude of an FID has decayed by a factor of $1/e$ (Fig. 1.21). The first term in (1.88) represents the natural line width caused by the spin–spin relaxation, whereas the second term is the contribution of field inhomogeneity to the spectral line width. When molecules in nonviscous liquids are moving very rapidly, $T_1 = T_2$ for nuclei with spin ½, which is called the extreme-narrowing limit. Therefore, T_2 is so long that the line widths can be narrower than 0.1 Hz. T_2 can be estimated from the determined T_1 relaxation time. For macromolecules or solid-state samples, $T_1 > T_2$, in which case the line width is broad.

When the system is not in the extreme-narrowing limit, T_2 is determined experimentally by the spin echo method which eliminates the effect of field inhomogeneity. The $T_{1\rho}$ experiment (Sect. 8.2.2) utilizes the spin lock method to lock the magnetization on a transverse axis, resulting in the measurement of the T_2 relaxation time.

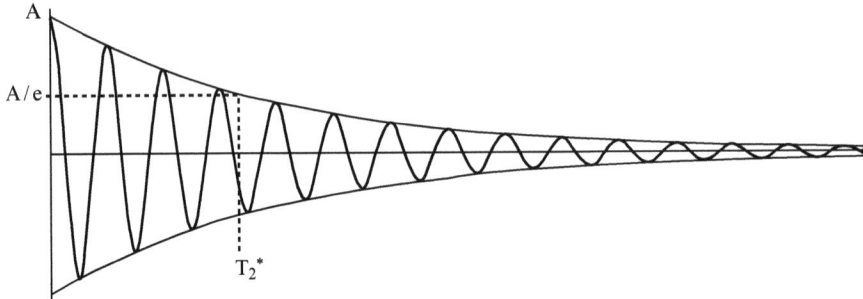

Fig. 1.21 Effective transverse relaxation time T_2^* of a FID, which is determined at the time when the amplitude of a FID has decayed by a factor of $1/e$

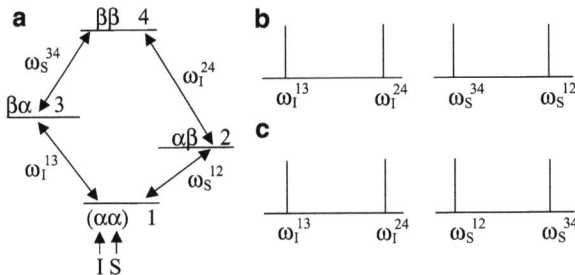

Fig. 1.22 (a) Standard energy diagram, (b, c) schematic spectra of IS spin system. The doublets of spin S with a negative gyromagnetic ratio in (c) are reversed relative to those with a positive gyromagnetic ratio in (b)

For proteins and nucleic acids, T_2 relaxation is dominated by the chemical shift anisotropy interaction and DD interaction with other spins (Brutscher 2000). A single spin interaction causes autocorrelated relaxation, whereas the interference between different nuclear interactions such as between CSA and DD, or between DD and DD interactions, gives rise to cross-correlated relaxation. Both autocorrelated and cross-correlated relaxations refer to the relaxation mechanism, which differ from cross relaxation, which refers to the relaxation pathway. In a weakly coupled two-spin-½ system, the contribution to T_2 relaxation from the cross-correlated relaxations by CSA and DD interactions (also called cross-correlation, see Sect. 8.1) leads to different relaxation rates for individual multiplet components in the spectrum. The CSA has the same effect on the T_2 relaxation of both components of the doublet. However, the influence of DD coupling on the T_2 relaxation of the doublet components is antisymmetric: the effect on the α transitions (ω_I^{13} and ω_S^{12}, Fig. 1.22) is the same as for the CSA, whereas the influence of DD coupling on the β transitions (ω_I^{24} and ω_S^{34}) is opposite to that of CSA. When CSA and DD coupling are comparable in magnitude and the CSA principle symmetry axis is aligned collinearly to the vector of DD coupling, the line shapes of the resonances at ω_I^{24} and ω_S^{34} are narrowed because

1.11 Selection of Coherence Transfer Pathways

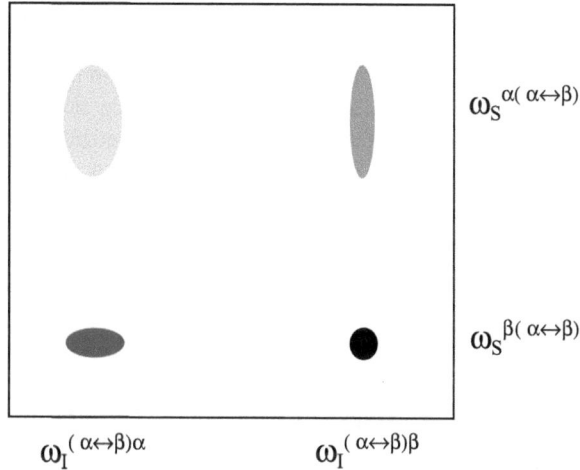

Fig. 1.23 Schematic representation of a 2D correlation spectrum of coupled IS spin system in the presence of cross-correlated relaxation of CSA and DD coupling. The line shape of the cross peak corresponding to the two β transitions is not broadened by the cross-correlated relaxation due the cancelation of their opposite effects on T_2 relaxation when CSA and DD coupling are aligned collinearly with comparable magnitude. Note that $\omega_I^{(\alpha\leftrightarrow\beta)\alpha}$, $\omega_I^{(\alpha\leftrightarrow\beta)\beta}$, $\omega_S^{\alpha(\alpha\leftrightarrow\beta)}$, and $\omega_I^{\beta(\alpha\leftrightarrow\beta)}$ are the same as ω_I^{13}, ω_I^{24}, ω_S^{12}, and ω_S^{34} in Fig. 1.22, respectively

the effects on T_2 relaxation from CSA and DD coupling cancel each other at the resonances (Fig. 1.23). This property of cross-correlated relaxation has been used to reduce transverse relaxation of large proteins by TROSY experiments (Chap. 5).

1.11 Selection of Coherence Transfer Pathways

Coherence is a term representing the transverse magnetization. In a coupled two-spin system, single-quantum coherence involves one spin changing its spin state ($\alpha \rightarrow \beta$, or $\beta \rightarrow \alpha$), whereas double-quantum coherence arises from a transition in which two spins alter their states at the same time ($\alpha\alpha \rightarrow \beta\beta$, or $\beta\beta \rightarrow \alpha\alpha$). The zero quantum coherence is referred to as a transition of $\Delta m = 0$ with two spins changing their states in opposite directions ($\alpha\beta \rightarrow \beta\alpha$, or $\beta\alpha \rightarrow \alpha\beta$). The transition rule governs such that only a transition with $\Delta m = \pm 1$ between the spins is allowed, meaning that single-quantum coherence is the only directly observable coherence. The type of transition, that is, the value of Δm, is known as coherence order, p. A coherence order of $p = \pm 1$ represents single-quantum coherence, whereas a coherence order of $p = 0$ is zero-quantum coherence or z magnetization.

Because coherence order corresponds to a transition between spins, only RF pulses cause a change in coherence order from one level to another, which is referred to as coherence transfer. Delays without an RF field conserve the coherence orders. A diagram is used to describe the coherence transfer at the different stages of a pulse sequence, called the coherence transfer pathway (Bain 1984; Bodenhausen et al. 1984). In order to use the diagram for coherence selection, the pathway must originate at the equilibrium state in which the magnetization is along the magnetic field direction (z magnetization) possessing a coherence order of zero, $p=0$. The first 90° pulse applied on the equilibrium z magnetization only generates single-quantum coherence, $p=\pm 1$, and the last pulse in a pulse sequence must bring the coherence to the coherence level of $p=-1$ for the use of quadrature detection to observe the complex signals (Sørensen et al. 1983). A noninitial 90° pulse (a 90° pulse other than the first one) generates higher order coherence along with single-quantum coherence.

1.12 Approaches to Understanding NMR Experiments

During an NMR pulse sequence, the equilibrium magnetization is manipulated to generate detectable signals. Several formalisms have been developed to describe the behavior of nuclear magnetization during a pulse sequence. In this section, three theoretical treatments (vector model, product operator formalism and density matrix) used to describe how magnetization is transferred during NMR experiments are briefly discussed. Simple pulse sequences are used as examples to help understand the approaches.

Questions to be addressed include the following:

1. What are the advantages and limitations of each approach?
2. How are the approaches used to describe how experiments work?

1.12.1 Vector Model

For a variety of experiments [such as the one-pulse experiment (Fig. 1.7), T_1 measurement, spin echoes, polarization transfer experiments and composite pulses] the vector model remains a useful approach for analysis of the experiments because of its visualization and simplicity. However, the vector model is inadequate to analyze certain experiments such as DEPT (distortionless enhancement by polarization transfer, Doddrell et al. 1982; Bendall and Pegg 1983) because it is unable to describe multiple quantum coherence.

Shown in Fig. 1.24 is the vector representation of magnetization during the pulse sequence of INEPT (insensitive nuclei enhanced by polarization transfer, Morris and Freeman 1979) which is used to transfer magnetization from one nucleus (usually ^{1}H) to a heteronucleus through J coupling. It is the most utilized building block of many heteronuclear NMR pulse sequences. The first 90° ^{1}H pulse rotates the ^{1}H magnetization onto the y axis at point a. After the first $1/(4J_{CH})$ delay, each of the doublets caused by the J_{CH} coupling is 45° away from the y axis. The simultaneous 180° pulses on both ^{1}H and ^{13}C refocus the chemical shift but allow the

1.12 Approaches to Understanding NMR Experiments

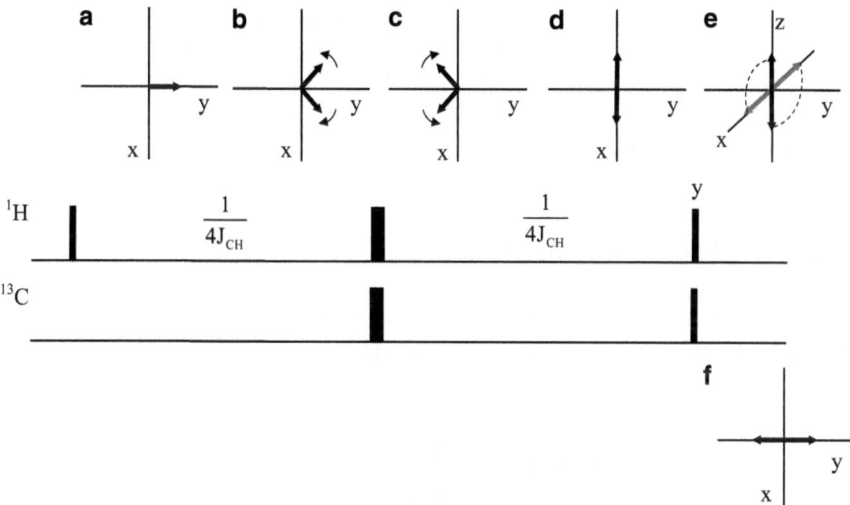

Fig. 1.24 Vector representation of INEPT experiment. The *narrow* and *wide bars* are 90 and 180° pulses, respectively. All pulses are x phase except the last ^{1}H 90°_y. After a 90°_x pulse (*point a*), the ^{1}H magnetization is on the y axis. After the first $1/(4J_{CH})$ delay, each of the doublets caused by the J_{CH} coupling is 45° away from the y axis (*b*). The simultaneous 180° pulses on both ^{1}H and ^{13}C refocus the chemical shift but allow the coupled ^{1}H vectors to continue to diverge (*c*). After the second $1/(4J_{CH})$ delay, the two vectors are 180° out of phase and aligned on the x axis (*d*). The following ^{1}H 90° pulse rotates the ^{1}H magnetization components back to the z and $-z$ axis (*e*), and the ^{13}C 90° pulse brings the coupled magnetization components back to the xy plane, which results in detectable antiphase magnetization (*f*)

coupled ^{1}H vectors to continue to diverge. This can be understood by considering the effect of ^{1}H and ^{13}C 180° pulses individually. After the ^{1}H 180° pulse flips the doublets about the x axis, the ^{13}C 180° inverts the populations of ^{13}C states of the coupled doublets but has no effect on uncoupled chemical shift. Consequently, the magnetization in the ^{13}C α state becomes that in the β state, and vice versa. It is represented in the vector diagram at point c that the slower and faster vectors exchange places, resulting in the vectors rotating in the other direction (c). After the second $1/(4J_{CH})$ delay, the two vectors are 180° out of phase and aligned on the x axis (d). The following ^{1}H 90° pulse rotates the ^{1}H magnetization components back to the z and $-z$ axes (e), and ^{13}C 90° pulse brings the coupled magnetization components back the xy plane, which results in detectable antiphase magnetization (f).

1.12.2 Product Operator Description of Building Blocks in a Pulse Sequence

Product operator formalism has become a popular approach for the theoretical description of NMR experiments because it combines the simplicity and visualization of the vector representation with quantum mechanics. The approach utilizes a linear combination of base operators to express the density matrix. Shorthand notation can be used to describe the operator matrices (Sørensen et al. 1983). Therefore, it does not require an understanding of quantum mechanics to utilize the formalism. The standard notation of the product operators for a two-spin system with their physical interpretation is explained in Appendix A.

1.12.2.1 Spin Echo of Uncoupled Spins

The spin echo experiment consists of two delays separated by a 180° pulse after the transverse magnetization is produced (Carr and Purcell 1954):

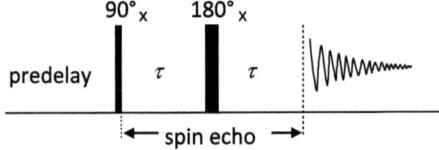

For an isolated spin, the magnetization starting at the equilibrium state proportional to I_z will undergo a series of changes during the spin echo sequence. Upon the application of the first $90°_x$ pulse [denoted by $(2/\pi)I_x$], the equilibrium magnetization I_z is converted to $-I_y$:

$$I_z \xrightarrow{(\pi/2)I_x} -I_y \tag{1.89}$$

During the first period τ of free precession of the spin echo, the magnetization evolves to:

$$-I_y \xrightarrow{\Omega_i I_z \tau} -I_y \cos(\Omega_i \tau) + I_x \sin(\Omega_i \tau) \tag{1.90}$$

in which Ω_i is the frequency of the chemical shift precession. Term $\Omega_i I_z \tau$ means that the magnetization of spin I precesses about the z axis for τ period at spin resonance frequency Ω_i, and $\Omega_i \tau$ represents an angle rotated by $\Omega_i I_z \tau$. The $180°_x$ pulse inverts I_y magnetization, but does not have any effect on I_x magnetization:

$$-I_y \cos(\Omega_i \tau) + I_x \sin(\Omega_i \tau) \xrightarrow{\pi I_x} I_y \cos(\Omega_i \tau) + I_x \sin(\Omega_i \tau) \tag{1.91}$$

In the final delay of the spin echo sequence, the free precession yields

$$I_y \cos(\Omega_i \tau) + I_x \sin(\Omega_i \tau) \xrightarrow{\Omega_i I_z \tau} [I_y \cos^2(\Omega_i \tau) - I_x \sin(\Omega_i \tau) \cos(\Omega_i \tau)]$$
$$+ [I_x \sin(\Omega_i \tau) \cos(\Omega_i \tau) + I_y \sin^2(\Omega_i \tau)] \tag{1.92}$$

1.12 Approaches to Understanding NMR Experiments

Because $\cos^2\theta + \sin^2\theta = 1$, (1.92) reduces to

$$I_y \cos(\Omega_i \tau) + I_x \sin(\Omega_i \tau) \xrightarrow{\Omega_i I_z \tau} I_y \tag{1.93}$$

Therefore, the net effect of a spin echo sequence on uncoupled spins is to change the sign of the transverse magnetization:

$$-I_y \xrightarrow{\tau \to \pi I_x \to \tau} I_y \tag{1.94}$$

There is no net evolution of chemical shift during the spin echo sequence because evolution of the chemical shift is refocused upon the completion of the spin echo sequence.

1.12.2.2 Spin Echo of Coupled Spins

For two coupled spins, the spin echo sequence can be applied to homonuclear spins or the following pulse sequence can be used for coupled heteronuclear spins.

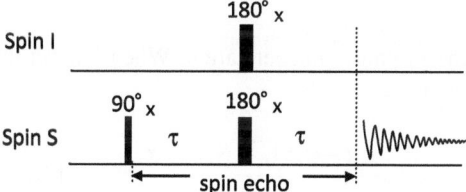

The two 180° pulses on the heteronuclear spins will have the same effect as the nonselective homonuclear 180° pulse on the coupled homonuclear spins. During the evolution period, scalar coupling is the only interaction to be considered because the evolution under the chemical shift interaction will be refocused by the spin echo sequence as described for the uncoupled spin. For heteronuclear magnetization after a $90°_x$ on the S spin, the evolution during the first τ period under the scalar coupling J_{IS} converts the in-phase magnetization of $-S_y$ to orthogonal antiphase magnetization, which is represented by the product operators:

$$-S_y \xrightarrow{\pi J_{IS}\tau} -S_y \cos(\pi J_{IS}\tau) + 2I_z S_x \sin(\pi J_{IS}\tau) \tag{1.95}$$

The 180° pulse on the S spin does not have any effect on the I spin and vice versa. Therefore, the 180° pulse on S spin changes the sign of S_y and the 180° pulse on the I spin inverts I_z:

$$-S_y \cos(\pi J_{IS}\tau) + 2I_z S_x \sin(\pi J_{IS}\tau) \xrightarrow{\pi S_x} S_y \cos(\pi J_{IS}\tau) + 2I_z S_x \sin(\pi J_{IS}\tau)$$

$$\xrightarrow{\pi I_x} S_y \cos(\pi J_{IS}\tau) - 2I_z S_x \sin(\pi J_{IS}\tau) \tag{1.96}$$

The conversion of the magnetization during the second evolution is described by:

$$S_y \cos(\pi J_{IS}\tau) - 2I_z S_x \sin(\pi J_{IS}\tau) \xrightarrow{\pi J_{IS}\tau} [S_y \cos^2(\pi J_{IS}\tau) - 2I_z S_x \cos(\pi J_{IS}\tau) \sin(\pi J_{IS}\tau)]$$

$$+ [-2I_z S_x \sin(\pi J_{IS}\tau) \cos(\pi J_{IS}\tau) - S_y \sin^2(\pi J_{IS}\tau)] \quad (1.97)$$

Using $\cos^2\theta - \sin^2\theta = \cos 2\theta$ and $2\cos\theta \sin\theta = \sin 2\theta$, the equation is simplified to:

$$S_y \cos(\pi J_{IS}\tau) - 2I_z S_x \sin(\pi J_{IS}\tau) \xrightarrow{\pi J_{IS}\tau} S_y \cos(2\pi J_{IS}\tau) - 2I_z S_x \sin(2\pi J_{IS}\tau) \quad (1.98)$$

Therefore,

$$-S_y \xrightarrow{\tau \rightarrow \pi(I_x + S_x) \rightarrow \tau} S_y \cos(2\pi J_{IS}\tau) - 2I_z S_x \sin(2\pi J_{IS}\tau) \quad (1.99)$$

If τ is set to $1/(2J_{IS})$, then

$$-S_y \xrightarrow{\tau \rightarrow \pi(I_x + S_x) \rightarrow \tau} -S_y \quad (1.100)$$

which gives the same in-phase magnetization. When τ is set to $1/(4J_{IS})$, antiphase coherence is generated:

$$-S_y \xrightarrow{\tau \rightarrow \pi(I_x + S_x) \rightarrow \tau} -2I_z S_x \quad (1.101)$$

For initial magnetization from spin I, $-I_y$, the result can be obtained by interchanging the I and S operators in the equations for the magnetization initiated at $-S_y$:

$$-I_y \xrightarrow{\tau \rightarrow \pi(I_x + S_x) \rightarrow \tau} -2I_x S_z \quad (1.102)$$

The above result is for the real part of the FID. If considering the complex data observed by quadrature detection, the operator of I_x in (1.101) or (1.102) is replaced by raising operator I^+ and lowering operator I^-

$$I^+ = I_x + iI_y$$

$$I^- = I_x - iI_y \quad (1.103)$$

in which I_y yields the observable imaginary magnetization.

1.12 Approaches to Understanding NMR Experiments

1.12.2.3 Insensitive Nuclei Enhanced by Polarization Transfer

The pulse sequence (Fig. 1.24) in INEPT before the final two 90° pulses is the same as the heteronuclear echo sequence. Hence, the antiphase magnetization of $-2I_xS_z$ is generated when $\tau = 1/(4J_{IS})$. In next step of INEPT, a $90°_y$ pulse is applied to spin I and $90°_x$ is applied to spin S simultaneously, which produces an antiphase observable S magnetization given by:

$$-2I_xS_z \xrightarrow{(\pi/2)(I_y + S_x)} -2I_zS_y \qquad (1.104)$$

or

$$-I_y \xrightarrow{\text{INEPT}} -2I_zS_y$$

The sensitivity of spin S in the INEPT is enhanced by spin I for a factor of γ_I/γ_S. In the antiphase signal, one component of the doublet has negative intensity, whereas the other is positive. The INEPT sequence is used in multidimensional heteronuclear experiments to transfer magnetization through scalar coupling between heteronuclei.

Because of the antiphase character of the magnetization after INEPT, ^{1}H decoupling, which is used to increase sensitivity, cannot be applied during the acquisition. A refocused INEPT sequence is used to convert antiphase magnetization in the INEPT sequence to in-phase coherence by appending additional spin echo sequence after the INEPT:

$$-2I_zS_y \xrightarrow{\pi J_{IS}\tau} -2I_zS_y \cos(\pi J_{IS}\tau) + S_x \sin(\pi J_{IS}\tau)$$
$$\xrightarrow{\pi(I_xS_x)} -2I_zS_y \cos(\pi J_{IS}\tau) + S_x \sin(\pi J_{IS}\tau)$$
$$\xrightarrow{\pi J_{IS}\tau} -2I_zS_y \cos(2\pi J_{IS}\tau) + S_x \sin(2\pi J_{IS}\tau) \qquad (1.105)$$

When $\tau = 1/(4J_{IS})$, the cosine term equals zero. Hence, the magnetization after the last spin echo sequence is given by:

$$-2I_zS_y \xrightarrow{\tau \to \pi(I_x + S_x) \to \tau} S_x \qquad (1.106)$$

which is an observable in-phase magnetization. Therefore, for the refocused INEPT, decoupling can be applied during acquisition to increase sensitivity by collapsing the I–S coupling doublet (Fig. 1.25b).

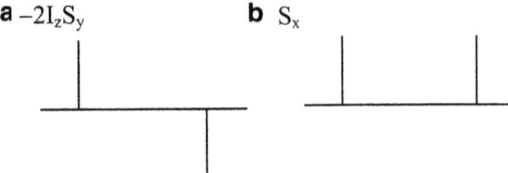

Fig. 1.25 The doublets of spin S corresponding to the product operator of (**a**) antiphase and (**b**) in-phase coherence

1.12.3 Introduction to Density Matrix

The density matrix approach (Fano 1957; Howarth et al. 1986; Hore et al. 2001) is used to treat a more complicated case involving two or more spins by solving density matrices for the operators involved in NMR experiments. For a system consisting of uncoupled spins of ½, the energy E of the two states for the Schrödinger equation is given by:

$$H(t)\Psi(t) = E\Psi(t) \tag{1.107}$$

The equation describes that the time-dependent Hamiltonian operator $H(t)$ governs the time dependence of the system wave function $\Psi(t)$. The system wave function can be represented in the Dirac notation (Dirac 1967) by a linear combination of the basis functions $\phi_1 = |\alpha\rangle$ and $\phi_2 = |\beta\rangle$ for a single-spin-½ system:

$$\Psi(t) = C_1|\alpha\rangle + C_2|\beta\rangle \tag{1.108}$$

$$H = -\gamma B_0 \hbar \hat{I}_z = \hbar \omega_0 \hat{I}_z \tag{1.109}$$

in which $\omega_0 = -\gamma B_0$ is the Larmor frequency, and

$$H|\alpha\rangle = \frac{1}{2}|\alpha\rangle, \quad H|\beta\rangle = -\frac{1}{2}|\beta\rangle \tag{1.110}$$

in which the eigenvalues are represented in units of $\hbar\omega_0$ for simplicity. Populations of the energy states are given by:

$$N_\alpha = \frac{e^{-(E_1/kT)}}{\sum_{i=1}^{2} e^{-(E_i/kT)}} = \frac{1}{2} - \frac{E_1}{2kT} = \frac{1}{2}\left(1 - \frac{\hbar\omega_0}{2kT}\right) \tag{1.111}$$

$$N_\beta = \frac{e^{-(E_2/kT)}}{\sum_{i=1}^{2} e^{-(E_i/kT)}} = \frac{1}{2} - \frac{E_2}{2kT} = \frac{1}{2}\left(1 + \frac{\hbar\omega_0}{2kT}\right) \tag{1.112}$$

The eigenvalues (energy) can be use to represent the operator $\hat{I}_z$ in term of a matrix:

$$\mathbf{I}_z = \begin{pmatrix} \frac{1}{2} & 0 \\ 0 & -\frac{1}{2} \end{pmatrix} \tag{1.113}$$

1.12 Approaches to Understanding NMR Experiments

The operators of the transverse x and y components interconvert $|\alpha\rangle$ and $|\beta\rangle$ because $|\alpha\rangle$ and $|\beta\rangle$ are not the eigenstates of the operators:

$$\hat{I}_x|\alpha\rangle = \frac{1}{2}|\beta\rangle, \quad \hat{I}_x|\beta\rangle = \frac{1}{2}|\alpha\rangle \tag{1.114}$$

$$\hat{I}_y|\alpha\rangle = \frac{1}{2}i|\beta\rangle, \quad \hat{I}_y|\beta\rangle = -\frac{1}{2}i|\alpha\rangle \tag{1.115}$$

which can be understood by the raise and lower operators $\hat{I}_+$ and $\hat{I}_-$:

$$\hat{I}_x|\alpha\rangle = \frac{1}{2}(\hat{I}_+ + \hat{I}_-)|\alpha\rangle = \frac{1}{2}(0 + |\beta\rangle) = \frac{1}{2}|\beta\rangle \tag{1.116}$$

$$\hat{I}_y|\beta\rangle = \frac{i}{2}(\hat{I}_- - \hat{I}_+)|\beta\rangle = \frac{i}{2}(0 - |\alpha\rangle) = -\frac{i}{2}|\alpha\rangle \tag{1.117}$$

The corresponding matrices for the operators are given by:

$$\mathbf{I}_x = \begin{pmatrix} 0 & \frac{1}{2} \\ \frac{1}{2} & 0 \end{pmatrix} = \frac{1}{2}\begin{pmatrix} 0 & 1 \\ 1 & 0 \end{pmatrix} = \frac{1}{2}\sigma_x \tag{1.118}$$

$$\mathbf{I}_y = \begin{pmatrix} 0 & -\frac{1}{2}i \\ \frac{1}{2}i & 0 \end{pmatrix} = \frac{1}{2}\begin{pmatrix} 0 & -i \\ i & 0 \end{pmatrix} = \frac{1}{2}\sigma_y \tag{1.119}$$

$$\sigma_z = \begin{pmatrix} 1 & 0 \\ 0 & -1 \end{pmatrix} \tag{1.120}$$

in which σ_x, σ_y and σ_z are the Pauli matrices with the property of $\sigma_r^2 = \sigma_0$ for $r = x$, y, z:

$$\sigma_0 = \begin{pmatrix} 1 & 0 \\ 0 & 1 \end{pmatrix} \tag{1.121}$$

The bras and kets are represented by vectors:

$$\langle\alpha| = (1 \quad 0); \quad \langle\beta| = (0 \quad 1) \tag{1.122}$$

$$|\alpha\rangle = \begin{pmatrix} 1 \\ 0 \end{pmatrix}; \quad |\beta\rangle = \begin{pmatrix} 0 \\ 1 \end{pmatrix} \tag{1.123}$$

By combining the matrices $\mathbf{I}_x$ and $\mathbf{I}_y$ the matrices for the raise and lower operators can be obtained:

$$\mathbf{I}_+ = \mathbf{I}_x + i\mathbf{I}_y = \begin{pmatrix} 0 & 1 \\ 0 & 0 \end{pmatrix} \tag{1.124}$$

$$\mathbf{I}_- = \mathbf{I}_x - i\mathbf{I}_y = \begin{pmatrix} 0 & 0 \\ 1 & 0 \end{pmatrix} \tag{1.125}$$

Matrices $\mathbf{I}_x$, $\mathbf{I}_y$, and $\mathbf{I}_z$ represent the magnetization along the x, y, and z axis, respectively. Quantum mechanics states that the expectation value of an operator $\hat{A}$ depends on the products of coefficients, which is given by:

$$\langle \hat{A} \rangle = \langle \Psi(t) | \hat{A} | \Psi(t) \rangle = \sum_{k,l} C_k^* C_l \langle k | \hat{A} | l \rangle = \sum_{k,l} C_k^* C_l A_{kl} \tag{1.126}$$

in which C_k, C_l are defined in (1.108) for a single-spin-½ system. Therefore, it is useful to define a density matrix $\rho(t)$ with an individual matrix element in the term of:

$$\rho_{lk} = \langle k | \rho | l \rangle = C_k^* C_l \tag{1.127}$$

The density matrix for a single-spin-½ system at equilibrium state is given by

$$\rho = \begin{pmatrix} \rho_{11} & \rho_{12} \\ \rho_{21} & \rho_{22} \end{pmatrix} = \begin{pmatrix} N_\alpha & 0 \\ 0 & N_\beta \end{pmatrix} = \frac{1}{2}\begin{pmatrix} 1 & 0 \\ 0 & 1 \end{pmatrix} + \frac{1}{2}\begin{pmatrix} \delta & 0 \\ 0 & -\delta \end{pmatrix}$$
$$= \frac{1}{2}\sigma_0 + \delta \mathbf{I}_z \tag{1.128}$$

in which $\delta = -\hbar\omega_0/2kT$. The unit matrix σ_0 is not of NMR interest because it does not evolve during any Hamiltonian. The δ is a scaling factor. Thus σ_0 and δ are often omitted. The expectation value of $\hat{A}$ is represented by the trace of the product ρ and A:

$$\langle \hat{A} \rangle = \sum_{k,l} C_k^* C_l A_{kl} = \sum_{k,l} \rho_{lk} A_{kl} = \sum_l (\rho A) = \text{Tr}(\rho A) \tag{1.129}$$

Tr( ) is the trace of a matrix defined as the sum of the diagonal elements of the matrix (here, the product of the matrices). By solving the Liouville–von Neumann equation

$$i\frac{d\hat{\rho}}{dt} = [H, \hat{\rho}] \tag{1.130}$$

the time-dependent density matrix can be described in term of operator exponentials:

$$\rho(t) = e^{-iHt} \rho(0) e^{iHt} \tag{1.131}$$

For RF pulses in the rotating frame, the Hamiltonian is given by:

$$H = \omega_1 \hat{\mathbf{I}}_x \quad \text{or} \quad H = \omega_1 \hat{\mathbf{I}}_y \tag{1.132}$$

1.12 Approaches to Understanding NMR Experiments

The matrices for the Hamiltonian of x and y pulses and free precession can be obtained by using the expansion of the exponential:

$$e^{\pm iaA} = \sigma_0 \cos a \pm iA \sin a \tag{1.133}$$

When $a = \phi/2$, and $A = 2\hat{I}_x$, then

$$e^{\pm i\phi\hat{I}_x} = \sigma_0 \cos\tfrac{\phi}{2} \pm i2\mathbf{I}_x \sin\tfrac{\phi}{2}$$

$$= \begin{pmatrix} \cos\tfrac{\phi}{2} & \pm i\sin\tfrac{\phi}{2} \\ \pm i\sin\tfrac{\phi}{2} & \cos\tfrac{\phi}{2} \end{pmatrix} \xRightarrow{\phi=\pi/2} \frac{1}{\sqrt{2}}\begin{pmatrix} 1 & \pm i \\ \pm i & 1 \end{pmatrix} \tag{1.134}$$

Similarily,

$$e^{\pm i\phi\hat{I}_y} = \begin{pmatrix} \cos\tfrac{\phi}{2} & \pm\sin\tfrac{\phi}{2} \\ \mp\sin\tfrac{\phi}{2} & \cos\tfrac{\phi}{2} \end{pmatrix} \xRightarrow{\phi=\pi/2} \frac{1}{\sqrt{2}}\begin{pmatrix} 1 & \pm 1 \\ \mp 1 & 1 \end{pmatrix} \tag{1.135}$$

$$e^{\pm i\omega t\hat{I}_z} = \begin{pmatrix} e^{\pm i\omega t/2} & 0 \\ 0 & e^{\mp i\omega t/2} \end{pmatrix} \tag{1.136}$$

in which $\phi = \omega_1 t$, is the pulse angle rotated by the RF field with strength $B_1 = \hbar\omega_1$, and ω is the resonance frequency of the spin in the rotating frame. When on-resonance, $\omega = 0$. For a one-pulse experiment with a $90°_x$ pulse applied to the initial magnetization I_z, the density matrix after the 90° pulse is given by:

$$\rho(t) = e^{-i(\pi/2)\hat{I}_x}\rho(0)e^{i(\pi/2)\hat{I}_x} \tag{1.137}$$

$$= e^{-i(\pi/2)\hat{I}_x}\mathbf{I}_z e^{i(\pi/2)\hat{I}_x} = \frac{1}{2}\begin{pmatrix} 0 & i \\ -i & 0 \end{pmatrix} = -\mathbf{I}_y \tag{1.138}$$

We now define a transform matrix as:

$$\mathbf{U}_H\rho(0) = e^{-iHt}\rho(0)e^{iHt} \tag{1.139}$$

$$\rho(t) = \mathbf{U}_{(\pi/2)x}\mathbf{I}_z = e^{-i(\pi/2)\hat{I}_x}\mathbf{I}_z e^{i(\pi/2)\hat{I}_x} = -\mathbf{I}_y \tag{1.140}$$

The above equation for $\rho(t)$ is equivalent to the product operator representation of:

$$\mathbf{I}_z \xrightarrow{(\pi/2)\mathbf{I}_x} -\mathbf{I}_y \tag{1.141}$$

The density matrix for the observed magnetization changing with time is given by:

$$\rho(t) = \mathbf{U}_{H_R}(-\mathbf{I}_y) = \frac{1}{2}\begin{pmatrix} e^{-i\omega t/2} & 0 \\ 0 & e^{i\omega t/2} \end{pmatrix}\begin{pmatrix} 0 & i \\ -i & 0 \end{pmatrix}\begin{pmatrix} e^{i\omega t/2} & 0 \\ 0 & e^{-i\omega t/2} \end{pmatrix}$$

$$= \frac{1}{2}\begin{pmatrix} 0 & ie^{-i\omega t} \\ -ie^{i\omega t} & 0 \end{pmatrix} \tag{1.142}$$

The observable magnetization is obtained by the trace of the product of the density matrix with the raise operator:

$$\overline{M(t)} = \mathrm{Tr}(\rho(t)\mathbf{I}_+) = -\frac{1}{2}ie^{i\omega t} = \frac{1}{2}(\sin\omega t - i\cos\omega t) \tag{1.143}$$

By now we have seen all of the basic concepts necessary to predict the behavior of an isolated spin system in the absence of relaxation.

Questions

1. In an inversion-recovery experiment, signal A has zero intensity when $\tau = 380$ ms. Estimate the spin–lattice relaxation time T_1 for the signal.
2. Assuming that a signal has an intensity of 100 observed using a 90° pulse, what is the intensity of the signal if a 45° pulse is used?
3. What is the longest dwell time one could use to collect ^{1}H data for a spectrum with 10 ppm on a 600 MHz spectrometer and the carrier frequency in the center of the spectrum?
4. What is the frequency of ^{13}C in the laboratory frame on a 500 MHz instrument? And what is its frequency in the rotating frame (on resonance)?
5. Assuming that we have a 100 % ^{13}C enriched sample and we have obtained ^{1}H and ^{13}C spectra with one transient, what is the intensity of ^{13}C likely to be relative to the ^{1}H spectrum?
6. Assuming that the DSS signal has a resonance frequency of 600.0123456 MHz, what is the reference frequency for ^{13}C and ^{15}N if ^{13}C and ^{15}N are chosen according to the ratio of their gyromagnetic ratios to that of ^{1}H (provided in Table 1.1)?
7. What is the frequency for 170 ppm and 58 ppm of ^{13}C, and for 118 ppm of ^{15}N using the reference frequencies in Question 6?
8. Assuming that the $^3J_{H^N H^\alpha}$ coupling constant of a residue in a protein is about 10 Hz, what is likely to be the value of torsion angle ϕ?
9. Where is the magnetization after applying a spin echo pulse sequence $90^\circ_x - \tau - 180^\circ_x - \tau$?
10. When is a high power rectangular RF pulse used? What are the four parameters needed to be specified?

Appendix A: Product Operators

A1. Uncoupled Spins

For an uncoupled spin system, the product operators can be applied to individual spins. For a spin-½ nucleus, there are three basic operators used to describe the spin magnetization during NMR experiments, I_x, I_y and I_z, which are the x, y, and z components of spin I magnetization. The transformations of these operators after a 90° ($\pi/2$) pulse are given by:

$$
\begin{aligned}
I_x &\xrightarrow{(\pi/2)I_x} I_x & I_x &\xrightarrow{(\pi/2)I_y} -I_z \\
I_y &\xrightarrow{(\pi/2)I_x} I_z & I_y &\xrightarrow{(\pi/2)I_y} I_y \\
I_z &\xrightarrow{(\pi/2)I_x} -I_y & I_z &\xrightarrow{(\pi/2)I_y} I_x
\end{aligned}
\quad (1.144)
$$

in which $(\pi/2)I_x$ and $(\pi/2)I_y$ are the 90° pulses applied along on the x and y axes, respectively. For a free precession during time t at resonance Ω, the transformations of the three operators are given by:

$$I_x \xrightarrow{\Omega t} I_x \cos \Omega t + I_y \sin \Omega t \quad (1.145)$$

$$I_y \xrightarrow{\Omega t} I_y \cos \Omega t - I_x \sin \Omega t \quad (1.146)$$

$$I_z \xrightarrow{\Omega t} I_z \quad (1.147)$$

The sign of the above rotations are shown in Fig. 1.26. An example of applying the product operators to describe the spin echo experiment of uncouple spins is discussed in Sect. 1.12.2.1.

A2. Two Coupled Spins

The spin system of two coupled spins I and S may be either homonuclear or heteronuclear spin system. For homonuclear spin system, non selective RF pulses act on both spins, whereas the RF pulses rotate only the specific heteronuclear spin because the Larmor frequencies of the heteronuclei are several tens of megahertz off-resonance to each other. In addition to the six operators for x, y, and z components of individual spin (i.e., I_x, I_y, and I_z for spin I, and S_x, S_y, and S_z for spin S), nine product operators represent the coupling between the two spins, which are summarized in Table 1.6, of which five product operators for the antiphase magnetization are illustrated in Fig. 1.27 in the vector representation.

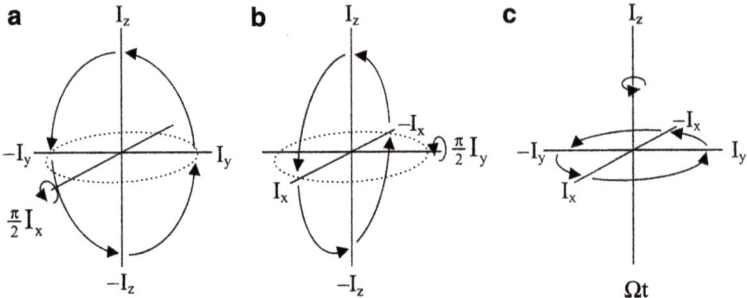

Fig. 1.26 The transformations of the product operators for uncoupled spin I (a) after a $90°_x$ pulse, (b) after a $90°_y$ pulse and (c) during free precession Ωt (Ziessow 1990; Hore et al. 2001)

Table 1.6 Product operators for a coupled two-spin IS system

	S_x	S_y	S_z
I_x	$2I_xS_x$	$2I_xS_y$	$2I_xS_z$
I_y	$2I_yS_x$	$2I_yS_y$	$2I_yS_z$
I_z	$2I_zS_x$	$2I_zS_y$	$2I_zS_z$

Note: The product operators in **bold** are the in-phase magnetization, the four product operators ($2I_xS_x$, $2I_xS_y$, $2I_yS_x$, $2I_yS_y$) are the double-quantum coherences, and the rest are the antiphase magnetization

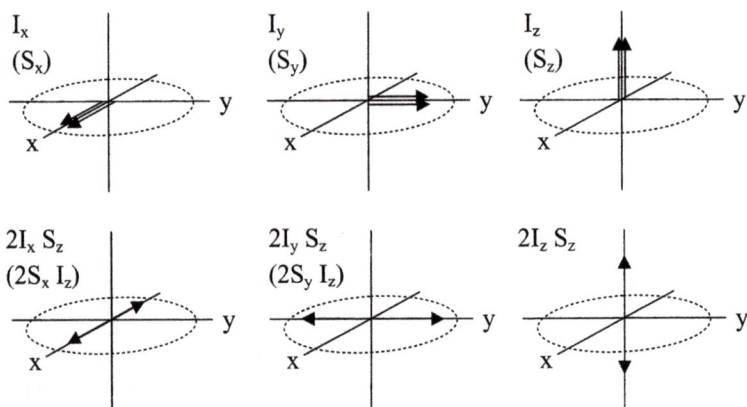

Fig. 1.27 Vector representation of product operators for in-phase and antiphase magnetization for spin I (or spin S) in a coupled two-spin system

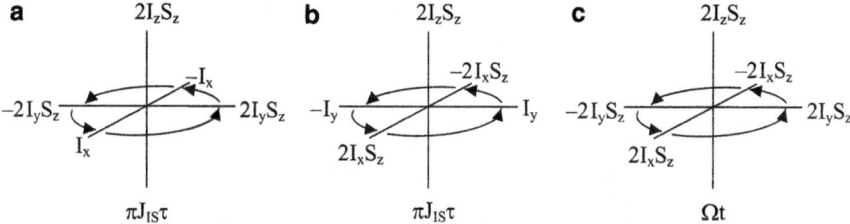

Fig. 1.28 The transformations of the product operators for two-coupled-spin IS system by scalar coupling J_{IS} (**a**, **b**) during the evolution under the influence of J_{IS} coupling for a period τ, and (**c**) during the free precession Ωt

The transformations of the product operators during the evolution under the influence of J_{IS} coupling for a period τ are given by:

$$I_x \xrightarrow{\pi J_{IS}\tau} I_x \cos \pi J_{IS}\tau + 2I_y S_z \sin \pi J_{IS}\tau \tag{1.148}$$

$$I_y \xrightarrow{\pi J_{IS}\tau} I_y \cos \pi J_{IS}\tau - 2I_x S_z \sin \pi J_{IS}\tau \tag{1.149}$$

$$I_z \xrightarrow{\pi J_{IS}\tau} I_z \tag{1.150}$$

$$2I_x S_z \xrightarrow{\pi J_{IS}\tau} 2I_x S_z \cos \pi J_{IS}\tau + I_y \sin \pi J_{IS}\tau \tag{1.151}$$

$$2I_y S_z \xrightarrow{\pi J_{IS}\tau} 2I_y S_z \cos \pi J_{IS}\tau - I_x \sin \pi J_{IS}\tau \tag{1.152}$$

$$2I_z S_z \xrightarrow{\pi J_{IS}\tau} 2I_z S_z \tag{1.153}$$

The product operators for spin S can be obtained similarly by interchanging I and S in (1.148)–(1.153). The above rules for the transformations of product operators are summarized in Fig. 1.28.

References

Abragam A (1961) Principles of nuclear magnetism. Clarendon Press, Oxford

Andrew ER, Bradburg A, Eades RG (1959) Removal of dipolar broadening of nuclear magnetic resonance spectra of solids by specimen rotation. Nature 183:1802–1803

Baber JL, Libutti D, Levens D, Tjandra N (1999) High precision solution structure of the C-terminal KH domain of heterogeneous nuclear ribonucleoprotein K, a c- myc transcription factor. J Mol Biol 289:949–962

Bain AD (1984) Coherence levels and coherence pathways in NMR: a simple way to design phase cycling procedures. J Magn Reson 56:418–427

Barkhuusen H, De Beer R, Bovee WMMJ, VanOrmondt D (1985) Retrieval of frequencies, amplitudes, damping factors, and phases from time-domain signals using a linear least-squares procedure. J Magn Reson 61:465–481

Bax A, Kontaxis G, Tjandra N (2001) Dipolar couplings in macromolecular structure determination. Meth Enzymol 339:127–174

Bendal MR, Pegg DT (1983) Complete accurate editing of decoupled 13C spectra using DEPT and a quaternary-only sequence. J Magn Reson 53:272–296

Bloch F (1946) Nuclear induction. Phys Rev 70:460–474

Bodenhausen G, Kogler H, Ernst RR (1984) Selection of coherence-transfer pathways in NMR pulse experiments. J Magn Reson 58:370–388

Bracewell RN (1986) The Fourier transformation and its application. McGraw-Hill, New York

Brutscher B (2000) Principles and applications of cross-correlated relaxation in biomolecules. Concepts Magn Reson 12:207–229

Carr HY, Purcell EM (1954) Effects of diffusion on free precession in Nuclear Magnetic Resonance experiments. Phys Rev 94:630–638

de Alba E, Tjandra N (2002) NMR dipolar couplings for the structure determination of biopolymers in solution. Prog Nucl Magn Reson Spectrosc 40:175–197

Dirac PAM (1967) The principles of quantum mechanics. Oxford University Press, New York

Doddrell DM, Pegg DT, Bendall MR (1982) Distortionless enhancement of NMR signals by polarization transfer. J Magn Reson 48:323–327

Fano U (1957) Description of states in quantum mechanics by density matrix and operator techniques. Rev Mod Phys 29:74–93

Friebolin H (1993) Basic one- and two-dimensional NMR spectroscopy, 2nd edn. VCH Publishers, New York

Harris J, Stocker H (1998) Handbook of mathematics and computational science. Springer, New York

Harris RK, Becker ED, De Menezes SMC, Goodfellow R, Granger P (2001) NMR nomenclature. Nuclear spin properties and conventions for chemical shifts. Pure Appl Chem 73:1795–1818

Hore PJ, Jones JA, Wimperis S (2001) NMR: the toolkit. Oxford University Press, Oxford

Howarth MA, Lian LY, Hawkes GE, Sales KD (1986) Formalisms for the description of multiple-pulse NMR experiments. J Magn Reson 68:433–452

Howarth O (1987) Multinuclear NMR. In Mason J (ed) Nuclear magnetic resonance spectroscopy. Springer, New York

Karplus K (1963) Vicinal proton coupling in nuclear magnetic resonance. J Am Chem Soc 85:2870–2871

Karplus M (1959) Contact electron-spin Interactions of nuclear magnetic moments. J Phys Chem 30:11–15

Lazzeretti P (2000) Ring currents. Prog Nucl Magn Reso Spectr 36:1–88

Lipstitz RS, Tjandra N (2004) Residual dipolar couplings in NMR structure analysis. Annu Rev Biophys Biomol Struct 33:387–413

Lowe IJ (1959) Free Induction decays of rotating solids. Phys Rev Lett 2:285–287

Markley JL, Bax A, Arata Y, Hilbers CW, Kaptein R, Sykes BD, Wright PE, Wüthrich K (1998) Recommendations for the presentation of NMR structures of proteins and nucleic acids – IUPAC-IUBMB-IUPAB Inter-Union Task Group on the standardization of data bases of protein and nucleic acid structures determined by NMR spectroscopy. J Biomol NMR 12:1–23

Mazzeo AR, Delsuc MA, Kumar A, Levy GC (1989) Generalized maximum entropy deconvolution of spectral segments. J Magn Reson 81:512–519

Morris GA, Freeman R (1979) Enhancement of nuclear magnetic resonance signals by polarization transfer. J Am Chem Soc 101:760–762

Moskau D (2002) Application of real time digital filters in NMR spectroscopy. Concepts Magn Reson 15:164–176

Neuhaus D, Williamson MP (1989) The nuclear overhauser effect in structural and conformational analysis. VCH Publishers, New York

References

Prestegard JH, Al-Hashimi HM, Tolman JR (2000) NMR structures of biomolecules using field oriented media and residual dipolar couplings. Quart Rev Biophys 33:371–424

Sibisi S, Skilling J, Brereton RG, Laue ED, Staunton J (1984) Maximum entropy signal processing in practical NMR spectroscopy. Nature (London) 311:446–447

Sibi MP, Lichter RL (1979) Nitrogen-15 nuclear magnetic resonance spectroscopy. Natural abundance nitrogen-15 chemical shifts of alkyl- and aryl-substituted ureas. J Org Chem 44:3017–3022

Solomon I (1955) Relaxation processes in a system of two spins. Phys Rev 99:559–565

Sørensen OW, Eich GW, Levitt MH, Bodenhausen G, Ernst RR (1983) Product operator formalism for the description of NMR pulse experiments. Prog Nucl Magn Reson Spectrosc 16:163–192

Spera S, Bax A (1991) Empirical correlation between protein backbone conformation and C.alpha. and C.beta. 13C nuclear magnetic resonance chemical shifts. J Am Chem Soc 113:5490–5492

Stern AS, Hoch JC (1992) A new, storage-efficient algorithm for maximum-entropy spectrum reconstruction. J Magn Reson 97:255–270

Tjandra N (1999) Establishing a degree of order: obtaining high-resolution NMR structures from molecular alignment. Structure 7:R205–R211

Tjandra T, Bax A (1997) Direct measurement of distances and angles in biomolecules by NMR in a dilute liquid crystalline medium. Science 278:1111–1114

Wang AC, Bax A (1996) Determination of the backbone dihedral angles φ in human ubiquitin from reparametrized empirical Karplus equations. J Am Chem Soc 118:2483–2494

Winder S (1997) Analog and digital filter design, 2nd edn. Elsevier Science, Woburn

Wishart DS, Bigam CG, Yao J, Abildgaard F, Dyson HJ, Oldfield E, Markley JL, Sykes BD (1995) 1H, 13C and 15N chemical shift referencing in biomolecular NMR. J Biomol NMR 6:135–140

Wishart DS, Richards FM, Sykes BD (1991) Relationship between nuclear magnetic resonance chemical shift and protein secondary structure. J Mol Biol 222:311–333

Wüthrich K (1986) NMR of proteins and nucleic acids. Wiley, New York

Zhu G, Bax A (1990) Improved linear prediction for truncated signals of known phase. J Magn Reson 90:405–410

Ziessow D (1990) Understanding multiple-pulse experiments - an introduction to the product-operator description. I. Starting with the vector model. Concept Magn Reson 2:81–100

Chapter 2
Instrumentation

2.1 System Overview

Questions to be answered in this section include:

1. What are the basic components in an NMR spectrometer?
2. What are their functions?

The basic components of an NMR spectrometer are shown in Fig. 2.1a, and include three major elements: magnet, console, and host computer. The working function of a NMR spectrometer is basically similar to a radio system. Some of the components are called by the terms used in radio systems, such as transmitter, synthesizer, and receiver. The magnet of an NMR spectrometer produces a stable static magnetic field which is used to generate macroscopic magnetization in an NMR sample. The linear oscillating electromagnetic field, B_1 field (see Chap. 1), is induced by a transmitter with a desirable B_1 field strength to interact with nuclei under study. The NMR signal, known as the free induction decay (FID), generated in the probe coil after irradiation by radio frequency (RF) pulses is first amplified by a preamplifier, then detected by a receiver. This detected signal is digitized by an analog-to-digital converter (ADC) for data processing and display, which is done on a host computer.

2.2 Magnet

In this section such questions about the NMR magnet will be addressed as:

1. What is the structure inside a magnet?
2. How is the magnetic field generated and how is the stability of the field maintained?
3. Why does the magnet need to be periodically filled with liquid nitrogen and liquid helium?

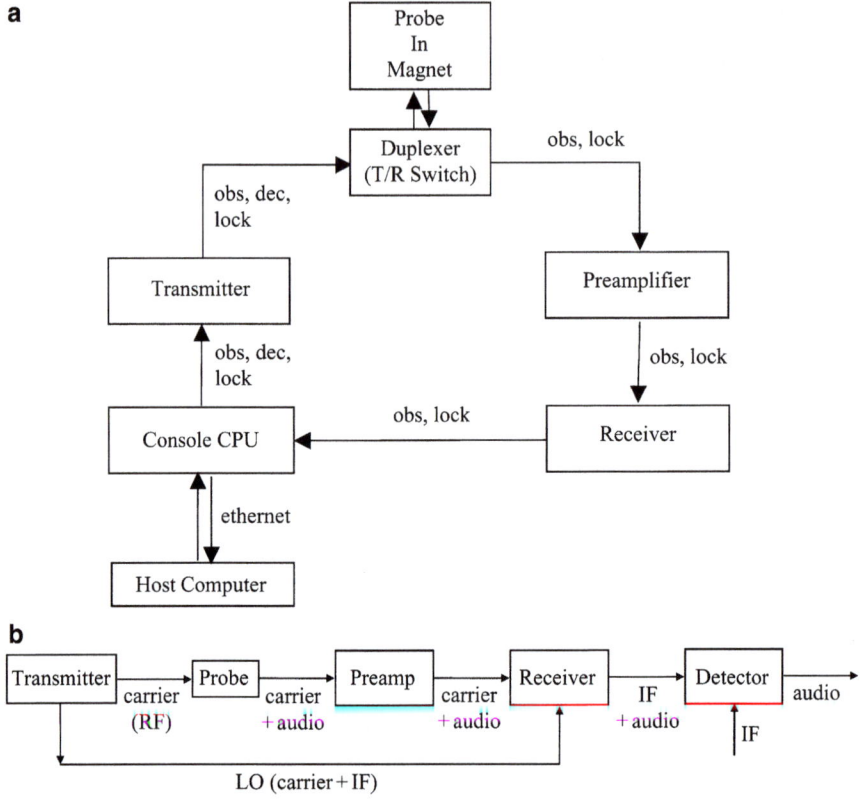

Fig. 2.1 Block diagram of NMR spectrometer.

4. What homogeneity of the field is required for NMR and how can it be obtained?
5. How is the sensitivity increased as the field strength increases?

Almost all high field NMR magnets are made of superconducting (SC) solenoids. In order to achieve superconductivity, an SC solenoid is enclosed in a liquid helium vessel (Fig. 2.2). Liquid nitrogen stored in a vessel outside the liquid helium vessel is used to minimize the loss of liquid helium because the cost of liquid nitrogen is about 40 times less than liquid helium. In addition, insulation of heat transfer between the vessels and the shell of the magnet is achieved by the use of high vacuum chambers. Vacuum is the most effective method of heat insulation, which prevents two of the three heat transfer processes: conduction and convection. The third process of heat transfer, radiation, is prevented through the use of reflective shields which are made of aluminum foil and surrounded by the high vacuum. Because of the efficient heat insulation, an NMR sample can be placed in the probe at room temperature or at other desired temperature and insulated from the liquid helium at 4.2 K just a few inches away. Liquid helium

2.2 Magnet

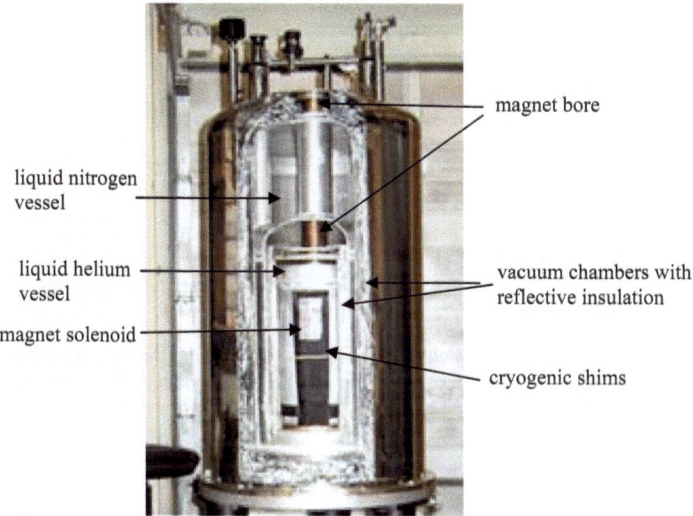

Fig. 2.2 Cutaway of a superconducting (SC) magnet. The magnet solenoid is in a liquid helium vessel, and contains approximately 12 miles of SC wire. The liquid nitrogen vessel is between the inner and outer vacuum chambers. The insulation in the outer vacuum chamber reflects heat radiation from the room temperature surface. The inner 20 K (Kelvin) radiation shield is used to prevent infrared heat radiation transfer from the liquid nitrogen vessel into the liquid helium vessel. The elimination of heat radiation reduces the liquid helium boil-off rate. Both radiation shields are made of aluminum foils (courtesy of JEOL USA, Inc.).

loss can be less than a liter per day for a modern 600 MHz NMR. Low helium loss magnets have a helium holding time longer than 1 year.

Once it is cooled down to operational temperature at or below that of liquid helium, the magnet is energized slowly by conducting DC current into the solenoid over a period of several hours to a few days. (For ultrahigh field such as 800 and 900 MHz, the magnet solenoid is kept below the temperature of liquid helium). When the magnetic field produced by the current reaches operational field strength, the two terminals of the solenoid are closed by a SC switch such as the one shown in Fig. 2.3. The SC switch is open during the entire energization process by turning on the heater nearby the SC wire in the SC switch. The heat causes the SC switch to lose superconductivity. Thus, the current passes through the magnet solenoid from the charging power supply. When the magnet reaches the operational field strength, the SC switch is closed by turning off the heater. As a result, the current passes through the loop formed by the SC switch and the solenoid, and stays inside the solenoid. Normally, an NMR magnetic field drifts less than 10 Hz h^{-1}. Quite frequently, a few months after installation, the field drifts less than 1 Hz h^{-1}.

High homogeneity of the magnetic field is an essential requirement for any NMR magnet. It is achieved with a set of SC shim coils, called cryogenic shims (or cyroshims), located just outside the magnet solenoid. The field homogeneity is

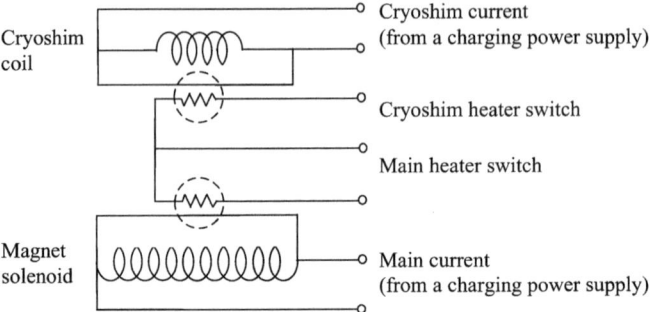

Fig. 2.3 Superconducting (SC) switch. When a heater switch is on, the SC wire inside the heater (the *dotted circle*) becomes a resistor due to the loss of superconductivity as the temperature is raised. The current flows from the power supply to the SC coil. After the heater is turned off, the current remains in the closed coil loops.

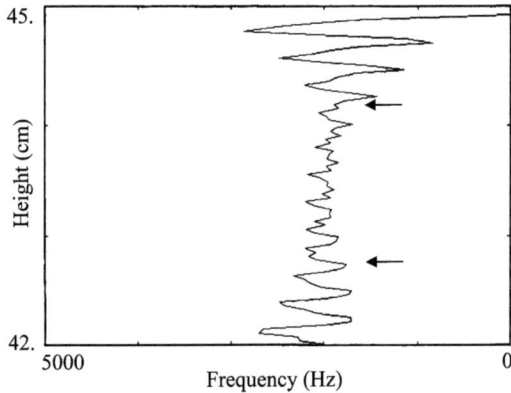

Fig. 2.4 Magnetic field mapping results of cryogenic shims for a 500 MHz magnet. High homogeneity of the magnetic field is obtained across a length over 1.6 cm (indicated by the *arrows*). The deviation of the magnetic field across the 1.6 cm length is about 500 Hz, which is equivalent to 1 ppm (500 Hz/500 MHz). The center of the probe coil is placed at the center of the field.

shimmed by a method called field mapping in which a tiny amount of sample (e.g., a drop of water) is used to obtain signal at different physical locations inside the magnet bore. The sample is moved spirochetically in the bore through the solenoid axis to record the magnetic field gradient. Then, cyroshims are adjusted according to a computer fitting for better field homogeneity. Figure 2.4 shows an example of field mapping results for an Oxford magnet during magnet installation. The results indicate that a field homogeneity of better than 1 ppm is obtained over 1.6 cm by cyroshims. During normal operation of the magnet, cyroshims need not to be changed.

For NMR experiments, a field homogeneity of 1 part per billion (ppb) or better is obtained by using a room temperature (RT) shim set which consists of as many as 40 shim gradients located in the area inside the magnet bore but outside probe.

High resolution NMR experiments require stability of the magnetic field in addition to field homogeneity. The fluctuation of the static magnetic field is corrected by monitoring a locking field frequency using a mini spectrometer, or lock system. The lock system has a lock transmitter (including a lock frequency synthesizer), a lock receiver, and a lock channel in the probe. It continuously observes the deuterium frequency of the NMR sample. The current of the z_0 coil of the RT shim coil assembly residing in the magnet bore is automatically adjusted to maintain the lock frequency at the correct value if the frequency changes. For this purpose, any NMR sample should be made from pure or partially deuterated solvent. ^2H_2O is the most common deuterated solvent used in biological samples. More details are discussed in Chap. 3.

Sensitivity and resolution of NMR signals are the fundamental reasons for the requirement of higher magnetic field strengths. Resolution of NMR spectra at a constant line width in hertz improves linearly with magnetic field strength (B_0). The sensitivity of an NMR signal is proportional to the population difference between two nuclear transition states. Because the energy gap of the two nuclear states is small (in the RF range), the population difference determined by Boltzmann distribution is also small. An increase in field strength will increase the population difference, and thus increases sensitivity (more details in Chap. 1). As a result, the sensitivity of the NMR signal increases in proportion to $B_0^{3/2}$ as the field strength increases and hence the time required to obtain the same signal-to-noise ratio is reduced in proportion to B_0^3.

2.3 Transmitter

Questions to be addressed about the transmitter include:

1. What is the function of the transmitter and what does it consist of?
2. How does a transmitter produce RF pulses with the desired pulse length and desired frequency (the carrier frequency)?
3. How can the amplitude of the pulses be attenuated?
4. What is the relationship between attenuated RF power and pulse length?

The function of a transmitter is to provide RF pulses to irradiate the samples with a desired pulse length (or pulse width) and frequency at the correct phase and power level. The transmitter channel consists of a frequency synthesizer, an RF signal generator, a transmitter controller, and an RF amplifier (Fig. 2.5). It provides RF pulses and quadrature phase generation. A frequency synthesizer provides a stable source of signal with the required frequencies using a standard reference frequency. The RF signal is gated by an RF controller to form pulses at a low amplitude level. A transmitter controller is used to create modulated phase, pulse power, and pulse

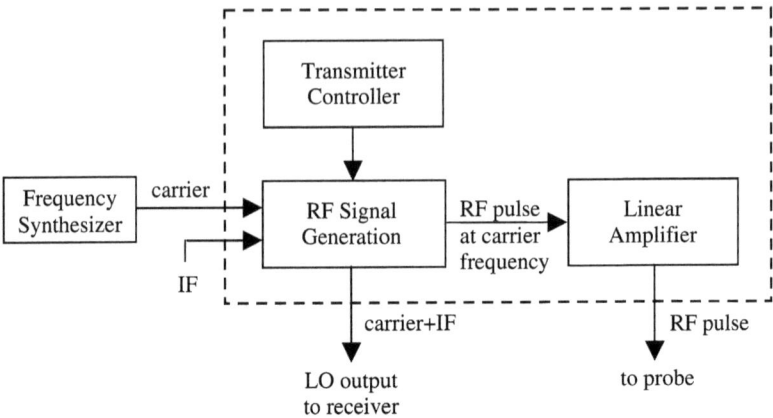

Fig. 2.5 Components of an NMR transmitter—block diagram.

gating (on and off). After it is routed through a computer-controlled attenuator to set the desired amplitude level, the RF signal then goes to the linear power amplifier to obtain the pulse power needed. The pulse from the amplifier is delivered to the probe where the NMR sample is irradiated. The output of the transmitter can be highly monochromic. Because the output power of an amplifier is attenuated linearly, the pulse length for a fixed pulse angle (for instance, a 90° pulse angle) is increased proportional to the power attenuation. The attenuation of the output amplitude is measured in a logarithmic unit, decibel or dB, which is a tenth of 1 Bel. By definition, the decibel of two signals in comparison is

$$dB = 20 \log \frac{V_2}{V_1} \tag{2.1}$$

in which V_1 and V_2 are two signal amplitudes, or voltage. A signal with twice the amplitude of the other is a 6 dB increase, whereas a signal of one half the amplitude is -6 dB (or a 6 dB attenuation). Twenty decibels represent a tenfold increase in signal amplitude. A signal amplitude V increased by N dB has a value given by

$$N(dB)V = 1.122^N V \tag{2.2}$$

Often the ratio of two signals is measured in terms of power levels:

$$dB = 10 \log \frac{P_2}{P_1} \tag{2.3}$$

in which P_1 and P_2 are the power levels of the signals and $P = V^2/R$ (R is resistance). In NMR, pulse "power" refers to the amplitude of the transmitter RF field in frequency units, rather than power in watts, because pulse length or pulse

2.3 Transmitter

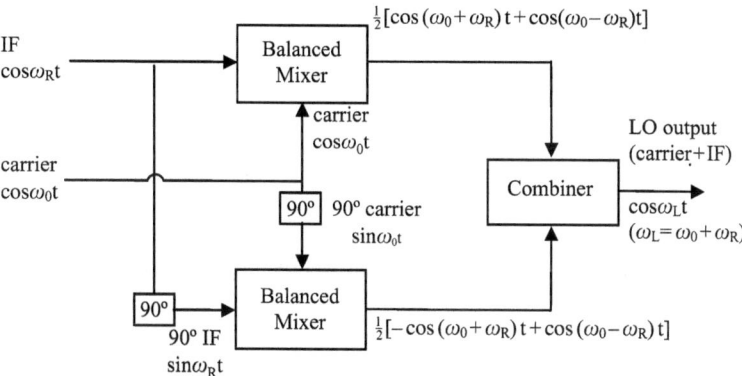

Fig. 2.6 Generation of LO frequency by a transmitter via SSB (single sideband) selection. When mixed at a balanced mixer (BM), two input frequencies are multiplied to produce a pair of sideband frequencies. The phase of the output is also dependent on the phases of the input signals. The output of a balanced mixer contains neither the carrier frequency nor the modulated intermediate frequency (IF) but only the sidebands.

angle is proportional to γB_1, in which γ is the gyromagnetic ratio and B_1 is the amplitude of the transmitter RF field. Therefore, the pulse length will increase to twice as long when attenuation is -6 dB.

One transmitter is required for each channel on an NMR spectrometer. Typically, a triple-resonance experiment requires separated proton, carbon, and nitrogen transmitter channels in addition to a lock channel. A four-channel NMR spectrometer may use the fourth channel for deuterium decoupling or for irradiation on other nuclei. Because of the low gyromagnetic ratios of heteronuclei, a heteronuclear channel has a longer pulse length for the same amplifier output power. A typical amplifier for high resolution NMR has an output power of a few hundred watts on each RF channel (see Sect. 2.8).

The local oscillator (LO) output of a transmitter which is used by a receiver to record the NMR signal (see discussion for receivers) is created by combining an intermediate frequency (IF) signal with the carrier frequency using the technique called single sideband (SSB) selection (Figs. 2.1b and 2.6). The IF is much lower than the carrier frequency, usually a few tens of MHz and is usually obtained from a fixed-frequency source. When the carrier and IF signals are mixed at a balanced mixer (BM is also called phase sensitive detectors, PSD), which is a device with two or more signal inputs that produces one signal output, the carrier multiplies the IF resulting in a pair of frequencies, carrier $-$IF and carrier $+$IF, known as a double sideband band suppressed carrier (DSBSC, Fig. 2.7). In order to convert DSBSC to SSB frequency, BMs are used to phase the signals. Quadrature IF and carrier frequencies (Quadrature means that two components of a signal differ in phase by $90°$) are met at two BMs whose output contains neither the carrier frequency nor the modulated IF but only the sidebands, resulting in two pairs of mixed IF with the carrier signals: a $90°$ phase-shifted pair in one path and a non-phase-shifted pair in the other. The output of a BM is a double sideband signal consisting of the

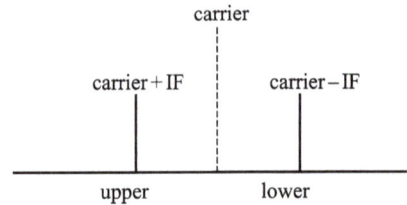

Fig. 2.7 Double sideband suppressed carrier frequency pair. The sidebands above and below the carrier frequency are called upper and lower sideband, respectively.

sum and difference of IF and the carrier frequencies produced by multiplying the two signal inputs:

$$\cos \omega_0 t \cos \omega_R t = \frac{1}{2}[\cos(\omega_0 + \omega_R)t + \cos(\omega_0 - \omega_R)t]$$

$$\sin \omega_0 t \sin \omega_R t = \frac{1}{2}[-\cos(\omega_0 + \omega_R)t + \cos(\omega_0 - \omega_R)t]$$

in which ω_0 and ω_R are the carrier frequency and intermediate frequency, respectively. Then, the double-band outputs of the mixers are combined at the combiner, where one sideband is enhanced and the other is canceled. The combination produces a single frequency output, an LO output, which usually is the frequency of the carrier +IF (it can also be designed to produce the frequency of the carrier −IF).

An alternative way to produce an LO is to use a synthesizer to generate an LO frequency. In this case, the carrier frequency of the transmitter is produced from the combination of LO and IF (Fig. 2.8). IF and LO quadrature frequency signals are mixed to produce the carrier frequency for RF pulses. The advantage of this configuration is that LO frequency to be used by the receiver is less noisy than the transmitter configuration represented in Fig. 2.6, and hence it potentially gives better sensitivity.

Since the transmitter provides the energy source for irradiating an NMR sample, it is wise to measure the output of the transmitter when the NMR spectrometer has problems such that no NMR signal can be observed. The convenient way to do this is to connect the transmitter output to an oscilloscope at the point just before the probe. The oscilloscope is set to measure voltage and appropriate attenuation should be used to protect the oscilloscope from damage by the high power of the transmitter amplifier (refer to Sect. 2.9 for operation of an oscilloscope). Attenuation can be done either by setting a transmitter attenuation parameter or by using an attenuator (e.g., 20 dB) between the oscilloscope and the transmitter output.

The dB value describes the relative power levels or amplitudes of two signals. Often, the amplitude of a signal is described relative to a reference power level. For instance, the term dB_m means dB relative to 1 mW into a given load impedance of a device, which is 50 Ω for an NMR instrument (we will assume that impedance is 50 Ω throughout this book unless specified), and dBW to 1 W:

$$1 \text{ mW} = 0.2236 V_{rms} = 0 \text{ dB}_m \tag{2.4}$$

$$dB_m = 10 \log P_{mw} \tag{2.5}$$

2.3 Transmitter

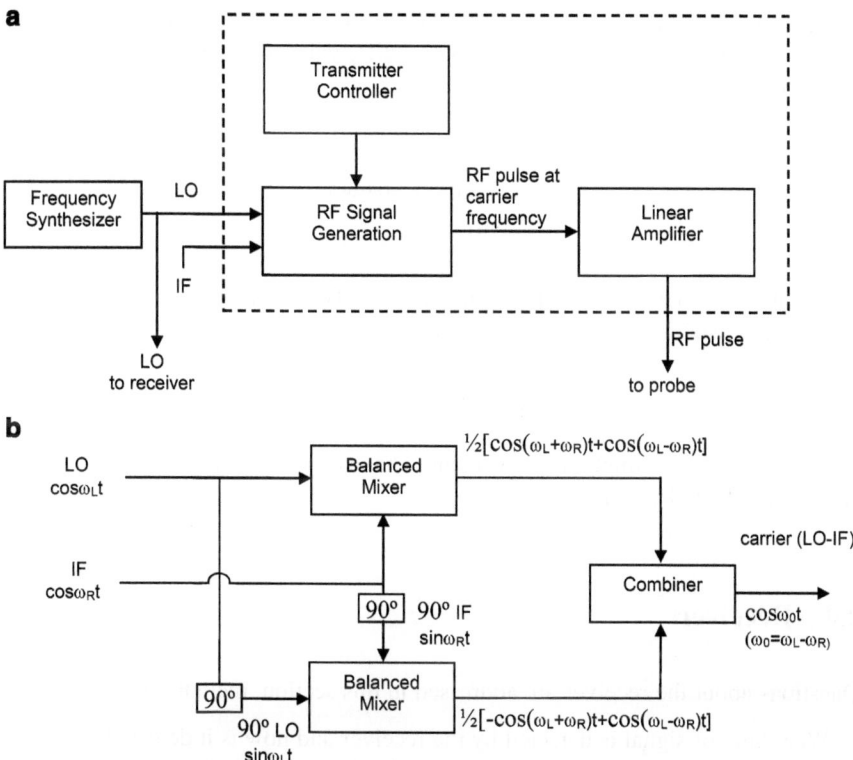

Fig. 2.8 An NMR transmitter using an LO to produce the carrier frequency. The carrier frequency of the transmitter is produced from the combination of LO and IF. IF and LO quadrature frequency signals are mixed to produce the carrier frequency for RF pulses.

$$P_{mw} = 10^{dB_m/10} \qquad (2.6)$$

in which P_{mw} is the power in mW. A signal into a 50 Ω impedance with 0 dB$_m$ amplitude has a voltage of 0.2236 V_{rms} [$V = (PR)^{1/2} = (10^{-3} \times 50)^{1/2}$]. The electric signal is also characterized by a peak-to-peak amplitude (V_{pp} which is twice the amplitude) and the root-mean-square amplitude (V_{rms}). For a sinusoidal signal, V_{rms} and V_{pp} have a relationship given by

$$V_{rms} = \frac{A}{\sqrt{2}} = \frac{V_{pp}}{2\sqrt{2}} = \frac{V_{pp}}{2.828} \qquad (2.7)$$

$$P_{mw} = 2.5 V_{pp}^2 \qquad (2.8)$$

in which A is the signal amplitude and V_{pp} is the peak-to-peak amplitude that corresponds to the voltage difference between the most positive and most negative points of a signal waveform (Fig. 2.22). It is two times the amplitude of a sine wave signal. A sine wave signal of 1 V_{pp} has a dB$_m$ value of 3.98, using one of the equations:

$$\begin{aligned} dB_m &= 3.98 + 20 \log V_{pp} \\ &= 13.01 + 20 \log V_{rms} \\ &= 30 + 10 \log P_{rms} \end{aligned} \quad (2.9)$$

in which P_{rms} is the power of the signal in watts. The value of V_{pp} for a given dB$_m$ value can be calculated by

$$V_{pp} = 10^{(dB_m - 3.98)/20} \quad (2.10)$$

When troubleshooting, it is convenient to have a table of V_{pp} vs. dB$_m$ although it can be calculated by (2.10).

2.4 Receiver

Questions about the receiver are addressed in this section, including:

1. What kind of signal is detected by the receiver and how is it detected?
2. How is the signal separated from the carrier frequency by the receiver?
3. How is quadrature detection achieved?

A receiver is used to detect the NMR signal generated at the probe and amplify the signal to a level suitable for digitization. Detection is the process of demodulating the NMR signal (in audio frequency, kHz range) from the carrier frequency (in RF, MHz range), and measures not only the amplitude or voltage of the signal, but also the phase modulation. Because the RF signal is very weak coming from the probe, it is amplified first by a preamplifier that is located near or inside the probe to reduce the loss of signal, before it is transferred to the receiver inside the console. The process of signal detection includes preamplification, several stages of RF signal amplification, quadrature detection (separation of the NMR signal from the carrier frequency), and amplification of the NMR (audio) signals.

In the simplest method, several stages of tuned RF amplification are used, followed by a detector. The frequencies of all amplification stages are tuned to a narrow range near the carrier frequency in order to amplify RF signals for detection. When signals pass through the amplifier, noise is also amplified along with the input signals which have very low amplitude. To reduce the effect of noise it is necessary to filter noise outside the signal frequency bandwidth and to only allow signals and noise with the same bandwidth as the signals to come through. For this reason, a bandpass filter is used with the center frequency tunable over a desired frequency range.

2.4 Receiver

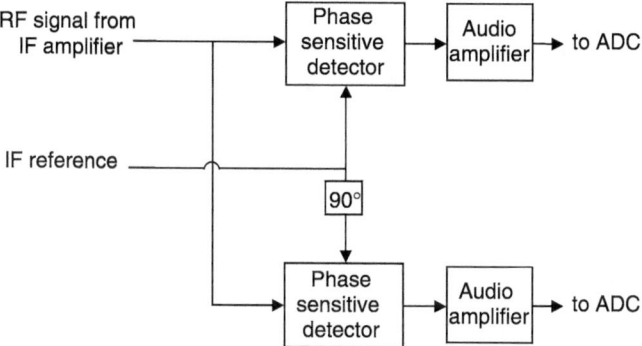

Fig. 2.9 Quadrature detection using two phase sensitive detectors (PSDs). The PSDs compare the frequencies of two input signals, and then generates an output. The output is the measure of phase and frequency differences of the input signals. When two signals with the same frequency are mixed at a PSD, the output is the measure of the phase difference of the two inputs.

Furthermore, all stages of amplification must have amplitude linearity over the full band frequency range. This configuration of NMR receiver is undesirable because it is difficult to construct amplifiers with linear response and accurate selectivity at all stages over the range of several hundred MHz. The tunable filters usually lack passband flatness over a wide frequency range. As a result, the resolution of the tuned receiver is dependent on frequency. This causes problems such as lack of sensitivity and resolution, and signal distortion.

The solution to the problem is the superheterodyne receiver (narrowband receiver). It differs from the tuned receiver in that the RF signals are adjusted to pass through fixed passband amplifiers and filters instead of tuning the amplifiers and filters for the RF signals. Unlike the transmitter which must use Larmor frequency RF pulses to irradiate sample in order to generate NMR signals, the receiver may be set to a fixed frequency to detect the signals. The incoming signals are amplified by a preamplifier (single stage tuned amplification), then mixed with an LO frequency to produce signals at a fixed IF. After the preamplifier, the signal at IF passes through a set of IF amplifiers and filters in the receiver. Finally, IF RF signals are terminated at a quadrature detector that subtracts the IF from the NMR signals using the reference IF, and NMR signals with audio frequency (kHz) are amplified by an audio amplifier for digitization. Tuning the IF receiver for different carrier frequencies is achieved by alternating the LO frequency so that an input carrier frequency gets mixed down to the IF frequency. Receivers that have one mixing stage are called single conversion receivers, whereas they are called multiple-conversion receivers if mixed in more than one stage. The single-conversion superheterodyne receiver has become very popular for modern NMR spectrometers. It offers higher sensitivity and better performance in the presence of interfering signals.

Detection of NMR signals is done by a quadrature detector, involving a phase detector shown in Fig. 2.9. The phase detector is a circuit that compares the

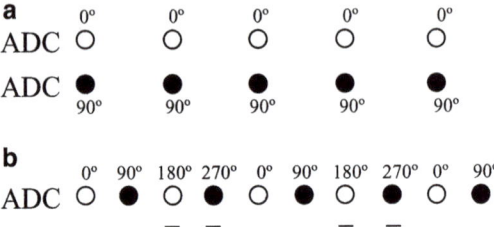

Fig. 2.10 Quadrature detection by (**a**) the simultaneous acquisition method and (**b**) the sequential method. The *open circles* represent the data points detected by the zero-phased detector (PSD) and the *filled circles* represent those detected by the 90°-phased detector in Fig. 2.9. The data points multiplied by −1 are indicated by the *minus sign* below the *circles* in (**b**). The receiver phases are shown above or below the data points.

frequencies of two input signals, and then generates an output. The output is the measure of phase and frequency differences of the input signals. The internal circuitry of a phase detector is actually a BM. When two signals with the same frequency are mixed at a phase detector, the output is the measure of the phase difference of the two inputs. The RF signal coming out of the IF amplifier is divided at a splitter. The two split signals are fed into separate phase detectors where they are mixed with quadrature IF reference signals generated by a phase shifter. Finally the output of each phase detector is amplified by an audio amplifier and digitized at the ADC (see the Sect. 2.7) as real and imaginary components of an FID.

Practically, quadrature detection in the observed dimension can be done either by two ADCs or by a single ADC (Fig. 2.10). The first method (known as simultaneous acquisition) uses one ADC for each PSD to simultaneously sample the data from two channels with one ADC acquiring the real part of the FID and the other recording the imaginary part of the data. Fourier transformation of the complex data produces a spectrum with the carrier in the center of the spectral window (SW). The second method (known as sequential acquisition or the Redfield method) uses a single ADC to sample the data from the two PSDs one after the other with the same time intervals set by the dwell time. The ADC digitizes the signal at a sample rate twice as fast as normal. The ADC switches between the two PSDs after sampling each point. Therefore, the odd number data points come from the first PSD and the even number from the second PSD, which is 90° out of phase to the first one. Additionally, every second pair of data points is multiplied by −1. The net result is that the phases of all the points are increased sequentially by 90° (=¼ cycle), which is known as time-proportional phase increment (TPPI). If a real Fourier transform is applied to the data, the sign of the frequency (the direction of the magnetization rotation) cannot be distinguished (−SW to SW), because the data does not contain an imaginary part. Since the sampling rate is twice as fast as in the simultaneous method, the spectral window now is −½SW to +½SW (real Fourier transformation produces a spectral window with 2SW from −SW to +SW and the ½ factor is caused by the doubled sampling rate). In addition, the effect of TPPI on the time domain is to increase the frequency by +½SW. This can be understood by considering that the spectral width is doubled because the real Fourier transformation cannot distinguish the sign of the spectrum, the 90° phase increment introduces a factor of ¼ because 90°/360° = ¼, and hence 2SW × ¼ = ½SW. Considering all the factors, the spectral window of the sequentially acquired data ranges from 0 to SW after the real Fourier transformation with the carrier in the center of the spectrum and the correct sign for all frequencies. The results obtained from the two methods are essentially identical. Some spectrometers (such as Bruker systems) allow users to use either of the acquisition methods, whereas others acquire the data simultaneously using two ADCs (such as Agilent or JEOL).

2.5 Probe

Probe circuits are usually characterized by three quantities: resonance frequency, total impedance at resonance, and the Q factor of the circuits. In the current section, simple circuits are discussed to illustrate the function of an NMR probe. Questions to be addressed include the following:

1. What are the electronic components inside a probe?
2. What are the inductor–capacitor (LC) parallel and series circuits and what are their resonance frequencies and impedance?
3. How is the quality factor or Q factor of the probe defined and what are the Q factors of the circuits?
4. What do probe tuning and matching mean and why must a probe be tuned before setting up experiments?
5. How are probe tuning and matching achieved?
6. What is a cryogenic probe and how is high sensitivity of a cryogenic probe obtained?
7. Why can a moderate salt concentration degrade the performance of cryogenic probes?
8. What is the radiation damping effect and what causes it?

NMR probes are basically resonant circuits (frequency dependent) in which capacitors and inductors are combined (Fig. 2.11). The sample coil in the probe

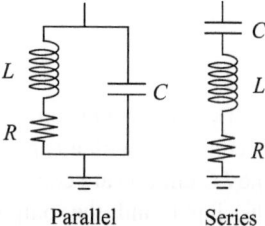

Fig. 2.11 Parallel and series LC circuits. C is the capacitance of the circuit and L is the inductance with resistance R.

circuit is used to generate a B_1 electromagnetic field to interact with the nuclei of the sample. Used with RF pulses, the probe circuit must have its impedance matched to the specific impedance of the cables, which means that the impedance of the cable terminated at the probe equals the characteristic impedance of the cable (50 Ω). This allows the RF pulses to be transferred to the probe without reflection so that all the power of the pulses is used by the probe without loss. In order to understand the function and working principle of probes, it is necessary to review the relationship between voltage and current, which are the two quantities characterized in electronic circuits. An important characteristic of capacitors and inductors is their frequency dependence. A device made from these components will produce an output waveform that is also frequency dependent, but maintains linearity in the amplitude of waveforms. The generalized Ohm's law is well used in analyzing the inductor–capacitor (LC) devices:

$$I = \frac{V}{Z} \tag{2.11}$$

$$V = IZ \tag{2.12}$$

in which Z is impedance in complex form, considered as a generalized resistor of a circuit, I is current and V is voltage. The capacitor with capacitance C and the probe coil with inductance L have impedances in the following terms:

$$\text{Capacitor: } Z_C = -\frac{j}{\omega C} = \frac{1}{i\omega C} \tag{2.13}$$

$$\text{Inductor: } Z_L = j\omega L \tag{2.14}$$

$$\text{Resistor: } Z_R = R \tag{2.15}$$

in which ω is the angular frequency of the waveform ($\omega = 2\pi \nu$) and j is the imaginary unit, $\sqrt{-1}$. Like resistors, impedance in parallel and series circuits has the formulas:

$$Z_p = \frac{1}{(1/Z_1) + (1/Z_2) + (1/Z_3) + \cdots} \tag{2.16}$$

$$Z_s = Z_1 + Z_2 + Z_3 + \cdots \tag{2.17}$$

The simplest LC circuits are parallel and series LC circuits, in which an inductor is combined with a capacitor in parallel and series, respectively (Fig. 2.11). Since the LC circuits are connected to the input in series where the current is the same for the input and the output at the junction and ground, the output voltage is

2.5 Probe

proportional to the total impedance of the LC circuit. For a parallel LC, the impedance is given by

$$Z = \frac{1}{(1/Z_L) + (1/Z_C)} = \frac{1}{(1/(j\omega L + R)) + j\omega C} = \frac{j\omega L + R}{1 - \omega^2 LC + j\omega RC} \quad (2.18)$$

By multiplying both the numerator and denominator of (2.18) by $(R - j\omega L)$, the total impedance is

$$Z = \frac{\omega^2 L^2 + R^2}{R - j\omega L(1 - \omega^2 CL - R^2 C/L)} \quad (2.19)$$

At resonance frequency ω_0, the circuit has real impedance. Therefore, the imaginary part of (2.19) must be zero:

$$1 - \omega_0^2 CL - R^2 C/L = 0 \quad (2.20)$$

Because R is always much smaller than L in the circuit, $R^2 C/L \approx 0$. Then

$$\omega_0 \approx 1/\sqrt{LC} \quad (2.21)$$

The impedance Z at resonance approximately equals [R^2 in (2.19) is negligible compared to L]:

$$Z = \frac{\omega^2 L^2}{R} \quad (2.22)$$

producing a sharp peak of output voltage as shown in Fig. 2.12.

The resonance condition is phase-resonance, meaning that the capacitance and inductance of the circuit are equal. The frequency function of the voltage ratio (V_{out}/V_{in}) in Fig. 2.12 shows that the output voltage of the parallel LC circuit is the same as the input voltage at the resonance frequency. Practically, the ratio is less than 1 due to imperfections in the electronic components.

The circuit is characterized by the quality factor of the circuit, Q, which is dependent on the resonance frequency:

$$Q = \omega_0 L/R \quad (2.23)$$

The practical significance of Q represents that the smaller the value of R, the greater the value of Q, resulting in a sharper resonance peak. In addition, the higher Q is, the more sensitive the probe. Changing C and L will alter the impedance of the circuit while tuning to the desired resonance frequency.

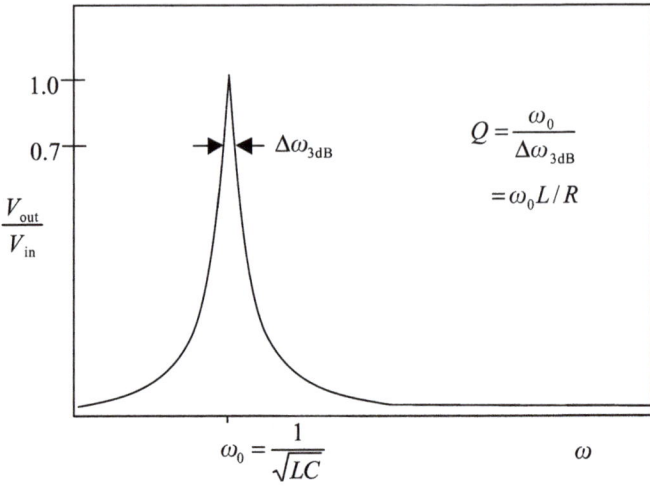

Fig. 2.12 Output voltage curve in a parallel LC circuit as a function of frequency. The maximum is at the resonance frequency ω_0 and is dependent on the capacitance and inductance of the circuit.

Another type of LC resonance circuit is the series LC circuit as shown in Fig. 2.11, which has impedance in the terms of

$$Z = Z_L + Z_C = R + j\omega L + \frac{1}{j\omega C} = \frac{\omega CR + j(\omega^2 LC - 1)}{\omega C} \quad (2.24)$$

By applying the resonance condition that the imaginary term of the impedance is zero, the resonance frequency has the same formula as that of parallel circuit, $\omega_0 = 1/\sqrt{LC}$. The series resonance circuit is different than the parallel in that it is a trap circuit which holds all input voltage at the resonance frequency (Fig. 2.13). There is no voltage through the circuit at the resonance condition, as if it is a short circuit. However, the individual components have voltage across them. In fact, the capacitor and inductor have the same amplitude and opposite voltages. In addition, they are larger than the input voltage and 90° out of phase with the input. The circuit has a Q factor of $\omega_0 L/R$ and the resonance impedance of the circuit equals the resistance R of the conductor (probe coil).

For the above circuits, to achieve the highest Q factor, L is chosen to be as large as possible and R as small as possible. The desired resonance frequency can be obtained by changing C for the given L. However, the impedance cannot be set to a desired value once L, C and R are selected for a resonance frequency and Q factor. As a result, matching the impedance is impractical for these kinds of circuits. The solution to the problem is to integrate an additional adjustable capacitor as shown in Fig. 2.14. For the series–parallel circuit the total impedance can be approximated to

2.5 Probe

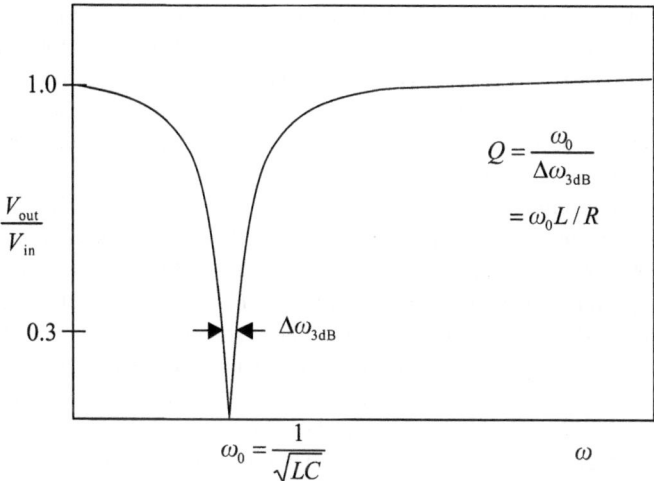

Fig. 2.13 Output voltage curve in a series LC circuit as a function of frequency. It becomes a short circuit at the resonance frequency ω_0.

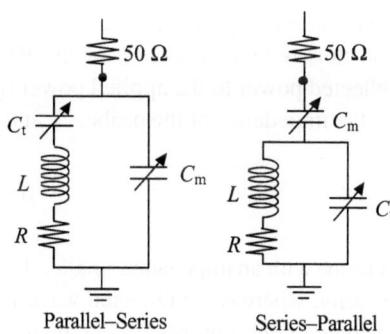

Fig. 2.14 Examples of probe circuits: parallel–series and series–parallel LC circuits. C_t and C_m are the adjustable capacitors for tuning and matching, R is the resistance of the probe coil, and 50 Ω is the impedance of the cable connected to the probe at the dot point.

$$Z_{LC} = \frac{1}{(1/Z_L) + (1/Z_{C_t})} + Z_{C_m} = \frac{1}{(1/j\omega L) + j\omega C_t} + \frac{1}{j\omega C_m}$$

$$= \frac{j(\omega^2 LC_t + \omega^2 LC_m - 1)}{(1 - \omega^2 LC_t)\omega C_m} \tag{2.25}$$

$$\omega_0^2 L(C_t + C_m) - 1 = 0 \tag{2.26}$$

$$\omega_0^2 = \frac{1}{\sqrt{LC}} \tag{2.27}$$

in which $C = C_t + C_m$. To obtain the Q factor and total impedance at resonance, the resistance R should be considered as treated earlier (refer to (2.18)). Q is the same at resonance as previously obtained for the resonance circuit, $Q = \omega L/R$.

The impedance at resonance is close to $Q\omega L/a$, which is the same as the parallel circuit except it is scaled by a factor of $a = (1 + C_m/C_t)^2$. Therefore, for a probe circuit with high Q obtained by large L, the impedance is brought down to 50 Ω by increasing C_m and simultaneously decreasing C_t to maintain the resonance frequency.

For a parallel–series circuit, the modification is obtained by adding a parallel capacitor to the series circuit. Using a similar treatment, the resonance frequency is proved to be approximately equal to $(LC_t)^{-1/2}$ for the situation of $C_m \gg C_t$ and the impedance at resonance is given by $Q\omega L C_t^2/(C_t + C_m)^2$. When such a circuit is used for an NMR probe, the resonance frequency is achieved by high L and small C_t to obtain high Q and to meet the condition of $C_m \gg C_t$. For such a probe, the matching capacitor has little effect on the tuning of the resonance frequency and the 50 Ω matching is achieved by adjusting the matching capacitor after the probe is tuned to ω_0.

Tuning the probe means adjusting the circuit capacitance and inductance to be on resonance at a desired frequency. For probe tuning, it is difficult and expensive to change the inductance of the probe circuit. Therefore, the frequency and impedance adjustment of the probe is achieved by changing the capacitance as described above. During the probe tuning, the impedance of the probe circuit is also adjusted to match impedance of the cable connected to the probe at 50 Ω. The probe acts as an RF load of the cable. In the case of mismatch, when the impedance of the probe circuit is not 50 Ω, the cable produces a reflected wave when an RF pulse is applied to the probe, and thus reflects a portion of the RF power delivered to the probe. The ratio of reflected power to the applied power (power loss due to the mismatch) is dependent on the impedance of the probe, Z_L and the characteristic impedance of the cable, Z_0:

$$\rho = \frac{Z_L - Z_0}{Z_L + Z_0} \tag{2.28}$$

A probe with an impedance smaller than 50 Ω produces a reflected wave with opposite polarity, whereas the reflected wave is not inverted if Z_L is larger than 50 Ω. At the matching condition ($Z_L = Z_0$), there is no power loss and hence all applied power remains in the probe, which in turn produces the shortest 90° pulse length.

As mentioned in Chap. 1, NMR spectroscopy is an insensitive technique owing to the small energy gap between the transition energy states. This insensitivity limits the application of NMR to samples with high concentration. Much effort has been carried out to develop more sensitive probes in parallel with the development of higher field magnets. The sensitivity of the probe is proportional to its Q factor, which means that the higher Q is, the higher the sensitivity:

$$\frac{S}{N} \propto \sqrt{\frac{\eta Q}{T}} \tag{2.29}$$

in which S/N is the signal-to-noise ratio, η is the filling factor of the probe coil and T is temperature in K. As discussed previously, the Q factor is inversely proportional to the resistance of the probe coil. Reduction in the resistance will significantly increase the Q value of the probe. Using high temperature superconducting material

2.5 Probe

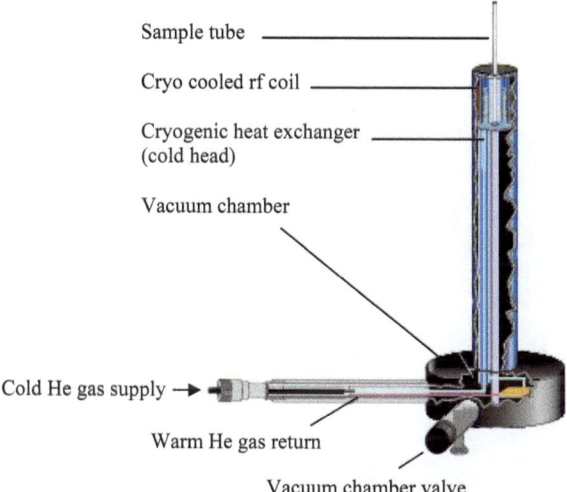

Fig. 2.15 Diagram example of a cryogenic NMR probe. The probe operates at cryogenic temperature (e.g., 25 K) and the temperature of the sample is regulated by a variable temperature (VT) control unit. The probe coil and preamplifier are cooled by the cold helium gas. After heat exchange at the cold head near the probe coil, the warm helium gas returns to helium compressor of a close-cycled cooling (CCC) system. The probe body is insulated by the vacuum chamber pumped continuously (courtesy of Agilent Technologies).

for the probe coil is an effective way to reduce the resistance. It has also been recognized that thermal noise generated at the probe coil limits the sensitivity of the probe. Cooling the probe coil made from the normal conductor and preamplifer to 25 K can significantly reduce the noise contribution and improve the sensitivity. For a cryogenic probe, the Q factor can be as high as 20,000 compared to 250 of conventional probes. In addition, a considerable amount of thermal noise in the probe is eliminated at the low temperature, which in turn increases the sensitivity of the probe. For this same reason, preamplifier circuits are integrated inside the cryogenic probe and cooled to the cryogenic temperature. An example diagram of a cryogenic probe for high resolution NMR is shown in Fig. 2.15. With the use of cryogenic probes, the sensitivity can be improved dramatically by a factor of 3–4-fold compared to a conventional probe as evidenced by the comparison of HNCA TROSY slices shown in Fig. 2.16. This leads to a reduction in experiment time of 9–16 fold or the ability to obtain data for more dilute samples.

Because of its high sensitivity, the performance of the cryogenic probe is more vulnerable to the salt concentration of the NMR sample. The sensitivity of a probe has a dependence on the conductivity of a sample according to the following relationship:

$$\frac{S}{N} \propto r_s \sqrt{\eta \sigma \omega_0} \qquad (2.30)$$

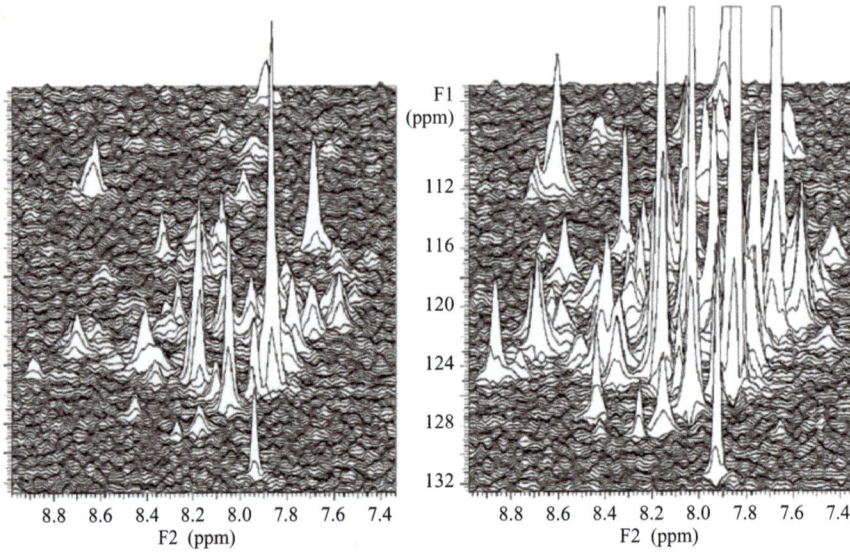

Fig. 2.16 HNCA TROSY slices of 2.3 mM ^{13}C, ^{15}N, ^{2}H DAGK (Oxenoid et al. 2004) obtained at 600 MHz field strength using a conventional triple-resonance probe (*left*) and a cryogenic probe (*right*) (courtesy of Agilent Technologies).

in which r_S is the radius of a cylindrical sample with conductivity of σ, η is the filling factor of the probe coil, and ω_0 is the resonance frequency. The high Q value of cryogenic probes is dramatically diminished by the increased resistance due to the presence of salts in the solution, whereas the function of a conventional probe is stable over a relatively wide range of salt concentrations. Even a moderate dielectric loss by a salt concentration of about 100 mM may substantially weaken the advantage of cryogenic probes. Therefore, careful attention must be paid when the sample is prepared with a buffer solution containing salts.

At high magnetic fields (>500 MHz), the radiation damping effect from water signal of an aqueous sample causes problems and artifacts such as artifacts and spurious harmonics in multidimensional spectra and distorted line shapes in T_1 and T_2 relaxation measurements. It has long been recognized that radiation damping is not signal dissipation but a process in which transverse magnetization is transformed to the longitudinal magnetization due to the coupling of water magnetization to the probe coil. The effect can be explained by considering the oscillating magnetic field produced by the water transverse magnetization. After an RF pulse, the water magnetization near the carrier frequency precesses in the xy plane of the laboratory frame (Augustine 2002). This rotating magnetization produces an oscillating magnetic field that induces an electromotive force (EMF) or a current flowing in the probe coil according to Faraday's law. The current will in turn produce an RF magnetic field inside the probe coil with the same frequency that rotates the water magnetization back to the z axis. The rate at which the water transverse magnetization generated by a 90° pulse returns to the z axis by the oscillating RF magnetic field can be described in terms of the radiation damping time constant T_{RD} (Bloembergen and Pound 1954), which is given by

$$R_{RD} = \frac{1}{T_{RD}} = 2\pi M_0 \gamma Q \eta \tag{2.31}$$

in which γ is the gyromagnetic ratio, Q and η are defined as in (2.29). For a high-Q NMR probe (specially a cryogenic probe), the water transverse magnetization can be transformed to longitudinal magnetization by the radiation damping effect on the order of milliseconds compared to the water ^{1}H T_1 relaxation times on the order of seconds (Lippens et al. 1995). Larger T_{RD} gives a slower rate (or a smaller R_{RD}).

Radiation damping can cause problems such as line width broadening, rapid sample repolarization, and solute signal distortion. Many methods have been developed to remove the radiation damping effects by either pulse sequences or probe hardware design. The active feedback-suppression method (Szoke and Meiboom 1959; Broekaert and Jeener 1995) uses hardware to feed the signal generated by the radiation damping back to the probe coil after the signal is phase shifted by 180°. As a result, the oscillating current in the probe coil is canceled in real time. Other methods include the overcoupling method, which uses overcoupled probe circuits (Picard et al. 1996) and hence increases T_{RD} and decreases the radiation damping rate, and Q-switching method which uses low Q during RF pulsing and switches to high Q during acquisition. As a result, T_{RD} is increased by decreasing the Q value.

The radiation damping effect has also been utilized to obtain information on the water/solute interactions and to achieve solvent suppression. For instance, the radiation damping was used to study the hydration of the protein BPTI without the feedback (Bockmann and Guittet 1995), to generate selective inversion pulse using the feedback to investigate the water/solute interaction (Abergel et al. 1996), and to suppress solvent signals in the measurement of the self-diffusion coefficients of biomolecules (Krishnan et al. 1999).

2.6 Quarter-Wavelength Cable

Questions to be addressed in the present section include the following:

1. What is a quarter-wavelength cable?
2. What are the functions of a quarter-wavelength cable?
3. What can it be used for?

If the load impedance of a cable matches the characteristic impedance of the cable, all applied power goes into the load and no power is reflected. This is true regardless of cable length or wavelength of the RF signal. However, when the impedance of the cable is mismatched, for a given cable length, the portion of the signal reflected back at the input terminal has a phase shift with respect to the input signal and the phase

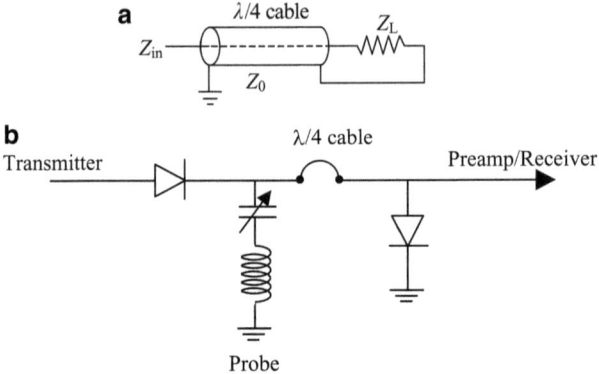

Fig. 2.17 (a) Quarter-wave coaxial cable whose input impedance is determined by (2.33) and (b) its application in a T/R (transmitter/receiver) switch.

depends on the frequency of the input signal (Parker et al. 1984). Consequently, the impedance at the input terminal will contain the reflected component and depends on the load impedance of the cable, the characteristic impedance, cable length ℓ and the wavelength λ corresponding to the applied frequency. The wavelength is 0.66 times the wavelength of light at a given frequency for a typical coaxial cable using solid dielectric spacing material (polyethylene). For a cable with length ℓ, characteristic impedance Z_0, and load impedance Z_L (Fig. 2.17), the input impedance is given by

$$Z_{in} = Z_0 \frac{Z_L \cos(2\pi\ell/\lambda) + jZ_0 \sin(2\pi\ell/\lambda)}{Z_0 \cos(2\pi\ell/\lambda) + jZ_L \sin(2\pi\ell/\lambda)} \qquad (2.32)$$

The equation describes the dependence of the impedance transformation on the cable length. If the cable length equals an odd number of quarter-wavelength:

$$\ell = n\frac{\lambda}{4} \quad (n = 1, 3, 5, 7, \ldots)$$

then the input impedance experienced by the cable is

$$Z_{in} = \frac{Z_0^2}{Z_L} \qquad (2.33)$$

For a short-circuited ¼ wavelength cable which has zero load impedance ($Z_L = 0$) such as by grounding, the input impedance becomes infinitely large according to (2.33), which means that the cable becomes open for its corresponding frequency. Thus, no signal with the frequency of the quarter-wavelength cable can pass through, and a signal with a different frequency will be attenuated by passing through the cable. This can be understood by considering that a shorted

¼ wavelength line must always have zero voltage and maximum current at the shorted end because $Z_L = 0$. At the input end which is a quarter-wavelength away from the shorted end, the voltage is maximum and the current is zero. Therefore, it looks like an open circuit for the signal with the corresponding frequency. This property of the quarter-wavelength cable (sometimes called quarter wave cable) is applied in a T/R (transmitter/receiver) switch to isolate the probe from the preamplifier during the transmitter pulse so that the RF power does not go into the preamplifier. A quarter-wavelength cable with actively shut diodes is connected to the receiver part of the T/R switch. When an RF pulse is applied by the transmitter, the diodes become one-way conductors (because of the high voltage of the pulse, Fig. 2.17b). The quarter-wavelength cable is shorted by the closed diode connected to it, becomes an open line for the specific frequency RF pulse, and hence separates the receiver from the transmitter during pulsing. Conversely, for an open-circuited ¼-wavelength cable (without transmitter pulsing), the input impedance becomes zero because of the infinite load impedance (according to (2.33)), and hence the line looks like a shorted circuit, resulting in attenuation of the signal with a frequency corresponding to that of the quarter-wavelength cable.

2.7 Analog/Digital Converters

Questions to be answered related to the topics of this section include the following:

1. What are analog-to-digital converters (ADC) and digital-to-analog converters (DAC)?
2. What are the basic principles used to make the devices?
3. What are their functions and applications to NMR instrumentation?

The signals generated at the probe coil and detected by the receiver are in continuous or analog form, meaning that their amplitudes change smoothly, such as in a sine wave. However, the signals to be processed by computers and other electronic devices in the NMR spectrometer are a digital or discrete type, which means that their amplitudes can only exist in certain levels or ranges, such as binary digits. On the other hand, the output controlled by the computer needs to be converted to analog form, for example, numbers for gradient pulse levels and RT shims must be converted to analog currents into gradient coil or RT shim coils. Therefore, for NMR spectrometers it is necessary to accurately convert an analog signal to a digital number proportional to its amplitude (ADC), and vice versa (DAC). These conversions are essential in a wide variety of processes in which the analog information is converted (ADC) for data processing and display such as the Fourier transformation of the time domain data, and the digital information is converted to analog (DAC) for a computer controlling the experimental setup such as shimming, gradient pulse amplitude, or waveform generation. The conversions are also necessary for measurement instruments such as signal generators as well as digital oscilloscopes. An ADC is a device that converts the information obtained in

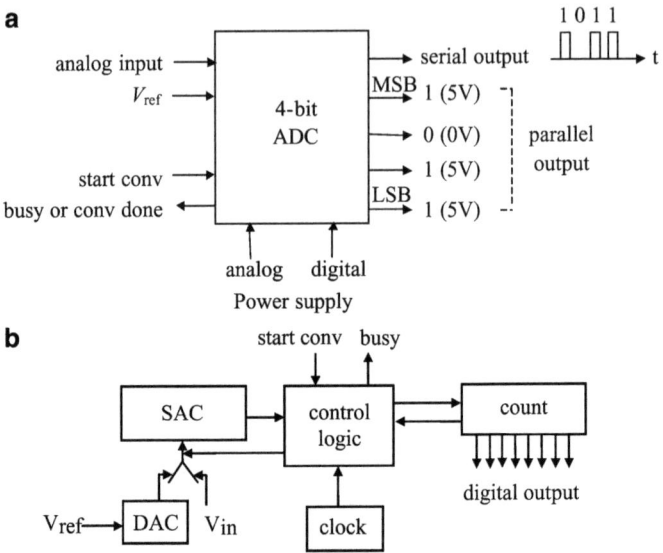

Fig. 2.18 Analog-to-digital conversion. (**a**) Block diagram of a 4-bit ADC converting an input signal with a amplitude of 11 to a parallel and serial output of 1011. The reference voltage is often produced within the converter. The ADC usually has two control lines to receive "start conversion" input and send status "busy" (conversion in progress) or "conversion done" output. The serial output is in the form of a pulse train with MSB first, whereas the parallel output is done simultaneously via four separate output lines. MSB and LSB mean most significant bit and least significant bit, respectively. (**b**) Successive approximation conversion (SAC). All bits of output are first set to zero. Then, each bit is compared to the DAC output, starting with the MSB. If the input signal voltage is larger than or equal to the DAC output, the register is set to 1, otherwise, it is set to 0. The process continues until the LSB is compared.

analog form such as the amplitude of the input signal to the information described in numerical values with respect to a reference signal, whereas the DAC is a device for the reverse conversion. They are integrated circuits and can have resolution higher than 16 bit and conversion rates faster than 50 MHz.

The ADC process includes quantizing and encoding. The analog input signal is first partitioned by a comparator unit during the quantization and then the partitioned signal is assigned to a unique digital code corresponding to the input signal during the encoding process. Usually, the binary number system is used in the conversion. For an n-bit converter, there are 2^n digital codes (numbers), resulting in a dynamic range of $2^{n-1}-1$ (which represents numbers between -2^{n-1} and $2^{n-1}-1$). The code is a set of n physical two-value levels (i.e., bits, 0 or 1). For example, a signal with a scale of 11 will be coded as 1011 by a 4-bit ADC as shown in Fig. 2.18a. Frequently, the signal is digitized by converting the electric voltage of the input signal into a set of coded binary electrical levels such as +5 or 0 V and the digitized signal is output in parallel (simultaneous) form or in series (pulse-train) form with the most significant bit first (MSB), and sometimes both.

2.7 Analog/Digital Converters

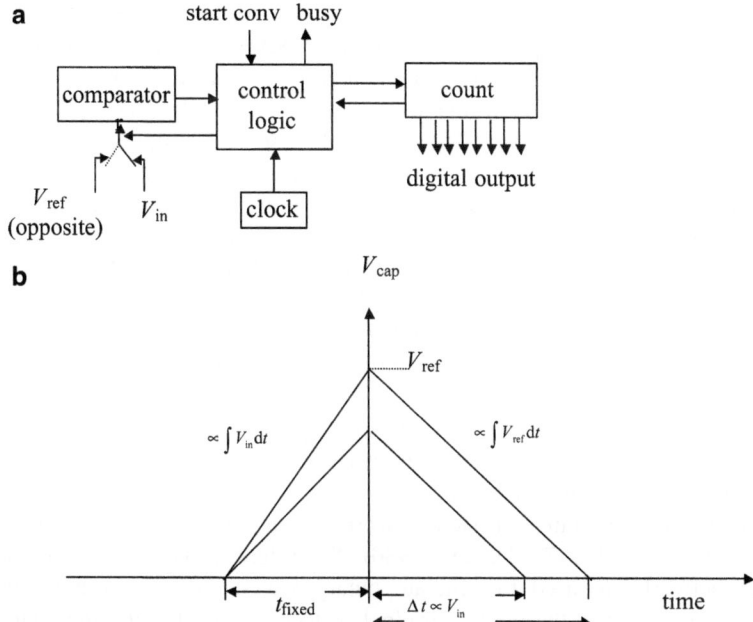

Fig. 2.19 Dual-slope integration conversion. (**a**) Block diagram of the ADC, (**b**) conversion cycle. The voltage of the reference is proportional to the input voltage, V_{in}, and hence the time to discharge the capacitor, Δt, is proportional to V_{in}. Because the current for discharging the capacitor is constant, the slope of the reference integration is unchanged for all V_{in}, while Δt is different for different V_{in}.

There are many techniques for analog-to-digital conversion, among which the successive-approximation (Fig. 2.18b) and dual-slope (Fig. 2.19) ADC remain popular because of their conversion speed and accuracy (Dooley, 1980; Sheingold, 1977). The dual-slope integration converters provide excellent accuracy with high sensitivity and resolution (Fig. 2.19). During the conversion, the input signal is integrated for a fixed time interval by charging a capacitor with a current accurately proportional to the input signal amplitude. The final value of the signal integral becomes the initial condition for integration of the reference in the reverse process, which is achieved by discharging the capacitor with a constant current. When the net integral is zero, as indicated by the voltage of the capacitor reaching zero again, integration of the reference stops. The time of reference integration (to discharge the capacitor) is counted by a counter driven from a clock, which is proportional to the input signal amplitude. Therefore, the result of the time count is a digital output proportional to the input signal amplitude. The drawback of the dual-slope integration conversion is the slow conversion rate.

The successive approximation conversion (SAC) is a popular high speed technique used primarily in data acquisition. The conversion is achieved by comparing the input signal with a reference set produced by a DAC, resulting in various output codes (Fig. 2.18b). Initially, all bits of output are set to zero. Then each bit is compared to the DAC output, starting with the MSB. If the input signal voltage is larger than or equal to the DAC output, the register is set to 1; otherwise, it is set to 0. It is a binary search starting from the middle of the full scale. The MSB is tried by the DAC output of ½ full scales. The MSB code is set to 1 or 0, respectively, if the input signal is at least equal to or does not exceed the DAC output. Then, the second bit is tried with ¼ full scale and assigned to 1 or 0 accordingly. The process continues until the least significant bit (LSB) is compared. An n-bit ADC has an n-step process. The maximum output is always 2^n-1, in which all bits are set to 1. The final digital output is usually provided in both the parallel form of all bits at once on n-separated output terminals and the series form of n-sequential output bits with the MSB first on one single output terminal. For NMR applications, the current typical conversion time is 500 kHz with a dynamic range of 16 bits.

The DAC is a device to convert an input signal representing binary numbers (or binary-coded decimals, BCDs) to information in the form of current or voltage proportional to the input signal. There are a variety of conversion methods, in which the reference voltage source, resistor network, and digital switches are the essential elements (Sheingold, 1977; Dooley, 1980). The reference voltage source and the resistor network are used to generate binary scaled currents, whereas digital switches are turned to the output terminal or to the ground under the control of the digital input code. The output signal voltage V (or current) is given by

$$V = V_{\text{ref}} \sum_{i=1}^{n} \frac{2^{n-i}}{2^n} \delta_i \qquad (2.34)$$

in which V_{ref} is the voltage of the reference source, δ_i is the input digital code which is equal to 0 or 1, and n is the bit of the converter. The MSB (i = 1) is converted first and the LSB (i = n) last. The maximum output voltage is limited to $V_{\text{ref}}(2^n - 1)/2^n$ because the maximum digital input is 2^n-1. For instance, if a digital input of 1011 is converted by the 4-bit DAC, the output has a voltage given by

$$V_{\text{out}} = V_{\text{ref}} \left(\frac{8}{16} + \frac{0}{16} + \frac{2}{16} + \frac{1}{16} \right) = \frac{11}{16} V_{\text{ref}} \qquad (2.35)$$

2.8 Instrument Specifications

In the current section, typical specifications of an NMR instrument are discussed which are useful to describe a desired NMR spectrometer. When purchasing an NMR spectrometer, there are certain specifications that must be considered and specified. The typical specifications to be discussed below are categorized based on the basic components of NMR spectrometer.

2.8 Instrument Specifications

Specifications for the NMR magnet include bore size, number of shims, actively vs. passively shielded, days between refills for liquid nitrogen and liquid helium, field drift rate and warranty period. NMR magnets are made with either a standard bore size (e.g., 51 mm diameter) or a wide bore (e.g., 69 mm). The wide bore magnets are usually used for micro imaging or solid state NMR because there is more space inside the bore, but they cost much more than the standard bore magnets due to the usage of more SC material. In recent probe development, solid state NMR probes have been built to fit in a standard bore magnet for solid state NMR research. The standard bore magnet may have as many as 40 RT shims for a field strength higher than 500 MHz whereas the wide bore type does not need more than 30 shims because of the large volume inside the magnet. An actively shielded magnet has a much shorter 5 Gauss line diameter than an unshielded magnet, which saves lab space. (A 5 Gauss line is the circle from the magnet center, where the fringe magnetic field strength outside the circle is less than 5 Gauss.) The time between refills should be >14 days for liquid nitrogen and >120 days for liquid helium. Although the drift rate is usually specified to <10 Hz h^{-1}, in most cases, the drift rate is in the range of 0.5–3 Hz h^{-1} for the magnets of 600 MHz or lower. For magnets of 500 MHz or higher, a set of antivibration posts should be included in the specifications. Homogeneity of the magnetic field is usually <1 ppm after cryogenic shimming.

Specifications for the console are more complicated than those for magnets, and are categorized based on the components of the console: RF channels (transmitter, amplifier, synthesizer, receiver, and digitizer), lock channel, and probes. The number of RF channels defines the spectrometer's capability of simultaneously delivering RF pulses to different nuclei. For consoles of 400 MHz or lower, the standard configuration has two RF channels with one full band and one low band frequency synthesizer, whereas three or four channel configuration with two full band and one low band synthesizer is the typical choice for 500 MHz or higher. An RF channel with a ^{1}H only frequency synthesizer is not a wise choice although it probably costs less than a full band RF channel. Full band is defined as the frequency range from ^{15}N resonance (or lower) up to ^{1}H resonance frequency and low band covers the frequency range from ^{15}N resonance up to ^{31}P resonance. Amplifier output power is >50 W for the frequency range of ±50 MHz about the ^{1}H resonance (~100 W for solid state NMR) and ~300 W for the heteronuclear frequency range. The additional specifications for transmitter include <500 ns event timing, >4,000 steps amplitude control over at least 60 dB range, <50 ns time constant for phase and amplitude change, and 0.1 Hz frequency resolution. The console should have at least two waveform generators with <50 ns pulse time resolution, and <200 ns minimum event time, and >1,000 linear steps. The lock channel should have the capability of automatic switching for ^{2}H gradient shimming and for ^{2}H decoupling. The frequency range of the lock is ~±5 MHz about the ^{2}H resonance frequency, which is necessary to adjust spectrometer frequency when needed (such as in the case of z_0 out of range due to field drift). An active T/R switch with <1.5 μs timing, a 16 bit ADC with 500 kHz speed and digital signal processing capability are the standard features of NMR spectrometers.

Specifications for the probe include signal-to-noise ratio, line width, gradient profile, gradient recovery time, 90° pulse lengths, and RF homogeneity. For a triple-resonance probe with a z-axis gradient, a typical 90° pulse length at 3 dB lower than the maximum pulse power is <7 μs for ^{1}H, <15 μs for ^{13}C, <40 μs for ^{15}N and <40 μs for ^{31}P. The gradient coil should be shielded with a strength >50 G/cm^{-1} (>20 G/cm^{-1} for 400 MHz or lower instruments) and a recovery time <0.1 ms. The sensitivity of a conventional (or room temperature, RT) triple-resonance probe is >1,000:1 for 500, >1,300:1 for 600 MHz and >1,800:1 for 800 MHz using the standard ^{1}H sensitivity sample (0.1 % ethylbenzene in CDCl$_3$), whereas cryogenic probes have a sensitivity of 3–4 fold higher. For instance, the cryogenic probe of a 600 MHz instrument should have a sensitivity of >4,500:1. RF homogeneity is >80 % for ^{1}H 450°/90° (which means that the intensity of the peak obtained by the 450° pulse is greater than 80 % the intensity obtained by the 90° pulse), >70 % for ^{1}H 810°/90°, 70 % for ^{13}C decoupler 360°/0°, and 55 % for ^{13}C decoupler 720°/0°. A typical ^{1}H non spinning line width should be better than 1/10/15 Hz at 50 %/0.55 %/ 0.11 % of peak amplitude using a 5 mm standard line shape sample for a RT probe and 1/10/20 Hz for a cryogenic probe. Spinning sidebands should be less than 1 % at a spin rate of 25 Hz. The variable temperature (VT) range is typically over −60 to 100 °C for a conventional probe and 0–40 °C for a cryogenic probe.

Additional specifications include quadrature image with one scan <0.4 %, with four scans <0.04 %, phase cycling cancelation (four scans) <0.25 %, and pulse turn-on time <0.05 μs.

2.9 Test or Measurement Equipment

The test equipment to be discussed in the present section are those routinely used in instrument setup or trouble-shooting, including the reflection bridge, oscilloscope, and spectral analyzer. Questions to be addressed about the test equipment are:

1. What is the test equipment needed for?
2. How are they operated?
3. What is the noise figure of a system?
4. How can it be measured?

2.9.1 Reflection Bridge

Although a reflection bridge (also known as duplexer or magic T, Parker et al. 1984) is not exactly a test instrument, it is a broadband device with four ports that is useful in tuning an NMR probe (Fig. 2.20). There is complete isolation (infinite impedance) between A and C or between B and D, but no isolation between the two terminals of any other combination. An RF signal fed into any port is equally split

2.9 Test or Measurement Equipment

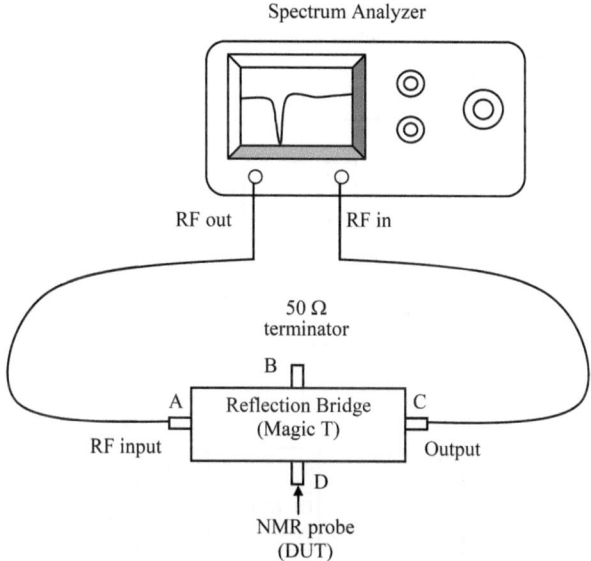

Fig. 2.20 Reflection bridge used for NMR probe tuning. There is complete isolation (infinite impedance) between A and C or between B and D, but no isolation (near zero impedance) between the two terminals of any other combination. An RF signal fed into port A is equally split into two output signals at ports B and D. If the impedances of the two output ports (B and D) are mismatched (unequally loaded), the reflected power is directed into the isolated port, resulting in an output at port C from port A. By monitoring the output RF signal (at C), a probe can be tuned for a desired resonance frequency at the desired impedance (50 Ω in this case). *DUT* device under test.

into two output signals at the closest ports with a specific phase shift (usually 0 or 180°). If the impedances of the two output ports (B and D) are mismatched (unequally loaded), the reflected power is directed into the isolated port, resulting in an output at port C from port A. By monitoring the output RF signal, a probe can be tuned for a desired resonance frequency at the desired impedance (50 Ω).

2.9.2 Oscilloscope

The two time-dependent physics quantities from electronic circuits we want to measure are current and voltage. An oscilloscope (Oliver and Cage 1971; Parker et al. 1984), or scope, is an essential and very useful test instrument because it measures the voltages or current (sometimes) in a circuit as a function of time and displays waveforms of the measured signals (Fig. 2.21). It is an electronic instrument which produces a graphical plot on its screen showing the relationship of two or more independent variables such as voltage vs. time. It can be adjusted for amplitude measurement or time measurement. For amplitude measurement, the scope

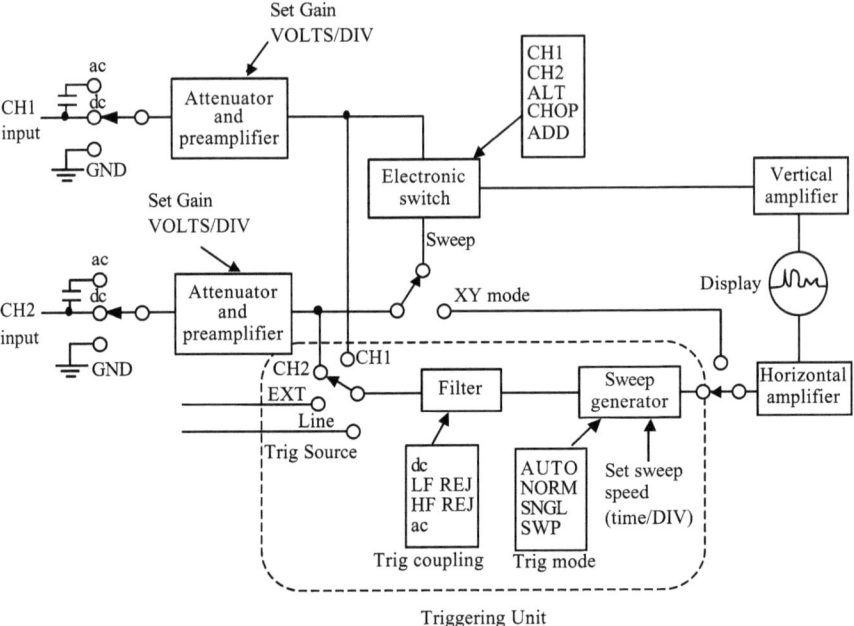

Fig. 2.21 Block diagram of an oscilloscope.

measures vertical deflection such as peak-to-peak voltage (V_{pp}) displayed on the oscilloscope screen (Fig. 2.22). If the effective voltage (V_{rms}) is needed to measure a sinusoidal signal, V_{pp} can be converted to V_{rms} according to (2.7). For time measurement, the time base setting is adjusted to observe time-dependent properties of the circuit, such as the frequency of the signal, the pulse rise time of the voltage step, or the phase difference of two signals.

Oscilloscopes usually have two or more input channels. Each channel has an input attenuation control knob labeled as VOLTS/DIVISION for vertical amplitude measurement (Fig. 2.21). Turning the knob increases or decreases the intensity of the measured signal in a calibrated condition. The knob is automatically rendered inactive if the channel related to it is set to ground input mode, GND, which lets the user observe the position of zero voltage on the scope screen. In ground mode, the signal is cut off from the scope input. The input of the scope is grounded, but the signal is not shorted to ground. There is also a VARIABLE control knob for each channel allowing the user to set the desired number of divisions. Turning the VAR knob adjusts the magnitude of a given signal, and the vertical deflection becomes uncalibrated as indicated on screen. The attenuation must be in the calibrated condition (VAR knob is not activated) when making an accurate measurement of signal voltage such as for the output of an amplifier.

There are other controls for vertical display, including input modes (DUAL, ADD and XY mode), Y POSITION control and an INVERT switch. Y POSITION (vertical position) allows one to change the vertical trace position. When there is no

2.9 Test or Measurement Equipment

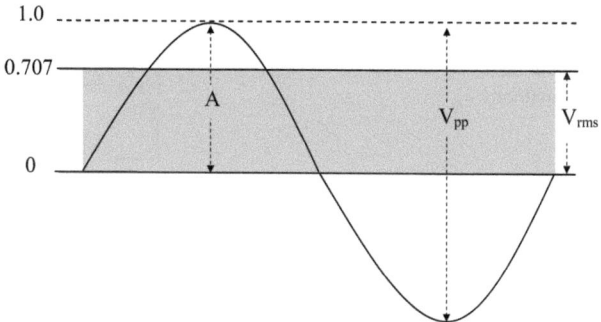

Fig. 2.22 Relationship between rms voltage, V_{rms}, signal amplitude, A, and peak-to-peak voltage, V_{pp}.

signal applied at the input, the vertical trace represents 0 V. The invert function is used to invert the signal display by 180°. This function is useful when looking at the difference of two signals in ADD mode. If the input mode is switched to DUAL mode, vertical signals from both channels are displayed on the screen either in ALTERNATE mode whereby the scope internally switches over from one channel to the other after each time base sweep or CHOPPED mode in which channel switching occurs constantly during each sweep. In XY mode, one channel is used for vertical (Y) deflection whereas the other causes horizontal (X) deflection (the amplitude change is displayed horizontally), which is useful for such measurements as frequency and phase comparisons of two signals. What is displayed on the screen is one signal vs. another (X–Y) rather than against time. The time unit controls the z axis and can be triggered internally from the vertical portion of the X–Y display.

Time related amplitude changes on an input signal are displayed in vertical mode as discussed above, deflecting the beam up and down whereas the time base generator moves the beam from left to right on the screen (time deflection). This gives a display of voltage vs. time. Similar to vertical attenuation control, calibrated TIME/DIV and VAR controls are used to change time deflection. Because test signals to be displayed are repetitive waveforms, the time base must accordingly repeat the time deflection periodically. To produce a stable display, the time base is triggered only if LEVEL and SLOPE (+ or −) on a waveform match with the previous time base. The slope is relative to rising or falling edge of the test signal. Triggering can be performed by measuring the signal itself (internal triggering) or by an externally supplied but synchronous voltage. In AUTO trigger mode, the sweep is free running without a trigger signal. A baseline will not disappear from the screen even if no signal is present. This is the best mode to use for all uncomplicated measuring tasks. The NORMAL trigger mode produces a waveform display by manually adjusting the trigger LEVEL control. When the trigger LELVEL is mismatched or signal is weak, no waveform is displayed.

Sometimes it is hard to get a signal to show on the screen. The following are tips for a quick start. Start by connecting the input to channel 1, setting the triggering on

Fig. 2.23 Measurement of phase shift using (**a**) the Lissajous figure and (**b**) alternate mode methods.

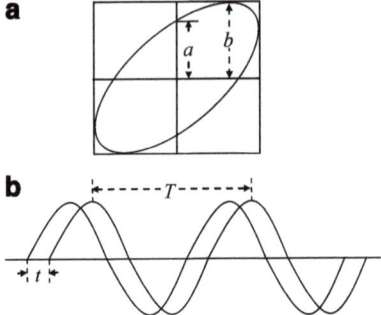

AUTO, DC, CH1 and setting time (horizontal) deflection at calibrated 1 ms per div with the X-magnifier off (1×). Next, ground the input signal by setting the input mode to ground input mode, GND, and adjust the display intensity and vertical position controls until the reference horizontal line appears. Now apply signal to the scope by ungrounding the input and adjust the time base switch TIME/DIV accordingly.

The peak-to-peak voltage of a signal can be directly measured by counting the amplitude scales on the scope. If V_{rms} is needed, V_{pp} can be converted according to the relationship shown in Fig. 2.22 or (2.7). The frequency of a sinusoidal signal may be measured by reading the time necessary for one full cycle and inverting the reading result. The relative phase of two waveforms is usually measured by means of a Lissajous figure as shown in Fig. 2.23a. Each of two signals is applied to each individual channel of the scope in XY mode. The phase angle can be determined from the dimensions of the ellipse according to the relationship:

$$\sin\theta = \pm\frac{a}{b} \qquad (2.36)$$

in which the minus sign is for a ellipse 90° rotated from the one in Fig. 2.23a. A more convenient method is to display both of the signals in alternate mode. Alter the full cycles of the waveforms are obtained by setting the appropriate time scale, the phase shift is determined by the quantities t and T:

$$\theta = 360\frac{t}{T} \qquad (2.37)$$

2.9.3 Spectrum Analyzer

A spectrum analyzer is another widely used test instrument, particularly in tuning an NMR probe. A scope observes signal voltage as a function of time, whereas a spectrum analyzer allows one to look at the signal voltage in the frequency domain, the graphical representation of signal amplitude as a function of frequency

2.9 Test or Measurement Equipment

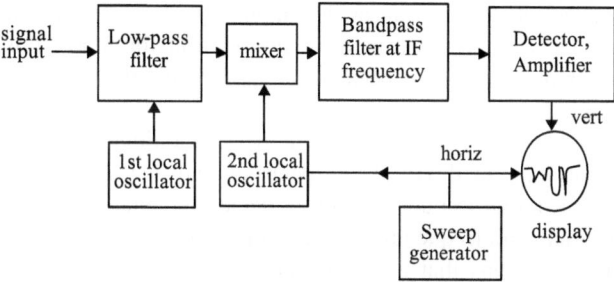

Fig. 2.24 Block diagram of a swept-tune (ST) spectrum analyzer. ST analyzers are tuned by a sweeping local oscillator (LO) of a superheterodyne receiver over its range of frequencies. The LO is mixed with the input signal to produce an intermediate frequency (IF). The signal frequency whose difference with LO frequency is equal to IF can pass through the IF amplifier and filter, and consequently is detected and displayed. As LO is swept through its frequency range, different input frequencies are successfully mixed to be observed.

(Coombs 1972; Parker et al. 1984). The time domain is used to view the relative timing and phase information of a characterized circuit. However, not all circuits can be appropriately characterized by time domain information. Circuit elements such as NMR probes, amplifiers, filters, receivers and mixers are best characterized by their frequency dependent information. In the time domain, all frequency components of a signal are overlapped together, whereas in the frequency domain, they are separated in frequency axis and voltage level at each frequency displayed. Therefore, a spectrum analyzer is useful in measuring resonance frequency, low level distortion and noise, etc.

There are two basic types of spectrum analyzers: swept-tuned (ST) and real-time (RT). ST analyzers are the most common type and they tuned by a sweeping LO of a superheterodyne receiver over its range of frequencies (Fig. 2.24). The LO is mixed with the input signal to produce an IF which can be detected and displayed on the analyzer screen. The signal frequency whose difference with LO frequency is equal to an IF can pass through the IF amplifier and filter, and consequently is detected and displayed. As the LO is swept through its frequency range, different input frequencies are successfully mixed to be observed. High sensitivity is obtained for this type of spectrum analyzers due to the use of IF amplifiers and filters, and it can be tuned up to a few gigahertz bandwidth. Since the input frequencies are sampled sequentially in time, only a small portion of the input signal is used at a given time. It is impossible to display transient responses on an ST analyzer. RT analyzers have lots of flexibilities in terms of sweep range, center frequency, filter bandwidth, display scale, etc. The instruments are able to simultaneously display the amplitudes of all signals in a wide frequency range. This preserves the time dependent relationship among signals, which allows one to analyze the phase change of signal vs. frequency. An RT analyzer can display transient events as well as random and periodic signals. A Digital analyzer is an RT analyzer which makes use of digital Fourier transformation. After the detection and filtering processes, it converts an analog input signal to digital using an ADC, and then generates a digital spectrum using Fourier transformation. It is particularly useful for low frequency signals because the sweep rate of swept analyzer is slow for practical use at low frequency.

Usually a tracking generator is used either in conjunction with a spectrum analyzer or as an integrated part of the spectrum analyzer. This is a special signal source whose RF output frequency tracks (checks) the analyzer signal with itself. It produces a signal with frequency precisely tracking the spectrum analyzer tuning. Precision tracking means that at any instant of time the tracking generator frequency is in the center of the spectrum analyzer passband. Certain spectrum analyzers have a tracking generator installed, whereas others require an external tracking source for accurate measurement.

Similar to an oscilloscope but with fewer controls, a spectrum analyzer has vertical (amplitude) and horizontal (frequency) controls. Attenuation control (dB/DIV) sets vertical scale unit per division, whereas SPAN/DIV adjusts the displayed spectral width of the signal. The center frequency is tuned by a dial "FREQUENCY." The tuning rate is dependent on the selected SPAN/DIV setting. The sweep rate is selected by TIME/DIV. For general operation, after turning the analyzer on, set attenuation to 0 dB, TIME/DIV to AUTO, SPAN/DIV to max, and adjust the center FREQUENCY control. Once the input signal is displayed, adjust SPAN/DIV to the desired spectral window. Figure 2.20 shows the connection of the probe to a spectrum analyzer using a reflection bridge for probe tuning.

2.9.4 System Noise Measurement

By definition, noise is the electrical interference which causes reduction of the signal being measured. Instrument sensitivity is affected by both the noise coming with the signal and the noise generated internally by the instrument. Generally, system noise is described by the amount of noise in dB, or the noise figure, which numerically equals the logarithm of the ratio of signal-to-noise ratios at the input and output of a system (Mazda 1987):

$$F = 10 \log \frac{SN_{in}}{SN_{out}} \quad (2.38)$$

in which F is in dB, SN is the input or output signal-to-noise ratio of the system. If the noise source of the system has excess power E, the noise figure is determined by the noise power N_c with noise source off (cold) and N_w with noise source on (warm):

$$F = 10 \log E - 10 \log \left(\frac{N_w}{N_c} - 1\right) \quad (2.39)$$

2.9 Test or Measurement Equipment

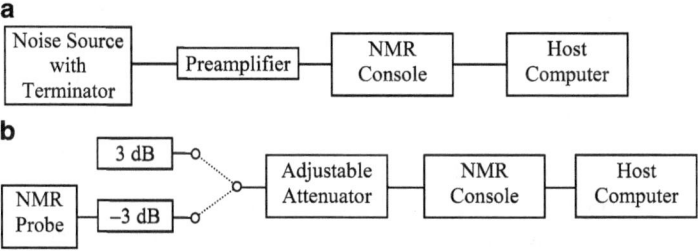

Fig. 2.25 Methods of noise figure measurement. (**a**) Cold/warm method separately measures the rms noise by placing the noise source at liquid nitrogen temperature and 20 °C. (**b**) Twice power method measures the rms noise with or without a probe. The source-off rms noise is measured with 3 dB attenuator and without the probe, whereas the probe and −3 dB attenuation are used for the measurement of the source-on rms noise.

E is also known as excess noise ratio and N_w/N_c is commonly called as Y factor. The noise figure can also be expressed in terms of noise temperature:

$$F = 10\log\left[\left(\frac{T_w}{290} - 1\right)\frac{N_c}{N_w} + \left(1 - \frac{T_c}{290}\right)\right] - 10\log\left(1 - \frac{N_c}{N_w}\right) \quad (2.40)$$

in which N_c and N_w are the noise measured at the cold T_c and warm T_w temperature, respectively. If the warm noise is measured at 290 K and cold noise measured in liquid nitrogen, the noise figure can be obtained by:

$$F = -1.279 - 10\log\left(1 - \frac{N_c}{N_w}\right) \quad (2.41)$$

In practice, the noise is measured as a V_{rms} value and hence $N = V_{rms}^2$. For an NMR system, the system noise figure should be less than 2 dB.

There are two methods to measure the noise figure based on (2.39) and (2.41), respectively. The cold/warm method measures the rms noise using a noise source in liquid nitrogen and at about 20 °C (Fig. 2.25a). Because the impedance of the NMR system is 50 Ω, the noise source for the cold measurement is constructed using the coaxial cable terminated with a 50 Ω resistor. After disconnecting the probe from the preamplifier, the noise source is connected to the preamplifier. The noise is measured with pulse length of 0, a ^{1}H SW of 50 ppm, maximum receiver gain and single scan. N_c is equal to the square of rms noise calculated after the Fourier transformation without any line broadening. N_w is measured in the same way except the noise source is warmed to 20 °C. Finally, the noise figure is calculated using (2.41).

The second method called twice-power measurement (Fig. 2.25b) is to make the noise ratio equal to 2 so that the noise figure is solely dependent on the first term of (2.39). The noise is first measured without the probe and the 3 dB attenuation using a pulse length of 0, an SW of 50 ppm, maximum receiver gain in the linear range and a single scan, which is N_c because switching off the noise source is equivalent to the cold condition. The probe is then connected to the preamplifier to allow the measurement of N_w with the -3 dB attenuation. The -3 dB attenuation can be achieved by decreasing the receiver gain according to the linear response of the instrument receiver gain. The next step is to adjust the inline attenuator to obtain the rms noise (N_w) at about the same level as the first measurement. In this condition, N_w has a value twice to N_c because of the -6 dB attenuation $[(-3\,\text{dB})_{N_w} - (+3\,\text{dB})_{N_c} = -6\,\text{dB}]$, resulting in the cancelation of the second term in (2.39). Therefore, the noise figure of the system is determined solely by the value of the first term in (2.39), which is equal to the value of the inline attenuator. This method does not require making noise sources but needs an adjustable attenuator. In addition, it may introduce error when using instrument receiver gain to attenuate the noise for the N_w measurement. The error of the twice-power measurement is in the range of 0.1–0.5 dB greater than that of the cold/warm method.

Questions

1. Which part of an NMR instrument generates NMR signals and which part detects? Where are they located?
2. How much are sensitivity and resolution of NMR signals on a 900 MHz instrument increased compared to a 500 MHz instrument? Assuming that the 900 MHz instrument has a cryogenic probe which has a gain in sensitivity by 3.5 fold, how much is the sensitivity increased compared to a 500 MHz with conventional probe? What field strength with a conventional probe is the sensitivity of the cryogenic probe on 500 MHz NMR equivalent to?
3. What is the function of ¼ wavelength cables? Where are they used in a NMR spectrometer? What could happen if the wrong ¼ wavelength cable is used during an experiment?
4. What is a T/R switch? Why does a NMR spectrometer have it?
5. What is the function of IF? And what is the value on an instrument you have used?
6. What part of an NMR spectrometer generates frequency? And what are the frequency ranges of the RF channels on an NMR spectrometer you have used?
7. If a 90° ^{1}H pulse length is much longer than the normal one (e.g., twice longer), what are the three things you should check before you conclude that something is wrong with the instrument?
8. Why does a magnet still have a magnetic field when the power is out?
9. If a 90° ^{1}H pulse length is 6.2 µs, what are the 90° ^{1}H pulse lengths after the RF field strength generated by a linear amplifier is reduced by 3 and 6 dB?
10. Why must the probe be tuned before the setup of an experiment? If a probe is tuned with the filters or without, which method gives the correct pulse length? Explain why.
11. What is the dynamic range (ratio of the largest to smallest signals) of a 16-bit ADC?
12. Where is a preamplifier located and what is its primary function?

13. Why is the ^{13}C sensitivity of a triple-resonance probe on a 600 MHz NMR spectrometer much lower than that of a broadband probe on a 400 MHz instrument?
14. Why can a cryogenic triple-resonance probe be used to directly observe ^{13}C?
15. Why is it necessary to fill two different cryogens in an SC magnet?
16. How is the heat insulation achieved in an SC magnet?
17. What is the function of LO in a NMR console?
18. How can a spectrum analyzer be used for probe tuning?

References

Abergel D, Louis-Joseph A, Lallemend JY (1996) Amplification of radiation damping in a 600-MHz NMR spectrometer: application to the study of water-protein interactions. J Biomol NMR 8:15–22

Augustine MP (2002) Transient properties of radiation damping. Prog Nucl Magn Reson Spectrosc 40:111–150

Bloembergen N, Pound RV (1954) Radiation damping in magnetic resonance experiments. Phys Rev 95:8–12

Bockmann A, Guittet E (1995) Water selective pseudo-3D NOESY-TOCSY experiment using 'radiation damping': application to the study of the effects of SCN- on the hydration of BPTI. J Chim Phys Chim Biol 92:1923–1928

Broekaert P, Jeener J (1995) Suppression of radiation damping in NMR in liquids by active electronic feedback. J Magn Reson 113A:60–64

Coombs CF (ed) (1972) Basic electronis instrument handbook. McGraw-Hill, New York

Dooley DJ (ed) (1980) Data conversion integrated circuits. IEEE Press, New York

Krishnan VV, Thornton KH, Cosman M (1999) An improved experimental scheme to measure self-diffusion coefficients of biomolecules with an advantageous use of radiation damping. Chem Phys Lett 302:317–323

Lippens G, Dhallium C, Wieruszeski JM (1995) Use of a water flip-back pulse in the homonuclear NOESY experiment. J Biomol NMR 5:327–331

Mazda FF (1987) Electronic instruments and measurement techniues. Cambridge University Press, New York

Oliver BM, Cage JM (1971) Electronic measurements and instrumentation. McGraw-Hill, New York

Oxenoid K, Kim HJ, Jacob J, Sonnichsen FD, Sanders CR (2004) NMR assignments for a helical 40 kDa membrane protein. J Am Chem Soc 126:5048–5049

Parker SP, Weil J, Richman B (1984) McGraw-Hill encyclopedia of electronics and computers. McGraw-Hill, New York

Picard L, von Keinlin M, Decorps M (1996) An overcoupled NMR probe for the reduction of radiation damping. J Magn Reson 117A:262–266

Sheingold D (ed) (1977) Analog-digital conversion notes. Prenctice-Hall, NJ

Szoke A, Meiboom S (1959) Radiation damping in nuclear magnetic resonance. Phys Rev 113:585–586

# Chapter 3
NMR Sample Preparation

3.1 Introduction

On many occasions while carrying out interesting NMR projects, researchers have found that protein sample preparation is the bottleneck and most time-consuming stage for the success of the planned studies. A similar situation occurs in crystallographic studies; however, a difference is that NMR samples usually require isotope-labeling. Because the cost of ^{13}C, ^{15}N, or ^{2}H source compounds is significantly higher than natural abundant sources, the isotopic labeling of the proteins is usually done in minimal growth media using bacterial expression systems. This chapter describes common steps of protein sample preparation for NMR studies. Most of these steps are common for regular protein preparation by molecular biologists and biochemists. Some steps are especially NMR-oriented and are emphasized in more detail. Some tips for sample preparation are also provided. Note that there is often a great deal of flexibility in the application of protocols. Hence, it is often possible to alter or adapt a technique to specific needs.

Questions to be addressed in the present chapter include the following:

1. How do you choose bacterial expression systems for high expression of target proteins?
2. How are protein expression and solubility optimized?
3. What are minimal media?
4. How much ^{13}C and ^{15}N source compounds are needed to obtain sufficient isotope-labeled samples for heteronuclear NMR experiments?
5. What are the brief steps in protein purification for NMR samples?
6. What buffers, protein concentration, pH, and temperature are suitable for NMR samples?
7. What is a typical procedure for NMR sample preparation?
8. How are alignment media prepared for residual dipolar coupling measurement?
9. How is protein–peptide complex sample prepared?
10. How is protein–protein complex sample prepared?
11. Examples of NMR sample preparation (complete protocols).

3.2 Expression Systems

3.2.1 *Escherichia coli Expression Systems*

NMR samples normally require large quantities of isotope-labeled proteins or protein complexes at mM concentrations (can be less if high sensitivity cryogenic probe or higher field spectrometers such as 800 or 900 MHz spectrometers are used, see Chap. 2). However, certain proteins or domains prepared using bacterial expression systems may have the following problems: (1) low expression level, (2) insolubility at high concentration, and (3) low stability. These problems require large efforts of exploration and optimization of various conditions. The first step is typically to develop good expression systems using recombinant genetic techniques. The recombinant technology can yield substantially high levels of target proteins compared to the low level of proteins from the natural source. Moreover, recombinant expression systems can be manipulated to produce protein domains or to attach tags for easy purification. The former is particularly important for NMR studies since many NMR projects focus on structures and dynamics of protein domains or domain–domain complexes. The common steps for protein subcloning can be found in many molecular biology books (such as "Short Protocols in Molecular Biology" by Ausubel et al. 1997) and are detailed here. What we focus on here are the steps in choosing the expression vectors for subcloning.

Compared to insect or mammalian cell expression systems, the prokaryotic *E. coli* expression systems are most commonly used to produce isotope-labeled proteins for NMR studies. One can start with many commercial *E. coli* expression vectors available such as the pET system from Novagen, Inc., the Impact system from New England Biolabs, etc. (Table 3.1), which are routinely used in most NMR laboratories. Because individual proteins may behave differently in different expression systems, it is often difficult to predict the best expression vector for a particular protein. For initial investigation, one usually starts with vectors encoding protease-cleavable N-terminal or C-terminal fusions such as poly(His)

Table 3.1 Common expression vectors

Vector name	Fusion partner	Source	Features
pET vectors	His-tag or none	Novagen, Inc.	Short His-tag allows quick NMR screening of uncleaved proteins
pGEX vectors	GST	Novagen, Inc.	Easy for purification but GST is a dimer that may interfere with target protein folding
pMal vectors	MBP	New England Biolabs, Inc.	Highly soluble MBP promotes expression and solubility of the target proteins but MBP has a low affinity for the resin, which reduces the yield
pTYB11 vectors	Intein	New England Biolabs, Inc.	Proteins purified without fusion attached

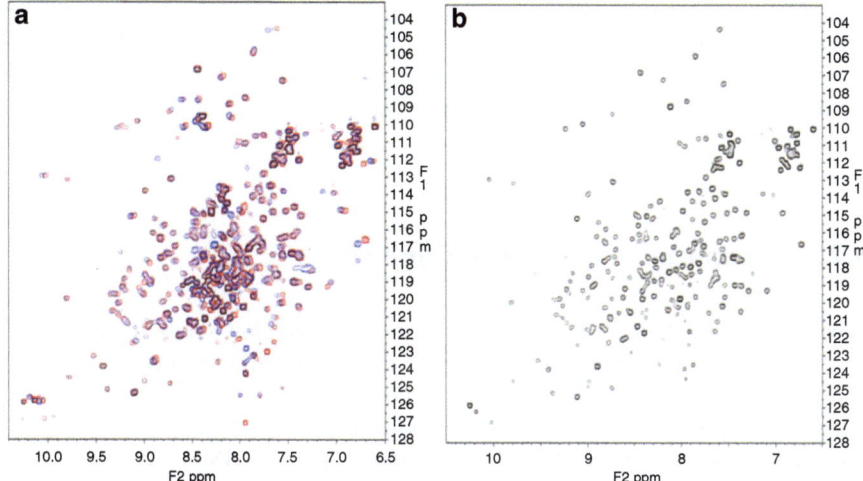

Fig. 3.1 HSQC of His-tag containing talin-HS and its interaction with β3 tail at 35 °C, pH 6.5 collected at 500 MHz. (**a**) Talin-HS (*blue*) and talin-HS in complex with β3 peptide (*red*). (**b**) Talin-HS in complex with the β3 peptide after cleavage of the His-tag from talin-HS. Comparison of spectrum (**a**) with spectrum (**b**) shows most of His-tag signals are in the center of the spectrum (**a**)

tag (or His-tag), GST (glutathione-*S*-transferase) tag, etc., which allows easy purification using an affinity column (see Sect. 3.4). A His-tag is a flexible linker containing only 6–20 residues. Hence, it is usually uncleaved after affinity purification since its short size imposes minimal complication of the NMR spectra of fogged proteins. For example, Fig. 3.1 shows the 2D ^{1}H-^{15}N HSQC of a fragment from the cytoskeletal protein talin with and without a His-tag. One can immediately conclude that the fragment is folded and the signals are well dispersed, which means that the protein is suitable for further investigation. It is clear that although the spectrum has an improved resolution after cleavage, the signals of those unstructured His-tag residues do not detract from the feasibility of the project.

3.2.2 Fusion Proteins in the Expression Vectors

In many cases, different fusion proteins lead to substantially different expression levels and solubility of the target proteins. Thioredoxin, protein B G1 domain, and maltose-binding protein (MBP) are known to promote the high expression as well as high solubility of the target proteins. Protein B G1 domain is small (~56 residues) and does not cause severe complication of spectral analysis and hence can be used as a sample solubility enhancer of the target proteins or protein complexes. However, caution must be paid since fusion proteins may prevent the folding or induce misfolding of the target proteins, particularly when the N-terminal regions of the proteins are important interior components of protein structures. In these cases, proteins are usually found in insoluble inclusion bodies or look unfolded as judged from the NMR spectra. To avoid this problem, vectors without fusions such as pET3a (Table 3.1) can be exploited.

3.2.3 Optimization of Protein Expression

When proteins or domains have low expression level and low solubility, a number of expression vectors need to be explored for systematic and sometimes time-consuming optimization. The first thing to do to increase protein expression is to choose appropriate cell lines. The most common cell lines (strains) for bacterial protein expression are BL21(DE3), BL21(DE3) pLysS, HMS, etc, which are commercially available. The cDNA containing the target protein is usually transformed into the above strains on Day 1. The next day, one colony is picked from each transformation and grown in a 5 mL culture for each strain. Cell density ($OD_{600\ nm}$) is checked after a few hours and 1 mM IPTG (isopropyl-1-thio-β-D-galactoside) is added (typical for the first time but can be varied, see below) when the OD is approximately 0.6 to induce protein expression (1 mL of culture is sampled before addition of IPTG). The cells are harvested after 3–4 h and spun down at $10,000 \times g$. The pellets are lysed by sonication using a standard PBS (phosphate buffered saline) buffer (100 μL). To check the protein expression, 2 μL of the lysate from each pellet (before and after IPTG) are run on an SDS-PAGE gel. This will give some idea of which strain gives the highest expression. The next step is to check protein solubility by spinning down the lysate containing IPTG and taking the supernatant to check if there is soluble protein. Hence, contained in the SDS-PAGE will normally be three lanes for each strain: lane 1, uninduced lysate; lane 2, lysate induced with IPTG; lane 3, supernatant for the induced lysate. If proteins are expressed in inclusion bodies, refolding protocols may be used (see Sect. 3.4) but many proteins cannot undergo reversible unfolding/refolding process. To maximize protein expression and to increase the soluble fraction in bacterial lysates, one usually needs to vary a series of expression conditions:

1. Prepare a cell culture growth curve to decide the best induction point. Each cell line may have different growth curve. The growth curve can be made by measuring cell density at 600 nm as a function of time in hours.
2. Induce protein expression at the middle of the log-phase derived from step 1 with different IPTG concentrations typically from 0.1 to 2 mM.
3. Vary the induction time between 1 h and overnight at different temperatures. Lower temperatures such as 16 °C are often useful to produce more soluble proteins than higher temperatures such as 37 °C.

Sometimes, protein solubility is still low after the above procedures due to inappropriate conditions for protein folding in bacteria. In such a case, insect and

3.3 Overexpression of Isotope-Labeled Proteins

After optimization of protein expression, the next step is to overexpress in a large scale isotope-labeled proteins for NMR studies. Heteronuclear multidimensional NMR experiments for structure determination of medium-sized protein requires that the target proteins or domains be uniformly ^{15}N and/or ^{13}C-labeled. This is done by growing cell cultures in minimal media in which $^{15}NH_4Cl$ and/or ^{13}C glucose are the only sources for nitrogen and carbon atoms. The standard recipe for the minimal media is shown in Table 3.2 and the recipe for trace element solution in Table 3.3. The most common cell lines for protein expression in minimal media are

Table 3.2 Recipe for minimal media[a]

Compound	Concentration	Comments
K_2HPO_4	10.0 g L^{-1}	
KH_2PO_4	13.0 g L^{-1}	
Na_2HPO_4	9.0 g L^{-1}	
K_2SO_4	2.4 g L^{-1}	
$^{15}NH_4Cl$	1–2 g L^{-1}	~\$30 g^{-1}
^{13}C Glucose	2–5 g L^{-1}	Needs to be optimized to reduce the cost (~\$80 g^{-1})
Trace element solution[b]	10 mL L^{-1}	See Table 3.3
1 M $MgCl_2 \cdot 6H_2O$	10 mL L^{-1}	
Thiamine (Vitamin B_1)	5 mg mL^{-1}	
Antibiotics	~0.1 mg L^{-1}	

[a] The media have to be sterilized
[b] Trace element solution is a combination of trace elements shown in Table 3.3

Table 3.3 Recipe for trace element solution[a]

Compound	Gram per 100 mL[b]
$CaCl_2 \cdot 2H_2O$	0.600
$FeSO_4 \cdot 7H_2O$	0.600
$MnCl_2 \cdot 4H_2O$	0.115
$CoCl_2 \cdot 6H_2O$	0.080
$ZnSO_4 \cdot 7H_2O$	0.070
$CuCl_2 \cdot 2H_2O$	0.030
H_3BO_3	0.002
$(NH_4)_6Mo_7O_{24} \cdot 4H_2O$	0.025
EDTA	0.500

[a] Add ingredients one at a time, waiting 5–10 min before they fully dissolve. After adding EDTA and stirring for a few hours, the color of the solution should be golden brown (if it is greenish, then leave stirring overnight). Sterilize afterwards by filtering through a 0.2 μm filter
[b] A fresh stock of 100–200 mL is usually made

BL21(DE3), BL21(DE3), pLysS, HMS, etc, which are commercially available. It is important to note that protein expression levels are usually lower in minimal media as compared to rich LB media. Hence, it is recommended that different cell lines be used to explore the optimum expression. Due to the expensive cost of ^{13}C glucose, it is recommended to optimize the glucose usage at small scale by measuring the growth curve as a function of different concentrations of unlabeled glucose. Protein expression levels should be checked and compared on SDS-PAGE by taking 1 mL of culture before and after the IPTG induction. Although a typical induction time for protein expression is 2–4 h, induction time is 2–3 times longer when the proteins are expressed in partial/full ^{2}H$_2$O due to the slow growth rate. Growing cultures in ^{2}H$_2$O is required to prepare deuterium labeled proteins for NMR studies when the proteins are relatively large (>20 kDa).

3.4 Purification of Isotope-Labeled Proteins

Purification of isotope-labeled proteins is a key step and probably the most time-consuming step for NMR sample preparation. The procedures and tips for purifying the isotope-labeled proteins are the same as nonlabeled proteins described in many textbooks and the literature. If the labeled proteins contain fusions, fusion-targeted affinity columns will be the first step for purification followed by protease cleavage and gel filtration. This procedure typically works well if the protein behaves during the process; however, protease cleavage sometimes can be a tricky process. Excess amounts of protease or over digestion by protease can lead to nonspecific cleavage, and hence, optimization is usually required. When proteins are not fused, the chemical structure and physical properties of the proteins are the two key parameters used to develop the most efficient purification protocols. Isoelectric point (pI), pH stability, and charge density are important properties to be exploited during purification. Several steps of different ion-exchange and hydrophobic chromatography are often used for large scale purification of nonfused proteins followed by a final step of gel-filtration. Note that 90 % pure proteins are usually sufficient for heteronuclear NMR studies if the proteins are stable in the presence of the impurities. Some important tips for purification are summarized below:

1. A French press is often better than sonication in lysing the cells by producing more soluble fractions of proteins.
2. A Cocktail of protease inhibitors (Table 3.4) is recommended in the cell lysis buffer. A typical lysis buffer for soluble proteins consists of 20–50 mM phosphate, pH 7.4, or 0.1 M Tris-HCl, pH 7.4, 0.1 M NaCl, 1–5 mM EDTA (assuming that the protein of interest has no metal), 5–20 mM β-mercaptoethanol (assuming that the protein of interest has no disulfide bonds), sucrose and the cocktail of protease inhibitors. Note that the conditions of lysis buffer vary significantly depending on the properties of the proteins. For example, low salt or no salt is used for Tris buffer if the protein or domain of interest behaves poorly in the presence of salt. The yield can differ by 2–10 fold between two different buffer conditions.

Table 3.4 Cocktail of protease inhibitors

Inhibitor	Target protease	Final concentration ($\mu g\ mL^{-1}$)
Leupeptin	Broad spectrum	0.5–10
EDTA-Na$_2$	Metalloproteases	5–10
Pepstatin A	Acidic proteases	0.7–10
Aprotinin	Serine proteases	50
PMSF	Serine proteases	0.2–2
Benzamidine HCl	Serine proteases	100
Soybean	Trypsin-like trypsin inhibitor	100

PMSF phenylmethanesulfonylfluoride or phenylmethylsulfonyl fluoride

3. Try different purification protocols to optimize the yield. To develop the best purification protocol, one should always try different protocols for initial screening. For example, if most of proteins go into inclusion bodies, it may be beneficial to try a refolding protocol. As mentioned above, some proteins are not reversible in folding/unfolding, but some proteins may be refoldable.

3.5 NMR Sample Preparation

3.5.1 General Considerations

The last step for NMR sample purification is to choose a good buffer in which the protein is concentrated to ~1 mM. Phosphate buffer at pH 5–7 (20–50 mM) with or without salt (e.g., KCl, NaCl) is often used for many NMR samples. It is recommended to try a series of conditions in small scales and then decide which condition is the best. Quite frequently, a protein/domain itself is not very stable in the buffer but becomes very stable after mixing with the target protein/peptide. High quality NMR tubes with appreciate specifications should be used for protein samples, which are usually 5 mm tube diameter containing 0.5 mL 95 % H$_2$O/5 % ^{2}H$_2$O for aqueous samples. If the volume of the sample is limited, microtubes whose susceptibility matches that of ^{2}H$_2$O are chosen for a total sample volume of ~200 μL, such as Shigemi microtubes (Shigemi Inc., Allison Park, PA). Because of the small sample volume, the buffer contains 7 % ^{2}H$_2$O used for ^{2}H lock. In addition, the samples are usually required to be degassed by blowing high purity argon or nitrogen gas into the samples to remove oxygen, whose paramagnetic property will broaden the line shapes of protein resonances.

3.5.2 Preparation of Protein–Peptide Complexes

The contact surface contributing to the interactions of high affinity and specificity often involves 30 or less amino acid residues from each protein of the complex (de Vos et al. 1992; Song and Ni 1998). Often this contact surface is located in a single continuous fragment of one of the proteins, which can be identified by mutation and deletion experiments. Therefore, fragments can be chemically synthesized in large amount and studied by 2D ^{1}H NMR experiments due to their small molecular size (Wüthrich 1986). Samples for protein–peptide complexes are commonly prepared from isotopically labeled protein and unlabeled peptide according to the following procedure, since the availability of labeled peptide is often prohibited by the expense of chemical synthesis from labeled amino acids and the difficulty of biosynthesis due to peptide instability during its expression and purification (Huth et al. 1997; Newlon et al. 1997).

Preparation of the complexes is done by titrating synthetic peptide (unlabeled) into the isotope-labeled target protein (Breeze 2000; Qin et al. 2001). The stoichiometry of association can be best determined by monitoring a ^{1}H-^{15}N HSQC spectrum of the target protein in different protein–peptide ratios. Because of the high sensitivity of the ^{1}H-^{15}N HSQC experiment, only a low concentration of ^{15}N labeled protein (0.1–0.2 mM) is needed for the titration experiments. Once the stoichiometry is determined, the labeled protein is preferably mixed with unlabeled peptide at dilute concentration and then concentrated to the 0.5–1 mM, required for most NMR experiments. A higher field magnet or high sensitivity probe such as a cryogenic probe makes it possible to study dilute samples if aggregation occurs at high concentrations. The pH of the samples is preferably kept below 7.0 to reduce the amide exchange rate. Further purification of the tightly associated complex by gel-filtration may be necessary to improve the sample quality.

3.5.3 Preparation of Protein–Protein Complexes

Protein–protein interactions play an essential role at various levels in information flow associated with various biological processes, such as gene transcription and translation, cell growth and differentiation, neurotransmission, and immune response. The interactions frequently lead to changes in shape or dynamics as well as chemical or physical properties of proteins involved. Solution NMR spectroscopy provides a powerful tool to characterize these interactions at the atomic level and at near physiological conditions. With the use of isotopic labeling, structures of many protein complexes in the 40 kDa total molecular mass regime have been determined (Clore and Gronenborn 1998). The development of novel NMR techniques and sample preparation has been further increasing the mass size available for the structural determination of protein complexes. Furthermore, NMR has been utilized to quickly identify the binding sites of the complexes based on the results of chemical shift mapping or hydrogen-bonding experiments. Because it is particularly difficult and sometimes impossible to crystallize weakly bound protein complexes ($K_d > 10^{-6}$), the chemical mapping method is uniquely suitable to characterize such complexes. The binding surfaces of small to medium sized isotopically labeled proteins with molecular mass less than 30 kDa to large target proteins (unlabeled, up to 100 kDa) can be identified by solution NMR (Mastsuo et al. 1999; Takahashi et al. 2000). As discussed in Chap. 6, the structures of small ligands weakly bound to the proteins can be determined by transferred NOE experiments (Clore and Gronenborn 1982, 1983). The structures of peptides or small protein domains of weakly bound protein complexes can also be characterized by the NMR technique, which may beneficial to the discovery and design of new drugs with high affinity. In addition to the structural investigation of protein complexes, NMR is a unique and powerful technique to study the molecular dynamics involved in protein–protein reorganization (Kay et al. 1998; Feher and Cavanagh 1999). Furthermore, protein binding site often contain residues from different parts of the protein or domain. Structure determination of protein–protein (domain) complexes is necessary for understanding the specificity.

Protein–protein complexes can be prepared as follows (Qin et al. 2001; Breeze 2000). For a complex A–B, A and B components are separately expressed and purified. The isotope-labeled A is mixed with unlabeled B or vice versa to simplify the NMR spectra. Because of the large sizes of complexes, partial or full deuteration may be necessary to reduce the line widths of signals. Isolated domains are sometimes unfolded or partially folded and may also undergo aggregation, which makes the purification difficult. Purification can be performed in the denatured condition and the unfolded domains can be refolded in the presence of target proteins. Fusion proteins are often needed to help for solubilization of domains for purification. Fusion proteins are cleaved after the domain is mixed with the target protein for stabilization. The complex is further purified by gel-filtration that removes impurities including the fusion protein and protease.

3.5.4 Preparation of Alignment Media for Residual Dipolar Coupling Measurement

Various media are available for moderately aligning macromolecules in solution in the magnetic field, of which two liquid crystalline media are most commonly used at the present time: dimyristoylphosphatidylcholine–dihexanoylphosphatidylcholine (DMPC–DHPC) bicelles and filamentous phage. It is the interaction of the magnetic field with the anisotropic susceptibility of liquid crystalline media that aligns these particles in the magnetic field. When particles with a nonspherical shape, such as disks or rods dissolved in solution, are placed in the magnetic field, the anisotropic distribution of the electron density leads to an orientational dependence of this interaction. If the anisotropic interaction is large enough to overcome the thermal energy of the particles, the degree of orientation order of the media in the magnetic field becomes significant enough to be measurable, which is usually in the range of 0.5–0.85. As discussed previously (Sect. 1.8.4), the order for macromolecules described by the magnitude of the alignment tensor is in the range of 10^{-3}, meaning that the interaction between aligned particles and macromolecules must be very weak.

The weakness of the interaction is necessary so that it does not perturb the native structure of macromolecules under study or broaden the resonance line shapes due to a high degree of order or a change of relaxation properties.

Disk-shaped bicelles (bilayered micelles) were the first liquid crystalline medium used to achieve weak alignment of macromolecules (Tjandra and Bax 1997), and were originally developed by Prestegard, Sanders and cowokers (Sanders and Prestegard 1990, 1991; Sanders et al. 1994). The medium contains a mixture of saturated lipids DMPC and DHPC in low concentration to form planer bicelles in which DMPC constitutes the planar bilayer region and DHPC stabilizes the rim of the bicelles. The bicelles of the medium align in the presence of magnetic field at 35 °C whereas they are isotropic at 25 °C.

The order of alignment media is transferred to the macromolecules by rapid random collisions and electrostatic interaction between medium particles and the molecules dissolved in the medium. The formation of liquid crystals by the DMPC–DHPC mixture depends on a number of factors including temperature, concentration, molar ratio of the mixture, and ionic strength. The presence of other charged amphiphilic compounds can also influence the phase transition from isotropic to the liquid crystalline phase. To maintain the weak interaction, the liquid crystalline medium is limited to a very dilute concentration, typically, about 10 % w/v. In such dilute lipid concentration, the molar ratio of DMPC/DHPC plays an important role in the formation of bicelles. If the ratio is too low, the lipids form small size disks that are too small to generate measurable alignment. On the other hand, a high ratio causes the oversized bicelles to collapse to form spherical micelles by DMPC. The upper limit of the ratio at which stable planar bicelles can be formed is 5. Usually, the ratio is maintained in the range of 3.0–3.5, which corresponds to bicelle diameters of 200–250 Å with a thickness of approximately 40 Å.

At the low concentration the DMPC/DHPC bicelles are unstable in the lower range (25–30 °C) of liquid crystalline phase temperature (25–45 °C). The stability of the bicelles can be improved by adding a small amount of charged amphiphile (Losonczi and Prestegard 1998) such as cetyl (hexadecyl) trimethyl ammonium bromide (CTAB, positively charged) or sodium dodecyl sulfate (SDS, negatively charged). Best results are obtained with a molar ratio of 0.003–0.01 relative to DMPC (e.g., DMPC–DHPC–CTAB = 3.5:1.0:0.005). The addition of these detergents widens the temperature range of liquid crystalline media, resulting in stabilizing the bicelles in the above original temperature range. Charging the bicelles with the detergents also produce an electrostatic potential which attracts and repels groups with opposite and like charges, respectively, resulting in a change in orientation and magnitude of the alignment tensor. This change is often enough to yield an independent alignment tensor and reduce the ambiguity of orientation mentioned early (also see Sect. 1.8.4).

The procedure for the preparation of bicelle samples includes a number of straightforward steps (Ramirez et al. 2000). It starts with weighing the appropriate amount of dry powder DMPC and DHPC according to the desired molar ratio, typically 3.5:1. The weighing should be done in a dry box to obtain an accurate amount of the material due to the hygroscopic property of DHPC. The buffer is prepared with the required amount of salt and sodium azide (highly toxic) at a concentration greater than or equal to 1 mM as an antibacterial agent, which is then added into DHPC. Dry powder DMPC is then added into the sample. In order to completely dissolve DMPC, cooling and heating cycles are repeated until the solution becomes clear by freezing the solution in liquid nitrogen and thawing it while vortexing at 35 °C. An alternative method is to leave the mixture solution at 20 °C for a few days to let it turn into a clear solution. It should be noted that a protein sample cannot be recovered from the liquid crystalline medium once it is dissolved in it. The alignment media may be frozen or lyophilized (Bax et al. 2001).

Filamentous phage is also utilized to achieve weak alignment of macromolecules in the magnetic field (Clore et al. 1998; Hansen et al. 1998a, b) and is commercially available (Asla Labs). Because of their rod shape (1–2 μm long, 6.5 nm diameter), filamentous phage with a certain concentration in solution (as low as a few milligram per milliliter) can form liquid crystals. When the sample is placed in the magnetic field, the particles are aligned with their long axis parallel to the field direction due to their very anisotropic nature. In order to obtain the liquid crystalline phase, the solution should be prepared at a pH higher than neutral to maintain a negatively charged environment so as to prevent glutamate and aspartyl side chains of the phage coat protein from protonating. In addition, a high salt concentration should be avoided at low phage concentration, because it can prevent alignment. Because of the relatively strong electrostatic interactions between phage and macromolecules relative to those between micelles and macromolecules, phage liquid crystallines generate different alignment tensors than in micelle media, resulting in additional alignment tensors for application to the reduction of orientational degeneracy. Phage samples are easily prepared by dissolving 3–10 mg mL^{-1} *pf1* (or *pd*) phage in the buffer solution (pH=8.0) at 4 °C (keep phage and the buffer on ice throughout the preparation). The liquid crystalline phase is formed over a wide range of temperatures (5–50 °C). However, the phage medium cannot be frozen or lyophilized. Protein in the sample is usually recovered by precipitating the phage through centrifugation.

3.6 Examples of Protocols for Preparing ^{15}N/^{13}C Labeled Proteins

3.6.1 Example 1: Sample Preparation of an LIM Domain Using Protease Cleavage

3.6.1.1 Background

PINCH LIM1 domain is a double zinc finger involved in cell adhesion. The protein was subcloned into several expression vectors including pET3a, pGEX-4T, and pMAL-C2X. Only pMAL-C2X gives good expression and hence was used for sample preparation.

3.6.1.2 Protein Expression

Expression plasmid pMAL-C2X, encoding a MBP fused to the N-terminus of residues 1–70 of human PINCH protein via a Factor X_a (FX_a) cleavable linker, was used for preparation of the NMR sample. Residues 1–70 contain the entire LIM1 domain. Due to cloning artifacts, the C-terminus of LIM1 had three additional residues (WIL), whereas N-terminus contained four (ISEF). BL21 (DE3) cells harboring plasmid were grown in LB medium or in M9 minimal medium (Table 3.2) in the presence of 100 μg mL^{-1} ampicillin. For isotope labeling, M9 contained 1.1 g L^{-1} ^{15}N-NH$_4$Cl and 3 g L^{-1} unlabeled or ^{13}C-labeled glucose. Three liters of culture were induced at OD$_{600\,nm}$ approximately equal to 0.5 for 4 h at 37 °C with 1 mM IPTG.

3.6.1.3 Protein Purification and Sample Preparation

Cells were lysed with a French Press and the cleared lysates were fractionated on a DEAE-sepharose column (50 mM Tris-HCl, pH = 8.0, gradient of NaCl 0.0–0.8 M). MBP-LIM1-containing fractions were concentrated, the buffer was exchanged into the optimal one for cleavage (50 mM Tris-HCl, 100 mM NaCl, 3.5 mM CaCl$_2$, pH = 8.0) and subjected to FX_a treatment. Cleaved LIM1 was further purified on a Superdex® 75 gel-filtration column. Fractions containing LIM1 were pooled and concentrated to ~0.5 mM with a buffer at pH = 7.5 containing 50 mM Na$_2$HPO$_4$, 100 mM NaCl, 0.5 mM β-mercaptoethanol.

3.6.2 Example 2: Sample Preparation Using a Denaturation–Renaturation Method

3.6.2.1 Background

The double-stranded RNA (dsRNA) activated protein kinase (PKR) contains a dsRNA binding domain (dsRBD), which was subcloned into pET15b vector with a thrombin-cleavable His-tag linker. However, the protein is expressed in inclusion bodies, and hence, the refolding method was used for protein purification.

3.6.2.2 Protein Expression

Four liters of E. coli BL21(DE3) pLysS cells transformed with pET15b encoding the dsRBD of human PKR were grown at 37 °C in minimal media containing 0.4 % glucose/0.1 % ^{15}NH$_4$Cl or 0.4 % [U-^{13}C] glucose/0.1 % ^{15}NH$_4$Cl in order to obtain ^{15}N- and ^{15}N/^{13}C-labeled proteins, respectively. Cells were grown in log phase to

$OD_{600\ nm} = 0.6$–0.8 in the presence of 50 μg mL^{-1} carbenicillin, and protein expression was induced for 4 h with 1 mM IPTG. The cells were harvested by centrifugation for 20 min at 6,000 × g at 4 °C, and the pellets were drained and stored at −80 °C.

3.6.2.3 Protein Purification and Sample Preparation

The pellets were resuspended in ice-cold lysis buffer (6 M guanidine HCl and 50 mM Tris-HCl, pH 8.0), sonicated 4 × 30 s at full power, and centrifuged at 20,000 × g for 20 min at 4 °C. The supernatant was passed over an Ni^{2+}-agarose metal affinity column, and the histidine-tagged protein was eluted according to the manufacturer's instructions. The denatured dsRBD protein was further purified by gel filtration chromatography on a Superdex®-75 column and refolded by dialysis against 50 mM phosphate buffer at 4 °C, pH 6.5, and 1 mM DTT. The refolded dsRBD binds to dsRNA with the same affinity as wild type PKR and was concentrated by a Amicon® Centriplus®-10 ultrafiltration device (Millipore, Billerica, MA). The dsRBD is mostly monomeric (>90 %) as judged by gel-filtration column, and its purity and concentration were determined by SDS-PAGE with Coomassie staining and UV spectrometry. The final ^{15}N-labeled or ^{15}N/^{13}C-labeled protein was purged with argon and adjusted to 7 % ^{2}H$_2$O, and transferred to 250 μL microcell NMR tubes (Shigemi Inc., Allison Park, PA) for NMR experiments.

Questions

1. Why do proteins require isotope-labeling for structure and dynamics studies?
2. Why are minimum media used to isotope-label the proteins and what are the differences between minimum media and rich media?
3. What is His-tag and can it affect the quality of the NMR spectrum?
4. When is a micro-NMR tube needed for NMR sample preparation?
5. What are the temperature ranges of the alignment media prepared using bicelles and filamentous phage?
6. For what purpose is the charged amphiphile such as CTAB added to the bicelles?

References

Ausubel F, Brent R, Kingston RE, Moore DD, Seidman JG, Smith JA et al (eds) (1997) Short protocols in molecular biology, 3rd edn. Wiley, New York

Bax A, Kontaxis G, Tjandra N (2001) Dipolar couplings in macromolecular structure determination. Meth Enzymol 339:127–174

Breeze AL (2000) Isotope-filtered NMR methods for the study of biomolecular structure and interactions. Prog Nucl Magn Reson Spectrosc 36:323–372

Clore GM, Gronenborn AM (1982) Theory and applications of the transferred nuclear overhauser effect to the study of the conformations of small ligands bound to proteins. J Magn Reson 48:402–417

Clore GM, Gronenborn AM (1983) Theory of the time dependent transferred nuclear Overhauser effect: applications to structural analysis of ligand-protein complexes in solution. J Magn Reson 53:423–442

Clore GM, Gronenborn AM (1998) Determining the structures of large proteins and protein complexes by NMR. Trends Biotech 16:22–34

Clore GM, Starich MR, Gronenborn AM (1998) Measurement of residual dipolar couplings of macromolecules aligned in the nematic phase of a colloidal suspension of rod-shaped viruses. J Am Chem Sco 120:10571–10572

de Vos AM, Ultsch M, Kossiakoff AA (1992) Human growth hormone and extracellular domain of its receptor: crystal structure of the complex. Science 255:306–312

Feher VA, Cavanagh J (1999) Millisecond-timescale motions contribute to the function of the bacterial response regulator protein SpoOF. Nature 400:289–293

Hansen MR, Mueller L, Paridi A (1998a) Tunable alignment of macromolecules by filamentous phage yields dipolar coupling interactions. Nat Struct Biol 5:1065–1074

Hansen MR, Rance M, Paridi A (1998b) Observation of long-range 1H−1H distances in solution by dipolar coupling interactions. J Am Chem Sco 120:11210–11211

Huth JR, Bewley CA, Jackson BM, Hinnebusch AG, Clore GM, Gronenborn AM (1997) Design of an expression system for detecting folded protein domains and mapping macromolecular interactions by NMR. Prot Sci 6:2359–2364

Kay LE, Muhandiram DR, Wolf G, Shoelson SE, Forman-Kay JD (1998) Correlation between binding and dynamics at SH2 domain interfaces. Nat Struct Biol 5:156–163

Losonczi JA, Prestegard JH (1998) Improved dilute bicelle solutions for high-resolution NMR of biological macromolecules. J Biomol NMR 12:447–451

Mastsuo H, Walters KJ, Teruya K, Tanaka T, Gassner GT, Lippard SJ, Kyogoku Y, Wagner G (1999) Identification by NMR spectroscopy of residues at contact surfaces in large, Slowly Exchanging Macromolecular Complexes. J Am Chem Sco 121:9903–9904

Newlon MG, Roy M, Hausken ZE, Scott JD, Jennings PA (1997) The A-kinase anchoring domain of Type IIα cAMP-dependent protein kinase Is highly helical. J Biol Chem 272:23637–23644

Qin J, Vinogradova O, Gronenborn AM (2001) Protein-protein interactions probed by nuclear magnetic resonance spectroscopy. Meth Enzmol 339:377–389

Ramirez BE, Voloshin ON, Camerini-Otero RD, Bax A (2000) Solution structure of DinI provides insight into its mode of RecA inactivation. Prot Sci 9:2161–2169

Sanders CR, Prestegard JH (1990) Magnetically orientable phospholipid bilayers containing small amounts of a bile salt analogue, CHAPSO. Biophys J 58:447–460

Sanders CR, Prestegard JH (1991) Orientation and dynamics of.beta.-dodecyl glucopyranoside in phospholipid bilayers by oriented sample NMR and order matrix analysis. J Am Chem Sco 113:1987–1996

Sanders CR, Hare B, Howard KP, Prestegard JH (1993) Magnetically-oriented phospholipid micelles as a tool for the study of membrane-associated molecules. Prog Nucl Magn Reson Spectrosc 26:421–444

Song J, Ni F (1998) NMR for the design of functional mimetics of protein-protein interactions: one key is in the building of bridges. Biochem Cell Biol 76:177–188

Takahashi H, Nahanishi T, Kami K, Arata Y, Shimada I (2000) A novel NMR method for determining the interfaces of large protein−protein complexes. Nat Struct Biol 7:220–223

Tjandra T, Bax A (1997) Direct measurement of distances and angles in biomolecules by NMR in a dilute liquid crystalline medium. Science 278:1111–1114

Wüthrich K (1986) NMR of proteins and nucleic acids. Wiley, New York

Chapter 4
Practical Aspects

The quality of NMR spectra is critically dependent on numerous factors, among which are the condition of the instrument at the time the data are collected, the homogeneity of the magnetic field, the type of experiments chosen, how the data are acquired and subsequently processed, etc. In this chapter, practical aspects related to the experiments are discussed in detail.

Key questions to be addressed in the current chapter include the following:

1. How is a probe tuned correctly?
2. What is a simple and correct way to do shimming?
3. How are transmitter (or observe) hard pulses calibrated?
4. How are decoupler (nonobserve) hard pulses calibrated?
5. How are the pulse lengths of decoupling calibrated?
6. What is the procedure to calibrate selective pulses?
7. How are the offset frequencies of the transmitter and decoupler calibrated and how often are they calibrated?
8. How can off-resonance pulses ($^{13}C'$, $^{13}C^{\alpha}$, $^{13}C^{\alpha,\beta}$) be calibrated?
9. How is the temperature calibration of a probe done?
10. What is the decoupling efficiency of different decoupling methods?
11. How is pulsed field gradient used to remove unwanted magnetization?
12. What are coherence and coherence transfer?
13. What is the coherence transfer pathway?
14. How are they selected by phase cycling or gradient methods?
15. What are the common water suppression techniques?
16. What are the advantage and drawbacks of the techniques?
17. How are the water suppression experiments set up?
18. How is a 2D (or 3D, 4D) experiment formed and what is the intrinsic difference between 1D and multidimensional experiments?
19. What are the typical procedures used to set up an experiment?
20. What is the typical procedure for processing data?

4.1 Tuning the Probe

The general procedure for setting up routine NMR experiments is described in the flow chart of Fig. 4.1. Tuning probe is the initial step before NMR data can be acquired. As discussed in Chap. 1, in order to move the equilibrium magnetization away from the magnetic field direction, an oscillating magnetic field B_1 generated by an RF pulse during NMR experiments, is needed to interact with the nuclei at Larmor frequency. In order to efficiently use the RF energy, the probe must be tuned to the carrier frequency (center frequency of the spectrum) for the sample to be measured and the impedance of the probe must be matched to the instrument impedance of $50\,\Omega$ at resonance (Chap. 2). Incorrect tuning or matching of the probe will cause an increased $90°$ pulse length and decreased sensitivity of detection.

Probe tuning and matching are carried out by adjusting the capacitors of the coil circuits in the probe. Probe tuning refers to the adjustment of both the probe resonance frequency to the carrier frequency, and the probe impedance to match $50\,\Omega$. A common triple-resonance solution NMR probe consists of two coils, of which the inner is double-tuned to ^1H and ^2H and the outer to ^{13}C and ^{15}N for a HCN probe or to ^{13}C and other heteronuclei over a wide range for an HCX probe, in which X represents a heteronucleus in the broadband range. Usually, each channel has a pair of adjustable capacitors, one for tuning and the other for matching. Some of the probes have auto-matched ^{15}N and ^{13}C channels, in which only tuning capacitors need to be adjusted. There are no adjustable matching capacitors for heteronuclear channels in this kind of probe. A typical procedure for tuning the probe involves turning the rods (or sliding bars) of the two adjustable capacitors alternatively and detecting the change of probe resonance frequency and impedance after setting up the carrier frequency for each channel being tuned.

Tuning should always be performed first for the nucleus with the lowest Larmor frequency, then for those with higher resonance frequency because the higher frequency is more sensitive to small changes in capacitance of the probe circuits, which can be caused by tuning neighboring channels. For instance, the ^{15}N channel

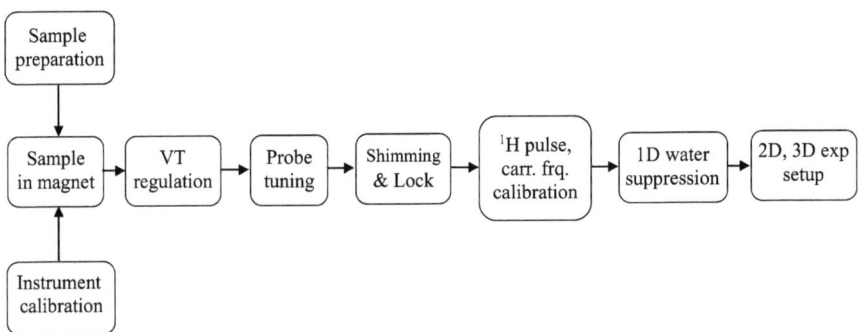

Fig. 4.1 Flowchart of a typical procedure for NMR experiment setup. The instrument must be properly calibrated before the experiment can be performed (see text for details). Each of the steps is discussed in the current chapter except sample preparation (previous chapter) and setup of 3D experiments (following chapter)

4.1 Tuning the Probe

is tuned first, then ^{13}C, and the ^{1}H channel is tuned last. In high field spectrometers the dielectric field of sample solvents is sufficiently high to influence the capacitance of probe circuits, resulting in a shift of probe resonance frequency when changing samples. Therefore, the probe must be tuned every time a sample is placed in the magnet. Since the circuits are temperature dependent, the probe must also be tuned after it reaches the selected temperature when the temperature is regulated.

There are two common methods used in probe tuning and matching: detecting the reflection power or detecting the resonance frequency of the probe during tuning and matching. Both methods are standard capabilities of NMR spectrometers. The reflection power method observes the RF power reflected by the probe and is more frequently used in routine tuning. When the spectrometer is in the tuning mode, RF pulses at the carrier frequency of the nucleus being tuned are delivered to the probe channel. This tuning method aims to minimize the amount of power being reflected by the probe. At the resonance condition with the input RF pulses, the probe circuit holds most of the power it receives. Thus, when a probe is well tuned, it reflects a minimum portion of the power being delivered to it and uses the RF power more efficiently. Consequently, a 90° pulse length is shorter for a given amount of RF power, compared to a mistuned probe.

The reflection power tuning method uses a directional coupler and a voltage meter to display the portion of the RF power reflected by the probe after a RF pulse with a specific frequency is sent to the probe. The RF pulses are set to the carrier frequency of the nuclei being tuned. Usually, the repetition rate is short, which is about five 20 µs pulses per second. The receiver does not need to be on during probe tuning (use a large number of steady-state scans or dummy scans). Most spectrometers setup the tuning parameters and start pulsing automatically, when in probe tuning mode. Practically, the capacitance of the probe circuits is adjusted by first turning the tuning capacitor (or sliding a bar) in a direction to reduce the amount of reflected RF power, until the reflection reaches a minimum. Keep turning the capacitor in the same direction, which makes the reflected power pass the minimum and increase by just a small amount. The next step is to adjust the matching capacitor in a direction so that the reflected power is back to a new minimum. If the new minimum is smaller than the previous one, keep turning the matching capacitor to pass the minimum by a small amount. If the new minimum is higher than the previous one, start with the adjustment of the tuning capacitor again but in the opposite direction so that a lower minimum can be found. Then, the above procedure is repeated by alternatively adjusting tuning and matching capacitors until the reflection power is best minimized. Usually, the ^{1}H channel of the probe can be tuned so well that the reading of power reflected from probe is zero. Normally, the reflection power reading for the heteronuclear channel is a little higher but the 90° pulse lengths are still in the range of the specifications (<40 µs for ^{15}N, <15 µs for ^{13}C).

Tuning using a spectral analyzer and an RF bridge to display the resonance frequency of the probe is another common method, referred as the wobbler method. The majority of NMR spectrometers have the capability to do the wobbler tuning as a standard feature (for instance, "qtune" on Agilent instruments and "wobb" on Bruker instruments). The basic procedure is the same as the reflection method except that the resonance frequency of the probe is monitored when adjusting the capacitors. The first step is to adjust the tuning capacitor so that the dip on the display moves from one side of the setup frequency to the other, approximately by 50 kHz. Adjusting the matching capacitor then makes the dip pass the setup frequency back to the original side. If the dip of the signal is lower, repeat the above process until it reaches a minimum. If the dip is higher than the initial value, change the direction. The depth of the dip indicates how well the impedance of the probe circuit matches to 50 Ω. The best tuning position is indicated by the lowest dip of the probe resonance frequency at the setup frequency.

The probe should be tuned every time the sample is changed, as noted above, especially for those with different solvents or salt concentrations because the dielectric property of the solvents affects the capacitance of probe circuits. However, the tuning of heteronuclear channels is not sensitive to different samples provided the sample does not contain a large amount of salts because the probe coil for heteronuclear channels is the outer coil. Thus, the tuning of ^{15}N and ^{13}C channels does not change much from sample to sample. It should also be noted that the correct way to tune the probe is without any filters attached because probe tuning aims to tune the resonance frequency of the probe. Filters are not integrated parts of the probe circuits and have their own resonance frequencies, which may alter the power fed into them. Therefore, the measurement of the reflected power should be performed directly at the probe output, rather than at any later stage.

4.2 Shimming and Locking

Because superior homogeneity of the magnetic field across the sample volume is necessary in order to obtain a sharp line shape in NMR spectra, the static magnetic field, B_0, must be perfected by adjusting the current to a series of coils inside the magnet (more details in Chap. 2). The process is called "shimming" and the coils called shimming coils. The number of the coils can be as many as 40 sets, which produce a magnetic field across the sample volume. Manually shimming the magnet is a rather complicated process and very time-consuming. It is difficult to manually shim higher order z shims (z shims only alter the magnetic field along the z direction and are also called spinning shims) such as z^5, z^6 or z^7 because they are dependent on each other, which means that a change in one of them alters the others. Shimming nonspinning shims (xy shims that only change the field in the xy plane) is relatively easy because most of them are independent of the others. Fortunately, gradient technology has been used in magnet shimming, called gradient shimming, and greatly reduces the amount of work in shimming. Gradient shimming of spinning shims requires a z gradient capability of the instrument, whereas gradient shimming of nonspinning or xy shims requires x and y gradients. Usually, solution NMR instruments do not have x, y pulsed field gradients (PFGs), and, hence, gradient shimming normally applies only on z shims.

Under normal operating conditions, all NMR magnets have a constantly fluctuating field strength due to the environment. The magnetic field is stabilized

4.2 Shimming and Locking

by a mini spectrometer operated at the ^{2}H frequency, which continuously observes the ^{2}H signal frequency through the dedicated lock channel to maintain a stable magnetic field. If the observed ^{2}H frequency deviates from the set value, the spectrometer changes the magnetic field strength in the opposite way through a coil inside the magnet to compensate for the fluctuation. This process is called "lock" and is carried out in real time. Four parameters are important to have correct operation of locking: lock field (also known as z_0 field), which is the ^{2}H resonance frequency of the solvent (or the offset of the ^{2}H lock frequency); lock power, which controls the ^{2}H RF power used to generate the ^{2}H signal; lock gain, which is used to amplify the signal received by lock receiver; and lock phase, which sets the correct dispersion phase of the ^{2}H signal used to monitor the deviation of the ^{2}H lock frequency caused by the B_0 magnetic field fluctuation. They should all be optimized for lock. The common procedure is to adjust z_0 first to be on resonance for the ^{2}H solvent. After locking on the ^{2}H, the lock power is adjusted to the highest possible level without saturating the lock signal. Saturation is the point at which the lock level is decreased when lock power increases. Then the lock power is slightly reduced to surely avoid the saturation (e.g., by 10%). The lock gain is adjusted to bring the lock level to 80%, which is the most sensitive level. Finally, the lock phase is optimized to maximize the lock level.

The lock level is monitored during manual shimming, which is the observation of the ^{2}H signal. The better the homogeneity of the field is, the higher the lock level, provided the lock phase is correctly adjusted. The FID of water may also be used to monitor the quality of shimming. When using the lock level for shimming, the lock phase must be properly adjusted because a poorly phased lock signal distorts the maximum intensity of the signal, which is used to monitor the homogeneity of the magnetic field. A distorted lock intensity (lock level) may not represent the field homogeneity linearly.

If the shimming is off significantly after changing samples, it is important to first adjust the z shims while spinning the sample. If z gradient auto shimming will be used, only z_1–z_3 need to be shimmed manually. The usual procedure is to adjust z_1 and z_2 alternatively to obtain a lock level as high as possible. Then z_3 needs to be optimized before z_1 and z_2 are revisited. The above procedure is repeated until the maximum lock level is reached. After initial adjustment on the z shims, the nonspinning shims must be shimmed without spinning the sample. The procedure usually is the following order:

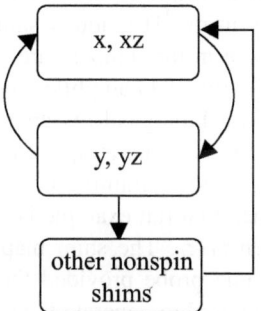

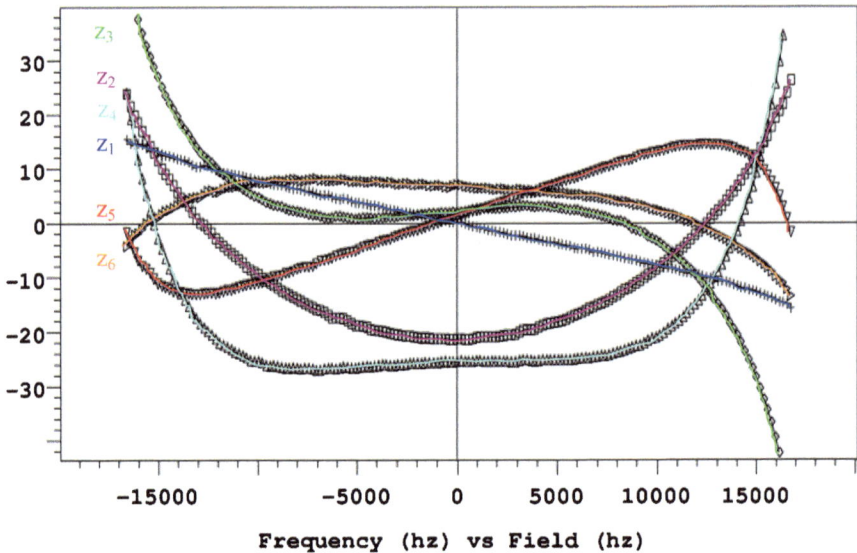

Fig. 4.2 Profiles of shimming gradients z_1–z_6 indicated by the shim map. The x-axis corresponds to the sample position along the z axis of the NMR tube with the zero frequency corresponding to the center of the sample. For this plot, the negative frequency is from the sample in the bottom of the NMR tube

where x and xz are shimmed first, then y and yz to optimize the lock level. The steps are repeated until the lock level does not increase. The rest of the nonspinning shims are checked one by one. If any of them changes the lock level, shims x, xz, y, and yz need to be checked again.

Gradient shimming, which provides rapid automatic optimization of room-temperature shims, is one of the profound applications of PFGs. It not only provides reliable shimming results on high-order z shims (and nonspinning shims) within a minute, which otherwise would take many hours if shimming manually, but also allows one to shim on nuclei besides ^2H, such as ^1H. It is critically useful for samples without ^2H solvent. For instance, a methanol sample is commonly used for calibration of variable temperature control. Without ^1H gradient shimming, it is impossible to shim on the sample because it does not contain ^2H solvent. It is usually done by shimming a deuterated methanol sample with the same volume as the calibration sample. By using ^1H gradient shimming, one can shim on the actual calibration sample (more in the temperature calibration section). Gradient shimming requires PFG capability (PFG amplifier and probe), which is a standard feature of NMR instruments for biological research, although it can be done by homospoil gradient through the z_1 room-temperature shim coil. It is a standard practice to always apply z gradient shimming on all NMR samples. It involves generating a gradient shim map first (an example is shown in Fig. 4.2), shimming z_1–z_4 using the map and then z_1–z_6. The shim map can be repeatedly used for different samples with the same probe provided the sample volume is approximately the same as the sample used to generate the map. Nonspinning shims must be optimized before the shim map is generated. By no means, is what has been described above the only way to perform shimming. There are different shimming methods available in different laboratories. However, in our hands, the above procedure has not failed.

"Deuterium lock frequency" and "lock field" are two terms that are sometimes easily confused. The lock frequency is the instrument's ^{2}H frequency at the magnetic field, whereas the lock field—as mentioned earlier—is the ^{2}H resonance frequency of the solvent. The function of the z_0 field (or lock field) is twofold. Different solvents have different ^{2}H resonance frequencies. The slightly different ^{2}H frequencies are adjusted by changing the z_0 field. The other situation that requires changing z_0 is when the magnetic field drifts. On the other hand, the lock frequency is unchanged as long as z_0 is adjustable, that is, z_0 is not out of range. Therefore, even if the magnetic field drifts (field strength decreases due to loss of current in the SC solenoid of the magnet), a constant field strength is maintained by changing the value of the z_0 field. In other words, the offset frequencies (see below), the chemical shift reference, and other field dependent qualities are unchanged when the instrument lock frequency is unchanged even if the magnetic field strength drifts as the magnet ages because the drifted field is compensated for through the adjustment of z_0.

4.3 Instrument Calibrations

4.3.1 Calibration of Variable Temperature

For each probe, the actual temperature vs. the setting value must be calibrated before data are collected for real samples. A typical method for temperature calibration over the range of -20 to $50°C$ is to observe the chemical shift difference ($\Delta\delta$ in ppm) between the two peaks of 100% methanol. The empirical expression in the relationship of the observed $\Delta\delta$ with temperature is given by:

$$T(°C) = 130.00 - 29.53\Delta\delta - 23.87\Delta\delta^2 \quad (4.1)$$

For the calibration over $50°C$, 100% ethylene glycol is used, whose chemical shift difference of the two peaks has a temperature dependence given by:

$$T(°C) = 193.0 - 101.6\Delta\delta \quad (4.2)$$

Spectrometer software may be used to calculate the temperature after the $\Delta\delta$ is measured. Since the accuracy of the measurement is greatly dependent on the line shape of the peaks, the sample should be well shimmed before the calibration. It is impossible to shim using the ^{2}H lock level by the conventional shimming method because the calibration samples do not contain any portion of deuterated solvents. Gradient shimming on ^{1}H is the best way to do the z shimmings. A sample containing about 1:1 methanol-d_4 in methanol, which has the same volume as

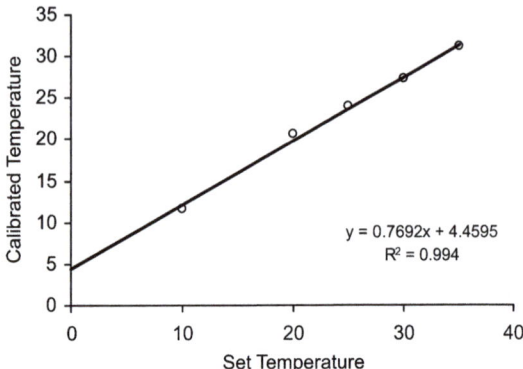

Fig. 4.3 Plot of set temperature vs. calibrated temperature (°C) using a methanol sample

the calibration sample, is first used for shimming purpose. After well shimming the methanol-d_4 sample, especially the nonspinning shims, the 100% methanol sample is placed in the probe and ^{1}H auto gradient shimming z_1–z_6 is applied. Now the sample is ready for temperature calibration. Change the temperature to the lowest point to be calibrated. It takes about 10 min for the temperature to reach equilibrium. Collect data with a single transient and calculate the temperature according to the above equation or by using the appropriate command of the instrument's software. Repeat the experiment and average the two calculated values of the calibrated temperature. Figure 4.3 shows a plot of the set temperature vs. the calibrated ones. A good VT control unit always has a linear relationship for all probes although the slope for individual probes may be different.

4.3.2 Calibration of Chemical Shift References

Tetramethylsilane (TMS) is the IUPAC standard compound used for ^{1}H and ^{13}C references for samples in organic solution (see Chap. 1). However, its insolubility in aqueous solutions makes it a poor reference sample for protein samples. For aqueous samples, DSS (2,2-dimethyl-2-silapentane-5-sulfonic acid) in 90% H_2O/10% ^2H_2O is a better standard sample for ^{1}H chemical shift reference because of its properties of high water solubility, far-upfield resonance, and insensitivity to pH and temperature. The water resonance is sometimes also used as a secondary (or indirect) ^{1}H reference. Although it is a convenient standard, the dependences of temperature and pH as well as the relatively broad line shape of water make it a controversial reference. The chemical shift of water at neutral pH is obtained by its temperature dependence in the range of 10–45°C:

$$\delta_{H_2O}(\text{ppm}) = 4.76 - (T - 25.0)0.01 \qquad (4.3)$$

in which temperature T is in °C. Before the calibration is performed using any reference sample, the ^{2}H signal of ^{2}H$_2$O is adjusted to be on-resonance. The chemical shift of DDS is set to zero frequency. For heteronuclei such as ^{13}C and ^{15}N, the best method to set chemical shift reference is to indirectly reference to DSS ^{1}H frequency via the frequency ratio Ξ at 25°C (Table 1.1) rather than using universal reference standards. There is a small temperature dependence $\Delta\Xi$ in the range of 5–50°C, which can be added to Ξ according the relationships:

$$\Delta\Xi^N(T) = (T - 25°C)2.74 \times 10^{-10} \quad (4.4)$$

$$\Delta\Xi^C(T) = (T - 25°C)1.04 \times 10^{-9} \quad (4.5)$$

As the equations indicate, the effect of temperature on the chemical shift for solution samples is extremely small. Therefore, it is generally ignored for most chemical and biological solution samples (Harris et al. 2001). If one prefers to use individual standard samples for heteronuclear chemical shift reference, the calibration can be done using DSS for ^{13}C and liquid ammonia for ^{15}N. The ^{13}C signal of DSS is directly observed and set to zero frequency. However, it is difficult to observe the ^{15}N signal of liquid ammonia because the sample is not in a liquid state at laboratory conditions of ambient temperature and pressure. Consequently, ^{15}N urea is commonly used as a secondary ^{15}N reference standard, whose chemical shift is referenced to liquid ammonia. ^{15}N urea with a concentration of 1 M in DMSO gives a ^{15}N peak at 77.6 ppm relative to liquid ammonia. Although the urea sample is in DMSO, the calibration must be done with the lock field on ^{2}H$_2$O resonance. First, the lock field is adjusted to the ^{2}H$_2$O resonance with a water sample in the probe because the real sample is in aqueous solution. Nonspinning shims should be checked before changing the sample to the urea sample. After gradient shimming with z_1–z_6 gradients, the ^{15}N spectrum is acquired without changing the lock field. It is crucial to make sure that the lock field is on resonance for ^{2}H$_2$O in order to obtain the accurate ^{15}N calibration.

The chemical shift reference is calibrated only periodically unless the magnetic field drift is substantial (which requires changing the lock frequency to correct for loss of magnetic field). How often the calibration needs to be performed depends on the stability of the instrument and magnet. Usually it is repeated twice annually if the lock frequency has not been changed.

4.3.3 Calibration of Transmitter Pulse Length

Once probe tuning and shimming have been completed, it is necessary to calibrate the pulse angle produced by a given amount of transmitter RF power, which is achieved by calibrating the 90° pulse length. The ^{1}H observe 90° pulse length should be calibrated for every sample. There are many different methods to

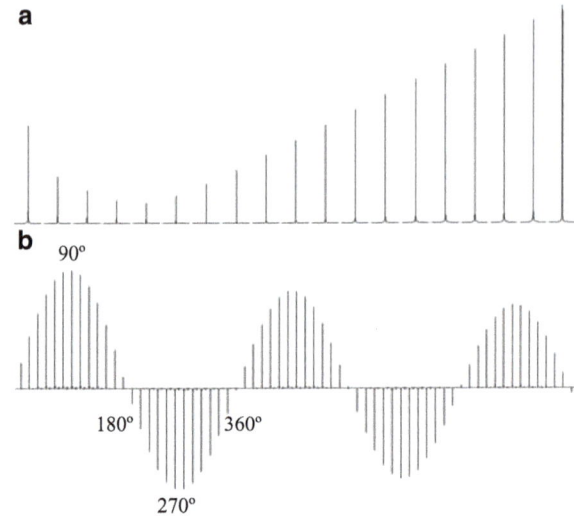

Fig. 4.4 Calibration of transmitter pulse length and offset frequency. (**a**) The offset frequency is arrayed with increment of 1 Hz using the PRESAT sequence. The minimum is obtained when the offset is on water resonance. (**b**) The pulse length is arrayed and the first minimum corresponds to a 180° pulse length

calibrate the observe pulse length. The easiest one is to calibrate a 180° pulse using a water or solvent signal. Because the maximum intensity of the observed signal by the 90° pulse is relatively insensitive to deviations of the pulse angle around 90°, the null intensity produced by a 180° pulse is commonly used to calibrate the 90° pulse length.

The calibration is performed using a one-pulse sequence with a relaxation delay longer than $5T_1$ and a single transient. The carrier frequency is set about 100 Hz from the water resonance to avoid interference of the carrier frequency with the water peak. The pulse length is arrayed by 2 μs for 15 or more FIDs to values sufficiently long to include the estimated 180° pulse length. After the first FID is transformed and phased, all FIDs are then transformed using the same phase correction. The next step is to fine array the range of pulse lengths that give signal intensity near the null value. Because of the inhomogeneity of the B_1 RF field generated at the probe coil, the signal is not completely nulled by a 180° pulse. Instead, a dispersive signal with similar amplitude of positive and negative peaks symmetric about the null position will be observed at the 180° condition (Fig. 4.4b). Sometimes a broad hump at the water resonance appears in the 180°-pulse spectrum, which is caused by the signal from the sample outside the probe coil. In the above "array" method, one is able to visualize the relationship of signal intensity with pulse length. The intensity increases with the pulse length and decreases after the pulse angle passes 90°. It passes through a null at 180° and reaches a negative maximum at 270°. This calibration can also provide information on RF field homogeneity, which is generally specified by the ratios of signal intensities at 450°/90°, and 810°/90°. However, this takes a longer time to finish.

For a quick calibration, a different method is used, which "targets" the pulse length in the range close to a previously calibrated 180° pulse. Acquire a pair of FIDs with pulse lengths of 5 μs and the estimated 180°. After Fourier transformation of the first FID followed by phase correction, the second FID is transformed with the same phase correction. A positive intensity of the water signal in the second spectrum indicates that the pulse angle is less than 180°, whereas a negative intensity is caused by a pulse length greater than 180°. Change the pulse length accordingly to get an evenly distributed dispersive signal obtained using the same phase correction. It is good practice to check the phase correction parameters compared to the first spectrum when the calibrated 180° pulse length is unusually long or short. Similarly, a 360° pulse length can be used to calibrate the 90° pulse length by the above procedure. The 180° pulse should be checked after the 360° is calibrated, to avoid mistaking the 180° as the 360°.

An alternative way to calibrate the 90° pulse is to observe real signals of an aqueous sample rather than the water signal, using a PRESAT sequence (see solvent suppression). Because the concentration of the sample is usually very low, this method needs multiple transients, and thus takes a longer time but gives a more accurate calibration. The normal setup includes setting the saturation frequency on water resonance, number of transients to 8, steady-state transients (or dummy scans) to 4, presaturation time to 3 s and predelay of 2 s or longer, and acquisition time to 1 s. Because a full range array will take a considerably longer time, the target method is more practical to use. Once the 180° pulse length is determined, the pulse calibration is checked again with a longer predelay to ensure the accuracy of the calibration because signal accumulation with an insufficient delay can cause an incorrect measurement of intensity.

^{1}H pulses for spin locking in TOCSY and ROESY, and decoupling from ^{15}N should be applied with longer pulse length, usually approximately 25–30 μs for TOCSY, and 40 μs for ROESY and the decoupling (Table 4.1). Therefore, it is necessary to calibrate these pulses with a lower power. As discussed previously, a reduction of the power level by 6 dB results in a pulse length that is twice as long. Therefore, if the pulse length for the hard 90° pulse is approximately 7 μs, the power needs to be attenuated by 12 dB for the TOCSY spin lock pulses and by 15 dB for 40 μs pulses. The calibration can be done with any aqueous sample by arraying every 1 μs in the range near the desired pulse length.

4.3.4 Calibration of Offset Frequencies

The offset frequency (v_{offset}) has the following relationship with the carrier frequency v_{carr}:

$$v_{\text{carr}} = v_{\text{inst}} + v_{\text{offset}} \tag{4.6}$$

in which v_{inst} is the base frequency of the instrument for the nucleus, which is the fixed frequency for a specific nucleus. For instance, v_{inst} for ^{1}H, ^{13}C, and ^{15}N of a 600 MHz instrument is 599.5200497 MHz, 150.7747596 MHz, and 60.7557335 MHz, respectively.

Table 4.1 Common pulses used for 500 and 600 MHz instruments[a]

Pulses	On/off frequency	500 MHz Pulse length[b]	Est. Pwr Att[c]	600 MHz Pulse length[b]	Est. Pwr Att[c]
^{1}H					
^{1}H hard 90°	On	7		7	
TOCSY spin lock	On	28	12	28	12
WALTZ dec	On	40	15	40	15
^{13}C					
All ^{13}C 90° hard	On	15		15	
^{13}C GARP	On	70	14	70	14
DIPSI-3 spin lock	On	30	6	30	6
$C^{\alpha,\beta}$ 90°, C' null	On	58.4	11–12	48.6	10
C^{α} 90°, C' null	On	64.7	12–13	53.9	11
C' 90°, C^{α} null	On	64.7	12–13	53.9	11
C' 90, C^{α} null	Off	64.7	12–13	53.9	11
$C^{\alpha/\beta}$ 180°, C' null	On	53.0	10–11	44.3	9
C^{α} 180°, C' null	On	57.9	11–12	48.3	10
C' 180°, C^{α} null[d]	Off	57.9 + 0.8	11–12	48.3 + 0.8	10
C^{α} 90° SEDUCE[e]	On	310.5	20	310.5	20
C^{α} 90° SEDUCE[e]	Off	310.5	20↑6	310.5	20↑6
C' 90° SEDUCE[e]	Off	310.5	20↑6	310.5	20↑6
C' 180° SEDUCE[f]	Off	200 + 4	16↑6	200 + 4	16↑6
^{15}N					
All ^{15}N hard 90°	On	40		40	
^{15}N GARP dec	On	250	15	250	15
^{15}N Waltz16 dec	On	200	14	200	14

[a]The pulse lengths listed in the table are used as starting values. Actual pulse lengths and powers should be calibrated accurately for the instruments
[b]The pulse lengths are calculated according to (4.14) and (4.17) using $C^{\alpha/\beta} = 45$ ppm, $C^{\alpha} = 58$ ppm, and C' = 177 ppm, and ^{13}C frequency of 125.68 and 150.86 MHz for 500 and 600 MHz instruments, respectively.
[c]The power attenuation for the 90° pulse calibration is the decreased power in dB from the power for the hard pulse (increase the power setting value for Bruker, or decrease the power setting value for Agilent instruments). "↑6" means increase the power by 6 dB (decrease the power value by 6 for a Bruker instrument). It is estimated relative to a hard 90° of 15 μs, assuming linearity of the amplifier. The reduction of power by 1 dB increases the pulse length by a factor of 1.122
[d]After the calibration, 0.8 μs is added to the off-resonance 180° pulse length
[e]The SEDUCE 90° pulses must be divisible by 50 ns and 45°. For off-resonance SEDUCE decoupling pulses: after the calibration, the power is increased by 6 dB (decreased for a Bruker instrument) because the off-resonance SEDUCE decoupling covers a double band width
[f]After the calibration, 4 μs is added to the pulse length of the off-resonance 180° SEDUCE and the power is increased by 6 dB (decreased for a Bruker instrument)

4.3.4.1 Calibration of Transmitter Offset Frequency

Because common biological samples are in aqueous solution, the transmitter ν_{offset} (e.g., o1 on a Bruker instrument and tof on an Agilent instrument) is required to be set on the water resonance for experiments with water suppression. If the transmitter ν_{offset} is off by even 1 Hz, it will significantly affect the result of water suppression experiments. Therefore, ν_{offset} calibration is performed for each individual sample. It is first estimated from the water peak using a one-pulse sequence by setting it on the center of water peak. It is then arrayed by 0.5 Hz in the range of ± 3 Hz of the setting value using a PRESAT sequence (see solvent suppression) with 2 transients for steady state and 8 transients per FID. The correct ν_{offset} gives the lowest intensity of the water peak (Fig. 4.4a). If the ^{2}H signal of ^{2}H$_2$O is adjusted to be on-resonance prior to ^{2}H locking, ν_{offset} should be within the range of a few hertz for different samples at a given temperature. It may have very different values if the lock is not adjusted to be on-resonance prior to the calibration.

4.3.4.2 Calibration of Decoupler Offset Frequency

The correct decoupler offset frequency (e.g., o2, o3 on a Bruker instrument or dof, dof2 on an Agilent instrument) needs to be known before pulse calibration for the decoupler channel. A simple one-pulse experiment with continuous wave (CW) heteronuclear decoupling is used for this purpose. The power of the CW decoupling is set very low so that only a narrow frequency range (a few tens of hertz) is decoupled in order to accurately determine the value of decoupler ν_{offset}. Arraying the decoupler ν_{offset} in the experiment provides a series of spectra with modulated signal intensity (Fig. 4.5). As the ν_{offset} becomes closer to the resonance frequency of the heteronuclear signal, the doublets become closer, resulting in an intensity increase of the peak. The maximum intensity is obtained when the ν_{offset} is on or near resonance. The lower the CW power, the narrower the frequency bandwidth it decouples. The spectra shown in Fig. 4.5 were obtained with a CW RF field strength of 70 Hz. The maximum pulse power of a 300 W heteronuclear amplifier was attenuated by -42 dB, which reduced the pulse power by a factor of 128 (or 2^7).

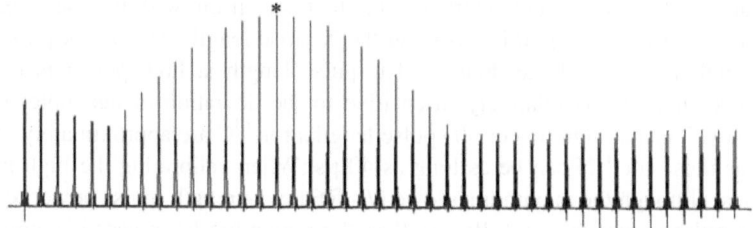

Fig. 4.5 Arrayed spectra for calibrating decoupler offset frequency using ^{15}N urea in DMSO. The decoupler offset frequency is arrayed with increments of 1 Hz as stated in the text. The spectrum labeled with an *asterisk* has the offset frequency on urea ^{15}N resonance

Fig. 4.6 Pulse sequence for calibrating the pulse length of the heteronuclear decoupler. The delay τ is set to $1/(2J_{HX})$. When the X pulse is 90°, the doublet has minimum intensity

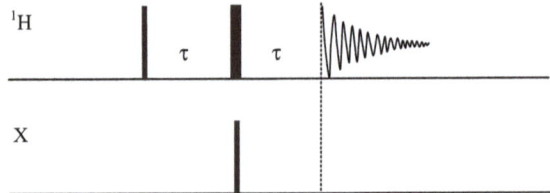

(As a reminder, attenuation by −6 dB decreases RF field strength by a factor of 2.) If the bandwidth of the CW decoupling is wider than the increment of the decoupler ν_{offset} array, there will be several spectra with the same maximum intensity at the peak. The average value of the ν_{offset} in the spectra with the same maximum intensity should be taken to get the calibrated decoupler offset frequency.

4.3.5 Calibration of Decoupler Pulse Length

Decoupler pulse length is calibrated only periodically because it usually does not change from sample to sample. The calibration is performed indirectly, meaning that the effect of the decoupler pulse on the observed nuclei is calibrated. In principle, any sample enriched with ^{15}N and/or ^{13}C can be used for ^{15}N and ^{13}C decoupler pulse calibration. Samples used for the calibration include (^{15}N, ^{13}C)-NAcGly in 90% H_2O/10% ^2H_2O, 1 M ^{15}N urea in DMSO-d_6 (Dimethyl Sulfoxide-d_6), 0.1% ^{13}C methanol in 99% ^2H_2O/1% H_2O (Agilent autotest sample), or other samples for indirect calibration. The pulse sequence shown in Fig. 4.6 is one of those commonly used to calibrate ^{15}N and ^{13}C decoupler pulses. A delay τ is set to $1/(2J_{HX})$ for calibrating the decoupler pulse of heteronucleus X. J_{HX} is measured by the frequency difference between the doublets. It is important to use the correct ^1H 90° pulse length and the decoupler offset frequency must be set on the center frequency of the heteronuclear doublets in order to get an accurate calibration at low decoupler power.

To calibrate the decoupler pulse length, the decoupler offset frequency is set to the calibrated value which is in the center of the coupled doublets. When the angle of the heteronuclear X pulse is zero, a doublet is observed. Similar to the ^1H pulse calibration, the phase parameters of the first spectrum will be used for the subsequent spectra acquired by arraying the X pulse length. The 90° X pulse will give a null intensity of the doublet. The pulse length at high power is usually calibrated first, and is relatively insensitive to the deviation of decoupler offset frequency. Typical high power 90° pulse lengths for ^{13}C are approximately 15 µs, whereas those for ^{15}N can be as long as 40 µs. When calibrating the high power X pulses, although the maximum allowed B_1 RF field strength (pulse power) can be used, a pulse power that is 3 dB less than the maximum RF power is commonly chosen because it is in the linear range of the amplifier. However, a properly chosen pulse power is the one that gives the above short pulse length.

The 90° pulses for heteronuclear decoupling use much less power and, hence, have a longer pulse length. For example, the desired ^{13}C 90° pulse for GARP

4.3 Instrument Calibrations

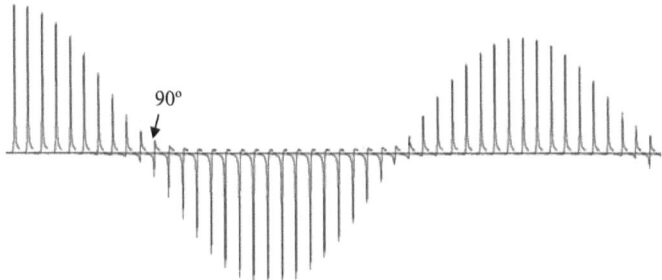

Fig. 4.7 Arrayed spectra for decoupler 90° pulse calibration. The 90° pulse gives minimal intensity

decoupling is 70 and 300 μs for SEDUCE decoupling, and ^{15}N 90° decoupling pulses are 200–250 μs for WALTZ-16 or GARP. Therefore, the decoupling pulses are calibrated with a lower decoupler power. The linearity of the amplifier can be used to estimate the power for the desired pulse length, based on the relationship given by:

$$dB = 20 \log_{10}(pw_1/pw_2) \quad (4.7)$$

in which pw_1 and pw_2 are the pulse lengths for the hard 90° pulse and the 90° decoupling pulse, respectively. If pulse power is decreased by 1 dB, the pulse length is increased by a factor of 1.122 linearly:

$$1 \text{ dB} = 10^{1/20} \approx 1.122 \quad (4.8)$$

Therefore, N dB reduction in pulse power lengthens the pulse by a factor of $(1.122)^N$. This equation is equivalent to (2.2) in Chap. 2 (on Instrumentation) because the pulse length is proportional to the B_1 field strength, or the voltage. Specification test data and previous calibration are also reasonable references to estimate the decoupler power setting. For the calibration of low power decoupling pulses, the pulse length is set to the desired value, for example, 70 μs for ^{13}C or 250 μs for ^{15}N (Table 4.1). Then the decoupler power is arrayed by 1 dB from lower power to higher in a range of 10 dB (lower power means a smaller value of the power setting for an Agilent spectrometer and a larger value for a Bruker spectrometer). When the power is lower, the selected pulse length is less than that of a 90° pulse, resulting in a doublet with positive phase. Phase the first spectrum obtained at the lowest power so that it has a doublet with correct phase. Then all FIDs are transformed using the same phase parameters as the first one. The intensity of the doublet will decrease as the power increases. The power corresponding to the intensity closest to null is used to calibrate the pulse length. The next step is to array the pulse length with the calibrated power in the range of 70 ± 10 μs by a step of 1 μs for ^{13}C, or 250 ± 50 μs by a step of 4 μs for ^{15}N, to get the 90° pulse length at the calibrated power. When calibrating the pulses, the relaxation delay must be sufficiently long and the decoupler ν_{offset} must be calibrated to be in the center of the doublet. Figure 4.7 shows the spectra obtained with correct delay and decoupler ν_{offset}.

4.3.6 Calibration of Decoupler Pulse Length with Off-Resonance Null

In certain triple resonance experiments, it is desirable to excite one region of ^{13}C resonances but not others. The range of resonances is usually centered at 58 ppm for C^α, 177 ppm for carbonyl carbon C', and 45 ppm for $C^{\alpha/\beta}$. In HNCO experiment (see the following chapter), for example, the carrier of the ^{13}C is set on C'. A 90° pulse on C' with null at C^α has a pulse length given by:

$$\tau_{90,\text{null}}(\mu s) = \frac{0.96825}{\Omega} \times 10^6 \quad (4.9)$$

in which Ω is the frequency difference in Hz between the center of C' and C^α:

$$\Omega = \nu_{C'} - \nu_{C^\alpha} = (\delta_{C'} - \delta_{C^\alpha})\nu_{\text{ref}} \quad (4.10)$$

in which $\nu_{C'}$ and ν_{C^α} are the center frequencies of C' and C^α in Hz, respectively, $\delta_{C'}$ and δ_{C^α} are the spectral centers in ppm, and ν_{ref} is the reference frequency of the ^{13}C chemical shifts in MHz. Assuming that a 600 MHz spectrometer has a ν_{ref} of 150.86 MHz for ^{13}C, $\Omega = (177 - 58)*150.86 \approx 17{,}950$ Hz, which gives a pulse length of 53.9 μs.

The effect of a hard 90° pulse on the spins is different when on-resonance and off-resonance. If on resonance, the 90° pulse length is solely produced by B_1 field because nuclear spins experience only the B_1 field:

$$\gamma B_1 \text{pw} = \frac{\pi}{2} \quad (4.11)$$

in which pw is the pulse length, $\gamma B_1 = \omega_1$ which is the frequency of B_1 field. As the resonance frequency moves away from the carrier, the effective field is the vector sum of B_1 and the field along the z direction with the field strength $2\pi\Omega/\gamma$ as a function of the frequency difference Ω and nuclear gyromagnetic ratio γ:

$$\gamma B_{\text{eff}} = \sqrt{(\gamma B_1)^2 + \left(\gamma \frac{2\pi\Omega}{\gamma}\right)^2} \quad (4.12)$$

The simplest way for an on-resonance 90° pulse to be a null pulse at Ω Hz off-resonance is to generate a 360° (or 2π) pulse at the off-resonance frequency. Under this condition, pw must satisfy the following relationship in the same way as the on-resonance pulse does:

$$\sqrt{(\gamma B_1)^2 + (2\pi\Omega)^2}\,\text{pw}_{90} = 2\pi \quad (4.13)$$

4.4 Selective Excitation with Narrow Band and Off-Resonance Shaped Pulses

in which $\gamma B_1 = \pi/(2\ \text{pw}_{90})$ according to (4.11) and pw_{90} is the on-resonance 90° pulse length. By substituting γB_1 with $\pi/(2\ \text{pw}_{90})$ and rearranging the equation, the pulse length that gives 90° on-resonance and null (360°) off-resonance has a value determined by:

$$\text{pw}_{90} = \frac{\sqrt{15}}{4\Omega} \tag{4.14}$$

For the case that the frequency difference Ω between C' and C^α is 17,950 Hz, pw_{90} is 53.9 μs.

Similarly, a 180° pulse is generated by the B_1 field on resonance:

$$\gamma B_1 \text{pw}_{180} = \pi \tag{4.15}$$

When the 180° pulse is needed to give a 360° pulse angle at Ω Hz off-resonance it must satisfy the following condition:

$$\sqrt{\left(\frac{\pi}{\text{pw}_{180}}\right)^2 + (2\pi\Omega)^2} \text{pw}_{180} = 2\pi \tag{4.16}$$

After the equation is rearranged, the pw_{180} can be determined by:

$$\text{pw}_{180} = \frac{\sqrt{3}}{2\Omega} \tag{4.17}$$

With a Ω of 17,950 Hz, pw_{180} is 48.3 μs. In practice, if pw_{180} is used for off-resonance excitation, it is set to be 0.8 μs longer than the calibrated value, which is 49.1 μs.

The 90° and 180° hard pulses mentioned above are used to rotate the magnetization at the frequencies on resonance and do not change the magnetization (360° rotation) at the off-resonance region Ω Hz away from the carrier frequency. In the cases when perturbation of the off-resonance frequencies is desired, which generates 90° or 180° rotation of the magnetization at off-resonance frequencies without exciting the region at the carrier frequency, shaped hard pulses are used, which are generated using the calibrated pw_{90} or pw_{180} (see Sect. 4.4). The desired pulse lengths for decoupler off-resonance pulses are calibrated by arraying the pulse power using the pulse sequence for decoupler pulse calibration. In most cases, it is necessary to calibrate fine attenuation of the decoupler power in order to obtain the desired pulse length.

4.4 Selective Excitation with Narrow Band and Off-Resonance Shaped Pulses

In many experiments, a certain frequency region must be excited without disturbing the magnetization at other frequencies. Selective pulses that have a narrow excitation range are utilized to selectively excite a group of nuclei, such as protons of water molecules or a kind of carbons of proteins. They are referred to as soft pulses as opposed to hard pulses that have a wide excitation range and cover all the nuclei of the same isotope, such as ^{1}H, ^{13}C and ^{15}N. Applications of selective pulses include improving solvent suppression, reducing spectral width so as to increase digital resolution in multidimensional experiments, and selectively exciting frequency regions of different spins such as C' and C$^\alpha$. A nonselective pulse, or hard pulse, is generated by a high power for a short time. The shorter the hard pulse, the wider the excited frequency range (see Sect. 1.5.1). As an approximation, the bandwidth of a pulse can be considered as the frequency range of its Fourier transformation, and it is inversely proportional to the pulse length. The Fourier transformation of an infinitely long time function gives a delta function in frequency domain. Therefore, a narrow excitation bandwidth can be achieved by reducing the pulse power and increasing pulse length. Such a selective pulse is utilized in solvent suppression by presaturation to saturate the transition of water spins. However, a desirable selective pulse should have properties such as a relatively short pulse length, a narrow excitation region covering the desired frequencies, a uniform excitation band, and linear phase dependence of transverse magnetization on the frequency offset. Amplitude-modulated and/or phase-modulated pulses, called shaped pulses, are used to generate selective pulses with the desired characteristics.

Although the pulse shape cannot be directly determined from the desired excitation profile due to the nonlinearity of the NMR response governed by the Bloch equations, Fourier transformation is a convenient method to design the pulse shape based on the desired excitation profile (Emsley 1994). The Gaussian function has a unique characteristic that the Fourier transformation of a Gaussian is also a Gaussian. It is for this reason that the Gaussian has been used as a shaped pulse for selective excitation (Bauer et al. 1984). Another example is a sinc (sine x/x) function that has a semirectangular frequency response, which is another application of the Fourier similarity theorem. A 90° Gaussian shaped pulse can be used to rotate the magnetization by 90° within the excitation bandwidth of the shaped pulse. However, the rotated off-resonance magnetization gradually dephases during the pulse, which produces a significant phase difference between the on-resonance and off-resonance magnetization (within the excitation bandwidth). When a Gaussian 90° pulse is used for selective excitation of solvent magnetization, this phase error does not cause any problem because the bandwidth of the pulse is narrow and the solvent magnetization is not desired to be observed. The problem arises if the Gaussian 90° is used to selectively excite a region of particular spin magnetization (e.g., C' or C$^\alpha$).

A simple solution to the phase error is to use a 270° Gaussian pulse that rotates the magnetization by 270° rather than 90° (Emsley and Bodenhausen 1989). The 270° pulse improves the phase response significantly compared to the 90° pulse. The improvement can be understood as the following. When applying a 90° Gaussian pulse on y axis, the on-resonance magnetization is rotated onto the x axis, whereas the magnetization at an offset from the resonance lands near the y axis, resulting in a nearly 90° phase difference between the magnetizations of on- and

4.4 Selective Excitation with Narrow Band and Off-Resonance Shaped Pulses

off-resonance. On the other hand, if the on-resonance magnetization is rotated by a 270° Gaussian pulse to pass the x axis, the $-z$ axis, and reach on $-x$ axis, the off-resonance magnetization is, through a different route, rotated to a position very close to $-x$ axis. The net result is that both the on- and off-resonance magnetizations are rotated to the $-x$ axis and phase error is reduced. Thus, a 270° Gaussian pulse produces an approximate in-phase excitation, which is sufficient for common applications in NMR spectroscopy of biological macromolecules. Other popular selective shaped pulses that provide improved in-phase excitations include half Gaussian (Friedrich et al. 1987; Xu et al. 1992), e-BURP-2 (Geen and Freeman 1991), and G^4 Cascade (a sum of Gaussians, Emsley and Bodenhausen 1990). The functions require a number of variable parameters to be defined, as many as 12 or more. The parameters can be optimized by software using pulse power and pulse length to produce the desired shaped pulse with desired excitation bandwidth and offset.

In order for NMR spectrometer hardware to utilize the shaped pulse function, the shaped pulse consists of a series of short rectangular pulses (a few microseconds) with different amplitudes. The duration of the shaped pulse is the sum of the pulse lengths of the element pulses. The shape is approximated by the amplitudes of the element pulses. A typical text file of a shaped pulse profile lists sequentially the individual width and power for each element pulse.

Like other pulses, the shaped pulse must be calibrated every time the sample is placed in the magnet if it is a transmitter pulse or periodically if it is a decoupler pulse. If the pulse length changes, the shaped pulse needs to be regenerated using the correct pulse length. This means that the transmitter shaped pulse will be regenerated more frequently than we prefer. One way to avoid the regeneration of the shaped pulse is to calibrate the pulse power for the fixed pulse length used by the shape function so that the existing shaped pulse function can be used. If the pulse length used in the existing shaped pulse is known, the conventional method is to calibrate the pulse power for the desired duration of the rectangular pulse. If the pulse length used to generate the shaped pulse is unknown, the shaped pulse itself must be used for the calibration.

For the calibration of the shaped pulse, the pulse length and pulse power used to generate the shaped pulse profile would be good reference values. For a 180° shaped pulse, the calibration can be performed using a one-pulse sequence in which the shaped pulse is defined as the same shape pulse profile to be used in the experiment. The correct pulse power for the desired pulse length is determined by arraying the fine power attenuation, which gives a dispersive residual signal. In general, a 90° shaped pulse cannot be obtained by dividing the calibrated 180° shaped pulse length or by doubling the power. A pulse sequence containing one hard 90° pulse and one shaped pulse (or two identical 90° shaped pulses) with the same phase is used for the 90° shaped pulse calibration. The sequence achieves 180° rotation of the magnetization around the axis. Arraying the pulse power using the same procedure in the 180° pulse calibration is performed to determine the power for the 90° shaped pulse.

4.5 Composite Pulses

RF pulses are used to rotate magnetization at desired angle, frequency, and duration. For certain applications, pulses are combined together to form a pulse train with or without delay in between pulses to accomplish specific functions. The pulse trains are called composite pulses. The applications of composite pulses range from improving imperfections of single pulses, spin lock for magnetization transfer to off-resonance excitation, and spin decoupling.

4.5.1 Composite Excitation Pulses

To compensate for the effects of off-resonance and RF field inhomogeneity, and to produce effective magnetization rotation by a specified angle, numerous composite pulses have been developed. Among them, the composite pulses for 90° and 180° excitation are more frequently utilized in biological NMR spectroscopy. The most commonly used composite pulses consist of a few 90° and/or 180° pulses varying in phase. The pulse sequence $90_x 90_y 90_{-x} 90_{-y}$ is a popular version of a composite 90° pulse used as a read out pulse (last pulse in the sequence) in certain NMR experiments with solvent suppression. The composite 180° pulse containing a sequence of $90_x 180_y 90_x$ is sometimes used to improve the performance in inversion of z magnetization in a pulse sequence.

4.5.2 Composite Pulses for Isotropic Mixing

The Hartmann–Hahn matching condition (Hartmann and Hahn 1962) is required to be satisfied during the coherence transfer between spins by isotopic mixing in TOCSY experiments (see TOCSY for detail). When the scalar coupled spins have different frequency offset from the B_1 carrier frequency, the efficiency of the magnetization transfer produced by the continuous RF field is extremely low. The required power of the RF field to cover a frequency range to satisfy the Hartmann–Hahn matching condition would produce a tremendous amount of heat during the period of mixing, which would overheat the probe and damage the NMR samples. Several pulse sequences for Hartmann–Hahn mixing have been developed to achieve effective coherence transfer. MLEV17 (Bax and Davis 1985a, b) and WALTZ16 (Shaka et al. 1983a, b) are the early versions, which utilize phase-modulated pulses. WALTZ16 combines basic elements of the sequences into supercycles to improve the transfer efficiency. Later, delays were added into the MLEV17 sequence in clean-TOCSY experiments to improve the performance (Griesinger et al. 1998). It has been found that the clean-TOCSY mixing scheme produces more heat during the mixing period and can cause overheating problem, which is not suitable for experiments with a long mixing time.

The mixing sequences commonly used in multidimensional NMR spectroscopy are DIPSI sequences that consist of pulses with different pulse angles (Shaka et al. 1988; Rucker and Shaka 1989). The DIPSI-2 and DIPSI-3 sequences provide much better efficiency of coherence transfer than phase-modulated sequences, MLEV17 and WALTZ16. For all of the composite pulse sequences, only the 90° pulse length needs to be calibrated. The sequences take the 90° pulse length and convert it into other pulse angles if necessary. A pulse length of the 90° pulse in the range of 25–30 μs is sufficient for the majority of applications to homonuclear magnetization transfer in multidimensional NMR spectroscopy.

4.5.3 Composite Pulses for Spin Decoupling

Nuclear scalar coupling can be removed by applying continuous RF irradiation. This technique is commonly used in a 1D ^{1}H homonuclear decoupling experiment. However, because of the off-resonance effect, it requires very high power to obtain effective decoupling over the broadband of heteronuclear frequencies using a continuous RF field. The decoupling efficiency of a continuous RF field is described in term of the scaling factor λ as:

$$\lambda = \frac{2\pi\Omega}{\sqrt{(\gamma B_1)^2 + (2\pi\Omega)^2}} \quad (4.18)$$

in which Ω is the offset frequency in Hz from the carrier frequency, γB_1 is the frequency of B_1 field. The value of $\lambda = 0$ represents complete decoupling, whereas $\lambda = 1$ provides no decoupling effect. In order to obtain complete decoupling using a continuous wave RF field, the field must be set on resonance, that is, $\Omega = 0$. The decoupling power of CW would be required to be higher to decouple a frequency range which is offset Ω from the carrier. Only a small portion of the scalar coupling is decoupled when the offset is about the same amplitude as the applied field ($\lambda = 0.707$ when $2\pi\Omega = \gamma B_1$).

One of the major applications of composite pulses is to perform spin decoupling. The spin decoupling is achieved by the inversion of spin magnetization, based on the assumption of instant flip approximation that the spin magnetization being decoupled is inverted instantly when the frequency of the RF pulse matches its Larmor frequency. To minimize the RF power and increase the decoupling bandwidth, composite pulse sequences including some of the sequences used for isotropic mixing are utilized for heteronuclear decoupling, of which WALTZ16, GARP (Shaka et al. 1985), SEDUCE (Coy and Mueller 1993) and DIPSI-3 are popular decoupling sequences. All of them use supercycles to improve the

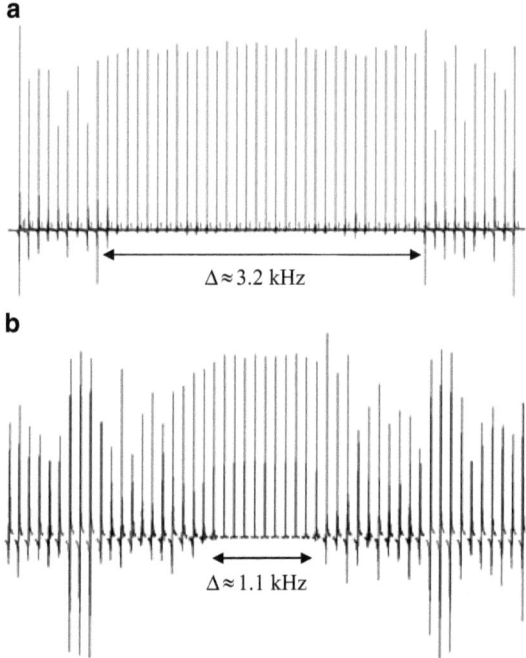

Fig. 4.8 Bandwidth of (a) GARP and (b) WALTZ16 decoupling. The ^{1}H–^{15}N doublets of ^{15}N urea in DMSO were observed using 51 decoupler offset frequencies with an increment of 100 Hz. (a) The decoupler 90° pulse length of 198 μs used in the experiments, which gives a B_1 field strength of 1.26 kHz. (b) 275 μs was used for the decoupler 90° pulse length, which corresponds to a B_1 field strength of 0.9 kHz. The observed decoupling bandwidths are indicated below the spectra (obtained using (4.19)). WALTZ16 has a decoupling efficiency of 1.2, whereas the observed bandwidth of GARP indicates a decoupling efficiency of 2.5

decoupling efficiency. Often the decoupling efficiency of the sequences is described by the factor of the decoupling bandwidth Δ (in units of Hz) to the strength of the applied field:

$$f_d = \frac{2\pi\Delta}{\gamma B_1} = \frac{\Delta}{1/(4\text{pw}_{90})} \tag{4.19}$$

in which pw_{90} is the 90° pulse length at the field strength $1/(4\,\text{pw}_{90})$ (in Hz) and f_d is the decoupling efficiency. It has theoretically been demonstrated that GARP produces the decoupling bandwidth by a factor of 2.5 over the field power it uses, whereas WALTZ16 gives the same bandwidth as the field power used. The decoupling profiles of the sequences observed experimentally are shown in Fig. 4.8. In multidimensional NMR spectroscopy, GARP is usually used for ^{13}C or ^{15}N broadband decoupling, SEDUCE for selective decoupling such as C$^\alpha$ decoupling during the experiment, and WALTZ16 is used for ^{15}N decoupling.

4.6 Adiabatic Pulses

The above composite pulses are regarded as linearly polarized waveforms which are phase-modulated RF pulses. There is another type of RF pulses whose waveforms are described as circularly polarized, referred to as adiabatic pulses. They are much more efficient for inversion of spin magnetization than the phase-modulated pulses. The advantages of adiabatic pulses are their wider band width with a minimum power and their insensitivity to inhomogeneity of B_1 field. The basic idea behind them is that if the change in the orientation of the effective field with time is sufficiently slow, the tilt angle θ of the magnetization with respect to the moving direction of the effective field is very small, with the duration of the pulses shorter than any relaxation process of the system (Abragam 1961; Kupče 2001):

$$\frac{1}{T_2^*} \ll \frac{\partial \theta}{\partial t} \ll \omega_{\text{eff}} \quad (4.20)$$

This means that the change in direction of the B_1 field must be sufficiently slower than the rotation of magnetization around the effective field. Initially, the B_1 field is turned on with a frequency far away from the resonance; the effective field B_{eff} is virtually parallel to the B_0 and the magnetization is along the effective field. The frequency of the B_1 field is then changed to approach the resonance at a rate sufficiently slow so that the magnetization changes its direction to follow the direction of B_{eff} (Fig. 4.9).

A dimensionless factor Q, called the adiabaticity factor, has been introduced to quantitatively describe the adiabatic condition (Baum et al. 1985):

$$Q = \frac{\omega_{\text{eff}}}{\dot{\theta}} \quad (4.21)$$

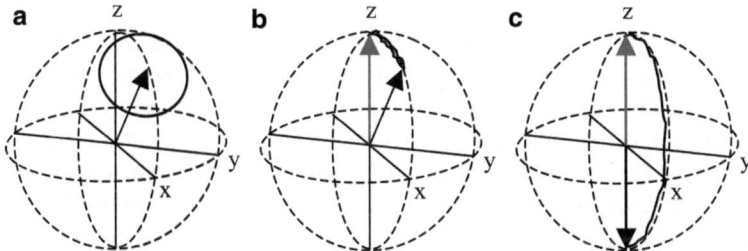

Fig. 4.9 Magnetization trajectories when RF pulses are applied (from Kupče 2001). (**a**) When the RF pulse is applied suddenly, the magnetization is rotated around the B_{eff} (*arrow*). (**b**) When the RF field is turned on adiabatically, the magnetization follows the B_{eff}, stays locked along the B_{eff} for most of time and then is returned back to equilibrium. (**c**) When the RF field is swept through the resonance frequency adiabatically, the magnetization follows the B_{eff} to the $-z$ axis and becomes inverted (reproduced with permission from Kupče 2001, Copyright © 2001 Elsevier)

in which $\dot{\theta} = \partial\theta/\partial t$. Equation (4.21) can be rewritten in terms of pulse field strength ω_1 and frequency offset Ω:

$$Q = \frac{(\omega_1^2 + \Omega^2)^{3/2}}{\omega_1 \dot{\Omega} + \dot{\omega}_1 \Omega} \tag{4.22}$$

in which the dot represents the rate of changing. Therefore, the adiabatic condition requires $Q \gg 1$, meaning that ω_{eff} is much greater than θ in (4.21).

Numerous adiabatic pulses whose amplitudes, phases, or both can be modulated have been developed for applications to spin decoupling, magnetization inversion, refocusing, and selective excitation. For instance, WURST (wideband uniform rate and smooth truncation, Kupče and Freeman 1995a) is an adiabatic pulse generated by adiabatically modulating the amplitude waveform of the B_1 field with the frequency offset swept linearly, according to the relationship:

$$B_1(t) = B_1^{\max}[1 - |\sin(\beta t)|^n] \quad -\frac{\pi}{2} < \beta t < \frac{\pi}{2} \tag{4.23}$$

The higher the value of n, the higher the bandwidth. WURST-20 has an n value of 20, and has been used for the inversion of magnetization to yield a uniform profile over a wide frequency range. On the other hand, both the amplitude and frequency sweep can be modulated in a coherent fashion, such as in hyperbolic secant (or sech) pulses (Kupče and Freeman 1995b; Tannus and Garwood 1996). The amplitude waveform and frequency sweep of B_1 are modulated in a hyperbolic secant pulse with the forms of:

$$B_1(t) = B_1^{\max} \operatorname{sech}(\beta t) \quad -\frac{\pi}{2} < \beta t < \frac{\pi}{2}$$
$$\Delta\omega(t) = -\mu\beta \tanh(\beta t) \tag{4.24}$$

The excitation profile and bandwidth ($\Delta\omega_{\text{BW}} = 2\mu\beta$) of the pulse is determined by the parameter μ in combination with β. As the value of μ increases, both the shape and bandwidth of the excitation are improved significantly (Fig. 4.10). The advantage of the adiabatic pulse for spin inversion is that the inversion is insensitive to the B_1 field strength and can have much more tolerance to the missetting of the B_1 field strength. Consequently, it is not affected by the inhomogeneity of either the B_1 or the B_0 fields during the experiment. Therefore, adiabatic pulses have found many applications in heteronuclear experiments in which inversion of the heteronuclear magnetization is needed such as HSQC and HMQC (see Chap. 5). They have become more and more popular in biological NMR spectroscopy.

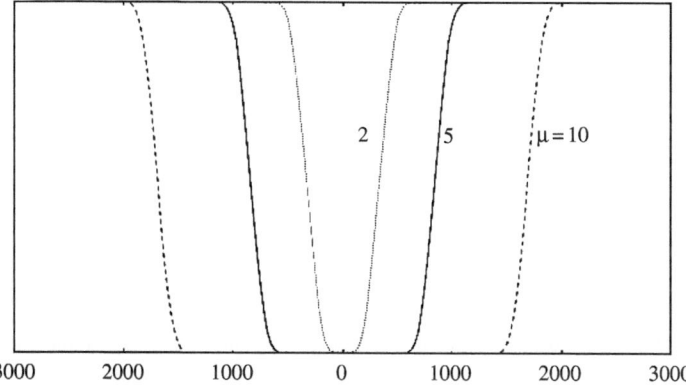

Fig. 4.10 Inversion profiles of hyperbolic secant pulses with different value of parameter μ as indicated. Higher μ generates wider bandwidth and flatter profile (reproduced with permission from Norris 2002, Copyright © 2002 Wiley Publishers)

4.7 Pulsed Field Gradients

In NMR experiments RF pulses are utilized to rotate the magnetization and create coherence transfers. Different coherence transfer pathways can be produced in the same pulse sequence. In order to extract a specific type of information, certain pathways must be separated from others. Phase cycling is a method used to select the desired coherence transfer pathway, in which the phases of pulses are systematically altered from one FID to another so that the unwanted pathway will be removed at the end of phase cycling during data acquisition. The method can be incorporated in a majority of experiments and provides satisfactory results (see Sect. 4.10.3). However, in this method, all types of magnetization (wanted and unwanted) must be collected first and then the unwanted coherence is removed via the differentiation of the data. In order to achieve this, the instrument is required to be perfect in stability and dynamic range.

PFGs have been chosen as an alternative method to select a desired coherence transfer pathway in NMR experiments. They provides several advantages over the phase cycling method in that only the desired magnetization is observed during data acquisition in each experiment, resulting in an increase in acquisition dynamic range, reducing artifacts caused by instrument instability, and shortening experimental time. Gradients also provide new methods for solvent suppression, sensitivity enhancement, and diffusion study.

When a gradient varying linearly, for example on the z axis, is applied, it causes spatial inhomogeneity of the magnetic field along the axis. The spins in different z locations experience different magnetic field strengths, which causes the spin to precess at different frequencies for chemically equivalent spins. This means that the transverse magnetization of the spins will have different precession frequencies for

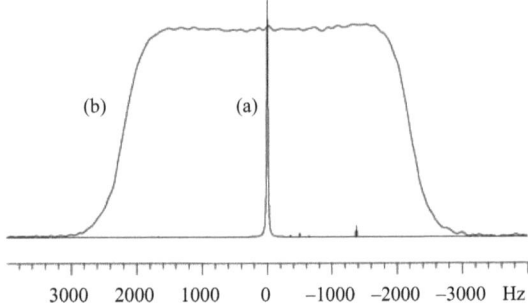

Fig. 4.11 ^{1}H spectra of water (a) without and (b) with gradients during acquisition. The intensities of the two spectra are not proportional. When the gradient is turned on, water across the sample experiences different magnetic field strengths due to the linear z gradient. Therefore, the chemical shifts are different as a function of the z position of the sample (see (4.30)). The center of the sample has zero frequency shift

different physical locations along the z axis in the sample volume. Some will precess faster, and some slower than the resonance frequency. The transverse magnetization of the spins across the sample volume will be decreased with the duration of the gradient irradiation, because a certain portion of the magnetization will be canceled out due to dephasing of the magnetization. At some time during the gradient irradiation, the transverse magnetization becomes zero. Therefore, the field inhomogeneity created by a gradient (z gradient) eliminates the magnetization on the plane (xy plane) perpendicular to the gradient axis. The process may be reversed to recover the dephased magnetization by applying the gradient with an opposite direction, or negative to the previous gradient with the same duration. When the gradient reverses the direction of the gradient field, the spins precessing slower under the previous gradient field will now precess faster, whereas those that were slower will now precess faster. After the same duration that caused the magnetization to disappear, the magnetization will regain its original amplitude, if ignoring relaxation effects. The former process is called dephasing, while the latter is called refocusing.

If the z gradient in a one-pulse experiment is applied during acquisition of signal from a single resonance, the observed FID is a sinc-type function. The pulse sequence is utilized by gradient shimming method. After Fourier transformation, a spectrum with a flat step shape is obtained (Fig. 4.11). The frequency range is symmetric about the resonance frequency in the absence of the gradient (the center of the spectrum) because the origin of the gradient field is in the center of the probe coil. The magnetic field, B_g, across the sample created by a gradient pulse on the z axis is a linear function of the position along the z direction (Keeler et al. 1994):

$$B_g = zG \quad (4.25)$$

in which G is the gradient field strength in units of Gauss per centimeter, Gcm^{-1}, or tesla per meter, Tm^{-1}. The convenient and commonly used unit is Gcm^{-1} because

4.7 Pulsed Field Gradients

gradient coils in NMR probes are a few centimeters in length. During the experiments, the field strength of the gradients is defined in terms of digital units such as DAC (digital-to-analog conversion). In order to set the gradient to specific strength in units of Gcm^{-1}, the maximum gradient strength corresponding to the maximum DAC unit must be known. For instance, if the maximum gradient strength of a probe is 70 Gcm^{-1} and the setting range of the gradient is from 32k DAC to −32k DAC, 8,000 DAC represent a gradient strength of 17 Gcm^{-1}. The field strength at a given position z of the sample is described by:

$$B = B_0 + B_g \tag{4.26}$$

Because B_0 will contribute evenly over the sample, we will drop the term and only consider in the rotating frame the effect of the gradient on the sample at different positions. The precession frequency of spins at any specific sample position z under the interaction of the gradient field is given by:

$$\omega_g = \gamma B_g = \gamma z G \tag{4.27}$$

in which ω_g is the precession frequency of the spins in the presence of B_g. The phase of the precession at the sample position z after the gradient field G is applied for a period of time t is determined by:

$$\phi(z) = \gamma z G t \tag{4.28}$$

Considering the above example in which a single peak is present in the spectrum, x magnetization of spins at any specific sample position z evolves after applying a z gradient for time t:

$$M_x(z) \xrightarrow{\gamma z G t} \cos(\gamma z G t) M_x(z) + \sin(\gamma z G t) M_y(z) \tag{4.29}$$

$M_x(z)$ is the transverse x magnetization at sample position z, which is dependent on the gradient strength and pulse time t. The appearing x magnetization comes from all of the sample slices along the z direction, and is the average over all sample slices. Because the origin of the gradient field is at the center of the sample, the integration (summation) of the $M_x(z)$ is from $-\frac{1}{2}r_{max}$ to $\frac{1}{2}r_{max}$ for a length of r_{max} (Fig. 4.12):

$$M_x = \frac{1}{r_{max}} \int_{-\frac{1}{2}r_{max}}^{\frac{1}{2}r_{max}} \cos(\gamma z G t)\, dz = \frac{2\sin\left(\frac{1}{2}r_{max}\gamma G t\right)}{r_{max}\gamma G t}$$

$$= \mathrm{sinc}(\frac{1}{2}r_{max}\gamma G t) \tag{4.30}$$

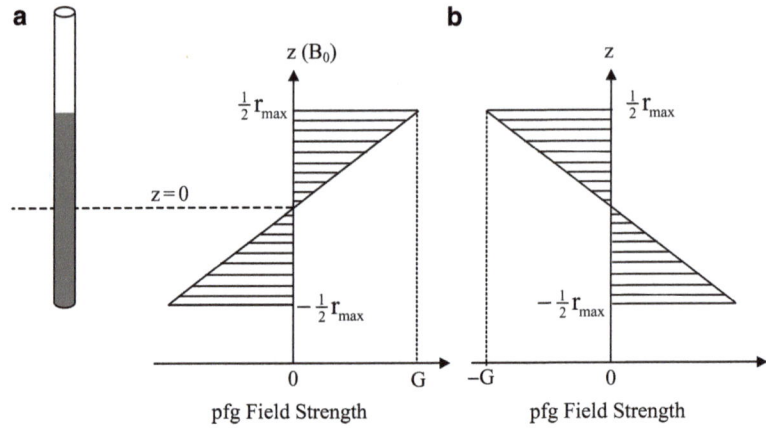

Fig. 4.12 The z gradient field strength across the sample volume. The origin of the gradient field is in the center of the sample (precisely, it is in the center of probe coil). The gradient field strength is a linear dependence of the z positions of the sample volume, and the total length along z axis is r_{max}. (**a**) The field strength distribution when a G gradient is applied. (**b**) The gradient field strength when a $-G$ gradient is applied. The sign of the gradients is relative

If the gradient pulse is applied during the entire acquisition time, the observed FID is a sinc function as described in (4.30). Fourier transformation of the sinc FID gives a spectrum with a broad range of frequencies caused by the gradient (spectrum b in Fig. 4.11). The width of the frequency band depends on the strength of the gradient. Because the origin of the gradient is at the center of the sample, the frequency distribution is symmetric about the resonance frequency as shown in the spectrum.

In multipulse experiments, a 180° pulse is frequently used to invert z magnetization $M_z \to -M_z$ (for instance, 180° decoupling pulse) or to generate a spin echo, that is, change the sign of the coherence order, $p \to -p$ (see Sect. 4.10.3). An imperfect 180° refocusing pulse is a noticeable source of pulse artifacts that can be removed by the phase cycling method EXORCYCLE (Bodenhausen et al. 1977). Gradient pulses can also be used to eliminate artifacts from an imperfect 180° refocus pulses. For a 180° inversion pulse, the first gradient pulse is applied before the 180° pulse to dephase the transverse magnetization (Fig. 4.13a), whereas the longitudinal magnetization is not influenced by z gradient pulses. After the 180° the second gradient pulse with opposite sign is applied to dephase any transverse magnetization produced by an imperfect 180° excitation. The opposite sign is used to avoid refocusing the transverse magnetization dephased by the first gradient pulse so that it is continuously dephased during the second gradient.

To remove artifacts during the inversion of coherence order by an imperfect 180 refocusing pulse, the coherence of order p is selected by a pair of gradient pulses (Fig. 4.13b). The 180° pulse inverts the order of the coherence after the first

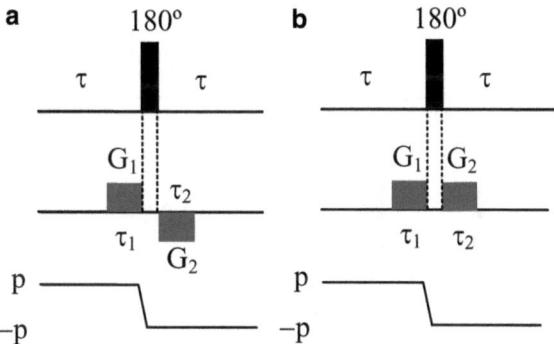

Fig. 4.13 Elimination via gradient pulses of artifacts due to imperfect 180° pulses. (**a**) Gradient-enhanced spin inversion sequence in which the artifacts are removed by a pair of z gradient pulses with opposite sign. The gradient pulses sufficiently dephase transverse magnetization while the longitudinal magnetization is not disturbed by the z gradient pulses. The opposite sign of the second gradient pulse is used to avoid refocusing any transverse magnetization dephased by the first one. (**b**) Gradient-enhanced spin echo sequence using gradient pulses to select the coherence echo so that the artifacts from the imperfect 180° pulse are removed. The gradient pulses are applied with the same strength and duration ($G_1\tau_1 = G_2\tau_2$) to accomplish the selection of the coherence with order of $-p$. Any coherence of order other than -1 generated by the 180° pulse is removed after the second gradient pulse

gradient pulse dephases the coherence. The second gradient pulse is applied to refocus the coherence according to the coherence selection rule:

$$G_1\tau_1 p + G_2\tau_2(-p) = 0 \tag{4.31}$$

$$G_1\tau_1 = G_2\tau_2 \tag{4.32}$$

The second gradient pulse is applied with the same strength, duration and sign as the first one. Any coherence with order other than $-p$ will sufficiently be removed during the second gradient pulse. For both methods, the artifacts are effectively eliminated by a single transient without any phase cycling. The reduction of phase cycling translates to faster data acquisition or higher digital resolution through the collection of more increments in the indirect dimensions. The selection of coherence transfer pathways by gradients is discussed in more detail in Sect. 4.10.3.

4.8 Solvent Suppression

In order to observe amide protons which contain rich structure information, it is necessary to dissolve protein samples in aqueous buffer solution containing only approximately 5% ^2H_2O for ^2H lock. It is difficult to observe the signals from the solute in the presence of the intense signal from water protons since water protons in 95% aqueous solution with a concentration of approximately 105 M have bulk

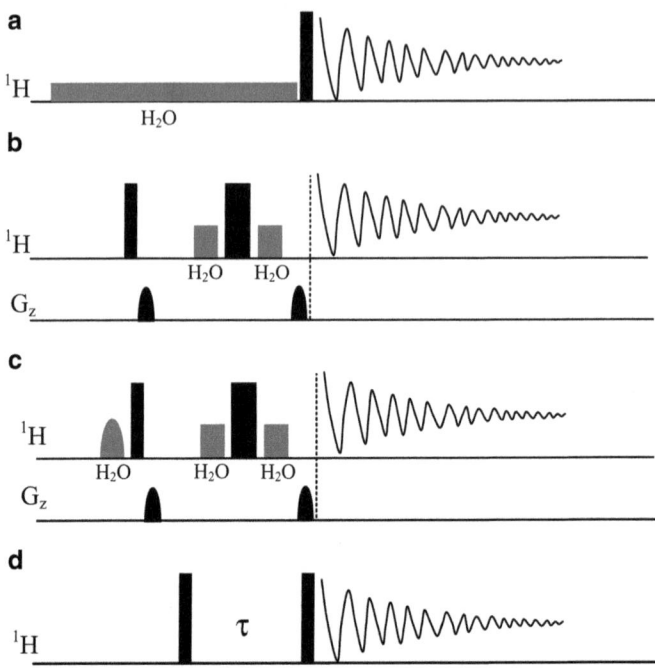

Fig. 4.14 Water suppression experiments. (**a**) PRESAT. Usually, the water saturation pulse is approximately 3 s. (**b**) WATERGATE. The 90° water selective pulses in gray can be shaped pulses (e.g., 3 ms sinc shape) or 1.7 ms rectangular pulses. (**c**) Water flip-back. The initial 90° water selective pulse is used to ensure water magnetization on the z axis after the hard 90° pulse to avoid radiation damping, which can be a 3 ms sinc or a 7.5 ms eburp1 shaped pulse. The two gradient pulses dephase and then refocus the solute magnetization, while the magnetization of residual water is continuously dephased by the gradient pulses because the pulse train formed by the two selective pulses sandwiching the hard 180° pulse keeps the water coherence order unchanged. The gradient strength is approximately 20 Gcm^{-1}. (**d**) Jump-return. The delay τ is set to $1/(4\Delta)$ in which Δ is the frequency difference between the water resonance and the center frequency of the interested region

magnetization approximately 10^4 greater than that from a type of protons in a protein. Therefore, the first task to do before acquiring data is to remove the solvent signal by a solvent suppression sequence. In the current section, some commonly used water suppression techniques are discussed.

4.8.1 Presaturation

There are various methods for water suppression, of which presaturation is the most basic and classical technique and is simple to use. A long and low power selective rectangular pulse is applied to saturate the water signal during the relaxation time (or predelay time) in the PRESAT pulse sequence shown in Fig. 4.14a.

4.8 Solvent Suppression

Because the quality of the water saturation experiment depends greatly on the homogeneity of the magnetic field, the quality of shimming will substantially influence the result of the water suppression. In the experiment, the frequency of the long pulse is set on the water resonance to saturate the water proton population. Because the chemical shift of water is near the center of the ^{1}H chemical shift range, the carrier frequency is also placed on the water resonance though it can be set to any frequency within the spectrum. The pulse is set to longer than 3 s for a 1D experiment or to the longest time allowed for 2D experiments (usually 1.0–1.5 s). The pulse power must be set sufficiently low for important reasons, one of which is because the lower the power, the better the selectivity of the excitation. A high power will saturate signals in the region close to the water resonance. Additionally, for such a long pulse, the probe could be damaged if the pulse power were not sufficiently low. Usually, the pulse power is set in the range of 100–150 Hz, approximately 2 ms for the 90° pulse. For instance, the correct setting of the power should be attenuated about −48 dB from the power for a hard 90° pulse length of approximately 7 μs, which is typically in the range of −59 to −55 dB lower than the maximum pulse power of a 50 W ^{1}H amplifier. A typical setup procedure includes calibration of the 90° pulse, setting the carrier on the water resonance frequency and adjustment of the receiver gain. The hard pulse is calibrated first using a one-pulse sequence according to the procedure described in the calibration section. The carrier frequency offset is set to the water peak as described previously. After the receiver gain is adjusted using a single transient, the data are collected with the proper number of transients for the desired signal-to-noise ratio. Occasionally, a composite 90° pulse is used to replace the 90° pulse so as to minimize the effect of radiation damping (Chap. 2). It has also been noted that solvent signals originating outside the main sample volume (outside the probe coil) degrade the solvent suppression.

4.8.2 Watergate

The PRESAT is a simple method and is easy to set up. However, it reduces or partially saturates signals from exchangeable protons (amide protons) due to the rapid chemical exchange and spin diffusion between amide protons and saturated water protons and signals from protons resonating near the water resonance (α protons), causing partial saturation of H^N spins and reducing the intensity of amide protons (saturation transfer). The exchange rate of amide protons with water protons linearly increases with the pH value of the solution over a pH range of 3–9 (Connelly et al. 1992). To overcome this drawback, more sophisticated methods have been developed to perform water suppression without saturating the water magnetization.

The WATERGATE sequence (Piotto et al. 1992) shown in Fig. 4.14b is a popular technique used for water suppression of aqueous samples. It uses gradient pulses to dephase magnetization of both water and solute, and then to refocus only the signals from the solute and further dephase the water magnetization. After the first hard 90° pulse brings the magnetization of all protons to the transverse plane, a 1 ms gradient pulse with an amplitude of approximate 20 Gcm^{-1} is applied to dephase the magnetization. The two 90° water selective pulses, which are either shaped or rectangular, are set on the water resonance. The net effect of the hard 180° pulse sandwiched by the two 90° water selective pulses is to keep the coherence order of water magnetization unchanged. Consequently, not only is the water magnetization dephased by the first gradient not refocused by the second, but the second gradient also further dephases the water magnetization. On the other hand, the second gradient refocuses the solute magnetization and produces a gradient echo because the hard 180° pulse changes the coherence order of the solute magnetization. The gradient echo used to dephase the water magnetization can sufficiently suppress water signal intensity compared to the presaturation method. However, it may still reduce the sensitivity of water exchangeable protons due to the partial saturation of water magnetization. Since the water magnetization is along the transverse plane at the time when acquisition starts, it needs much more time to completely relax back to its equilibrium state than do protein resonances due to its long T_1 relaxation time. Thus, the water will not fully relax back during the predelay time and will be partially saturated. The partial saturation will be transferred to the exchangeable protons by the saturation transfer, resulting in partial saturation of the H^N protons as reflected in a reduction of sensitivity for the desired resonances.

4.8.3 Water Flip-Back

An alternative version of Watergate, called water flip-back (Grzesiek and Bax 1993), uses a selective pulse on the water resonance before the Watergate sequence (Fig. 4.14c). The combination of the selective 90° and hard 90° brings the water magnetization back to the z axis before the first gradient pulse in the Watergate sequence. Because of the inhomogeneity of the 180° pulse, some magnetization from residual water is left on the transverse plane. The water flip-back sequence will suppress the magnetization from the residual water, whereas the majority of water magnetization stays on the z axis and is not disturbed to avoid radiation damping (Redfield et al. 1975). Consequently, the sensitivity for the exchangeable amide protons is increased because of the reduction of saturation transfer. Additionally, the amplitude of the required gradient pulses is considerably reduced to approximate 5 Gcm^{-1} for the gradient echo, which also increases the sensitivity of the experiment due to reduced effect of gradient pulses on the solute magnetization. In order to maximize the suppression of the water magnetization, the selective pulses must be well calibrated. To simplify the experimental setup, the flip-back selective pulse has a different shape and pulse length than the two Watergate selective pulses.

For instance, the first selective 90° pulse is a 90° eburp-1 pulse of 7.5 ms, whereas the last two can be either 1.7 ms rectangular 90° pulse or 3 ms sinc 90° pulses. The selective pulses are calibrated using the sequence described in the calibration section and the flip-back selective pulse is retuned in the water flip-back sequence by adjusting the fine power attenuation of the pulse.

In some experiments, distortion of the baseline occurs due to the fact that a substantial fraction of the FID is lost during the waiting time between the last pulse and acquisition (called the dead time). The time period is needed for the RF power of the high power pulse to go below the observable amplitude of the FID so the receiver is not saturated. A spin-echo sequence can be cooperated into any pulse sequence to prevent the FID from substantial loss during the dead time. The idea is to start the acquisition at the top of the first FID echo so that the full information of the FID can be retained. A typical value for the echo delays is 500 µs in the water suppression sequences, whereas the last delay is adjusted within 500 ± 50 µs to correct any phase errors in the spectrum.

4.8.4 Jump-Return

Although they provide sufficient water suppression, the gradient-based water suppression techniques and spin echo sequence may not produce observable signal for certain dilute samples because of partial loss in sensitivity during the experiment. For such samples, a better result can be achieved using the jump-return sequence shown in Fig. 4.14d (Plateau and Guéron 1982). The sequence is rather simple: a two-pulse sequence with two 90° pulses is separated by a delay. Like all other water suppression techniques, the carrier frequency is placed on the water resonance. The time of the delay is dependent on the spectral region of interest, which is determined by the frequency difference between the center of the region and the water resonance according to:

$$\tau = \frac{1}{4(v_c - v_w)} \quad (4.33)$$

in which v_c is the center frequency of region of interest and v_w is the water resonance. For example, the signals in the region of 7–10 ppm are expected to be observed. With the delay set to 111 µs for a 600 MHz spectrometer, the resonance at 8.5 ppm will precess 90° out of phase from the water resonance after the delay, whereas the water is on resonance. After the second 90° pulse that has a 180° phase shift from the first 90° pulse, the water is back to the z axis, whereas resonances in the region are detected with scaled intensity about the center frequency, v_c. Normally, the second 90° pulse length is required to be optimized to achieve the best solvent suppression.

4.9 NMR Data Processing

For data analysis, an FID recorded during data acquisition is Fourier transformed to generate a spectrum. Prior to and post FT, the data are treated through a series of data processing steps to optimize sensitivity or resolution. A typical data processing includes DC offset correction (or drift correction), application of a solvent suppression filter, linear prediction, apodization, Fourier transformation, phase correction, and spectral baseline correction (if necessary).

4.9.1 DC Drift Correction

An FID collected by quadrature detection contains DC (direct current) offset error originating from the receiver. A DC offset in the time domain data produces a spike at the carrier frequency after Fourier transformation. The DC correction is automatically applied by NMR data processing programs without user's attention.

4.9.2 Solvent Suppression Filter

Although the water signal is suppressed during experiments, the suppression does not completely eliminate the unwanted water peak and sometimes the residual water signal is sufficiently high to interfere with nearby resonances. Digital filtering the FID will further suppress the water peak. There are two types of popular methods for solvent suppression by digital filtering, namely low frequency suppression and zero frequency suppression. In the low frequency method, a low passband digital filter with the desired bandwidth and resonance frequency on the water frequency is applied to the acquired FID (Marion et al. 1989a, b). The filter only allows the water signal and signals within the bandwidth of the filter to pass through. This filtered FID is subsequently subtracted from the original FID. Fourier transformation of the treated FID gives the water suppressed spectrum. The suppression filtering leaves a flat line in the bandwidth region if the profile of the digital filter fits the water line shape. Three parameters need to be specified for the filtering: the bandwidth in Hz (50–200 Hz), the resonance frequency position offset (in Hz) from the carrier frequency (the center of the spectrum) and the number of coefficients for the digital filter, which define the passband profile of the filter. More coefficients produce a digital filter with a steeper passband (brick wall type).

The zero frequency method first filters the FID using the same digital filtering as in the low frequency method. Then the filtered FID is fit with a polynomial function using a specified order. The polynomial is subtracted from the original FID to eliminate the water contribution to the FID. This method only subtracts the water signal residing exactly on resonance. The water peak is removed from the spectrum after Fourier transformation. In addition to the parameters used in low frequency water suppression, the polynomial order needs to be specified, usually less than five. A brick wall type digital filter (with a higher number of coefficients) should be avoided in order to obtain reasonably fitting by the polynomial.

4.9.3 Linear Prediction

The first few points of the acquired time domain data are often distorted due to the imperfect condition of instrument hardware, such as tailing of RF pulses. Fourier transformation of the FID causes problems in the frequency domain such as baseline distortion. In addition, multidimensional data are always recorded with fewer points in the indirectly observed dimensions due to the long experimental time required. Collecting fewer data points in the indirectly observed dimension and more scans in the acquired dimension translates to higher S/N ratio for a given time. However, Fourier transformation of the truncated time domain data will severely distort the spectrum and reduce both sensitivity and resolution. Linear prediction is utilized to repair both types of FID distortions (Barkhuijsen et al. 1985). The algorithm analyzes the FID using the correct data points of the FID to obtain information on the frequencies present in the FID. Based on the analysis, it extrapolates more data points in the forward direction (forward prediction) to eliminate or reduce the truncation problem, or in the backward direction (back prediction) to replace the distorted initial points (in the observe dimension) or to extend a data point (in the indirect dimension). Mirror image linear prediction (Zhu and Bax 1990) uses the observed positive-time points to extrapolate negative-time points using the relationship of $f(-t) = f^*(t)$, in which * represents complex conjugation, resulting in doubling the size of the time-domain data points.

For the acquired dimension it may not be necessary to extend the data because it does not take long time to acquire the desired number of data points. However, backward prediction is sometimes used to correct the first few points that are influenced by the power of the last pulse. Forward predictions are always used for indirect dimensions because it provides more data points without actually acquiring the data so as to save experimental time (a more efficient way to acquire NMR data). The data points are usually doubled by the prediction (twofold prediction) because overextended data may introduce artifacts after Fourier transformation. The maximum number of points that can be used is the acquired data points for extending data points in either direction and the total data points minus the number of predicted points (usually the first one or two points) for replacing the distorted points in the backward predication.

4.9.4 Apodization

It is often necessary to manipulate time domain data (without altering frequencies) to improve the sensitivity or resolution of the spectrum. This can be achieved by the application of weighting functions. The basic idea is that according to the FT similarity theorem, multiplying a FID by a function results in changing the shape of spectrum, but not the frequency:

$$F'(\omega) = H(\omega)F(\omega)$$
$$F'(\omega) = Ft\{h(t)f(t)\} = Ft\{f'(t)\} \quad (4.34)$$

in which $h(t)$ is a weighting function (or window function). The process used to change the appearance of the data is called data apodization. Because the FID trends to decay to zero, the S/N ratio (amplitude of FID to noise level) in the tail part of the FID is much lower than that at the beginning. By applying a weighting function that smoothes the tail region of the FID, the noise level can be zeroed out so that the sensitivity of the spectrum can be improved. The signals in the tail region, if any, will also be reduced to zero. An exponential function is used to serve this purpose:

$$h(t) = e^{-t/a} \quad (4.35)$$

applying to the FID $f(t)$:

$$f(t) = e^{-t/T_2} \quad (4.36)$$

results in new time-domain data $f'(t)$:

$$f'(t) = h(t)f(t) = (e^{-t/a})(e^{-t/T_2}) = e^{-t/(T'_2)} \quad (4.37)$$

in which the apparent T'_2 is given by:

$$\frac{1}{T'_2} = \frac{1}{a} + \frac{1}{T_2} \quad (4.38)$$

This states that after Fourier transformation the line shape is broadened by the application of the exponential function with the line width increased by a factor of 2 (i.e., $T'_2 = T_2/2$) when $a = T_2$. Therefore, the exponential function decreases the resolution of the spectrum but increases sensitivity by removing the noise level at the tail of the FID. On the other hand, the sensitivity enhancement of the exponential function is reduced when the line broadening parameter a becomes larger. Exponential function is usually applied as a matched filter with $a = T_2$ which has a decay profile matched to the FID. Alternatively, if the exponential function can be applied with a negative a, the line width should be reduced as the reverse effect to the line broadening. The line shape of the spectral peak can be narrowed by a factor of 2 (i.e., $T'_2 = 2T_2$) when $a = -2T_2$. However, as its profile indicates, the function with a negative a amplifies the noise level at the end part of the FID and reduces the signal amplitude at the beginning part of FID, resulting in a significant loss in sensitivity. Resolution enhancement by applying these types of weighting functions will certainly degrade the sensitivity of the spectrum because the functions reduce the amplitude of the FID at the beginning. On the other hand, to increase sensitivity by smoothing the end part of FID, both signal and noise level are reduced and the

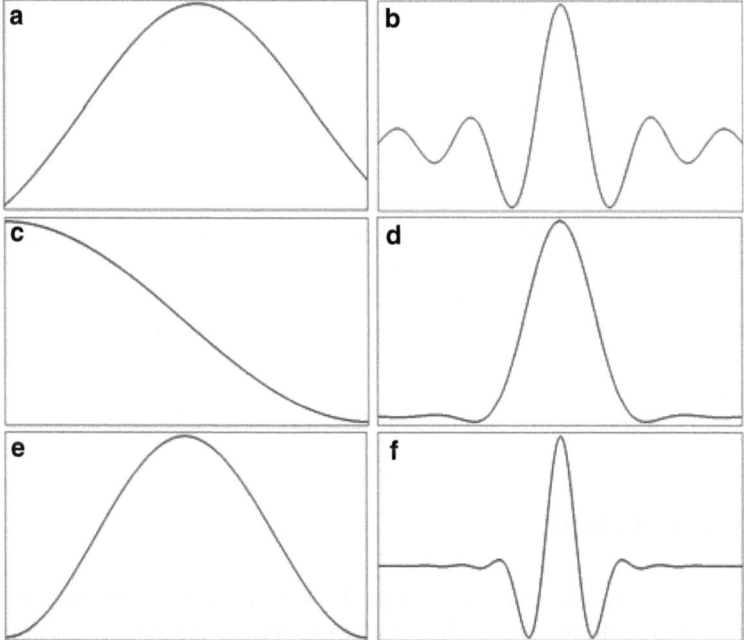

Fig. 4.15 Common weighting functions for apodization. (**a**) Lorentz–Gauss function with the parameters $a = -0.5$, $b = 1.5$, and (**b**) its Fourier transformation. (**c**) Squared 90° shifted sine bell function and (**e**) squared sine bell function, and their Fourier transformations (**d**) and (**f**), respectively

apparent T'_2 is decreased. As a consequence, the resolution of the spectrum is sacrificed. In general, sensitivity and resolution of an NMR spectrum cannot be improved simultaneously by means of applying a weighting function (or window function). Enhancement in one of the aspects has to be achieved at the expense of the other.

A usual solution to the problem is to combine the negative exponential function with another sensitivity enhancement function such as Gaussian function:

$$h(t) = e^{-t/a} e^{-t^2/b} \qquad (4.39)$$

Compared to the natural line shape of NMR spectral peaks (Lorentzian line shape) with the same half height width, a Gaussian function has a much sharper line shape. While negative a gives resolution enhancement, b (always positive) provides sensitivity enhancement to compensate for the loss of sensitivity caused by a being less than zero. The optimization of parameters a and b of the Lorentz–Gauss function can increase resolution with a minimum loss of sensitivity and transforms the Lorentzian line shape to a narrow Gaussian line shape.

The Lorentz–Gauss function has a profile shown in Fig. 4.15 with the maximum in meddle of the function. This leads to the application of other functions with a similar profile but less reduction to the beginning of FID. Among them, the phase shifted or unshifted sine-bell and squared sine-bell functions are commonly used for multidimensional data to achieve the desired sensitivity or resolution enhancement. Application of weighting functions such as Lorentz–Gauss, unshifted sine-bell, or unshifted squared sine-bell functions (Fig. 4.15) increases the amplitude of middle part, reduces the beginning, and smoothes the tail of the FID, producing an FID with longer decay (flatter FID). The corresponding spectral peak has a narrow line shape and hence the spectral resolution is increased. For 1D data, Lorentz–Gauss is commonly used if resolution needs to be improved, otherwise, an exponential function with $a = T_2$ is applied. For multidimensional data, an unshifted sine-bell or unshifted squared sine-bell function is utilized in the COSY to gain resolution enhancement, whereas the shifted weighting functions are used for all other 2D, 3D, and 4D data to improve the sensitivity. A 90°-shifted function provides maximum sensitivity enhancement.

4.9.5 Zero Filling

Data points collected in an FID are necessarily limited to just enough points to cover the FID to avoid collecting noise. However, a spectrum needs more data points to be able to characterize the individual peaks with the correct digital resolution. Digital resolution in the frequency domain is defined as SW/f_n, in which SW is the spectral window in Hz and f_n is the number of frequency domain points (a spectrum consists of f_n data points in the real part and f_n data points in the imaginary part). Additionally, $f_n = n_p/2$, in which n_p is the number of time domain points. By adding more points with zero amplitude to the end of the FID after data are collected (called zero-filling), the digital resolution is improved because f_n becomes larger after the zero-filling. The digital resolution without zero-filling is $SW/(½n_p) = 1/(DW \times n_p) = 1/at$, in which at is acquisition time and DW is dwell time.

Exchanging the order between apodization and zero-filling (i.e., the data are zero-filled first and then the apodization is applied) does not affect the result of the data processing because the amplitudes of the zero-filled data points are still zero after the apodization. However, the apodization after the zero-filling takes more computing time. Although this extra computing time is unnoticeable in 1D and 2D data processing, it may be substantial during 3D and 4D data processing. Therefore, the apodization is usually applied before zero-filling.

4.9.6 Phase Correction

Fourier transformation of an FID gives a spectrum which can be represented by:

$$F(\omega) = R(\omega)\cos(\theta) + I(\omega)\sin(\theta) \qquad (4.40)$$

4.9 NMR Data Processing

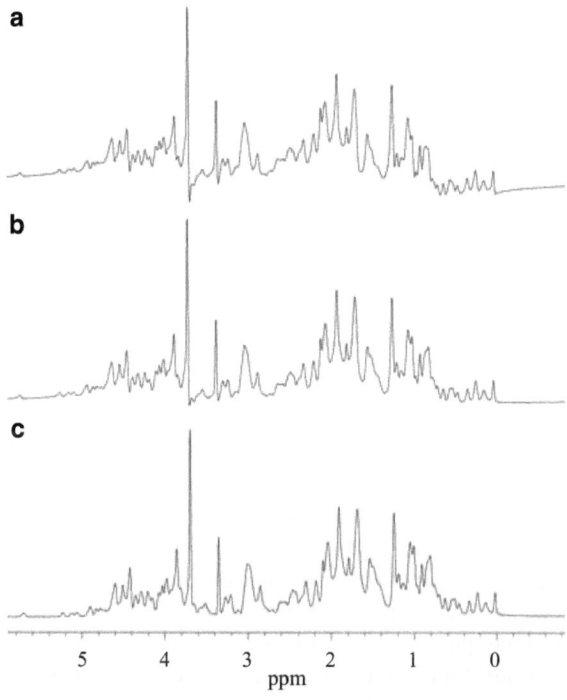

Fig. 4.16 Phase correction of spectrum. (**a**) The spectrum contains a zero order phase error of 40°. All peaks are distorted by the same amount. (**b**) The spectrum contains a first order phase error of 40°. The degree of distortion of the spectrum is frequency dependent and peaks at downfield are off more than those in the upfield region. (**c**) Spectrum with correct phase

in which R and I are the real and imaginary parts of the spectrum $F(\omega)$, respectively, and θ is the phase error of the spectrum. Ideally, $\theta = 0$, that is, real part is pure absorption and the imaginary part is pure dispersion. In practice, $\theta \neq 0$ for most cases, because the FID does not start with the maximum amplitude. There are many factors causing the problem, one of which is that quadrature detection has a phase error and does not acquire pure cosine and sine components of the FID. This causes an equal amount of phase error on all resonances of the spectrum and is called zero order phase error (Fig. 4.16a). The other type is first order phase error typically introduced by the delay between the end of last pulse and the start of acquisition (dead time, or receiver gating time). This delay is unavoidable because of tailing problem of the RF pulse. During the dead time, the amplitudes of time domain signals with different frequencies are reduced by different fractions, leading to a frequency-dependent phase error (Fig. 4.16b). The total phase error is the linear combination of the two types of phase errors:

$$\theta(\omega) = \phi_0 + (\omega - \omega_0)\phi_1 \qquad (4.41)$$

in which ϕ_0 and ϕ_1 are the zero order and first order phase error, respectively, and can be corrected by the phase correction function of a data processing programs.

It should be noted that an experiment can be set up in such a way that the FID is acquired without any phase error (both ϕ_0 and ϕ_1 are zero). Adjusting the dead time to ensure all signals start at a maximum amplitude eliminates the first order phase error.

Table 4.2 Methods of quadrature detection in the indirect dimension

Methods	DW	Exp. per incr.	Phase shift	τ	Phase correction	References
States	1/SW	2	90° for same incr. 0° per incr.	DW/2	90°, −180°	States et al. (1982)
TPPI	1/(2 × SW)	1	90° per incr.	DW	90°, 0°	Marion and Wuthrich (1983)
States-TPPI	1/SW	2	90° for same incr. 90° per incr.	DW/2	90°, −180°	Marion et al. (1989a, b)

This can be done by simply optimizing the dead time to the point that the first order phase error is zero, which is usually longer than the value set by the instrument software (rof2 and alfa on an Agilent spectrometer, de on a Bruker spectrometer). The drawback is loss in sensitivity if the delay is long. Alternatively, in a better method, a spin echo sequence is used during the normal delay period which refocuses all coherence at any desired time so that the FID is always starts with the maximum amplitude. The zero order phase error can be corrected by setting the receiver phase after the first order phase error is eliminated in the above procedure. The adjustment of the receiver phase can be done by the spectrometer software or by adding an adjustable phase parameter to the receiver phase table in the pulse sequence. Note that every time probe tuning is adjusted, zero-order phase is changed. After the above adjustment of both zero order and first order phases, the acquired FID does not contain any phase error. Therefore, phase correction after Fourier transformation is not necessary.

The phase error for the indirect dimension can be corrected after the t_1 Fourier transformation (see the following section). However, the first order phase error in the indirect dimension causes such problems as baseline distortion and dispersive folded cross-peaks, which significantly degrade the sensitivity of the spectrum. To avoid these problems, the phases can be set before the acquisition according to the relationships:

$$\phi_1 = -\tau \times SW \times 360°$$
$$\phi_0 = -\frac{1}{2}\phi_1 \quad (4.42)$$

in which SW is the spectral window of the indirect dimension, and τ is the sampling delay, which is the sum of all delays and 180° pulses in the evolution time. The 90° pulses that flank the evolution time are also included in the sampling delay in terms of the 90° pulse length multiplied by $2/\pi$. The sampling delay τ is set to $\tau = 0$ or $\tau = DW/2$ (DW is the dwell time of the indirect dimension. See Table 4.2 for the relationship between SW and DW in the t_1 dimension), resulting in $\phi_0 = 0°$, $\phi_1 = 0°$ for $\tau = 0$, and $\phi_0 = 90°$ and $\phi_1 = -180°$ for $\tau = 1/(2\ SW)$.

4.9 NMR Data Processing

The conditions can be met by adjusting the initial value of t_1 as described in the following pulse sequence elements:

(a) Homonuclear evolution period $90° - t_1 - 90°$
The sampling delay is given by:

$$\tau = t_1(0) + \frac{4 \times pw_{90}}{\pi} = \frac{DW}{2} \quad \text{for } \phi_0 = 90° \text{ and } \phi_1 = -180° \quad (4.43)$$

The initial t_1 value is calculated according to:

$$t_1(0) = \frac{DW}{2} - \frac{4 \times pw_{90}}{\pi} \quad (4.44)$$

in which $t_1(0)$ is the initial value of t_1, and pw_{90} is the 90 pulse length. The effect of the phase variation during the flanking 90° pulses can be removed from the sampling delay in the pulse sequence element $90° - t_1 - \delta - 180° - \delta - 90°$, in which δ is a fixed delay. Because the 180° pulse is placed symmetrically in the initial evolution period ($t_1 = 0$) and refocuses the effect of the phase variation, the sampling delay is zero. Consequently, $t_1(0)$ can be set to $t_1(0) = DW/2$ or $t_1(0) = 0$. For $t_1(0) = 0$, the first point of the interferogram is usually scaled by a factor of 0.5 prior to Fourier transformation (Otting et al. 1986).

(b) HMQC element

- With spin decoupling $90°(X) - \frac{1}{2}t_1 - 180°(^1H) - \frac{1}{2}t_1 - 90°(X)$
  The sampling delay is given by:

$$\tau = t_1(0) + pw_{180(^1H)} + \frac{4 \times pw_{90(X)}}{\pi} = \frac{DW}{2} \quad (4.45)$$

$$t_1(0) = \frac{DW}{2} - pw_{180(^1H)} - \frac{4 \times pw_{90(X)}}{\pi} \quad (4.46)$$

in which $pw_{180(^1H)}$ and $pw_{90(X)}$ are the pulse lengths of a ^1H 180° pulse and a heteronuclear 90° pulse, respectively.
- Without spin decoupling $90°(X) - t_1 - 90°(X)$
  The initial $t_1(0)$ is set to:

$$t_1(0) = \frac{DW}{2} - \frac{4 \times pw_{90(X)}}{\pi} \quad (4.47)$$

- Constant time $90° - \frac{1}{2}t_1 - T - 180° - (T - \frac{1}{2}t_1) - 90°$.
  Because the refocusing 180° pulse is placed symmetrically in the initial evolution period ($t_1 = 0$), the sampling delay is zero. As mentioned earlier, $t_1(0)$ can be set to either $DW/2$ or zero.

4.10 Two-Dimensional Experiments

4.10.1 The Second Dimension

The signals of protons in a protein sample recorded in 1D NMR experiments are not able to be completely assigned due to severe overlapping of the signals. To improve the resolution for spectral assignment, a second frequency dimension is introduced to disperse the signals over two frequency dimensions, forming a 2D NMR spectrum. In a 1D experiment, the FID is acquired after an RF pulse or pulses (called the preparation period). In a 2D experiment, one additional period called the evolution time, which contains a variable time delay t_1, is introduced into the experiment between the preparation and acquisition periods (Fig. 4.17). The evolution delay increases systematically by the same amount of time for each increment during a 2D experiment from zero to the final value determined by the number of increments. An FID is acquired for one increment with the starting value of the t_1 evolution time. For next increment, an FID is acquired with an increased t_1 and all other parameters the same as the previous FID, and so on for the required number of increments. After the experiment is done, all the FIDs are transformed with the same phase parameters. The phase of the signals after Fourier transformation is modulated as a function of the t_1 evolution time. As a result, the amplitudes of the signals after Fourier transformation vary sinusoidally as a function of the t_1 evolution time, which forms an interferogram similar to a FID when the

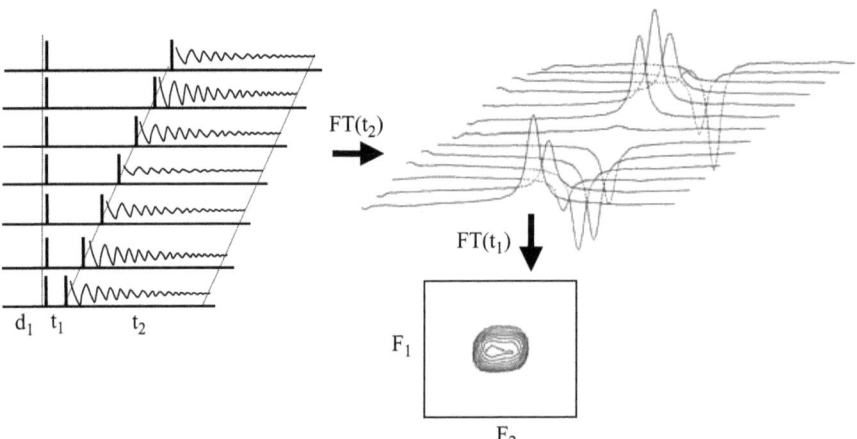

Fig. 4.17 Two-dimensional experiment and 2D spectrum. The evolution time t_1 is increased systematically for each of the FIDs. The FIDs acquired during t_2 are Fourier transformed to obtain a set of 1D spectra. The second Fourier transformation is applied along t_1, that is, each point of the spectra, resulting in a cross peak coordinated by both frequencies on F_2 and F_1. The F_2 dimension has better digital resolution because t_2 is always longer than t_1. The parameter d_1 is the relaxation delay (or predelay) and t_2 is the acquisition time

amplitude of a peak is plotted as a function of t_1 (Fig. 4.17). Fourier transformation of the second FID obtained by the t_1 evolution time generates another frequency domain as the transformation of the acquired FIDs. The result of the two Fourier transformations is a 2D NMR spectrum with two frequency axes and an intensity axis on the third dimension that is usually plotted as contours. The second dimension can be frequency for ^{1}H, which gives a square spectrum with diagonal peaks, or a heteronucleus which gives an asymmetric spectrum. Two dimensional experiments will also contain other periods in addition to the evolution time such as a mixing period. However, it is the evolution time that introduces the additional dimension into a spectrum. Similarly, more dimensions can be formed by introducing additional evolution time periods. The widely accepted terminology is that for nD data, $t_1, \ldots, t_{n-1}$ are evolution delays, and t_n is the acquisition period, whereas $F_1, \ldots, F_{n-1}$ are the indirect dimensions generated by the corresponding evolution times and F_n is the direct dimension from the observed FID. For instance, in a 3D spectrum the F_3 dimension is obtained from the transformation of the observed FID or t_3, which is the direct dimension, whereas the F_1 and F_2 dimensions are from the evolution times t_1 and t_2, respectively, and hence the indirect dimensions.

Before setting up a 2D experiment, calibrations of pulses, water resonance frequency, and optimization of the spectral window should first be performed. The spectral window is set to the value optimized in the 1D spectrum. The acquisition is generally set to about 200 ms for homonuclear experiments or 64 ms for heteronuclear experiments with ^{15}N or ^{13}C decoupling during acquisition. If each FID is acquired with the number of points, n_p and number of increment is n_i, the final size of the data file is $n_p \times n_i$. Assuming that each transient of a FID takes n seconds to finish (n is called the recycle time) and n_t transients are taken for each FID with n_i increments, the total experiment time will be $n_t \times n \times n_i$. A 2D experiment can take as little time as 2–15 min to finish (e.g., gCOSY, gHSQC). Instrument software can be used to calculate the total experiment time before hand.

4.10.2 Quadrature Detection in the Indirect Dimension

In order to correct the phase distortion of cross peaks, quadrature detection in the indirect dimension is necessary to obtain a phase sensitive 2D spectrum. This is achieved by hypercomplex or TPPI methods similar to quadrature detection in the observed dimension (see Sect. 2.4). In the hypercomplex method (also known as the States method; States et al. 1982), two FIDs are recorded for each t_1 increment. In the second FID, the phase of the preparation pulse or pulses (the pulse or pulses prior to the t_1 evolution period) shifts by 90° phase (Fig. 4.18). Two sets of data are obtained: one with 0° phase and the other with 90° phase. After the first Fourier transformation, the t_1 interferogram is constructed by taking the real part of the 0°-phase data (open circles in Fig. 4.18) as the real part of the interferogram, and the real part of the 90°-phase data (filled circles) as the imaginary part of the interferogram.

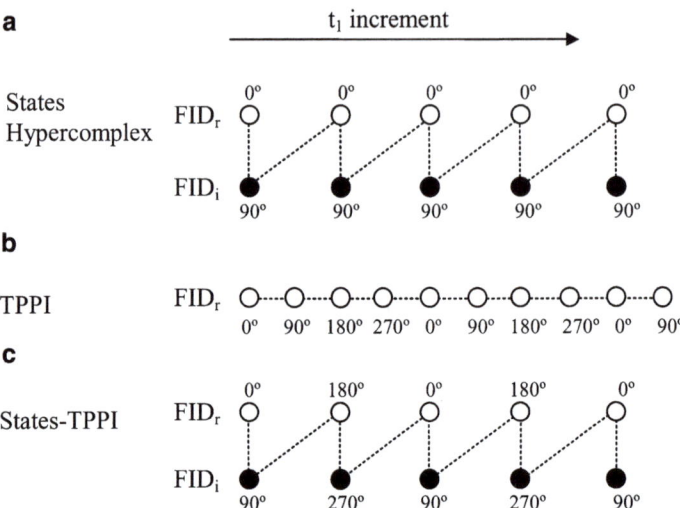

Fig. 4.18 Quadrature detection methods in the indirect dimension. Each *circle* represents an FID: the *open circles* represent the FIDs used for the real part (FID$_r$) of the t_1 interferogram. The *filled circles* represent the FIDs used for the imaginary part (FID$_i$) of the t_1 interferogram. The values above or below the circles represent the relative phases of the pulse (or pulses) before the t_1 evolution period in a pulse sequence. The *dotted lines* represent the order of FIDs being acquired. (a) The States or hypercomplex method. For each t_1 increment, two FIDs are acquired. After t_2 Fourier transformation (FT), the real parts of the open-circled FIDs form the real part of the t_1 interferogram and the real parts of the filled FIDs form the imaginary part of the t_1 interferogram. Complex FT is then applied to transform t_1 to the F_1 dimension. (b) TPPI method. The t_1 is incremented for every FID along with a 90° shift of the relative phase of the pulse (or pulses) before the t_1 evolution period. The real parts of the FIDs form the real part of the t_1 interferogram, which is then transformed by a real Fourier transformation. (c) States-TPPI method. The data are acquired in the same way as the States method except the relative phase is increased by 90° after each FID is acquired. As a result, the axial peaks are suppressed by shifting them to the edge of the spectrum. Complex FT is used in both dimensions

Therefore, both parts of the interferogram are from absorptive data. A phase sensitive spectrum is obtained after the interferogram is transformed by the complex Fourier transformation. The t_1 increment is set to $1/(2 \times SW_1)$, in which SW_1 is the spectral window in the F_1 dimension.

In the TPPI method (Fig. 4.18b; Marion and Wuthrich 1983), the phase of the preparation pulse shifts 90° for every t_1 increment. After the first Fourier transformation, the t_1 interferogram is constructed by taking the real part of all FIDs as real part and the interferogram is transformed by the real Fourier transformation, resulting in a phase sensitive spectrum. Similar to the sequential acquisition method, the TPPI method requires twice as fast t_1 increment as the hypercomplex method. Therefore, the t_1 increment is set to the $1/SW_1$. The advantage of TPPI relative to the hypercomplex method is that axial peaks appear at the edge of F_1 dimension ($F_1 = 0$), whereas axial peaks are located in the center of the 2D spectrum with the States method, which overlap with the signals of interest.

However, TPPI has the disadvantage of undesirable folding properties (Marion and Bax 1988). To overcome this disadvantage while still suppressing the axial peaks, a combination of the States and TPPI methods was developed (Marion et al. 1989a, b).

The States-TPPI method combines the two methods together as shown in Fig. 4.18c. The data are acquired in the same way as the States method except that the phase of the preparation pulse shifts 90° after each FID is acquired. As a result, the axial peaks are suppressed and the spectrum does not have the TPPI folding properties. The t_1 increment is set to $1/(2 \times SW_1)$, the same as in the States method. Complex Fourier transformations are used in both dimensions.

4.10.3 Selection of Coherence Transfer Pathways

When quadrature detection is used, only coherence with the order of either $+1$ or -1 can be observed (single quantum coherence), but not both at the same time. The common choice assumes that the coherence order of -1 is observed in all experiments by quadrature detection. An initial non-90° pulse generates higher order coherence in addition to the single quantum coherence. Therefore, there exist many other coherence transfer pathways along with the desired pathway at the end of a pulse sequence, which produce observable signals during acquisition time. To select the specific coherence transfer pathway and eliminate the unwanted signals from the other pathways, either phase cycling or gradient pulses are usually utilized.

The change in the phase of the coherence caused by an RF pulse or gradient pulse depends on its coherence order p. The selection of the coherence pathway is based on this property of the coherence phase. Phase cycling uses changes in the phase of RF pulses which lead to different phase shifts of the coherence. If the phase of an RF pulse is set to ϕ, the phase shift gained by the coherence experiencing a coherence order change Δp by the pulse is given by $-\phi \Delta p$. A particular coherence transfer pathway can be selected by phase cycling using N steps with increments of $\phi = 360°/N$. Phase cycling is the process in which the phases of pulses in a pulse sequence are shifted after each acquisition during the signal averaging (detailed treatment of phase cycling methodologies can be found in Bodenhausen et al. 1984; Bain 1984). The pathway selection by phase cycling requires precise phase shifts of RF pulses and repeated acquisitions with different phase shifts of pulses. Usually, it is not the method of choice. Coherence selection by gradient pulses is more commonly used in multidimensional NMR experiments to reduce experimental time in addition to other advantages.

The coherence selection by gradient pulses is achieved by applying a gradient pulse to dephase and then refocus a specific coherence, while allowing the unwanted coherence to continue dephased. Unlike the phase cycling method, phase changes by gradient pulses only depend on the coherence order at the time the gradient pulse is applied, not the difference of two coherence orders.

The spatially dependent phase induced by a gradient pulse with field strength G applied on a coherence of order p for a duration τ is given by:

$$\phi = G\tau\gamma p \tag{4.48}$$

in which γ is the gyromagnetic ratio of nuclear isotope. In the case of a heteronuclear system and a shaped gradient, the phase is described by:

$$\phi = sG\tau\gamma p \tag{4.49}$$

in which s is the shape factor that describes the amplitude profile of gradient pulse.

Considering a pair of simple dephasing-refocusing gradient pulses to select a specific homonuclear coherence transfer pathway from p_1 to p_2, the dephasing gradient pulse of duration τ_1 prior to the RF pulse induces a phase of coherence with order p:

$$\phi_1 = s_1 G_1 \tau_1 p_1 \tag{4.50}$$

Similarly, the refocusing gradient pulse of duration τ_2 following the pulse generates a phase:

$$\phi_2 = s_2 G_2 \tau_2 p_2 \tag{4.51}$$

The condition for the selected coherence to be refocused at the end of the refocusing gradient pulse is that the overall phase change ($\phi_1 + \phi_2$) after the gradient pulses becomes zero. Because usually the gradient pulses are applied with the same shape in a pulse sequence, the shape factor s may only be different by the sign of the gradient, $+$ or $-$. For the use of two gradient pulses, the selection condition can be rearranged as:

$$s_1 G_1 \tau_1 p_1 = -s_2 G_2 \tau_2 p_2 \tag{4.52}$$

This states that the desired coherence transfer pathway is achieved by adjusting the duration, amplitude and the sign of the gradient pulses and the unwanted pathways can be removed when the condition is not met for other coherences, provided the amplitude of the gradient pulses is sufficiently high to completely dephase the unwanted coherences.

When more than two gradient pulses are used for the coherence selection, the refocusing condition is also the same that the overall phase induced by all gradients for the selected pathway is zero:

$$\sum_i \phi_i = \sum_i s_i G_i \tau_i p_i = 0 \tag{4.53}$$

This condition applies to all combinations of gradients used in the coherence selection. When more than two gradients used, the combination for dephasing

4.10 Two-Dimensional Experiments

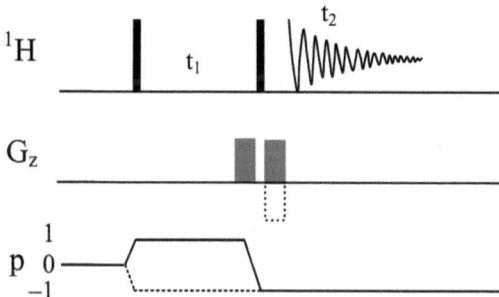

Fig. 4.19 COSY pulse sequence with coherence transfer pathway diagram. The general points for coherence selection using coherence transfer pathway are described as follows. The coherence transfer pathway must start at the equilibrium state in which the magnetization is along the field direction, z magnetization, represented by coherence order of $p = 0$. The first pulse acting on the equilibrium z magnetization gives rise to only single quantum coherence $p = \pm 1$. Delays without an RF field do not change the coherence order. Finally, the pathway must terminate at $p = -1$ as quadrature detection is used to observe the complex signals. The *solid line* in the coherence transfer pathway diagram and the solid gradient represents N-type selection, whereas the *dotted* coherence pathway and the open gradient is P-type selection

and refocusing can be different. For instance, the coherence can be continuously dephased by many gradients until the final gradient right before acquisition refocuses the desired coherence. Alternatively, the set of gradients can be used to select one part of the preferred pathway, whereas others select the other part of the pathway. More examples are discussed in detail for 2D experiments in the following sections.

4.10.4 COSY

Shown in Fig. 4.19 is the gradient version of the 2D *C*orrelation *S*pectroscop*y* experiment, abbreviated as COSY, which is one of the earliest 2D experiments. It has the simplest 2D pulse sequence consisting of two 90° pulses separated by the evolution time, t_1. The first 90° pulse rotates the equilibrium magnetization to the transverse plane and generates single-quantum coherence. During the evolution time t_1, the single-quantum coherence evolves, resulting in F_1 frequency labeling of the detected coherence. The last pulse transfers the magnetizations between spins via the scalar coupling between them. Finally, the correlated coherence is detected as the FID. The coherence selection is achieved by the gradients pulses following each RF pulse.

There are two possible pathways for acquiring COSY spectra, as shown in Fig. 4.19. A pathway through $p = -1$ is known as P-type (or anti-echo) in contract to that through $p = +1$ known as N-type (echo). The frequency modulations of the coherence during t_1 and t_2 have the same sign for the P-type pathway, whereas they have opposite sign for the N-type pathway. The N-type COSY is recorded using the last gradient with the same sign as the first one because the refocusing condition is obtained by $G_1 p_1 = -G_2 p_2$, in which $p_1 = 1$ and $p_2 = -1$, assuming that both gradient pulses have the same amplitude, shape, and duration. The P-type spectra can be similarly obtained by setting the refocusing gradient pulse with the sign opposite to the first one with the same duration and amplitude. The selection of either the N- or P-type pathway by gradient pulses results in frequency discrimination in the F_1 dimension without using any of the quadrature detection methods. However, the two pathways are not usually selected simultaneously.

Both pathways give rise to phase-twisted line shapes containing the superposition of absorptive and dispersive signals, which is caused by the use of the gradient during the evolution period. A phase sensitive spectrum can be constructed by acquiring both types of data separately and then combining them in such a way that the axis-reversed N-type spectrum is added to P-type. In the resulting spectrum, the dispersive portion due to the phase twist is canceled and complete absorptive line shapes are obtained (see quadrature detection in the indirect dimension). The phase sensitive spectrum acquired by this method has a reduction in sensitivity by a factor of $2^{1/2}$ compared to that by States, States-TPPI or TPPI methods due to the cancelation of the absorptive and dispersive signals. Alternatively, N-type and P-type data can be combined in a different way to obtain a phase sensitive spectrum. Addition of the two data sets generates cosine-modulated data, whereas subtraction of the two yields sine-function data. Hypercomplex data are formed by using the constructed cosine and sine portion of the data, which can be processed in the same way as the States method.

Digital resolution in the F_1 dimension also has a great influence on the sensitivity of the COSY spectrum owing to the antiphase character of the cross peaks. Because the cross peaks contain scalar coupling, the last t_1 increment must have a t_1 value ($t_{1,\max}$) greater than $1/(4J)$, whereas t_2 is set to a value comparable to $1/(2J)$. Usually, $t_{1,\max}$ is chosen in the range of 50–80 ms and t_2 is approximately 200 ms. For example, the data collected with 2,048 complex points and 600 increments yields $t_{1,\max}$ of 50 ms and t_2 of 170 ms for spectral window of 10 ppm on a 600 MHz instruments. Linear predication is used to add sufficient data points for both the F_1 and F_2 dimensions.

4.10.5 DQF COSY

The double quantum filtered (DQF) COSY is sometimes preferred over the COSY experiment because the intense peaks from uncoupled singlets are substantially attenuated and resolution in the region adjacent to the diagonal peaks is dramatically improved. The advantages of DQF COSY also include correction of the phase distortion present in COSY and absorptive line shapes for both cross peaks and diagonal peaks. The drawback of DQF COSY is reduction in sensitivity, thus requiring longer experimental time. The basic pulse sequence of DQF COSY consists of three 90° pulses (Fig. 4.20). The single quantum coherence generated

4.10 Two-Dimensional Experiments

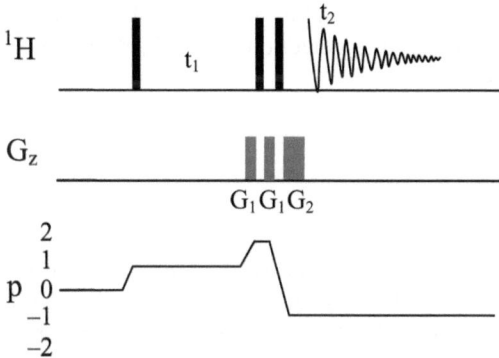

Fig. 4.20 Double-quantum filtered COSY pulse sequence with coherence transfer pathway diagram. The RF pulses are 90° pulses. The double-quantum coherence is selected by the second gradient with the coherence order $p_2 = 2$. The ratio of the gradients is $G_2 = 3G_1$. The N-type coherence pathway is shown in the coherence transfer pathway diagram

by the 90° pulse evolves during t_1. After the second 90° pulse transfers the coherence into double quantum coherence, the last 90° pulse produces observable single quantum coherence. Only double quantum coherence will be selected by the gradient pulses. The last gradient pulse refocuses the double quantum coherence dephased by the first two gradient pulses, leaving other coherence to continue dephased. According to the selection rules, the overall phase created by the gradient pulses must be zero. Assuming that the gradient pulses have the same duration, then

$$p_1 G_1 + p_2 G_2 + p_3 G_3 = 1G_1 + 2G_1 - 1G_2 = 0 \tag{4.54}$$

$$G_2 = 3G_1 \tag{4.55}$$

The spectrum may contain phase errors caused by the application of the gradient pulse during t_1 evolution. Higher order multiquantum filter experiments can also be implemented using the coherence selection rule. For instance, the detectable triple quantum filtered coherence is refocused by the condition:

$$G_1 + 3G_1 - 1G_2 = 0 \tag{4.56}$$

$$G_2 = 4G_1 \tag{4.57}$$

The above selection of double quantum coherence chooses the N-type coherence pathway ($p = +2$), whereas the P-type pathway with $p = -2$ is not selected by the gradient pulses. As a result, half of the initial magnetization does not contribute to the detected magnetization, causing a loss of signal compared to the data acquired by the phase cycling method. An absorption mode spectrum can be obtained by acquiring P- and N-type data and then combining the data as

4.10.6 TOCSY

In a 1D ^{1}H spectrum of a protein, the side chain region (up-field region) is very crowded with severe spectral overlap due to the large number of side chain protons. In the contrast, the chemical shifts of H^N protons are less crowded, especially for small proteins, and the peaks are sometimes well resolved. It is useful in resonance assignment to transfer the correlations of side chain protons to the region of H^N proton chemical shifts. TOCSY (also known as HOHAHA; Braunschweiler and Ernst 1983; Bax and Davis 1985a, b) experiments were developed to obtain, through coherence transfer via ^3J scalar couplings, the relayed correlations between spins within a spin system, a network of mutually coupled spins. COSY experiments provide the information about the correlation between spins, whereas TOCSY transfers the coherence to other coupled spins through the ^3J coupling (through molecular bonds) by isotropic mixing. During the mixing period, the coherence along an axis on xy plane is transferred throughout the spin system under the interaction of scalar coupling, at the same time the multiquantum coherence is dephased by the inhomogeneity of RF pulses in the isotropic mixing pulse train. Several methods for the isotropic mixing have been used in different types of TOCSY experiments (Sect. 4.5.2). Much effort has been spent on the development of mixing pulse sequences (also known as spin lock sequences) that use minimal RF power to produce wide isotropic frequency ranges. DIPSI series mixing sequences are most commonly used now, among others such as Mlev17, Waltz16, and clean-TOCSY.

During the mixing time, the magnetization of amide protons is transferred to other protons within amino acid residues through three-bond scalar coupling. Because there is no through-bond scalar coupling between the interresidual amide proton H^N and α proton H^α, the magnetization cannot be transferred across the peptide bond. The transfer distance (the number of protons the H^N magnetization can reach) depends on the efficiency of the spin lock sequence, length of mixing time, the coupling constants, and the relaxation rate of the molecule. Protons further away from the H^N proton require a longer mixing time to reach. Ideally, it is preferred to obtain magnetization transfer to every proton within the spin system. However, it is practically difficult to acquire a single TOCSY with a mixing time to optimize magnetization transfer from H^N to all other protons. The magnitude of transfer to neighboring protons decreases quickly as the mixing time exceeds than 30 ms. For remote protons, the magnetization transfer is achieved with the mixing time as long as 100 ms. The correlations between H^N and H^α, H^β can be observed in a TOCSY with a mixing time of 30 ms, whereas a 60 ms or longer mixing time is required to achieve magnetization transfer from H^N to other side chain protons. Normally, a TOCSY experiment is first acquired with a mixing time in the range of 30–60 ms to efficiently transfer H^N magnetization to the neighboring side chain protons. Then, the experiment is repeated with a longer mixing time to favor the long range magnetization transfer of H^N magnetization to the end protons of the side chain, but is usually less than 100 ms to avoid magnetization loss caused by relaxation.

TOCSY data are collected in a phase sensitive mode in the F_1 dimension, for example. States-TPPI, with 512 increments which yield 256 complex points in the F_1 dimension. Trim pulses are normally set to 2 ms. The pulse length of spin lock is dependent on the spectral window of the 2D data. Usually, a 90° spin lock pulse is about 25–30 µs, which covers a bandwidth close to 8–10 kHz, and a 90° excitation pulse is less than 10 µs. It creates sample heating problem when 90° spin lock pulse is shorter than 20 µs. The problem becomes severe for short 90° spin lock pulses in the case of the clean TOCSY. A spin lock time in the range of 30–60 ms is required for protein samples, which can only be selected in an integer number of spin lock cycles. A typical set up involves calibrating a 90° pulse at high power, a 90° spin lock pulse at lower power, and optimizing the spectral window. After the data are acquired with 200 or more FIDs, linear prediction should be utilized to double data points for the F_1 dimension. TOCSY data are processed with a square sine-bell function shifted by 30–60° before Fourier transformation. Note that spin lock power can be calculated based on the power and length of hard 90° pulse within the linear range of the RF amplifier.

4.10.7 NOESY and ROESY

COSY and TOCSY use scalar coupling (through bond nuclear interaction) to correlate the spins within a spin system. These types of experiments provide orientational information about the molecules via three-bond J coupling in addition to the correlation used in resonance assignment. The through-space nuclear interaction (dipolar interaction) is also used in multidimensional experiments to obtain distance information between spins for structural and dynamic studies of molecules. NOESY (nuclear Overhauser effect spectroscopy) and ROESY (rotating frame Overhauser effect spectroscopy, also known as CAMELSPIN for cross-relaxation appropriate for minimolecules emulated by locked spins) experiments utilize the dipolar interaction in the form of cross-relaxation to correlate spins that are close in distance (Bothner-By et al. 1984; Bax and Davis 1985a, b; Griesinger and Ernest 1987).

During the mixing time τ_m of the pulse sequences shown in Fig. 4.21b, c correlations between spins that are close in space occur via cross-relaxation, which give rise to cross peaks in NOESY and ROESY spectra. In a NOESY experiment, the intensities of the cross peaks not only are proportional to the cross-relaxation rate but also come from spin diffusion. Although the contributions to the cross-peak intensity from both mechanisms increase as the mixing time τ_m increases, the spin diffusion effect can be minimized with a short mixing time. For protein samples, the mixing time is usually chosen in the range of 30–200 ms to avoid the spin diffusion effect. For a quantitative measurement of the internuclear

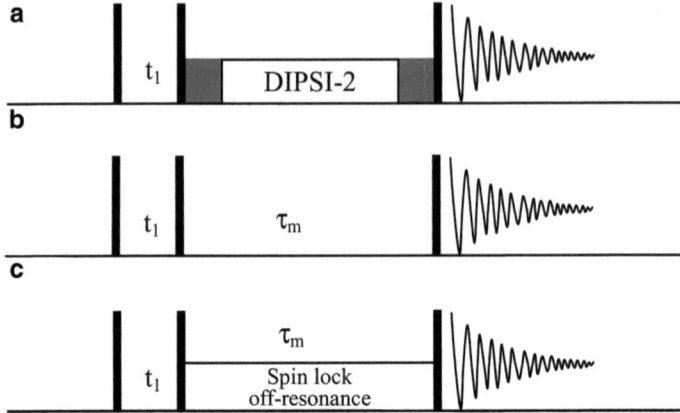

Fig. 4.21 Two-dimensional homonuclear pulse sequences. (**a**) In TOCSY, a DIPSI-2 pulse train is applied with low power during spin lock period. The trim pulses (*gray bars*) are usually set to 2 ms. (**b**) τ_m is the mixing time in the NOESY sequence. (**c**) Spin lock pulses during the mixing time τ_m in the ROESY experiment are set to off-resonance to avoid the generation of TOCSY cross-peaks that have an opposite sign to the ROESY cross-peaks. *Solid bars* represent 90° pulses

distance, a series of NOESY experiments with different mixing times is collected and an NOE buildup curve is obtained by plotting the cross-peak intensity vs. the mixing time for each individual resonance. The NOESY cross-peak intensity (or volume) is proportional to the length of the mixing time:

$$I_a = \xi \frac{\tau_m}{r_{ab}^6} \qquad (4.58)$$

in which ξ is a constant for a given sample at a specific magnetic field strength, τ_m is the mixing time, and r_{ab} is the distance between protons a and b. The initial linear range of the NOE buildup curve is used to calibrate and calculate the distance based on the cross-peak intensity relative to the intensity from a known distance. The constant ξ is first obtained by measuring the cross-peak intensity of a NOESY spectrum with a short mixing time for a known internuclear distance:

$$\xi = I_{ref} \frac{r_{ref}^6}{\tau_m} \qquad (4.59)$$

in which I_{ref} is the cross-peak intensity of a reference proton to another proton with a fixed and known distance r_{ref}. ξ is used to calculate the distance r_{ab} from I_a for all cross peaks. Alternatively, the cross-peak intensity I_{ref} from r_{ref} can be used directly to calculate the distances for other protons using the linear range of the NOE buildup curve:

$$r_{ab}^6 = \frac{I_{ref}}{I_a} r_{ref}^6 \qquad (4.60)$$

4.10 Two-Dimensional Experiments

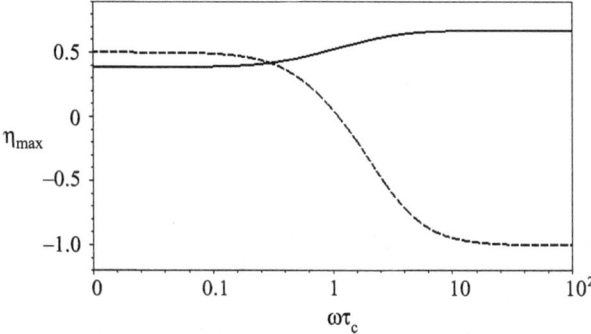

Fig. 4.22 NOE and ROE enhancement factor η_{max} as a function of $\omega\tau_c$. The steady-state NOE enhancement (*dashed line*) is close to zero when $\omega\tau_c \approx 1$, equals 0.5 for $\omega\tau_c \ll 1$ (extreme narrowing limit), and -1.0 for $\omega\tau_c \gg 1$ (spin diffusion limit). The ROE enhancement (*solid line*) is always positive in all regimes, and it is 0.38 for $\omega\tau_c \ll 1$ and 0.68 for $\omega\tau_c \gg 1$

The intensity ratio of each assigned cross peak is used to obtain the distance r_{ab}.

In addition to the spin diffusion effect, another limitation of the NOESY experiment is that the NOE enhancement is close to zero when $\omega\tau_c$ is approximately 1 (small and medium size molecules) as indicated in Fig. 4.22. To overcome these problems, the ROESY experiment has been developed. The pulse sequence of the ROESY experiment is similar to TOCSY although the cross peaks of a ROESY spectrum has an opposite phase to those in the TOCSY spectrum. The ROE enhancement is always positive in all regimes, and it is 0.38 for $\omega\tau_c \ll 1$ and 0.68 for $\omega\tau_c \gg 1$ (Fig. 4.22). The presence of TOCSY peaks in the ROESY spectrum will reduce the intensity of the ROESY cross-peaks. A typical solution is to use an inefficient spin lock scheme with an off-resonance B_1 to mismatch the Hartmann–Hahn matching condition that is required for the TOCSY experiment. Usually, the spin lock field strength for a ROESY experiment is in the order of 1–5 kHz and off-resonance from the carrier frequency by several kilohertz. Such a B_1 field is not sufficient for TOCSY isotropic mixing.

NOESY and ROESY data are collected in a phase sensitive mode in the F_1 dimension, for example, States-TPPI, with 512 increments that yield 256 complex points in the F_1 dimension. A mixing time in the range of 30–200 ms is usually used for NOESY, whereas the mixing time for the ROESY experiment is in the range of the rotating frame relaxation time $T_{1\rho}$, which can be longer than 500 ms for small molecules. Spin lock power for the ROESY is usually set to be less than 5 kHz and several kilohertz off-resonance from the carrier frequency. The data are acquired using 200 or more FIDs and linear prediction should be utilized to double data points for the F_1 dimension. The data are processed with a square sine-bell function shifted by 30–60° before Fourier transformation. The cross peaks of a NOESY spectrum have the same phase as the diagonal peaks, whereas the cross peaks of a ROESY spectrum have the opposite phase from the diagonal peaks.

Questions

1. For a 14.1 T magnet with a proton resonance frequency of 599.89836472 MHz at 0.0 ppm, what is the $^{13}C'$ 180° pulse length that gives a null at C^α?
2. For a 14.1 T magnet with a proton resonance frequency of 599.89836472 MHz at 0.0 ppm, what is the ^{13}C C' off-resonance 90° pulse length that does not disturb the on-resonance C^α?
3. When calibrating a shaped pulse (except a rectangular shaped pulse), which one of the two pulse parameters (pulse length and pulse power) is usually adjusted? And why?
4. What pulse length should be used for the GARP decoupling scheme if a 10 kHz decoupling field strength is desired for ^{13}C decoupling?
5. What is the difference between the watergate and water flip-back watergate sequences? Why does a water flip-back watergate generally yield superior water suppression for aqueous protein samples?
6. What happens to the spectrum if the receiver gain is set too high and how the gain is set for 2D and 3D experiments?
7. Why is a protein sample never span when acquiring data (water suppression, 2D and 3D data)?
8. Why must nonspinning shims be optimized first before gradient shimming on z shims?
9. Which nucleus do you shim when gradient shimming on a 90% H_2O/10% ^2H_2O sample? And explain why.
10. Why is there a splitting in ^2H spectrum of an alignment medium for dipolar measurement?
11. What is the modification of water flip-back made to the watergate solvent suppression sequence and for what purpose is the modification?
12. Why does a weighting function for resolution enhancement almost always decrease sensitivity of the spectrum?
13. Why is unshifted squared sine-bell function used for COSY data, but a 80–90°-shifted squared sine-bell function is used for 3D data?
14. Why is a 180° pulse used for ^1H pulse length calibration but 90° X pulse is used for heteronuclear decoupler pulse length calibration?
15. How many frequency axis or axes do 1D and 3D spectra have? Which parameter is present in any 2D and 3D experiment but not in any 1D experiment?
16. What are the typical ranges of gradient strengths used for suppressing water magnetization, perfecting a 180° pulse, and coherence selection?
17. Assuming that 32,000 DAC on an NMR spectrometer corresponds to the gradient strength of 70 Gcm^{-1}, what is the DAC value for a gradient of -30 Gcm^{-1}?

References

Abragam A (1961) Principles of nuclear magnetism. Clarendon Press, Oxford

Bain A (1984) Coherence levels and coherence pathways in NMR. A simple way to design phase cycling procedures. J Magn Reson 56:418–427

Barkhuijsen H, de Beer R, Bovée WMJ, van Ormindt D (1985) Retrieval of frequencies, amplitudes, damping factors, and phases from time-domain signals using a linear least-squares procedure. J Magn Reson 61:465–481

Bauer C, Freeman R, Frenkiel T, Keeler J, Shaka AJ (1984) Gaussian pulses. J Magn Reson 58:442–457

Baum J, Tycko R, Pines A (1985) Broadband and adiabatic inversion of a two-level system by phase-modulated pulses. Phys Rev A32:3435–3447

Bax A, Davis DG (1985a) Practical aspects of two-dimensional transverse NOE spectroscopy. J Magn Reson 63:207–213

Bax A, Davis DG (1985b) MLEV-17-based two-dimensional homonuclear magnetization transfer spectroscopy. J Magn Reson 65:355–360

Bodenhausen G, Kogler H, Ernst RR (1984) Selection of coherence-transfer pathways in NMR pulse experiments. J Magn Reson 58:370–388

Bothner-By AA, Stephens RL, Lee J, Warren CD, Jeanloz RW (1984) Structure determination of a tetrasaccharide: transient nuclear Overhauser effects in the rotating frame. J Am Chem Soc 106:811–813

Braunschweiler L, Ernst RR (1983) Coherence transfer by isotropic mixing: application to proton correlation spectroscopy. J Magn Reson 53:521–528

Connelly GP, Bai Y, Feng MF, Englander SW (1993) Isotope effects in peptide group hydrogen exchange. Proteins: Struct Funct Genet 17:87–92

Dykstra RW (1983) A broadband radiofrequency reflected-wave detector and its application to a nuclear magnetic resonance spectrometer. J Magn Reson 52:313–317

Emsley L, Bodenhausen G (1990) Gaussian pulse cascades: new analytical functions for rectangular selective inversion and in-phase excitation in NMR. Chem Phys Lett 165:469–476

Emsley L, Bodenhausen G (1989) Self-refocusing effect of 270° Gaussian pulses. applications to selective two-dimensional exchange spectroscopy. J Magn Reson 82:211–221

Emsley L (1994) Selective pulses and their applications to assignment and structure determination in nuclear magnetic resonance. Meth Enzymol 239:207–246

Ernst RR, Bodenhausen G, Wokaun A (1987) Principles of nuclear magnetic resonance in one and two dimensions. Oxford University Press, New York

Freeman R, Morris GA, Turner DL (1977) Proton-coupled carbon-13 J spectra in the presence of strong coupling. I. J Magn Reson 26:373–378

Friedrich J, Davies S, Freeman R (1987) Shaped selective pulses for coherence-transfer experiments. J Magn Reson 75:390–395

Geen H, Freeman R (1991) Band-selective radiofrequency pulses. J Magn Reson 93:93–141

Griesinger C, Ernest RR (1987) Frequency offset effects and their elimination in NMR rotating-frame cross-relaxation spectroscopy. J Magn Reson 75:261–271

Grzesiek S, Bax A (1993) The importance of not saturating water in protein NMR. Application to sensitivity enhancement and NOE measurements. J Am Chem Soc 115:12593–12594

Harris R, Becker ED, de Menezes SMC, Goodfellow R, and Granger P, NMR Nomenclature. Nuclear Spin Properties And Conventions For Chemical Shifts, Pure Appl. Chem., Vol. 73, pp. 1795–1818, 2001

Hartmann SR, Hahn EL (1962) Nuclear double resonance in the rotating frame. Phys Rev 128:2042–2053

Kay LE, Marion D, Bax A (1989) Practical aspects of 3D heteronuclear NMR of proteins. J Magn Reson 84:72–84

Keeler J, Clowes RT, Davis AL, Laue ED (1994) Pulsed-field gradients: theory and practice. Meth Enzymol 239:145–207

Kupče Ē (2001) Applications of adiabatic pulses in biomolecular nuclear magnetic resonance. Meth Enzymol 338:82–111

Kupče Ē, Freeman R (1995) Adiabatic Pulses for wideband inversion and broadband decoupling. J Magn Reson A115:273–276

Kupče Ē, Freeman R (1996) Optimized adiabatic pulses for wideband spin inversion. J Magn Reson A118:299–303

Marion D, Wuthrich K (1983) Application of phase sensitive two-dimensional correlated spectroscopy (COSY) for measurements of 1H-1H spin-spin coupling constants in proteins. Biochem Biophys Res Comm 113:967–974

Marion D, Ikura M, Bax A (1989a) Improved solvent suppression in one- and two-dimensional NMR spectra by convolution of time-domain data. J Magn Reson 84:425–430

Marion D, Ikura M, Tschudin R, Bax A (1989b) Rapid recording of 2D NMR spectra without phase cycling. Application to the study of hydrogen exchange in proteins. J Magn Reson 85:393–399

McCoy MA, Mueller L (1993) Selective decoupling. J Magn Reson A101:122–130

Norris D (2002) Adiabatic radiofrequency pulse forms in biomedical nuclear magnetic resonance. Concepts Magn Reson 14:89–101

Otting G, Widmer H, Wagner G, Wuthrich K (1986) Origin of τ_2 and τ_2 ridges in 2D NMR spectra and procedures for suppression. J Magn Reson 66:187–193

Piotto M, Saudek V, Sklenár V (1992) Gradient-tailored excitation for single-quantum NMR spectroscopy of aqueous solutions. J Biomol NMR 2:661–665

Plateau P, Guéron M (1982) Exchangeable proton NMR without base-line distorsion, using new strong-pulse sequences. J AM Chem Soc 104:7310–7311

Redfield AG, Kunz SD, Ralph EK (1975) Dynamic range in Fourier transform proton magnetic resonance. J Magn Reson 19:114–117

Rucker SP, Shaka AJ (1989) Broadband homonuclear cross polarization in 2D N.M.R. using DIPSI-2. Mol Phys 68:509–517

Shaka AJ, Lee CJ, Pines A (1988) Iterative schemes for bilinear operators; application to spin decoupling. J Magn Reson 77:274–293

Shaka AJ, Keeler J, Freeman R (1983) Evaluation of a new broadband decoupling sequence: WALTZ-16. J Magn Reson 53:313–340

Shaka AJ, Barker PB, Freeman R (1985) Computer-optimized decoupling scheme for wideband applications and low-level operation. J Magn Reson 64:547–552

States J, Haberkorn RA, Ruben DJ (1982) A two-dimensional nuclear overhauser experiment with pure absorption phase in four quadrants. J Magn Reson 48:286–292

Tannus A, Garwood M (1996) Improved performance of frequency-swept pulses using offset-independent adiabaticity. J Magn Reson A120:133–137

Xu P, Wu XL, Freeman R (1992) User-friendly selective pulses. J Magn Reson 99:308–322

Zhu G, Bax A (1990) Improved linear prediction for truncated signals of known phase. J Magn Reson 90:405–410

Chapter 5
Multidimensional Heteronuclear NMR Experiments

In order to study the structure and dynamics of proteins, the spectral resonances must be identified before the structural information and relaxation parameters can be utilized. This chapter addresses the general strategy and heteronuclear experiments required to achieve sequence-specific assignments of backbone and side-chain resonances. Practical aspects related to the experiments are also discussed in detail.

Key questions to be addressed in the current chapter include the following:

1. How do the heteronuclear single-quantum coherence (HSQC) and heteronuclear multiquantum correlation (HMQC) experiments work?
2. How does the transverse relaxation optimized spectroscopy (TROSY) experiment work?
3. How does the IPAP–HSQC experiment work?
4. What are variants of these experiments used for?
5. What are the experiments necessary for backbone and side-chain assignments?
6. How is the magnetization transferred during each of the experiments in terms of product operators?
7. What types of information can be obtained from each experiment?
8. What are the typical procedures to set up these experiments?
9. What is the general strategy to obtain complete assignments using the above experiments?
10. How are the multidimensional NMR data processed?

5.1 Two-Dimensional Heteronuclear Experiments

HSQC and HMQC experiments (Mueller 1979; Hurd and John 1991; Bodenhausen and Ruben 1980; Bax et al. 1990b) are the basic building blocks of multidimensional experiments. In the experiments, the heteronuclear correlations via one-bond scalar couplings are obtained by observing the ^{1}H signals because ^{1}H is more sensitive relative to the heteronuclei. The relative sensitivity of a heteronuclear isotope to ^{1}H is defined as:

$$\rho_i = \frac{S_X}{S_H} = \frac{N_X \gamma_X^3\, I(I+1)}{N_H \gamma_H^3\, \frac{1}{2}(\frac{1}{2}+1)} \quad (5.1)$$

in which S_X and S_H are the signal-to-noise ratios of heteronucleus X and ^{1}H, respectively, N is the number of the nucleus, which is corresponding to the natural abundance of the nucleus, I is the nuclear spin quantum number, and γ is the gyromagnetic ratio. For equal numbers of spin-½ nuclei, the relative sensitivity to ^{1}H is given by:

$$\rho_i = \left(\frac{\gamma_X}{\gamma_H}\right)^3 \quad (5.2)$$

The relative sensitivities of ^{13}C and ^{15}N are approximately 1.6×10^{-2} and 1.0×10^{-3}, respectively. Therefore, the sensitivity can be increased by observing ^{1}H as in HSQC and HMQC, compared to observing the heteronuclei. These experiments are also the most routinely acquired proton-detected heteronuclear correlation experiments used in structural and dynamic studies of proteins. Understanding how the experiments work, how they are carried out, and what types of information can be obtained will certainly aid in the understanding of more complicated multidimensional experiments (see below).

5.1.1 HSQC and HMQC

The pulse sequence of the HSQC (Fig. 5.1) uses the INEPT (insensitive nuclei enhancement by polarization transfer) sequence to transfer proton magnetization into HSQC. After it is frequency labeled according to the heteronuclear chemical shift during the following evolution period, the heteronuclear SQ coherence is transferred back into proton magnetization via the second INEPT sequence. The desired SQ coherence pathway is selected by the gradient pulses. The first gradient pulse dephases heteronuclear magnetization by producing a spatially dependent phase shift of $\phi_x = \gamma_x G_4$ at the end of the evolution period t_1. After the magnetization is transferred back to the ^{1}H spins, the coupled coherence is refocused by G_3 according to the refocusing condition:

$$\pm \gamma_X G_4 - \gamma_H G_3 = 0$$

or

$$G_4 = \pm \frac{\gamma_H}{\gamma_X} G_3 = \kappa G_3 \quad (5.3)$$

5.1 Two-Dimensional Heteronuclear Experiments

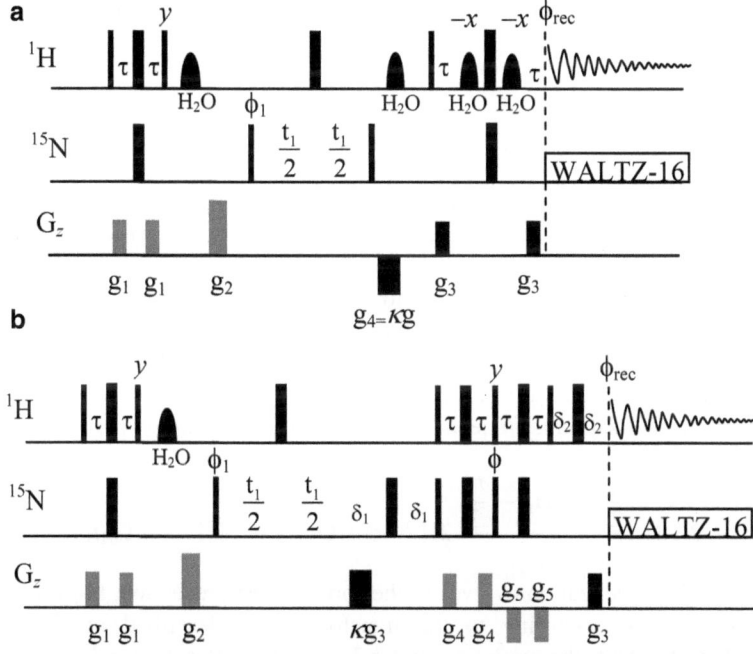

Fig. 5.1 Pulse sequences for gradient HSQC experiments. (**a**) Water flip-back using a 1 ms 90° Gaussian-selective pulse is combined with the water gate sequence for solvent suppression. (**b**) A PEP sensitivity- enhanced HSQC with water flip-back is acquired for a pair of data sets by inversing the sign of κ and the phase of ϕ. For both pulse sequences, $\phi_1 = x, -x$, +States–TPPI, $\phi_{rec} = x, -x$, and $\kappa = \pm 10$, $\phi = \pm y$ for PEP. The phases of all other pulses are x. Delays τ are set to 2.7 ms, δ_1 is the pulse length of gradient pulse κg_3, δ_2 gradient g_3. The black gradient pulses are used for coherence selection

in which G_3 and G_4 are the gradient strength of gradient pulses g_3 and g_4, respectively. Equation (5.3) assumes that the gradients have the same duration. Signs + and − produce P- and N-type spectra, respectively. Either type will give t_1 frequency discrimination type data with a phase twisted line shape as described in the COSY experiment (Chap. 4). The value of κ is set to ± 10 and ± 4 for ^{15}N and ^{13}C heteronuclear correlation, respectively. The opposite sign of G_3 to G_4 is to avoid refocusing any water magnetization that is dephased by the G_4. The selective pulse on water is to ensure the water magnetization is in the longitudinal direction during acquisition to avoid the effect of saturation transfer on amide proton signals. The gradient applied immediately after the selective pulse dephases the transverse magnetization of the residual water.

The magnetization during the gHSQC can be described by product operator analysis as shown for the INEPT since the experiment utilizes an INEPT sequence to transfer coupled H^N and N coherence. Staring at $^1H^N$ (H) coupled with ^{15}N (N), the magnetization transfer pathway is represented using the product operators as follows:

$$H_z \xrightarrow{\frac{\pi}{2}H_x} -H_y$$

$$-H_y \xrightarrow{\tau \to \pi(H_x+N_x) \to \tau \to \frac{\pi}{2}(H_y+N_x)} -2H_zN_y \quad (5.4)$$

Proton magnetization is transferred to nitrogen in the first INEPT sequence (see Sect. 1.12.2.3 for details). Note that during τ period, the NH magnetization evolves under the influence of $\pi J_{NH}\tau$, according to the transformations shown in Fig. 1.28a, b. In (5.4), $\pi J_{NH}\tau$ is simplified to τ.

The N magnetization is frequency labeled during t_1 evolution period:

$$-2H_zN_y \xrightarrow{\frac{t_1}{2} \to \pi H_x \to \frac{t_1}{2}} 2H_zN_y \cos(\Omega_N t_1) - 2H_zN_x \sin(\Omega_N t_1) \quad (5.5)$$

The AP NH magnetization evolves in the form of free precession, $\Omega_N t_1$, during t_1 evolution period, according to the transformations described in Fig. 1.28c. Throughout the text, the evolution under J coupling interaction, $\pi J_{NH}\tau$, and free precession during evolution period, $\Omega_N t_1$, are simplified to τ and t_1 for product operator description, respectively.

During the reverse INEPT of the pulse sequence, the magnetization is transferred back to proton:

$$\xrightarrow{\frac{\pi}{2}(H_x+N_x) \to \tau \to \pi(H_x+N_x) \to \tau} -H_x \cos(\Omega_N t_1) - 2H_yN_x \sin(\Omega_N t_1) \quad (5.6)$$

in which τ is set to $1/(4J_{NH})$, causing $\sin(2\pi 4J_{NH}\tau) = 1$. The 180° pulse on spin H in the middle of the t_1 period refocuses the coupling of spin H and N and hence removes the coupling J_{NH} in the F_1 dimension (^{15}N). Thus, the line shape on F_1 dimension does not include a contribution from J_{NH} coupling. The term $2H_yN_x$ represents multiquantum coherence that is unobservable magnetization.

5.1.2 HSQC Experiment Setup

A water suppression sequence such as the water-flip-back is usually integrated into a gradient HSQC pulse sequence (Fig. 5.1). The 90° ^{1}H pulse is calibrated first after the probe is tuned. The transmitter carrier frequency (^{1}H) is then calibrated by setting the offset frequency on the water resonance and acquiring a PRESAT experiment with the transmitter offset frequency arrayed over ±3 Hz in 0.5 Hz steps (see previous chapter). A water-selective shaped 90° pulse is calibrated by arraying the fine power of the pulse (refer to the section on shaped pulses in Chap. 4). The ^{15}N decoupler offset frequency is set to 118 ppm, which should be used for all ^{15}N double- and triple-resonance experiments. Timing parameters include a delay τ of 2.7 ms, a relaxation delay time of 1 s, and an acquisition time of 64 ms. The first FID of HSQC is recorded for a test run to optimize the receiver gain for better sensitivity. The dead time and receiver phase are adjusted to obtain a phase-correct spectrum without phase correction. Two-dimensional data are collected with 128 increments.

In the ^{1}H dimension, the data size is doubled with mirror image linear prediction followed by apodization with a 60°-shifted square sine-bell function, zero filling to 2,048 complex points and Fourier transformation. If the phase parameters are not optimized before acquisition, they are obtained using the first FID of the 2D data. Since only the amide protons are observed in the experiment, the upfield half of the ^{1}H spectrum (frequency range lower than 6 ppm) is eliminated before processing the ^{15}N dimension. After being transposed, the interferogram is processed in the same way as the ^{1}H dimension except with zero filling to 512 complex points.

A phase-sensitive gradient HSQC can be acquired using States, TPPI or States–TPPI methods to obtain a pure absorptive spectrum in both dimensions. It can also be acquired with a sensitivity-enhanced method (also known as PEP for preservation of equivalent pathway, and COS for coherence order selective) to improve the sensitivity of the experiment by a factor of $\sqrt{2}$ over the conventional gradient HSQC.

Water suppression schemes such as water-flip-back and Watergate can be implemented into the HSQC sequence to efficiently suppress water (Fig. 5.1b). The water-flip-back uses a water-selective pulse after the first INEPT to bring the water magnetization back to the z axis. The water magnetization remains along the z axis during the reverse INEPT and WATERGATE, while only the residual water magnetization resulting from pulse imperfection is efficiently dephased by the WATERGATE sequence. Utilization of the flip-back pulse prevents dephasing the majority of the water magnetization so as to avoid saturation transfer. As a result, the sensitivity is improved by 10–20 % over the sequence without the water-flip-back pulse (Grzesiek and Bax 1993).

5.1.3 Sensitivity-Enhanced HSQC by PEP

The PEP sensitivity enhancement method has been incorporated into the HSQC sequence to improve the sensitivity without increasing the experimental time (Kay et al. 1992; Cavanagh et al. 1991; Akke et al. 1994). In the conventional HSQC, the reverse INEPT transfers the magnetization back to ^{1}H (see (5.4)–(5.6)):

$$-2H_zN_y \xrightarrow{t_1 \to \left(\frac{\pi}{2}\right)(H_x + N_x) \to \tau \to \pi(H_x + N_x) \to \tau}$$

$$-H_x \cos(\Omega_N t_1) - 2H_y N_x \sin(\Omega_N t_1) \qquad (5.7)$$

The second term represents multiple-quantum (MQ) coherence that is not observable. Therefore, only the cosine component of the t_1 evolution will be detected. A sensitivity-enhanced HSQC using the PEP method (Fig. 5.1b) allows the detection of both components of the coherence.

The last period after the reversed INEPT in the pulse sequence is for the PEP sensitivity enhancement, in which the two terms of the coherence can be considered separately. The first FID is acquired with $\phi = x$ and $\kappa = 10$ and the second one is recorded with both the phase ϕ and gradient κ inverted (i.e., $-x$ and -10, respectively). During the PEP, the coherence for the first FID yields:

$$-H_x \cos(\Omega_N t_1) - 2H_y N_x \sin(\Omega_N t_1) \xrightarrow{\left(\frac{\pi}{2}\right)(H_y + N_y)} H_z \cos(\Omega_N t_1) + 2H_y N_z \sin(\Omega_N t_1)$$

$$\xrightarrow{\tau \to \pi(H_x + N_x) \to \tau} -H_z \cos(\Omega_N t_1) - H_x \sin(\Omega_N t_1)$$

$$\xrightarrow{\left(\frac{\pi}{2}\right) H_x} H_y \cos(\Omega_N t_1) - H_x \sin(\Omega_N t_1) \tag{5.8}$$

The last 180° ^{1}H pulse in Fig. 5.1b is used to invert ^{1}H magnetization for coherence selection by g_3 gradient and δ_2 delays are the same as the duration of g_3.

The second FID is obtained by inverting phase ϕ and the sign of gradient factor κ:

$$-H_x \cos(\Omega_N t_1) - 2H_y N_x \sin(\Omega_N t_1) \xrightarrow{\left(\frac{\pi}{2}\right)(H_y - N_y)} H_z \cos(\Omega_N t_1) - 2H_y N_z \sin(\Omega_N t_1)$$

$$\xrightarrow{\tau \to \pi(H_x + N_x) \to \tau} -H_z \cos(\Omega_N t_1) + H_x \sin(\Omega_N t_1)$$

$$\xrightarrow{\left(\frac{\pi}{2}\right) H_x} H_y \cos(\Omega_N t_1) + H_x \sin(\Omega_N t_1) \tag{5.9}$$

Therefore, the two FIDs acquired and stored in separated memory locations in the same data are given by:

$$\text{FID1} \propto H_y \cos(\Omega_N t_1) - H_x \sin(\Omega_N t_1) \tag{5.10}$$

$$\text{FID2} \propto H_y \cos(\Omega_N t_1) + H_x \sin(\Omega_N t_1) \tag{5.11}$$

Addition of the two FIDs gives rise to a data set containing observable magnetization described by:

$$2H_y \cos(\Omega_N t_1) \tag{5.12}$$

Similarly, subtraction of the two FIDs (FID$_2$–FID$_1$) yields a second data set:

$$2H_x \sin(\Omega_N t_1) \tag{5.13}$$

5.1 Two-Dimensional Heteronuclear Experiments

The two PEP data sets can be processed to obtain separate 2D spectra. The spectra are then combined to form a single spectrum with pure absorptive line shapes on both dimensions. Because the noise is increased by a factor of $\sqrt{2}$ and the signal intensity is increased by 2 after the spectral combination, the combined spectrum increases the sensitivity by a maximum of $\sqrt{2}$ compared to that in the conventional gradient HSQC spectrum without considering the relaxation factor (Palmer et al. 1991; Kay et al. 1992; Cavanagh and Rance 1993; Schleucher et al. 1994; Muhandiram and Kay 1994). For a ^{1}H–^{13}C HSQC experiment, the identical result will be obtained by replacing operator N with C for ^{13}C spins.

The water signal can be suppressed during the PEP sequence. In a conventional gradient-enhanced HSQC experiment, the water magnetization is fully dephased by the gradient echo immediately before the data acquisition. This causes reduction of NH magnetization through saturation transfer. In an seHSQC experiment, application of a selective water pulse combined with a two step phase cycle for the ^{1}H pulse ensures that the water magnetization stays along the z axis prior to the acquisition period while the residual water is suppressed during dephasing by the gradient echo pulses. Since the saturation transfer is minimized and both the echo and anti-echo coherences are selected in the experiment, the sensitivity is improved in the flip-back seHSQC compared to the conventional gradient phase-sensitive HSQC, in which quadrature detection in the F_1 dimension is obtained by using gradients to generate pure absorption spectra.

5.1.4 Setup of seHSQC Experiment

The general procedure to set up experiments for aqueous samples includes tuning the probe for all three channels in the order of ^{15}N, ^{13}C and then the ^{1}H channel; calibrating the 90° ^{1}H pulse; finding the resonance frequency of water; and calibrating the selective 90° pulse on water. The ^{15}N and ^{13}C pulses do not require recalibration for every experiment as long as they are calibrated regularly and the probe is well-tuned. The data are acquired with 1,024 t_2 and 128 t_1 complex points. The initial t_1 value is set according to

$$t_1(0) = \frac{1}{2SW} - pw_{180(H)} - \frac{4pw_{90(N)}}{\pi} \tag{5.14}$$

whereas for CT–HSQC, initial t_1 is chosen by

$$t_1(0) = \frac{1}{2SW} \tag{5.15}$$

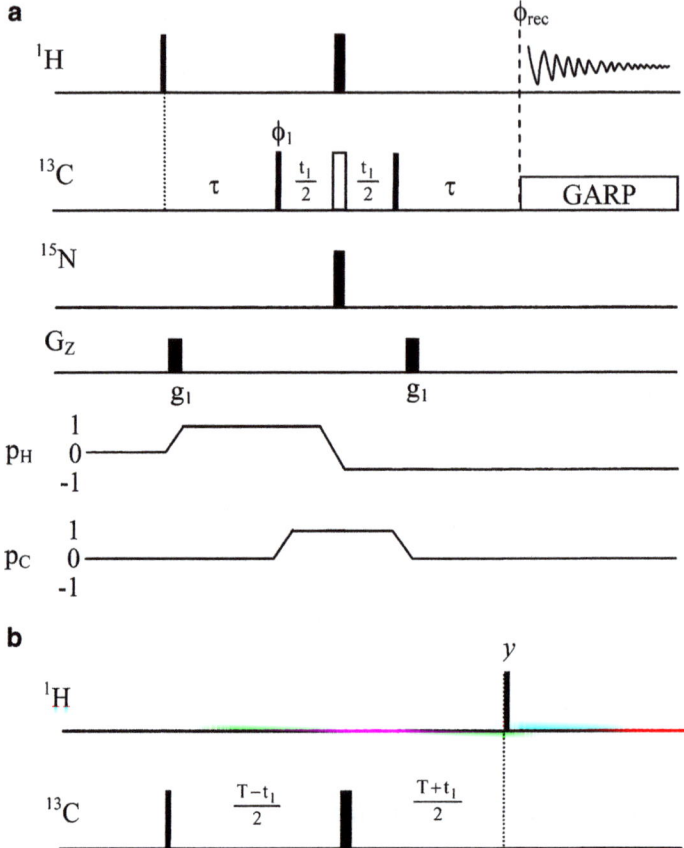

Fig. 5.2 Pulse sequence for HMQC experiments. The 90° (*narrow bars*) and 180° (*wider bars*) are x phase except as indicated. (**a**) Conventional HMQC with the coherence transfer pathways shown below the pulse sequence. The open 180° pulse is a $^{13}C^{\alpha}$- or $^{13}C^{\beta}$-selective pulse, which can be turned on for decoupling $^{13}C^{\alpha}$–$^{13}C^{\beta}$ coupling in the t_1 dimension since the frequency bands are well-separated with the exception of Gly, Ser, and Thr residues (Matsuo et al. 1996). Phase $\phi_1 = x, -x$, +States–TPPI and $\phi_{rec} = x, -x$. Delay $\tau = 1/(2J_{CH})$. Gradients are applied with duration of 0.5 ms and amplitude of $g_1 = 10$ Gcm^{-1}. P_H and P_C are ^1H and ^{13}C coherence transfer pathways, respectively. (**b**) Constant-time element (Ernst et al. 1987)

5.1.5 HMQC

Shown in Fig. 5.2a is an HMQC pulse sequence. The magnetization during the experiment can be described by product operator analysis. Staring at $^1H^C$ (H) coupled with ^{13}C (C), the magnetization transfer pathway is represented using the product operators as follows:

$$H_z \xrightarrow{\left(\frac{\pi}{2}\right)H_x} -H_y \xrightarrow{\tau} -H_y \cos(\pi J_{HC}\tau) + 2H_xC_z \sin(\pi J_{HC}\tau) \quad (5.16)$$

5.1 Two-Dimensional Heteronuclear Experiments

Delay τ is set to $1/(2J_{HC})$ to maximize the anti-phase (AP) coherence. The multiple-quantum (MQ) coherence is generated by the 90° carbon pulse and evolves during t_1 time:

$$2H_xC_z \xrightarrow{\left(\frac{\pi}{2}\right)C_x} -2H_xC_y \xrightarrow{\frac{t_1}{2} \to \pi H_x \to \frac{t_1}{2}} -2H_xC_y\cos(\Omega_C t_1) + 2H_xC_x\sin(\Omega_C t_1) \quad (5.17)$$

The 180° ^{1}H pulse in the middle of t_1 refocuses the evolution of ^{1}H chemical shift during the t_1 period and during the τ period. Therefore, the magnetization of ^{1}H does not evolve. In addition, the heteronuclear scalar coupling does not affect the evolution of the MQ coherence H_xC_y during t_1. As a result, the only evolution occurred during t_1 is at the ^{13}C chemical shift frequency. The MQ coherence is converted into observable single-quantum (SQ) magnetization by the 90° ^{13}C pulse combining with delay τ, while the zero-quantum term H_xC_x will not produce observable coherence:

$$-2H_xC_y\cos(\Omega_C t_1) \xrightarrow{\left(\frac{\pi}{2}\right)C_x} -2H_xC_z\cos(\Omega_C t_1) \xrightarrow{\tau} -H_y\cos(\Omega_C t_1) \quad (5.18)$$

In summary,

$$H_z \xrightarrow{\left(\frac{\pi}{2}\right)H_x} -H_y \xrightarrow{\text{HMQC-type}} -H_y\cos(\Omega_C t_1) \quad (5.19)$$

The above product operator treatment does not include the homonuclear scalar coupling between protons and between ^{13}C carbons, which also evolve during the t_1 period of ^{1}H–^{13}C HMQC. As a result, the F_1 dimension has the ^{13}C–^{13}C scalar-coupled multiplets as well as the contribution from ^{1}H–^{1}H homonuclear coupling.

A constant time evolution period (Fig. 5.2b) can be used to obtain F_1 decoupled spectrum (Ernst et al. 1987). In addition, studies have indicated that the relaxation rate of the ^{1}H–^{13}C MQ coherence is much slower than that of ^{1}H–^{13}C SQ coherence for non-aromatic methane sites in ^{13}C-labeled protein and in nucleic acids at slow tumbling limit. This property has been utilized to obtain better sensitivity in CT HMQC-type experiment than the CT HSQC-type experiment. During the CT evolution period the homonuclear scalar coupling J_{CC} evolves for a period of $(T - t_1)/2 + (T + t_1)/2 = T$. Therefore, after the t_1 evolution period the magnetization in (5.17) has a form of:

$$-2H_xC_y \xrightarrow{\frac{(T-t_1)}{2} \to \pi H_x \to \frac{(T+t_1)}{2}} [-2H_xC_y\cos(\Omega_C t_1) + 2H_xC_x\sin(\Omega_C t_1)]\cos(\pi J_{CC}T) \quad (5.20)$$

By setting the delay T to $1/(2J_{CC})$, the effect of the J_{CC} can be removed.

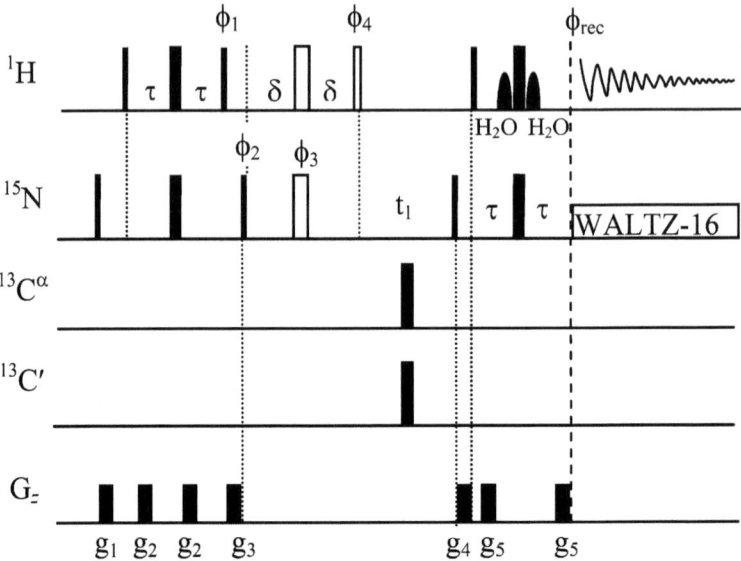

Fig. 5.3 Pulse sequence for IPAP (^{1}H, ^{15}N) HSQC experiment. The sequence element $\delta - 180°$ (^{1}H/^{15}N) $- \delta - 90°$ (open pulses) is only used in the experiment for generating the anti-phase (AP) spectrum and is omitted for generating the in-phase (IP) spectrum. The 90° (*narrow bars*) and 180° (*wider bars*) are x phase except $\phi_1 = -y, y$; $\phi_2 = 2(x), 2(-x)$, +States–TPPI for IP; $\phi_2 = 2(-y), 2(y)$, +States–TPPI for AP; $\phi_3 = 4(x), 4(y), 4(-x), 4(-y)$, +States–TPPI; $\phi_4 = 8$ $(x), 8(-x)$; $\phi_{rec} = x, 2(-x), x$ for IP; $\phi_{rec} = x, 2(-x), x, -x, 2(x), -x$ for AP. Delays $\tau = 2.5$ ms, $\delta = 2.7$ ms. The gradients are sine-bell shaped with amplitude of 25 Gcm^{-1} and durations of 2.0, 0.4, 2.0, 1.0, and 0.4 ms for g_1, g_2, g_3, g_4, and g_5, respectively. IP and AP spectra are recorded in an interleaved manner (from Ottiger et al. 1998)

5.1.6 IPAP HSQC

The IPAP HSQC (in-phase anti-phase) experiment is used to measure the residual dipolar coupling between amide ^{1}H and ^{15}N spins (Ottiger et al. 1998). The experiment records two HSQC data sets, one of which yields IP doublets after Fourier transformation while the other gives AP doublets. Addition and subtraction of the spectra produce a pair of individual spectra each containing one of the doublet components. The coherence transfer during the pulse sequence can be described by product operators. The first ^{15}N 90° combined with the gradient pulse is used to ensure the initial magnetization originates only from amide proton spins. After the first INEPT transfer, the evolution of the ^{1}H–^{15}N scalar coupling and ^{15}N chemical shift during t_1 evolution period is given by (Fig. 5.3):

$$H_z \xrightarrow{\left(\frac{\pi}{2}\right)H_x \to \tau \to \pi(H_x + N_x) \to \tau \to \left(\frac{\pi}{2}\right)(H_y + N_x)} 2H_zN_y \quad \text{(First INEPT)}$$

(5.21)

5.1 Two-Dimensional Heteronuclear Experiments

$$2H_zN_y \xrightarrow{t_1} 2H_zN_y \cos(\Omega_N t_1) \cos(\pi J_{NH} t_1)$$
$$- 2H_zN_x \sin(\Omega_N t_1) \cos(\pi J_{NH} t_1) + IP \quad (t_1 \text{ evolution period}) \quad (5.22)$$

in which term IP contains only IP coherence N_y and N_x with the coefficients, which cannot be converted into observable magnetization at the beginning of the acquisition. Therefore, the IP term is omitted during the following transfer path. For the first data set, the coherence transferred in the same path as for conventional HSQC is given by:

$$2H_zN_y \cos(\Omega_N t_1) \cos(\pi J_{NH} t_1) - 2H_zN_x \sin(\Omega_N t_1) \cos(\pi J_{NH} t_1)$$

$$\xrightarrow{\left(\frac{\pi}{2}\right)(H_x + N_x)} -H_yN_z \cos(\Omega_N t_1) \cos(\pi J_{NH} t_1) + 2H_yN_x \sin(\Omega_N t_1) \cos(\pi J_{NH} t_1)$$

$$\xrightarrow{\tau \to \pi(H_x + N_x) \to \tau} H_x \cos(\Omega_N t_1) \cos(\pi J_{NH} t_1)$$

$$(5.23)$$

in which $\tau = 1/(4J_{NH})$ and the MQ term H_yN_x cannot be converted to observable magnetization and thus is omitted.

The second data set is collected by inserting an ^{15}N refocusing period before the evolution time to obtain AP doublets. The magnetization after the first INEPT transfer with phase ϕ_2 decreased by 90° is given by:

$$H_z \xrightarrow{\left(\frac{\pi}{2}\right)H_x \to \tau \to \pi(H_x + N_x) \to \tau \to \left(\frac{\pi}{2}\right)(H_y - N_y)} 2H_zN_x \quad (5.24)$$

The coherence after the ^{15}N refocusing period is described by:

$$2H_zN_x \xrightarrow{\delta \to \pi(H_x + N_x) \to \delta \to \left(\frac{\pi}{2}\right)H_x} 2H_yN_x \cos(2\pi J_{NH}\delta) - N_y \sin(2\pi J_{NH}\delta)$$

$$(5.25)$$

in which the delay δ is optimized to $1/(4J_{NH})$, approximately 2.7 ms. Ignoring the MQ term, the evolution of scalar coupling J_{NH} and ^{15}N chemical shift yields:

$$-N_y \sin(2\pi J_{HN}\delta) \xrightarrow{t_1} 2H_zN_x \cos(\Omega_N t_1) \sin(\pi J_{NH} t_1) + 2H_zN_y \sin(\Omega_N t_1) \sin(\pi J_{NH} t_1) + IP$$

$$\xrightarrow{\left(\frac{\pi}{2}\right)(H_x + N_x)} -2H_yN_x \cos(\Omega_N t_1) \sin(\pi J_{NH} t_1) - 2H_yN_z \sin(\Omega_N t_1) \sin(\pi J_{NH} t_1)$$

$$\xrightarrow{\tau \to \pi(H_x + N_x) \to \tau} H_x \sin(\Omega_N t_1) \sin(\pi J_{NH} t_1)$$

$$(5.26)$$

Addition of the two FIDs produces:

$$H_x \cos(\Omega_N t_1) \cos(\pi J_{NH} t_1) + H_x \sin(\Omega_N t_1) \sin(\pi J_{NH} t_1)$$
$$= H_x \cos[(\Omega_N - \pi J_{NH}) t_1] \quad (5.27)$$

according to $\cos(\alpha)\cos(\beta) + \sin(\alpha)\sin(\beta) = \cos(\alpha - \beta)$, whereas subtraction of the data gives:

$$H_x \cos(\Omega_N t_1) \cos(\pi J_{NH} t_1) - H_x \sin(\Omega_N t_1) \sin(\pi J_{NH} t_1)$$
$$= H_x \cos[(\Omega_N + \pi J_{NH}) t_1] \quad (5.28)$$

because $\cos(\alpha)\cos(\beta) - \sin(\alpha)\sin(\beta) = \cos(\alpha + \beta)$. Fourier transformation of the combined data sets generates individual spectra with one of the doublets at either $\Omega_N - \pi J_{NH}$ or $\Omega_N + \pi J_{NH}$.

The above product operator treatment does not include the effect of relaxation that causes the signal loss in the second FID during the 2δ period, which is a factor $e^{2\delta/T_2}$. Deviation of J_{NH} from the selected value of δ also causes a signal loss, but the loss is identical for both IP and AP spectra. Thus, it is necessary to multiply a scaling factor before addition or subtraction, which can be adjusted during data processing. The quadrature detection in the t_1 dimension is achieved by the States–TPPI phase mode, because the IPAP does not provide frequency discrimination in the t_1 dimension as shown above. By altering the phase of the first 90° ^{15}N pulse (IP) and both the first 90° ^{15}N pulse and the 180° ^{15}N pulse during 2δ period (AP) in the States–TPPI manner, the two phase-sensitive IP and AP FIDs can be obtained:

$$H_x \cos(\pi J_{NH} t_1) e^{-i\Omega_N t_1} \quad (5.29)$$

$$iH_x \sin(\pi J_{NH} t_1) e^{-i\Omega_N t_1} \quad (5.30)$$

The data obtained after addition and subtraction are given by:

$$H_x e^{-i(\Omega_N - \pi J_{HN}) t_1} \quad (5.31)$$

$$H_x e^{-i(\Omega_N + \pi J_{NH}) t_1} \quad (5.32)$$

The IPAP method can also be implemented in triple-resonance experiments to resolve the overlapped signals.

5.1.7 SQ–TROSY

TROSY experiment utilizes the interference effect of cross correlated relaxations caused by CSA and dipolar interaction on the T_2 relaxation rate at the individual multiplet components to reduce the line width of heteronuclear correlation spectra (Pervushin et al. 1997). For a weakly coupled two spin–½ system isolated from other spins in a protein molecule, T_2 relaxation of the spin system is dominated by the CSA of each individual spin and the DD coupling between the two spins. The CSA has the same effect on the T_2 relaxation of all multiplet components, while the

5.1 Two-Dimensional Heteronuclear Experiments

effect of DD coupling on the T_2 relaxation of the resonances from the β transitions is opposite to the effect from CSA. This interference effect from CSA and DD coupling leads to different T_2 relaxations for the doublet peaks of the individual spins in the spin system. When the orientations of the two interactions are approximately collinear and their magnitudes are comparable, the line widths originating from the β transitions are reduced and those from the α transitions are broadened (Fig. 1.23, Pervushin et al. 1997).

The interference effect can be well-understood by considering the relaxation matrix containing SQ transition operators. For macromolecules in aqueous solution, the isotropic motion of the molecules is in the slow-tumbling limit. Only the spectral density, $J(0)$, contributes significantly to the relaxation rate. The change of the magnetization corresponding to the transitions as a function of time can be represented by the first-order relaxation matrix (Sørensen et al. 1983; Ernst et al. 1987; Pervushin et al. 1997):

$$\frac{d}{dt}\begin{bmatrix} I_{13}^\pm \\ I_{24}^\pm \\ S_{12}^\pm \\ S_{34}^\pm \end{bmatrix} = -\left\{ i \begin{bmatrix} \pm\omega_I^{13} \\ \pm\omega_I^{24} \\ \pm\omega_S^{12} \\ \pm\omega_S^{34} \end{bmatrix} + 4J(0) \begin{bmatrix} p^2 - 2C_{p,\delta_I}\, p\delta_I + \delta_I^2 \\ p^2 + 2C_{p,\delta_I}\, p\delta_I + \delta_I^2 \\ p^2 - 2C_{p,\delta_S}\, p\delta_S + \delta_S^2 \\ p^2 + 2C_{p,\delta_S}\, p\delta_S + \delta_S^2 \end{bmatrix} \right\} \begin{bmatrix} I_{13}^\pm \\ I_{24}^\pm \\ S_{12}^\pm \\ S_{34}^\pm \end{bmatrix} \quad (5.33)$$

in which $I_{ij}^\pm$ and $S_{ij}^\pm$ are the magnetization of spins I and S corresponding to the SQ transitions $i \leftrightarrow j$ in the standard energy diagram (Fig. 1.18) with the corresponding resonance frequencies:

$$I_{13}^\pm \Rightarrow \text{transition } 1 \leftrightarrow 3, \quad \omega_I^{13} = \omega_I + \pi J_{IS}$$

$$I_{24}^\pm \Rightarrow \text{transition } 2 \leftrightarrow 4, \quad \omega_I^{24} = \omega_I - \pi J_{IS}$$

$$S_{12}^\pm \Rightarrow \text{transition } 1 \leftrightarrow 2, \quad \omega_S^{12} = \omega_I - \pi J_{IS}$$

$$S_{34}^\pm \Rightarrow \text{transition } 3 \leftrightarrow 4, \quad \omega_S^{34} = \omega_S + \pi J_{IS} \quad (5.34)$$

$J(0)$ is the spectral density function at the zero-frequency, $C_{kl} = \frac{1}{2}(3\cos^2\Theta_{kl} - 1)$ and Θ_{kl} is the angle between the tensor axes of the interaction k and l, and p, δ_I, and δ_S are given by:

$$p = \frac{1}{2\sqrt{2}} \frac{\gamma_I \gamma_S \hbar}{r_{IS}^3}, \quad \delta_I = \frac{\gamma_I B_0 \Delta\sigma_I}{3\sqrt{2}}, \quad \delta_S = \frac{\gamma_S B_0 \Delta\sigma_S}{3\sqrt{2}} \quad (5.35)$$

in which γ_I and γ_S are the gyromagnetic ratios of spin I and S, respectively, $\hbar$ is reduced Plank's constant, r_{IS} is the distance between the two spins, B_0 is the magnetic field strength, $\Delta\sigma_I$ and $\Delta\sigma_S$ are the chemical shift difference between the axial and perpendicular principle components of the axially symmetric

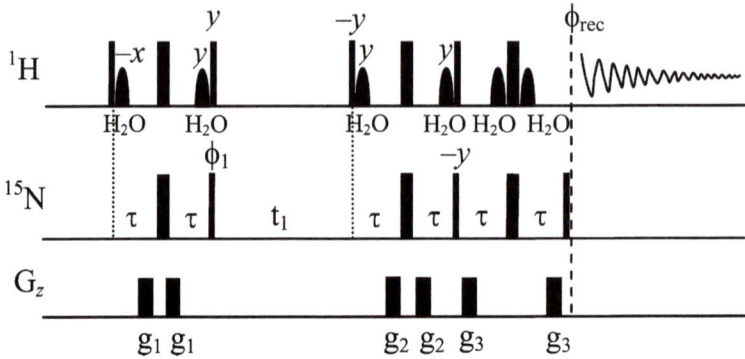

Fig. 5.4 Pulse sequence for TROSY experiment. The 90° (*narrow bars*) and 180° (*wider bars*) are x phase except $\phi_1 = y, x$, +States–TPPI, $\phi_{rec} = y, -x$, and others as indicated. The selective H$_2$O 90° pulses (shaped) are used to avoid the saturation of water by retaining water magnetization along z axis during the experiment. The delay $\tau = 2.7$ ms. Gradients are applied with duration of 1.0 ms and amplitude of $g_1 = 30$ Gcm^{-1}, $g_2 = 40$ Gcm^{-1}, and $g_3 = 48$ Gcm^{-1}

CSA of spin I and S, respectively. Equation (5.33) tells us that if the magnitudes of CSA and DD coupling are comparable ($p \approx -\delta_S$ and $p \approx -\delta_I$ for I = ^{1}H and S = ^{15}N), and the principle symmetric axis of the CSA tensor and the bond vector r_{IS} are approximately collinear, such as in the case of the backbone amide NH moiety of the proteins, the line widths at the resonance frequencies ω_I^{24} and ω_S^{34} are narrower than the other two due to the slow transverse relaxation even for large size proteins (Pervushin et al. 1999). For backbone NH, $\Delta\sigma_H = 15$ ppm, $\Theta_{p,\delta_H} = 10°$ (Gerald et al. 1993), $\Delta\sigma_N = -156$ ppm, and $\Theta_{p,\delta_N} = 17°$ (Teng and Cross 1989). Therefore, both ^{1}H and ^{15}N CSA tensors are almost axial symmetric and nearly collinear with the DD vector (the NH bond). It has been estimated using the above values that the transverse relaxation effect at ^{1}H frequencies can be completely canceled for one of the four multiplet components when the magnetic field strength is near 1 GHz (Pervushin et al. 1997).

Shown in Fig. 5.4 is the water-flip-back ^{1}H/^{15}N TROSY pulse sequence. The experiment specially correlates the ^{15}N 4 → 3 transition with the ^{1}H 4 → 2 through the SQ transition (SQ–TROSY, Pervushin et al. 1999). Saturation of the water magnetization is avoided during the experiment by the water-flip-back pulses at the beginning of the pulse sequence. The selective pulses on the water resonance during the first INEPT sequence are used to keep the water magnetization on the z axis during the t_1 evolution period. The two water-selective pulses during the SQ polarization transfer element ensure the water magnetization is on the z axis immediately before acquisition, resulting in a minimal saturation transfer from water to the exchangeable NH protons. The Watergate element at the end of the pulse sequence is used to suppress the residual transverse water magnetization before acquisition.

5.2 Overview of Triple-Resonance Experiments

The homonuclear NOE is a widely utilized technique in various types of calculation methods such as distance geometry, restrained molecular dynamics (or simulated annealing), and variable target function methods in structural characterization of proteins based on the information on approximate distances between protons obtained in NOESY or ROESY experiments. Before the information can be used, the origin of each resonance in the NMR spectrum must be linked to a nucleus in the molecular sequence. The process is called sequence-specific assignment. In order to obtain a high resolution structure, it is necessary to complete the assignment for a sufficient number of atoms in the sequence. Frequently, the heteronuclear isotopes (^{15}N and/or ^{13}C) are also used to establish sequential assignment as well as to increase the spectral resolution by spreading ^{1}H resonances on the heteronuclear dimensions so that the degeneracy of ^{1}H resonances can be reduced. Two types of nuclear interactions are used in NMR spectra for the assignment of the chemical shifts of the nuclei: through-bond interaction—scalar coupling, and through-space interaction—dipolar coupling via NOE.

For homonuclear correlations, the sequential assignments must be done by the inter-residue backbone NOEs due to the near-zero ^{1}H four bond scalar coupling. However, an inter-residue NOE may not occur between sequential correlations. In addition, for large proteins (MW>10 kDa) for which resonance degeneracy becomes severe, it is extremely difficult to assign the resonances even if ^{15}N edited experiments are employed. Consequently, it is necessary to make use of ^{13}C isotopes in the sequence-specific assignment. With the current availability of ^{13}C glucose, which is the most frequently used ^{13}C source in the preparation of ^{13}C-labeled proteins (details in Chap. 3), the cost of ^{13}C labeling has decreased to an affordable level, and it promises to be even less expensive in the future as isotopic labeling is becoming a standard procedure in NMR sample preparation for structure study. One of the advantages of heteronuclear isotopic labeling is the much larger scalar couplings than those between protons (Fig. 5.5). The larger scalar coupling constant means that the magnetization transfer between heteronuclear spins is more efficient, resulting in more intense cross peaks. By introducing a ^{13}C frequency dimension, the resonance ambiguity can be further reduced with higher spectral resolution, and hence the resonance assignment is significantly simplified.

A large number of triple-resonance experiments have been developed and optimized for structure determination of proteins using heteronuclear multidimensional NMR spectroscopy. These experiments make full use of one- and two-bond heteronuclear scalar couplings to correlate backbone and side-chain ^{1}H, ^{15}N, and ^{13}C spins of isotope-labeled proteins. Because the ^{1}J and ^{2}J couplings (shown in Fig. 5.5) generally are relatively large compared to the spectral line width, and independent of conformation, the coherence transfers through these couplings can efficiently compete with the loss of magnetization caused by short transverse relaxation times of macromolecules during the experiment. The nomenclature used for triple-resonance experiments is based on the coherence transfer pathway in the experiment. The name

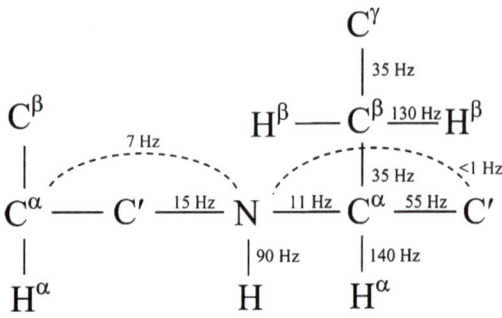

Fig. 5.5 J coupling constants between ^{1}H, ^{15}N, and ^{13}C along a polypeptide chain are used in triple-resonance NMR experiments for resonance assignments

of an experiment is formed by the spins involved in the coherence transfer in the order following the transfer pathway. Spins are given in parentheses if their chemical shifts do not evolve. The name is formed only by the first half of the coherence transfer when the magnetization of the proton spin is transferred to neighboring spins and then back to the proton by the same pathway. This type of experiments is called "out and back." For instance, in a 3D triple-resonance "out and back" type experiment, the magnetization of an amide proton (H^N) is transferred to the C^α carbon (CA) via the amide nitrogen (N) and then back to the amide proton via the amide nitrogen, and hence it is called HNCA. If the magnetization is transferred further from C^α to carbonyl carbon C' (CO) and the chemical shift of C' evolves instead of C^α, the experiment is named HN(CA)CO. The parentheses reflect that the chemical shifts of C^α carbons involved in the magnetization transfer do not evolve during the experiment.

Although a variety of triple-resonance experiments are available, only a certain number of experiments are frequently used to obtain backbone and side-chain assignments, Scalar couplings used in the experiments are summarized in Fig. 5.5. The combination of the 3D experiments HNCA (Ikura et al. 1990a, b; Kay et al. 1990; Grzesiek and Bax 1992c) and HN(CO)CA (Ikura et al. 1990a, b; Kay et al. 1991; Grzesiek and Bax 1992c) can be used to establish backbone sequential connectivities by connecting the resonance frequencies of spins with those of preceding residue. The HN(CO)CA provides correlations of H_i^N and N_i of residue i with C_{i-1}^α chemical shifts of the preceding residue $i-1$, whereas the HNCA correlates the chemical shifts of H_i^N and N_i with both C_i^α and C_{i-1}^α because the scalar coupling of 7 Hz has a similar size to $^1J_{NC\alpha}$ of 11 Hz. However, the correlation between N and C^α within a residue is not observed in HN(CO)CA due to the very weak $^2J_{NC'}$ (<1 Hz). These experiments are relatively sensitive and usually yield an excellent signal-to-noise ratio. Another pair of 3D experiments, CBCANH and CBCA(CO)NH (Grzesiek and Bax 1992a, b), are used to extend the connectivities from the backbone to C^β, which provide useful information on the type of amino acids. Assignment of C^α and C^β will also be used to establish side-chain connectivity in addition to the backbone assignment. For proteins with more than 130 residues, ambiguities in assignment sometimes still remain based on the data obtained from the above four experiments. Then, a pair of experiments HNCO (Ikura et al. 1990a; Grzesiek and Bax 1992c) and HN(CA)CO (Clubb et al. 1992; Kay et al. 1994; Engelke and Rüterjans 1995), which spread H–N correlations into C' chemical shifts, are generally sufficient to completely resolve the spectral overlap. In practice, these six experiments will provide the backbone assignments. Since these experiments are the most commonly used for backbone assignments, they are discussed in detail in the following sections. Note that the HN(CA)CO has a lowest sensitivity compared to the other five experiments. Therefore, it may require more transients to achieve a sizable S/N for spectral analysis and sometimes a subset of cross peaks may not be observable. However, with the use of a cryogenic probe, the problem of low sensitivity for the HN(CA)CO can be readily overcome.

Assignment of aliphatic side-chain proton and carbon resonances is necessary for high resolution structure determination using NOE distance constraints. Since the assignments of C^α and C^β have been obtained by the backbone assignment, the side-chain resonance can be assigned by transferring the magnetization of backbone amide protons to side-chain spins. (H)CC(CO)NH–TOCSY (Montelione et al. 1992; Grzesiek et al. 1993; Logan et al. 1993; Lyons and Montelione 1993) correlates the chemical shifts of H_i^N and N_i to all C_{i-1}^{aliph} via couplings of $^1J_{NC'}$ and $^1J_{CC}$, while the HCCH–TOCSY (Fesik et al. 1990; Bax et al. 1990a; Olejniczak et al. 1992; Majumdar et al. 1993) provides correlations among aliphatic protons and carbons within residues. A 3D ^{15}N HSQC–TOCSY experiment can also be used to confirm and obtain complete assignment of aliphatic ^{1}H resonances. For proteins larger than 30 kDa, the line widths of the aliphatic ^{1}H and ^{13}C resonances increase to the size comparable with the scalar couplings used for coherence transfer in the HCCH–TOCSY experiment, causing the significant reduction in sensitivity. An HCCH–NOESY experiment has been used to correlate the side-chain resonances of large proteins via NOESY in replacement of the TOCSY. TROSY-type experiments have successfully been used to establish backbone assignments for proteins as large as 110 kDa.

5.3 General Procedure of Setup and Data Processing for 3D Experiments

All pulses for the transmitter and decouplers should be properly calibrated (refer to instrument calibration). The spectral windows are set on a 500 MHz spectrometer to 6,500, 1,600, 7,500, and 1,750 Hz for ^{1}H, ^{15}N, aliphatic ^{13}C, and ^{13}C', respectively, or on a 600 MHz spectrometer, 8,000, 2,000, 9,000, and 2,100 Hz for ^{1}H, ^{15}N, aliphatic ^{13}C, and ^{13}C', respectively. The carrier frequency for ^{1}H is always set to H$_2$O resonance for aqueous samples, whereas the decoupler offset frequencies are set to the center of the chemical shift range of the indirectly observed heteronuclear nuclei. The center of chemical shift is commonly selected as 118, 177, 54, and 40 ppm for ^{15}N, ^{13}C', ^{13}C$^{\alpha/\beta}$, and C$^\alpha$, respectively. The data should be collected with minimal digital resolutions for all directions, which typically are 0.025, 0.10, 1.00, 1.70, and 0.35 ppm/point for ^{1}H, ^{1}H indirect, ^{15}N, aliphatic ^{13}C, and ^{13}C', respectively. If the data are collected with an acquisition time of 64 ms, 128 complex points the ^{1}H indirect dimension, 35 complex points for ^{15}N, and 40 for the ^{13}C dimension, this provides digital resolutions of 0.014, 0.10, 0.95, 1.50, and 0.35 ppm/point, respectively.

First, 1D ^{1}H and ^{1}H–^{15}N HSQC data are collected to check the condition of the sample. Then, a 1D trail spectrum of the 3D experiment is collected with 1 scan and the t_1 and t_2 increments set to 1. The receiver gain needs to be optimized using the 1D trail spectrum. Next, two 2D slices of the 3D experiment should be collected with 16 scans and the optimized gain to make certain that the setup is correct. Other parameters include a predelay set to 1.0–1.3 s and 32 steady state scans (or dummy scans). The number of transients is set to the minimum number needed for phase cycling, usually eight scans. If more scans are necessary, the data may be collected with fewer increments for each of the indirect dimensions.

The acquired data are converted to a specific format before they can be processed using NMRpipe (Delaglio et al. 1995, 2004), NMRview (Johnson 2004) or other software. The 2D versions of the 3D data are extracted and processed with the same procedure as in 3D processing (see following text). The spectra are phased and phase parameters are used for 3D data processing. In the observed ^{1}H dimension, all data are usually processed in the same way. Data are processed with linear prediction and a solvent suppression filter applied to the time domain data prior to apodization by a 70°-shifted squared sine-bell function, zero filling to 1,024 complex points, Fourier transform, and phasing. In the ^{13}C or indirect ^{1}H dimension, after size-doubling by mirror image linear prediction the data are apodized by 70° squared sine-bell function, zero filled to 256 complex points, Fourier transformed, and phased. In the ^{15}N dimension, the data sizes are doubled by mirror image linear prediction followed by apodization with a squared cosine-bell function, zero-filling to 128 complex points, Fourier transformation, and phasing.

5.4 Experiments for Backbone Assignments

Six 3D triple-resonance experiments are discussed; these were mentioned previously as being the most common experiments for backbone assignments using uniformly ^{15}N- and ^{13}C-labeled proteins. Since most of the experiments include the magnetization transfer of amide protons that are exchangeable with water under normal sample conditions, water-flip-back is used in all experiments for water suppression to avoid the saturation of amide proton magnetization, which provides superior

5.4 Experiments for Backbone Assignments

sensitivity along with the use of PEP sensitivity enhancement. In addition, the experiments utilize gradient echoes to select desired coherence pathways in combination with limited phase cycling for PEP sensitivity enhancement and quadrature detection in the indirect dimensions. The pulse sequences discussed here are for three-channel configuration of the spectrometer, that is, RF band-specific pulses for C′ and C$^\alpha$ are applied via the same RF channel of the spectrometer by applying off-resonance pulses for one of the carbon regions, which is denoted by "off" in Table 5.1. For a four-channel spectrometer, pulses for C′ and C$^\alpha$ may be delivered on separate channels with the carrier set at each carbon region. In that case, on-resonance-selective pulses can be applied. When the magnetization of ^{15}N is selectively transferred to C′ or C$^\alpha$ via INEPT, selective carbon pulses are used to avoid exciting unwanted carbon magnetization because the scalar couplings $^1J_{NC\alpha}$ and $^1J_{NC'}$ have a similar size (11 Hz and 15 Hz, respectively). However, nonselective carbon pulses can be applied if the magnetization transfer is from aliphatic protons to their attached carbons, because $^2J_{HC}$ and $^3J_{HC}$, if they are not zero, are much smaller than $^1J_{HC}$ (~135 Hz) and the INEPT optimized for $^1J_{HC}$ does not have magnetization transferred via $^2J_{HC}$ and $^3J_{HC}$.

5.4.1 HNCO and HNCA

If C$^\alpha$ pulses are exchanged for C′ pulses, the HNCO and HNCA experiments have identical pulse sequences as shown in Fig. 5.6. Of the six experiments mentioned previously, the HNCO has the highest sensitivity. The HNCO spectrum contains the correlations of H$_i^N$, N$_i$, and C′$_{i-1}$ but not C′$_i$ since $^2J_{N(i)C'(i)}$ has a value close to zero, whereas the HNCA gives two sets of backbone correlations, within the residue and with the preceding residue: H$_i^N$, N$_i$, and C$_i^\alpha$ as well as H$_i^N$, N$_i$, and C$_{i-1}^\alpha$ due to the fact that $^1J_{NC'}$ (11 Hz) and $^2J_{NC'}$ (15 Hz) are comparable in size.

The pulse sequences utilize the "out and back" transfer pathway to transfer the magnetization:

$$\text{HNCO}: H_N \xrightarrow{J_{NH}} N \xrightarrow{J_{NC'}} C'(t_1) \xrightarrow{J_{NC'}} N(t_2) \xrightarrow{J_{NH}} H_N(t_3) \qquad (5.36)$$

$$\text{HNCA}: H_N \xrightarrow{J_{NH}} N \xrightarrow{J_{NC\alpha}} C_\alpha(t_1) \xrightarrow{J_{NC\alpha}} N(t_2) \xrightarrow{J_{NH}} H_N(t_3) \qquad (5.37)$$

The magnetization originating from the amide proton H^N is transferred to the attached N via the $^1J_{NH}$ coupling during the INEPT sequence. The magnetization is then transferred to C′ in HNCO (or C$^\alpha$ in HNCA) in the next INEPT during which the coupling of H^N with N is removed by DIPSI-2 proton decoupling. The delay 2δ is set to 1/(2$^1J_{NC'}$) approximately 13.5 ms in HNCO [11.0 ms for HNCA, 2δ ≈ 1/(4$^1J_{NC\alpha}$)] for optimizing the refocus of the coupling. After the chemical shift of carbonyl C′ evolves during the t_1 evolution period, the magnetization is

Table 5.1 Parameters and correlations of the 3D experiments

	Experiments						
Parameter	HNCO	HNCA	HN(CO)CA	HN(CA)CO	CBCANH	CBCA(CO)NH	H(C)CH-TOCSY
^{13}C offset	C'	C^α	C^α	C'	$C^{\alpha,\beta}$	$C^{\alpha,\beta}$	C^{aliph}
"On" pulses	C' 90°, 180°	C^α 90°, 180°	C^α 90°, 180°	C' 90°, 180°	$C^{\alpha,\beta}$ 90°, 180°	$C^{\alpha,\beta}$ 90°, 180°	C^{ali} 90°, 180°
"Off" pulses	C^α 180°	C' 180°	C' 90°, 180°	C^α 90°, 180°	C' 180°	C' 90°, 180°	C' 180°
δ	13.5	11.0	13.5	11.0	11.0		
δ_1			7.0	3.4	1.8	1.8	1.8
δ_2							0.8
δ_3							1.1
τ	2.7	2.7	2.7	2.7		4.5	
τ_1					2.7	2.7	
τ_2					2.2	2.2	
τ_3					11.0	3.5	
T						13.5	
					3.6	3.6	
Correlations	H_i, N_i, C'_{i-1}	$H_i, N_i, C^\alpha_i, C^\alpha_{i-1}$	H_i, N_i, C^α_{i-1}	H_i, N_i, C'_i, C'_{i-1}	$H_i, N_i, C^{\alpha,\beta}_i, C^{\alpha,\beta}_{i-1}$	$H_i, N_i, C^{\alpha,\beta}_{i-1}$	$H^{aliph}_i, H^{aliph}_j, C^{aliph}_i$

"^{13}C offset" represents ^{13}C offset frequency (the center of ^{13}C spectral window) with values of 177, 54, 40, and 47 ppm for C', $C^{\alpha,\beta}$, C^α, and C^{aliph}, respectively
"On" or "Off" pulses are on-resonance or off-resonance pulses (see Chap. 4)
All delays are in milliseconds. Delay δ_2 is the duration of the last gradient (G_z), except for H(C)CH-TOCSY

5.4 Experiments for Backbone Assignments

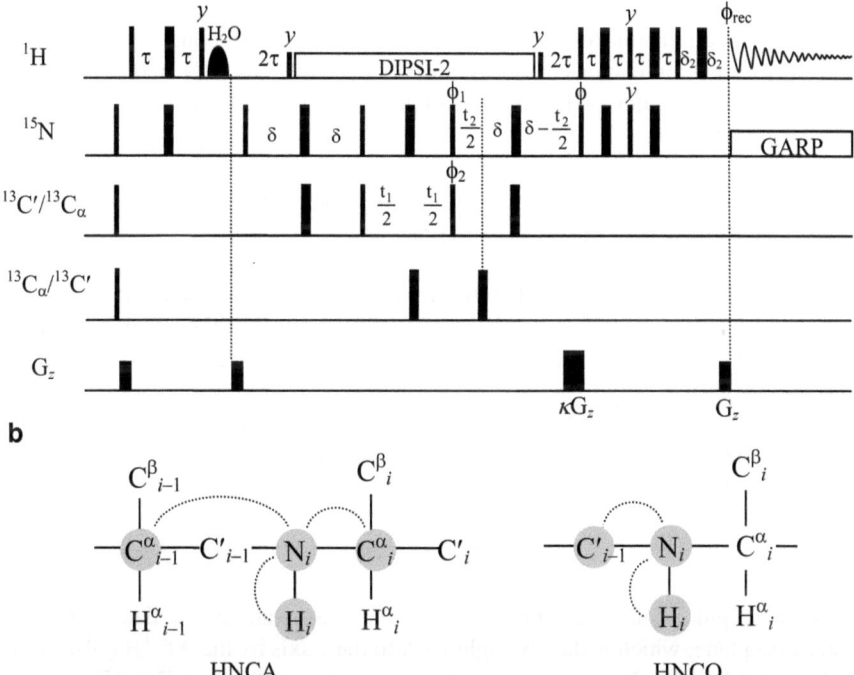

Fig. 5.6 HNCO and HNCA pulse sequences (**a**) and the corresponding resonance connectivities (**b**). (**a**) The delay $\tau = 2.7$ ms, $\delta = 13.5$ ms for HNCO, 11.0 ms for HNCA, δ_2 equals the G_z gradient pulse length. All pulses are x phased, except that $\phi_1 = x, x, -x, -x$; $\phi_2 = x, -x$, +States–TPPI, and $\phi_{rec} = x, -x, -x, x$. For PEP, the signs of κ and ϕ are inverted: $\kappa = \pm 10$, $\phi = \pm x$. (**b**) The *dotted lines* linking shaded nuclei indicate the observed correlations for the experiments: H_i, N_i, and C_{i-1}^α and C_i^α in HNCA or H_i, N_i, and C'_{i-1} in HNCO

transferred via the "back" pathway to N and then back to its origin, H. The chemical shift evolution of N proceeds during the "back" pathway to minimize the loss in magnetization caused by relaxation during the INEPT periods involving C', which have long delays due to the weak $^2J_{NC'}$ coupling. Spin decoupling is applied to suppress evolution under the scalar coupling interaction of J_{NH}. The other important role of ^1H decoupling is to ensure that IP coherence N_y (^{15}N spins) is generated throughout the INEPT sequences. DIPSI decoupling increases the experimental sensitivity compared to decoupling by refocusing 180° pulses because the IP coherence is not affected by T_1 relaxation in contract to the AP coherence N_xH_z that relaxes with both T_1 and T_2 relaxation times. The two 90° ^1H pulses next to the DIPSI-2 sequence are low-power water-selective pulses that restore the water magnetization along $+z$ at the end of the DIPSI-2 sequence. At the beginning of pulse sequence, the 90° N and C pulses combined with the gradients after them are used to dephase all N and C magnetizations so that the observed magnetization solely originates from ^1H (contributes to the FID).

5.4.1.1 Product Operator Description of the HNCO Experiment

The magnetization transfer can be described in terms of product operators. The operators for magnetizations of H^N, N, and C' are denoted as H, N, and C. After first 90° ^{1}H pulse generates $-H_y$ magnetization, the first INEPT sequence yields AP magnetization:

$$-H_y \xrightarrow{\tau \rightarrow \pi(H_x + N_x) \rightarrow \tau} -2H_xN_z \xrightarrow{\left(\frac{\pi}{2}\right)(H_y + N_x)} -2H_zN_y \qquad (5.38)$$

in which τ is set to $1/(4J_{NH})$ so that the coefficient $\sin(2\pi J_{NH}\tau)$ of the AP magnetization has a maximum. The selective water 90° pulse brings the water magnetization to the z axis and the gradient destroys any residual transverse water magnetization.

At the end of 2τ, the ^{15}N magnetization is refocused to be IP coherence with respect to the ^{1}H spin when $2\tau = 1/(2J_{NH})$, which leads to $\sin(\pi J_{NH}2\tau) = 1$ and $\cos(\pi J_{NH}2\tau) = 0$:

$$-2H_zN_y \xrightarrow{2\tau} N_x \qquad (5.39)$$

A 90° ^{1}H pulse before the DIPSI-2 sequence brings water magnetization back to transverse plane, which is then brought back to the z axis by the 90° ^{1}H pulse at the end of DIPSI-2. The N magnetization transfers to C' via the next INEPT:

$$H_x \xrightarrow{\delta \rightarrow \pi(N_x + C_x) \rightarrow \delta} 2N_yC_z \xrightarrow{\left(\frac{\pi}{2}\right)(N_x + C_x)} -2N_zC_y \qquad (5.40)$$

The delay δ is set to $1/(4J_{NC'})$ causing $\sin(2\pi J_{NC'}\delta) = 1$. During t_1 evolution both N and C^α spins are decoupled from C' by the refocusing 180° pulses in the middle of t_1. After the C' chemical shift evolves, the magnetization is transferred from C' back to N by the two 90° pulses:

$$-2N_zC_y \xrightarrow{t_1} 2N_zC_y \cos(\Omega_C t_1) \xrightarrow{\left(\frac{\pi}{2}\right)(N_x + C_x)} -2N_yC_z \cos(\Omega_C t_1) \qquad (5.41)$$

The sine-modulated MQ component generated during t_1 will not contribute to the observable magnetization and hence is omitted from consideration. During the constant time t_2 evolution period, ^{1}H is still being decoupled by DIPSI-2 decoupling and C^α is decoupled by the 180° C' refocus pulse during the initial period of t_2 and by 180° C^α and N pulses in the remaining t_2 period. The NC' coupling evolves for the entire 2δ ($(t_2/2) + \delta + \delta - (t_2/2) = 2\delta$) set to $\delta = 1/(4J_{NC'})$ to have $\sin(2\pi J_{NC'}\delta) = 1$, whereas the N chemical shift evolves for the period of t_2, resulting in:

$$-2N_yC_z \cos(\Omega_C t_1) \xrightarrow{2\delta} N_x \cos(\Omega_C t_1) \qquad (5.42)$$

5.4 Experiments for Backbone Assignments

$$N_x \cos(\Omega_C t_1) \xrightarrow{t_2} N_x \cos(\Omega_C t_1) \cos(\Omega_N t_2) + N_y \cos(\Omega_C t_1) \sin(\Omega_N t_2) \quad (5.43)$$

The N magnetization evolves during the 2τ period after DIPSI-2 under the influence of J_{NH} coupling, resulting in two AP coherences:

$$N_x \cos(\Omega_C t_1) \cos(\Omega_N t_2) + N_y \cos(\Omega_C t_1) \sin(\Omega_N t_2) \xrightarrow{2\tau}$$
$$2H_z N_y \cos(\Omega_C t_1) \cos(\Omega_N t_2) - 2H_z N_x \cos(\Omega_C t_1) \sin(\Omega_N t_2) \quad (5.44)$$

The last period in the pulse sequence is for the PEP sensitivity enhancement sequence in which the two terms of the coherence can be considered separately. The first FID is acquired with $\phi = y$ and $\kappa = 10$ and the second one is recorded with both the phase ϕ and gradient κ inverted. To simplify, the coefficients are temporarily dropped and will be retrieved later, since they are not changed by the PEP sequence. During the PEP, the evolution from the first term yields:

$$2H_z N_y \xrightarrow{\left(\frac{\pi}{2}\right)(H_x + N_x)} -2H_y N_z \quad (5.45)$$

$$-2H_y N_z \xrightarrow{\tau \to \pi(H_x + N_x) \to \tau} H_x \xrightarrow{\left(\frac{\pi}{2}\right)(H_y + N_y)} -H_z \quad (5.46)$$

$$-H_z \xrightarrow{\tau \to \pi(H_x + N_x) \to \tau} H_z \xrightarrow{\left(\frac{\pi}{2}\right)H_x} -H_y \quad (5.47)$$

The second term evolves:

$$-2H_z N_x \xrightarrow{\left(\frac{\pi}{2}\right)(H_x + N_x)} 2H_y N_x \quad (5.48)$$

Because $H_y N_x$ is MQ coherence, it does not evolve under the influence of scalar coupling.

$$2H_y N_x \xrightarrow{\tau \to \pi(H_x + N_x) \to \tau} 2H_y N_x \xrightarrow{\left(\frac{\pi}{2}\right)(H_y + N_y)} -2H_y N_z \quad (5.49)$$

$$-2H_y N_z \xrightarrow{\tau \to \pi(H_x + N_x) \to \tau \to \left(\frac{\pi}{2}\right)H_x} H_x \quad (5.50)$$

After retrieving the coefficients for both terms, the observable magnetization has a form of:

$$H_x \cos(\Omega_C t_1) \sin(\Omega_N t_2) - H_y \cos(\Omega_C t_1) \cos(\Omega_N t_2) \quad (5.51)$$

The last ^{1}H 180° pulse is used to invert ^{1}H magnetization for coherence selection by the gradient. The delay δ_2 is set to be long enough for the gradient plus gradient recovery time.

The second data set is recorded with inverted ϕ and gradient factor κ. For the first term,

$$2H_zN_y \xrightarrow{\left(\frac{\pi}{2}\right)(H_x - N_x)} 2H_yN_z \tag{5.52}$$

$$2H_yN_z \xrightarrow{\tau \to \pi(H_x + N_x) \to \tau} -H_x \xrightarrow{\left(\frac{\pi}{2}\right)(H_y + N_y)} H_z$$

$$\xrightarrow{\tau \to \pi(H_x + N_x) \to \tau} -H_z \xrightarrow{\left(\frac{\pi}{2}\right)H_x} H_y \tag{5.53}$$

For the second term,

$$-2H_zN_x \xrightarrow{\left(\frac{\pi}{2}\right)(H_x - N_x)} 2H_yN_x \tag{5.54}$$

which is the same as in the first FID. Therefore, this term reminds the same:

$$2H_yN_x \xrightarrow{\tau \to \pi(H_x + N_x) \to \tau} 2H_yN_x \xrightarrow{\left(\frac{\pi}{2}\right)(H_y + N_y)}$$

$$-2H_yN_z \xrightarrow{\tau \to \pi(H_x + N_x) \to \tau \to \left(\frac{\pi}{2}\right)H_x} H_x \tag{5.55}$$

The second FID has a form of:

$$H_x \cos(\Omega_C t_1) \sin(\Omega_N t_2) + H_y \cos(\Omega_C t_1) \cos(\Omega_N t_2) \tag{5.56}$$

The two obtained FIDs are:

$$\begin{cases} H_x \cos(\Omega_C t_1) \sin(\Omega_N t_2) - H_y \cos(\Omega_C t_1) \cos(\Omega_N t_2) \\ H_x \cos(\Omega_C t_1) \sin(\Omega_N t_2) + H_y \cos(\Omega_C t_1) \cos(\Omega_N t_2) \end{cases} \tag{5.57}$$

Addition of the two FIDs gives rise to a PEP data set which contains the observable magnetization described by:

$$2H_x \cos(\Omega_C t_1) \sin(\Omega_N t_2) \tag{5.58}$$

Similarly, subtraction of the two FIDs (FID$_2$–FID$_1$) yields another PEP data set:

$$2H_y \cos(\Omega_C t_1) \cos(\Omega_N t_2) \tag{5.59}$$

5.4 Experiments for Backbone Assignments

The two PEP data sets can be processed to obtain separate 3D spectra. The two spectra are then combined to form a single spectrum with pure absorptive line shapes in all three dimensions. Alternatively, the data sets can be combined before Fourier transformation.

5.4.1.2 HNCO Experiment Setup

The transmitter channel is set to the ^{1}H frequency with the carrier frequency on the water resonance. The first decoupler channel is used for ^{13}C with the offset frequency set to the middle of carbonyl C' (177 ppm) whereas the ^{15}N pulses are applied on the second decoupler channel with the offset frequency in the middle of the ^{15}N spectral window (118 ppm). The pulse calibration for ^{1}H includes a 90° hard pulse, a 90° pulse for broadband decoupling (WALTZ16 or DIPSI-2; ~100 μs), and a water-selective 90° pulse (see section on instrument calibration). Pulses for ^{13}C and ^{15}N are not required to be calibrated for every experiment setup, meaning that the pulse lengths can be used repeatedly after they are calibrated periodically. The hard 90° ^{15}N pulse is normally shorter than 40 μs and the 90° pulse length for GARP decoupling is about 250 μs. The ^{13}C pulses used in HNCO are C' 90° pulse nulling at C^α (64.7 μs for 500 MHz and 53.9 μs for 600 MHz), C' 180° pulse nulling at C^α (57.9 μs for 500 MHz and 48.3 μs for 600 MHz), shaped C^α 180° off-resonance pulse. Alternatively, the rectangular on-resonance C' pulses can also be replaced by selective pulses, such as eBURP (Geen and Freeman 1991), or G_4 (90° excitation), and G_3 (180° inversion) with a bandwidth of 50–60 ppm (Emsley and Bodenhausen 1990). The C^α 180° off-resonance decoupling pulses for decoupling are applied by SEDUCE-shaped pulses (Coy and Mueller 1993; see Table 4.1).

The delay τ is optimized to 2.7 ms which is used to cancel the coefficient cos $(2\pi J_{NH}\tau)$ with a J_{NH} of 90 Hz and delay δ is optimized to 13.5 ms, using $\delta = 1/(4^1 J_{NC'})$ with $^1 J_{NC'}$ of 15 Hz for effective refocusing of $^1 J_{NC'}$. These delays are optimized by recording a 1D experiment to obtain the most intense signals. The gradient echo pulses used for coherence selection are set to 2 ms and 200 μs with the strength of approximately 20 Gcm^{-1} for the dephasing and refocusing gradient pulses, respectively. The pulse length of the refocus gradient should be optimized to obtain the best sensitivity. For dephasing water, the gradient pulse is applied with a strength of approximately 15 Gcm^{-1} for a duration of 2 ms.

Before the acquisition can be started for 3D data collection, 2D ^{1}H/^{15}N and ^{1}H/^{13}C slices of the experiments are recorded first to make sure that the number of transients is sufficient, and all parameters are optimized to yield reasonable sensitivity. To collect a ^{1}H/^{15}N 2D slice, the t_1 increment is set to 1 and t_2 increment to 50 while the ^{1}H/^{13}C 2D slice is collected with 100 t_1 increments and a single t_2 increment. Once the 2D spectra indicate that the experiment works correctly, the 3D experiment is recorded with 35 complex points for ^{15}N and 40 for ^{13}C'. HNCO data are processed using the procedure described in Sect. 5.3 on General Procedure of Setup and Data Processing for 3D Experiments.

5.4.1.3 HNCA

After interchanging C' and C$^\alpha$ pulses, the basic setup of HNCA is identical to HNCO with a few changes. The off-resonance C' 180° pulse is generated with an offset of positive 122 ppm (downfield from the carrier). The delay δ is optimized to 11.0 ms using $2\delta = 1/(4^1 J_{NC\alpha})$, according to the coefficients for intra- and inter-residue coherence transfers:

$$\Gamma(^1 J_{NC\alpha}) = \sin(2\pi\ ^1 J_{NC\alpha}\delta) \cos(2\pi\ ^2 J_{NC\alpha}\delta)$$

$$\Gamma(^2 J_{NC\alpha}) = \sin(2\pi\ ^2 J_{NC\alpha}\delta) \cos(2\pi\ ^1 J_{NC\alpha}\delta) \tag{5.60}$$

In addition, the HNCA may require more transients to obtain a good S/N ratio because it is at least 50 % less sensitive than HNCO. Since $^1 J_{NC\alpha}$ (11 Hz) and $^2 J_{NC\alpha}$ (7 Hz) have a similar size, correlations of H$_i$, N$_i$ to both the intra-residue C$^\alpha_i$ and the C$^\alpha_{i-1}$ of the preceding residue are observed in the experiment.

In summary, the ^{1}H magnetization is the starting magnetization that is transferred to ^{15}N by INEPT in both sequences. The time period 2τ lets the AP magnetization evolve to IP magnetization that is further transferred to ^{13}C by the second INEPT. This period is an important step because it allows the coherence after evolving in the real time (RT) t_1 period to be transferred back to ^{15}N by a pair of 90° ^{15}N/^{13}C pulses, which requires fewer 180° pulses in the sequences. In the last step of the coherence transfer pathway, the ^{15}N coherence is transferred back to proton by the reverse INEPT after evolving in the constant time (CT) t_2 period. Note that the RT ^{13}C and CT ^{15}N evolutions provide optimal sensitivity compared to other combinations. The other 2τ period after DIPISI-2 converts the two IP coherences into AP ones that are ready to be manipulated by the PEP enhancement method. This delay–DIPSI–delay combination appears frequently in multidimensional NMR spectroscopy and will be met again in following pulse sequences.

5.4.2 HN(CO)CA

The HN(CO)CA correlates amide ^{1}H and ^{15}N chemical shifts (H$_i$ and N$_i$) with the ^{13}C chemical shift of preceding residue, C$^\alpha_{i-1}$, which is used to establish the backbone sequential connectivity across the peptide bond. By combining the information provided here with that from the HNCA, both intra-residue and inter-residue connectivities can be distinguished. Since the stronger one-bond spin couplings ($^1 J_{NC'}$ and $^1 J_{C'C\alpha}$) are utilized in the HN(CO)CA compared to couplings ($^1 J_{NC\alpha}$ and $^2 J_{NC\alpha}$) used in the HNCA, the magnetization transfer is more efficient. Therefore, the HN(CO)CA is more sensitive than the HNCA.

5.4 Experiments for Backbone Assignments

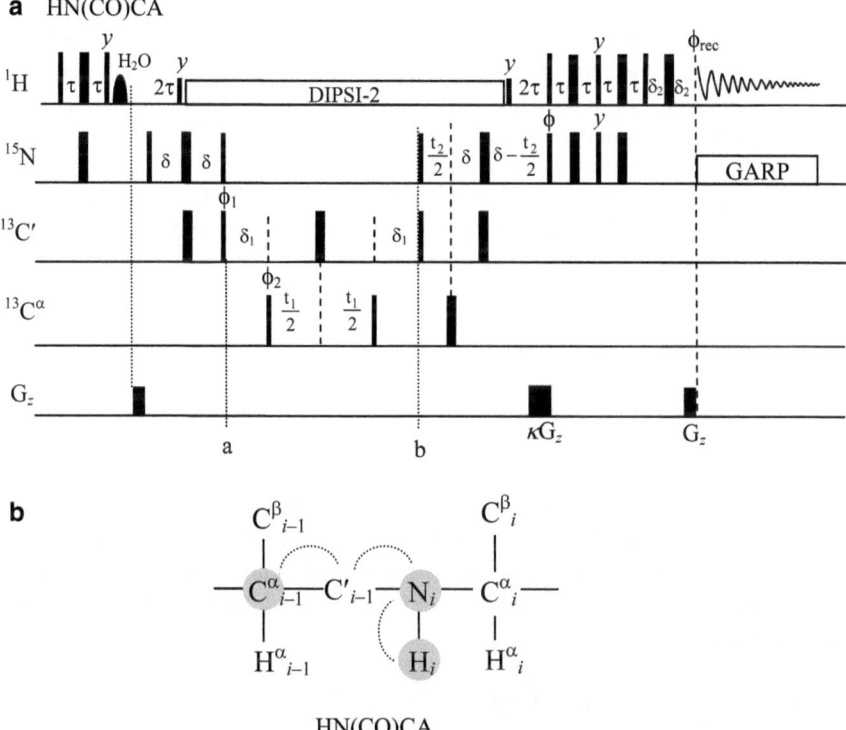

Fig. 5.7 Pulse sequence of HN(CO)CA (**a**) and the resonance connectivity (**b**). The experiment is derived from the HNCO. An HMQC-type sequence is used to transfer the magnetization from C′ to C^α and then back to C′. The phase cycles of ϕ_1 and ϕ_2 are same as in the HNCO. For PEP, the signs of κ and ϕ are inverted: $\kappa = \pm 10$, $\phi = \pm x$. The delays are set to $\tau = 2.7$ ms, $\delta = 13.5$ ms, $\delta_1 = 7.0$ ms, and δ_2 equals the G_z gradient pulse length. The ^{13}C offset frequency is set on C^α as in the HNCO experiment. Pulses on C^α are selective pulses that do not excite C′, while C′ pulses are off-resonance-selective pulses (shaped pulses, see Table 4.1). (**b**) The *dotted lines* indicate the magnetization transfer and the observed correlations are between the shaded H_i, N_i, and C^α_{i-1} relayed via C′$_{i-1}$

The HN(CO)CA pulse sequence shown in Fig. 5.7 is derived from the HNCO by transferring the magnetization from N to C′ and then from C′ to C^α via $^1J_{NC'}$ and $^1J_{C'C\alpha}$, respectively. The pulse sequence uses an "out and back" transfer pathway:

$$H \xrightarrow{J_{NH}} N \xrightarrow{J_{NC'}} C' \xrightarrow{J_{C'C\alpha}} C_\alpha(t_1) \xrightarrow{J_{C'C\alpha}} C' \xrightarrow{J_{NC'}} N(t_2) \xrightarrow{J_{NH}} H(t_3) \quad (5.61)$$

After the ^1H magnetization is transferred to N in the INEPT, the N magnetization is transferred to carbonyl C′ via the second INEPT sequence. The transfer of C′ magnetization to C^α is achieved by an HMQC-type sequence, which has been demonstrated to be superior for the $J_{C'C\alpha}$ magnetization transfer. After C^α chemical shift evolves during t_1, the magnetization is transferred back through the reverse pathway. The evolution of N chemical shift takes place during t_2 along the reverse pathway, before the coherence is transferred to H for detection. A PEP building block sequence combined with gradients is used to achieve sensitivity enhancement. The ^{1}H-selective pulse on water is to align the water magnetization along the z axis.

The magnetization transfer in the pulse sequence can be described by product operators which are denoted as H, N, C', and C$^\alpha$ for amide ^{1}H, ^{15}N, carbonyl ^{13}C, and ^{13}C$^\alpha$, respectively. The coherence transfer from ^{1}H to ^{15}N and then to ^{13}C' is in the same way as that in an HNCO experiment ((5.38)–(5.40)):

$$H_z \xrightarrow{\left(\frac{\pi}{2}\right)H_x \to \tau \to \pi(H_x+N_x) \to \tau \to \left(\frac{\pi}{2}\right)(H_y+N_x) \to 2\tau} N_x$$

$$\xrightarrow{\delta \to \pi(N_x+C'_x) \to \delta \to \left(\frac{\pi}{2}\right)(N_x+C'_x)} -2N_zC'_y \quad (5.62)$$

in which $\tau = 1/(4J_{NH})$, 2.7 ms, and $\delta = 1/(4J_{NC'})$, 13.5 ms. Note that the transverse ^{15}N magnetizations evolve for 2τ period under the influence of J_{NH}, because DIPSI-2 ^{1}H decoupling is turned off before the 2τ delay. The magnetization is then transferred to C$^\alpha$ and in turn back to C' after t_1 evolution during an HMQC-type sequence:

$$-2N_zC'_y \xrightarrow{\delta_1 \to \left(\frac{\pi}{2}\right)C^\alpha_x} -4N_zC'_xC^\alpha_y$$

$$\xrightarrow{\frac{t_1}{2} \to \pi C'_x \to \frac{t_1}{2} \to \left(\frac{\pi}{2}\right)C^\alpha_x \to \delta_1} -2N_zC'_y\cos(\Omega_Ct_1) \quad (5.63)$$

in which $\delta_1 = 1/(2J_{C'C\alpha})$. In practice, the delay δ_1 is set to a value in between $1/(3J_{C'C\alpha})$ and $1/(2J_{C'C\alpha})$, 7.0 ms. The magnetization transfer in the reverse INEPT sequence and sensitivity-enhancement PEP sequence in the HN(CO)CA, which are identical to those in HNCO, yield two FIDs.

$$-2N_zC'_y\cos(\Omega_Ct_1) \xrightarrow{\left(\frac{\pi}{2}\right)(N_x+C_x)} 2N_yC'_z\cos(\Omega_Ct_1)$$

$$\xrightarrow{2\delta} -N_x\cos(\Omega_Ct_1)$$

$$\xrightarrow{t_2} -N_x\cos(\Omega_Ct_1)\cos(\Omega_Nt_2) - N_y\cos(\Omega_Ct_1)\sin(\Omega_Nt_2)$$

$$\xrightarrow{2\tau} -2H_zN_y\cos(\Omega_Ct_1)\cos(\Omega_Nt_2) + 2H_zN_x\cos(\Omega_Ct_1)\sin(\Omega_Nt_2) \quad (5.64)$$

in which τ and δ are the same as in (5.62). The first FID after PEP is given by:

$$H_x\cos(\Omega_Ct_1)\sin(\Omega_Nt_2) - H_y\cos(\Omega_Ct_1)\cos(\Omega_Nt_2) \quad (5.65)$$

and the second FID is proportional to:

$$H_x\cos(\Omega_Ct_1)\sin(\Omega_Nt_2) + H_y\cos(\Omega_Ct_1)\cos(\Omega_Nt_2) \quad (5.66)$$

5.4 Experiments for Backbone Assignments

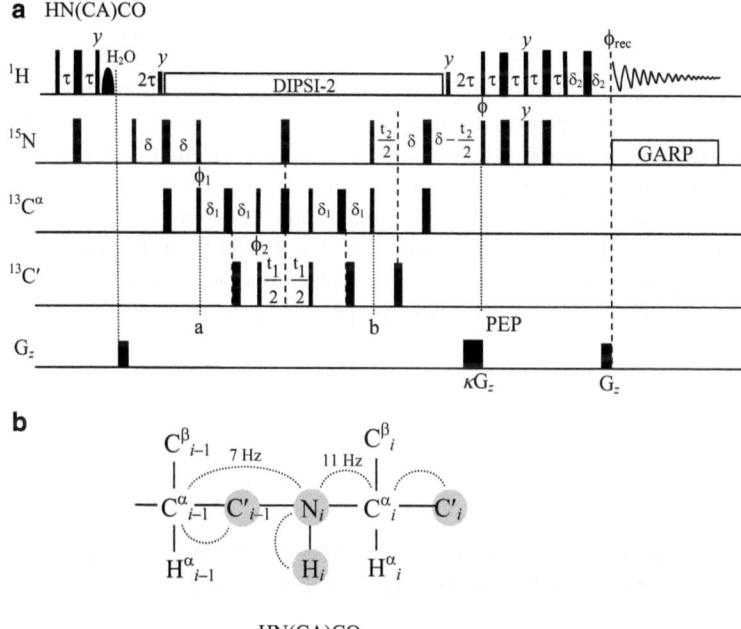

Fig. 5.8 Pulse sequence of HN(CA)CO (**a**) and the resonance connectivity (**b**). The experiments is derived from HNCA. An INEPT sequence is used for all steps of the out-back transfer. All phases are the same as in HN(CO)CA sequence. Delay are set to $\tau = 2.7$ ms, $\delta_1 = 3.4$ ms, and $\delta = 11.0$ ms. The ^{13}C offset frequency is set on C' as in HNCA. Pulses on C' are selective pulses that do not excite C^α, while C^α pulses are off-resonance-selective pulses (shaped pulses, see Table 4.1). (**b**) The *dotted lines* indicate the magnetization transfer pathways and the observed correlations are indicated by the shaded nuclei: H_i, N_i, C'_{i-1}, and C'_i relayed via C^α_{i-1}, and C^α_i

The two FIDs are acquired separately and stored in different memory locations. The addition and subtraction of the two FIDs yield two data sets that can be processed separately and combined to a single spectrum with absorptive phase.

5.4.3 HN(CA)CO

The HN(CA)CO experiment provides correlations of H^N_i, N_i, and C'_i chemical shifts. Similar to HNCA, the sequential connectivities from H^N_i, N_i to C'_{i-1} are also observed in the experiment owing to the comparable size of the scalar couplings $^1J_{NC\alpha}$ and $^2J_{NC\alpha}$. Because of the low sensitivity caused by the weak couplings, a fraction of the correlations may not be observed in the experiment.

A sensitivity-enhanced version of the HN(CA)CO experiment shown in Fig. 5.8 is derived from the HNCA experiment. The pulse sequence uses the "out and back" transfer pathway:

$$H \xrightarrow{J_{NH}} N \xrightarrow{J_{NC_\alpha}} C_\alpha \xrightarrow{J_{C_\alpha C'}} C'(t_1) \xrightarrow{J_{C_\alpha C'}} C_\alpha \xrightarrow{J_{NC_\alpha}} N(t_2) \xrightarrow{J_{NH}} H(t_3) \quad (5.67)$$

After the magnetization originating from amide protons is transferred to the N spins via the INEPT sequence, the amide ^{15}N magnetization is transferred to C^α via the next INEPT sequence. The C^α magnetization is further transferred to C', followed by the evolution of C' chemical shifts. During the reverse transfer path, the coherence is transferred back via C^α and amide ^{15}N spins to amide protons for detection. The amide ^{15}N chemical shifts evolve during the CT evolution t_2. The product operator terms leading to observable magnetization throughout the transfers at the indicated time points in the pulse sequence are given by:

$$H_z \xrightarrow{\begin{array}{c}\left(\frac{\pi}{2}\right)H_x \to \tau \to \pi(H_x+N_x) \to \tau \to \left(\frac{\pi}{2}\right)(H_y+N_x) \to 2\tau\end{array}} N_x$$

$$\xrightarrow{\begin{array}{c}\delta \to \pi(N_x+C_x^\alpha) \to \delta \to \left(\frac{\pi}{2}\right)(N_x+C_x^\alpha)\end{array}} -2N_zC_y^\alpha \quad (5.68)$$

in which $\tau = 1/(4J_{HN})$, 2.7 ms, and $\delta = 1/(8J_{NC\alpha})$, 11.0 ms, to optimize for both intra- and inter-residue coherence transfer [see (5.60)]. At point a:

$$-2N_zC_y^\alpha \xrightarrow{\begin{array}{c}\delta_1 \to \pi(C_x^\alpha + C'_x) \to \delta_1 \to \left(\frac{\pi}{2}\right)(C_x^\alpha + C'_x)\end{array}} -4N_zC_x^\alpha C'_y$$

$$\xrightarrow{\begin{array}{c}\frac{t_1}{2} \to \pi(C_x^\alpha + N_x) \to \frac{t_1}{2} \to \left(\frac{\pi}{2}\right)(C_x^\alpha + C'_x)\end{array}} -4N_zC_x^\alpha C'_z \cos(\Omega_{C'}t_1)$$

$$\xrightarrow{\begin{array}{c}\delta_1 \to \pi(C_x^\alpha + C'_x) \to \delta_1 \to \left(\frac{\pi}{2}\right)(N_x + C_x^\alpha)\end{array}} 2N_yC_z^\alpha \cos(\Omega_{C'}t_1) \quad (5.69)$$

in which $\delta_1 = 1/(4J_{CC\alpha})$ and is optimized to a value between $1/(4J_{CC\alpha})$ and $1/(6J_{CC\alpha})$, 3.4 ms. At point b:

$$2N_yC_z^\alpha \cos(\Omega_{C'}t_1) \xrightarrow{2\delta} -N_x \cos(\Omega_{C'}t_1)$$
$$\xrightarrow{t_2} -N_x \cos(\Omega_{C'}t_1)\cos(\Omega_N t_2) - N_y \cos(\Omega_{C'}t_1)\sin(\Omega_N t_2)$$
$$\xrightarrow{2\tau} -2H_zN_y \cos(\Omega_{C'}t_1)\cos(\Omega_N t_2) + 2H_zN_x \cos(\Omega_{C'}t_1)\sin(\Omega_N t_2) \quad (5.70)$$

in which τ and δ are set as in (5.68). The magnetization $-N_x$ and $-N_y$ evolve under the influence of J_{NH} coupling for 2τ period because DIPSI-2 ^{1}H decoupling ends before the 2τ delay. The PEP sequence yields two FIDs:

$$H_x \cos(\Omega_{C'}t_1)\sin(\Omega_N t_2) - H_y \cos(\Omega_{C'}t_1)\cos(\Omega_N t_2)$$
$$H_x \cos(\Omega_{C'}t_1)\sin(\Omega_N t_2) + H_y \cos(\Omega_{C'}t_1)\cos(\Omega_N t_2) \quad (5.71)$$

which are stored in different memory locations and are treated as described above.

5.4.4 CBCANH

The 3D CBCANH experiment correlates resonances of H_i^N and N_i with C_i^α and C_i^β, and C_{i-1}^α and C_{i-1}^β carbons. The correlation to residue $i - 1$ is caused by the similar values of the couplings of C_{i-1}^α and C_i^α to N_i. For each amide H or N resonance, there are four cross peaks in the spectrum, which provides information about the amino acid type of residues and reduces the effect of C^α–H degeneracy for resonance assignment. The experiment allows not only complete sequential assignments but also assignment of side-chain carbons, which is useful information for complete assignment of aliphatic resonances using 3D HCCH–TOCSY experiment. The experiment is a transfer type, which makes use of a relay-COSY sequence to transfer C^β to C^α before the coherence is transferred to amide N spins:

$$H_{\alpha,\beta} \xrightarrow{J_{C_{\alpha\beta}H}} C_{\alpha,\beta}(t_1) \xrightarrow{J_{C_\alpha C_\beta}} C_\alpha \xrightarrow{J_{C_\alpha N}} N(t_2) \xrightarrow{J_{NH}} H(t_3) \tag{5.72}$$

The product operators representing the observable coherence throughout the transfers in the pulse sequence (Fig. 5.9) are given by:

$$H_z \xrightarrow{\left(\frac{\pi}{2}\right)H_x \to \delta_1 \to \pi(H_x + C_x^{\alpha,\beta}) \to \delta_1 \to \left(\frac{\pi}{2}\right)(H_y + C_x^{\alpha,\beta})} -2H_zC_y^{\alpha,\beta} \tag{5.73}$$

in which $C_y^{\alpha,\beta} = C_y^\alpha + C_y^\beta$. Because of the ^{1}H decoupling, the CH coupling evolves for only t_1 period during the CT evolution period:

$$-2H_zC_y^{\alpha,\beta} \xrightarrow{\pi J_{C_{\alpha\beta}H}\tau_1} -2H_zC_y^{\alpha,\beta}\cos(\pi J_{C_{\alpha\beta}H}\tau_1) + C_x^{\alpha,\beta}\sin(\pi J_{C_{\alpha\beta}H}\tau_1) \tag{5.74}$$

The delay δ_1 is set to 1.8 ms and is optimized for $J_{C_{\alpha\beta}H}$ couplings, $1/(4J_{C_{\alpha\beta}H})$ and τ_1 is set to 2.2 ms to simultaneously optimize the CH_n coherence transfers of methine, methylene, and methyl groups (Fig. 5.10). The gradient pulse after the 90° ^{1}H pulse dephases all transverse magnetization. The C_α–C_β coupling evolves as follows:

$$C_x^{\alpha,\beta} \xrightarrow{t_1} (C_x^\alpha + C_x^\beta)\cos(\Omega_{C_{\alpha,\beta}}t_1) \tag{5.75}$$

$$\xrightarrow{2T} [C_x^\alpha \cos(2\pi J_{C_\alpha C_\beta}T) + 2C_y^\alpha C_z^\beta \sin(2\pi J_{C_\alpha C_\beta}T)$$
$$+ C_x^\beta \cos(2\pi J_{C_\alpha C_\beta}T) + 2C_z^\alpha C_y^\beta \sin(2\pi J_{C_\alpha C_\beta}T)]\cos(\Omega_{C_{\alpha,\beta}}t_1) \tag{5.76}$$

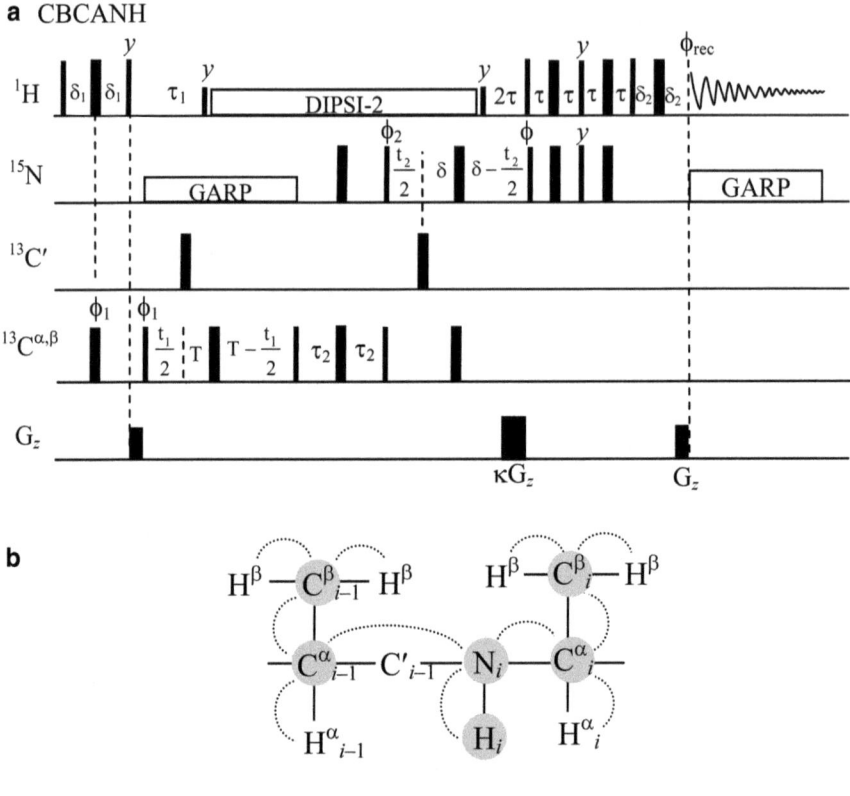

Fig. 5.9 Pulse sequence of CBCANH (**a**) and the resonance connectivity (**b**). The magnetization starts from the H^α and H^β protons and is finally transferred to HN protons and an INEPT sequence is used for all steps of the out-back transfer. The phase cycles of ϕ_1 and ϕ_2 are same as in the HNCO (Fig. 5.6). The delays are set to $T = 3.6$ ms, $\tau = 2.7$ ms, $\tau_1 = 2.2$ ms, $\tau_2 = 11.0$ ms, $\delta = 11.0$ ms, $\delta_1 = 1.8$ ms, and δ_2 equals the G_z gradient pulse length. (**b**) The *dotted lines* indicate the magnetization transfer pathways and the observed correlations are indicated by the shaded nuclei: H_i, N_i, C^α_{i-1}, C^β_{i-1}, C^α_i, and C^β_i

The time constant T is set to $1/(8J_{C_\alpha C_\beta}) = 3.6$ ms and only C^α magnetization terms, C^α_x and $2C^\alpha_z C^\beta_y$ will be transferred to N in the following steps, because $J_{C_\beta N}$ is negligibly small.

$$(C^\alpha_x + 2C^\alpha_z C^\beta_y)\cos(\Omega_{C_{\alpha,\beta}} t_1) \xrightarrow{\left(\frac{\pi}{2}\right) C^{\alpha,\beta}_x} (C^\alpha_x - 2C^\alpha_y C^\beta_z)\cos(\Omega_{C_{\alpha,\beta}} t_1)$$

$$\xrightarrow{2\pi J_{C_\alpha C_\beta} \tau_2} C^\alpha_x \cos(\Omega_{C_{\alpha,\beta}} t_1)[\cos(2\pi J_{C_\alpha C_\beta} \tau_2) + \sin(2\pi J_{C_\alpha C_\beta} \tau_2)] \quad (5.77)$$

Note that $J_{C_\alpha C_\beta}$ coupling evolves for $2\tau_2$ period. The first term $[\cos(2\pi J_{C_\alpha C_\beta} \tau_2)]$ is from C^α and the second term $[\sin(2\pi J_{C_\alpha C_\beta} \tau_2)]$ is from C^β. Then, the coherence is transferred to N via $^1J_{C_{\alpha\beta}N}$ and $^2J_{C_{\alpha\beta}N}$ during the INEPT sequence:

5.4 Experiments for Backbone Assignments

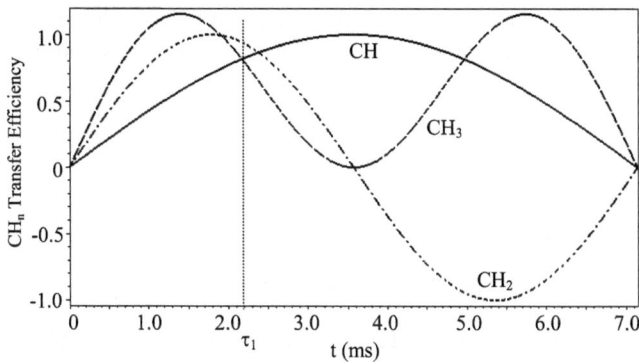

Fig. 5.10 Coherence transfer efficiency via INEPT for different CH_n groups with a J_{CH} coupling constant of 140 Hz. The delay τ_1 in CBCANH experiment is set to 2.2 ms (indicated by the *dotted vertical line*) to simultaneously optimize the CH_n coherence transfers of methine, methylene, and methyl groups

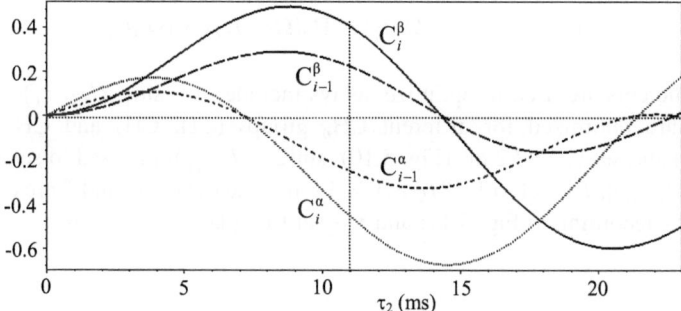

Fig. 5.11 Transfer amplitude as a function of τ_2 in the CBCANH experiment using (6.78). The *dotted vertical line* at 11.0 ms indicates the value of τ_2 optimized for all four coherence transfer pathways using $^1J_{C\alpha C\beta} = 35$ Hz, $^1J_{C\alpha N} = 11$ Hz, and $^2J_{C\alpha N} = 7$ Hz. The opposite sign of the C^α transfer amplitudes to the C^β indicates that C^α cross peaks of the CBCANH spectrum have opposite sign relative to the C^β cross peaks, except glycine

$$\xrightarrow{\tau_2 \to \pi(C_x^{\alpha,\beta} + N_x) \to \tau_2} 2C_y^\alpha N_z \cos(\Omega_{C_{\alpha,\beta}} t_1)[\sin(2\pi\,^1J_{C_\alpha N}\tau_2)\cos(2\pi\,^2J_{C_\alpha N}\tau_2)\cos(2\pi J_{C_\alpha C_\beta}\tau_2)$$
$$+ \sin(2\pi\,^1J_{C_\alpha N}\tau_2)\cos(2\pi\,^2J_{C_\alpha N}\tau_2)\sin(2\pi J_{C_\alpha C_\beta}\tau_2)$$
$$+ \cos(2\pi\,^1J_{C_\alpha N}\tau_2)\sin(2\pi\,^2J_{C_\alpha N}\tau_2)\cos(2\pi J_{C_\alpha C_\beta}\tau_2)$$
$$+ \cos(2\pi\,^1J_{C_\alpha N}\tau_2)\sin(2\pi\,^2J_{C_\alpha N}\tau_2)\sin(2\pi J_{C_\alpha C_\beta}\tau_2)]$$

(5.78)

in which the first two terms correspond to the coherence transfers of C_i^α and C_i^β, while the last two terms are for those of C_{i-1}^α and C_{i-1}^β, respectively. The delay τ_2 is optimized to 11 ms (Fig. 5.11) according to the coefficients in (5.78).

During the next constant time evolution period, the coherence evolves for $\delta - (\tau - \delta) + \tau = 2\delta$ under the influence of $J_{C_\alpha N}$, while the evolution caused by J_{NH} lasts for a period of 2τ:

$$2C_y^\alpha N_z \cos(\Omega_{C_{\alpha,\beta}} t_1) \xrightarrow{\left(\frac{\pi}{2}\right)(C_x^\alpha + N_x)} -2N_y C_z^\alpha \cos(\Omega_{C_{\alpha,\beta}} t_1)$$

$$\xrightarrow{2\delta} N_x \cos(\Omega_{C_{\alpha,\beta}} t_1)$$

$$\xrightarrow{t_2} N_x \cos(\Omega_{C_{\alpha,\beta}} t_1) \cos(\Omega_N t_2) + N_y \cos(\Omega_{C_{\alpha,\beta}} t_1) \sin(\Omega_N t_2)$$

$$\xrightarrow{2\tau} 2H_z N_y \cos(\Omega_{C_{\alpha,\beta}} t_1) \cos(\Omega_N t_2) - 2H_z N_x \cos(\Omega_{C_{\alpha,\beta}} t_1) \sin(\Omega_N t_2) \quad (5.79)$$

The delay δ is set to 11 ms according to Fig. 5.11 and τ is set to $1/(4J_{NH}) = 2.7$ ms. The coherence of the two FIDs obtained via PEP scheme can be described by:

$$H_x(\Omega_{C_{\alpha,\beta}} t_1) \sin(\Omega_N t_2) - H_y(\Omega_{C_{\alpha,\beta}} t_1) \cos(\Omega_N t_2)$$

$$H_x(\Omega_{C_{\alpha,\beta}} t_1) \sin(\Omega_N t_2) + H_y(\Omega_{C_{\alpha,\beta}} t_1) \cos(\Omega_N t_2) \quad (5.80)$$

The coefficients are used to optimize delays include: $n \times \sin(\pi J_{CH_n} \tau_1) \cos^{n-1}(\pi J_{CH_n} \tau_1)$ are optimized for different CH_n groups (CH, CH_2 and CH_3 groups; Fig. 5.10) and set $\tau_1 = 2.2$ ms (Fig. 5.10); $\sin(2\pi^1 J_{C^{\alpha\beta}H} \delta_1)$ is used to set $\delta_1 = 1.8$ ms $[1/(4J_{C^{\alpha\beta}H})]$; T is set to $1/(8J_{C_\alpha C_\beta}) = 3.6$ ms; the delays τ_2 and δ are optimized to 11.0 ms according to Fig. 5.11; and τ is set to $1/(4J_{NH}) = 2.7$ ms.

5.4.5 CBCA(CO)NH

The 3D CBCA(CO)NH experiment correlates the chemical shifts of both C_{i-1}^α and C_{i-1}^β carbons with H_i^N and N_i. By correlating both C^α and C^β simultaneously, the degeneracy of C^α–H resonances can be eliminated. The resonances of C^α and C^β provide information about the amino acid type of the preceding residue in addition to the sequential connectivity. The experiment is derived from the CBCANH, utilizing a relayed COSY sequence to transfer C^β to C^α before the coherence is transferred to the C' spins:

$$H_{\alpha,\beta} \xrightarrow{J_{C_{\alpha\beta}H}} C_{\alpha,\beta}(t_1) \xrightarrow{J_{C_\alpha C_\beta}} C_\alpha \xrightarrow{J_{C_\alpha C''}} C' \xrightarrow{J_{C'N}} N(t_2) \xrightarrow{J_{NH}} H(t_3) \quad (5.81)$$

The product operators representing the observable coherence throughout the transfers after the 90° $C^{\alpha\beta}$ pulse at the end of t_1 evolution period in the CBCA (CO)NH pulse sequence (Fig. 5.12) is the same as in the CBCANH:

5.4 Experiments for Backbone Assignments

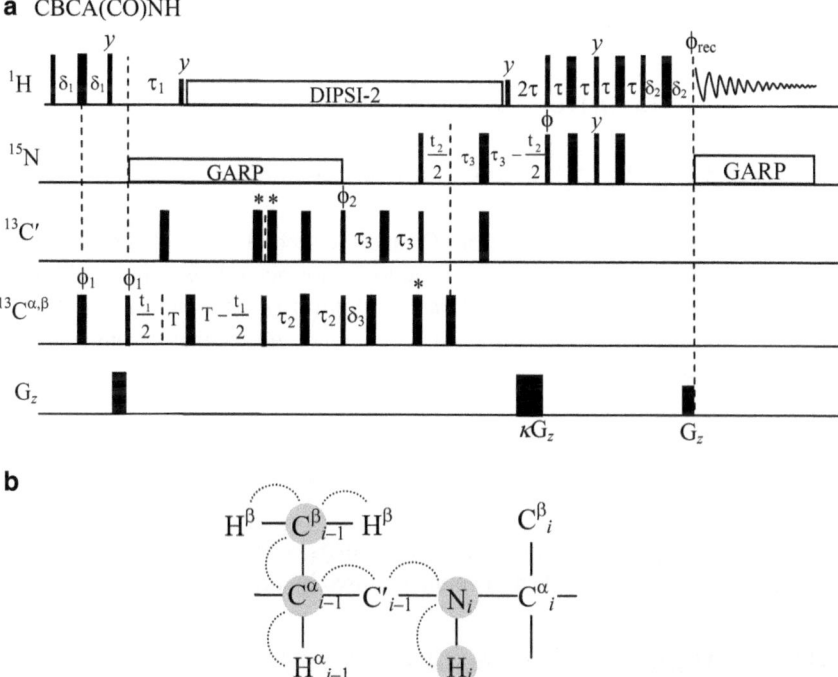

Fig. 5.12 Pulse sequence of CBCA(CO)NH (**a**) and the resonance connectivity (**b**). The magnetization starts from the H^α and H^β protons and is finally transferred to HN protons. The two 180° C′ pulses labeled by *asterisks* are used to compensate for the phase error introduced by the previous and subsequent off-resonance 180° C′ pulses, respectively, while the C^α 180° pulse labeled by *asterisks* is used to compensate for the phase error introduced by the previous off-resonance 180° C^α pulse. The phase cycle ϕ_1 and ϕ_2 are the same as in the HNCO (Fig. 5.6). The delays are set to $\tau = 2.7$ ms, $\tau_1 = 2.2$ ms, $\tau_2 = 3.5$ ms, $\tau_3 = 13.5$ ms, $T = 3.6$ ms, $\delta_1 = 1.8$ ms, $\delta_3 = 4.5$ ms, and δ_2 equals the G_z gradient pulse length. (**b**) The *dotted lines* indicate the magnetization transfer pathways and the observed correlations are indicated by the shaded nuclei: H_i, N_i, to C^α_{i-1}, C^β_{i-1}, relayed via C'_{i-1}

$$H_z \xrightarrow{\left(\frac{\pi}{2}\right)H_x \rightarrow \delta_1 \rightarrow \pi(H_x+C^{\alpha,\beta}_x) \rightarrow \delta_1 \rightarrow \left(\frac{\pi}{2}\right)(H_y+C^{\alpha,\beta}_x) \rightarrow \tau_1} -C^{\alpha,\beta}_x$$

$$\xrightarrow{t_1 \rightarrow 2T \rightarrow \left(\frac{\pi}{2}\right)C^{\alpha,\beta}_x} (C^\alpha_x - 2C^\alpha_y C^\beta_z)\cos(\Omega_{C_{\alpha,\beta}} t_1) \qquad (5.82)$$

The time constant T is set to $1/(8J_{C_\alpha C_\beta})$, 3.6 ms. The delay δ_1 is set to 1.8 ms and is optimized for $^1J_{C^{\alpha\beta}H}$ couplings and τ_1 is set to 2.2 ms to simultaneously optimize the CH_n coherence transfers of methine, methylene, and methyl groups (Fig. 5.10). The gradient pulse after the 90° ^1H pulse dephases all transverse magnetization. Only C^α magnetization terms, C^α_x and $2C^\alpha_z C^\beta_y$, will be transferred to C′ in the following steps, because $^2J_{C_\beta C'}$ is negligibly small.

$$(C^\alpha_x - 2C^\alpha_y C^\beta_z)\cos(\Omega_{C_{\alpha,\beta}}t_1) \xrightarrow{\tau_2 \to \pi(C^\alpha_x + C'_x) \to (2\pi J_{C_\alpha C_\beta}\tau_2 + 2\pi J_{C_\alpha C'}\tau_2)}$$

$$2C^\alpha_y C'_z \cos(\Omega_{C_{\alpha,\beta}}t_1) \cos(2\pi J_{C_\alpha C_\beta}\tau_2) \sin(2\pi J_{C_\alpha C'}\tau_2)$$

$$+ 2C^\alpha_y C'_z \cos(\Omega_{C_{\alpha,\beta}}t_1) \sin(2\pi J_{C_\alpha C_\beta}\tau_2) \sin(2\pi J_{C_\alpha C'}\tau_2) \quad (5.83)$$

in which the first term originates from C^α and the second from C^β coherence (the term $-2C^\alpha_y C^\beta_z$). The delay τ_2 is optimized to 3.5 ms according to the coefficients $\cos(2\pi J_{C_\alpha C_\beta}\tau) \sin(2\pi J_{C_\alpha C'}\tau)$ and $\sin(2\pi J_{C_\alpha C_\beta}\tau) \sin(2\pi J_{C_\alpha C'}\tau)$ with $J_{C_\alpha C'} = 55$ Hz, and $J_{C_\alpha C_\beta} = 35$ Hz. The 90°_x C^α pulse gives coherence $2C^\alpha_z C'_z$ that is converted into $-2C'_y C^\alpha_z$ by the 90°_x C' pulse:

$$2C^\alpha_y C'_z \cos(\Omega_{C_{\alpha,\beta}}t_1) \xrightarrow{\left(\frac{\pi}{2}\right)(C^\alpha_x + C'_x)} -2C'_y C^\alpha_z \cos(\Omega_{C_{\alpha,\beta}}t_1) \quad (5.84)$$

During the next period, the coherence evolves for $\delta_3 - (\tau_3 - \delta_3) + \tau_3 = 2\delta_3$ under the influence of the $J_{C_\alpha C'}$, while the evolution caused by $J_{NC'}$ lasts for a period of $2\tau_3$:

$$-2C'_y C^\alpha_z \cos(\Omega_{C_{\alpha,\beta}}t_1) \xrightarrow{2\delta_3} C'_x \cos(\Omega_{C_{\alpha,\beta}}t_1) \sin(2\pi J_{C_\alpha C'}\delta_3)$$

$$\xrightarrow{\tau_3 \to \pi(N_x + C^\alpha_x) \to \tau_3 \to \left(\frac{\pi}{2}\right)(N_x + C^\alpha_x)} -2N_y C^\alpha_z \cos(\Omega_{C_{\alpha,\beta}}t_1) \sin(2\pi J_{C_\alpha C'}\delta_3) \sin(2\pi J_{NC'}\tau_3) \quad (5.85)$$

The delay δ_3 is optimized to $1/(4J_{C_\alpha C'})$, which is approximately 4.5 ms and τ_3 to $1/(4J_{NC'})$, approximately 13.5 ms. During t_2 constant evolution, the coherence evolves for $2\tau_3$ under the interaction of $J_{NC'}$:

$$-2N_y C^\alpha_z \cos(\Omega_{C_{\alpha,\beta}}t_1) \xrightarrow{2\tau_3} N_x \cos(\Omega_{C_{\alpha,\beta}}t_1)$$

$$\xrightarrow{t_2} N_x \cos(\Omega_{C_{\alpha,\beta}}t_1) \cos(\Omega_N t_2) + N_y \cos(\Omega_{C_{\alpha,\beta}}t_1) \sin(\Omega_N t_2) \quad (5.86)$$

The delay τ_3 is the same as in (5.85) to optimize the $J_{NC'}$ coherence transfer. The ^{15}N magnetization is then transferred back to protons during the last INEPT:

$$\xrightarrow{2\tau} 2H_z N_y \cos(\Omega_{C_{\alpha,\beta}}t_1) \cos(\Omega_N t_2) - 2H_z N_x \cos(\Omega_{C_{\alpha,\beta}}t_1) \sin(\Omega_N t_2) \quad (5.87)$$

in which τ is set to $1/(4J_{NH}) = 2.7$ ms for maximizing the NH coherence transfer. The two FIDs obtained via the PEP sequence for the sensitivity enhancement are given by:

$$H_x \cos(\Omega_{C_{\alpha,\beta}}t_1) \sin(\Omega_N t_2) - H_y(\Omega_{C_{\alpha,\beta}}t_1) \cos(\Omega_N t_2)$$

$$H_x \cos(\Omega_{C_{\alpha,\beta}}t_1) \sin(\Omega_N t_2) + H_y(\Omega_{C_{\alpha,\beta}}t_1) \cos(\Omega_N t_2) \quad (5.88)$$

In the same way as described earlier to rearrange the data by addition and subtraction, the final spectrum has a maximum gain in sensitivity by a factor of $\sqrt{2}$, which comes from the factor-two gain of signal intensity reduced by the increase of noise by a factor of $\sqrt{2}$. The delays and ^{13}C pulses for the experiment are set up as described in Fig. 5.12.

5.5 Experiments for Side-Chain Assignment

5.5.1 HCCH–TOCSY

The HCCH–TOCSY experiment correlates all aliphatic ^{1}H and ^{13}C spins within residues, and is used to assign aliphatic ^{1}H and ^{13}C resonances and connect the side-chain chemical shifts with the backbone assignments. The experiment spreads a 2D TOCSY spectrum into a third or even fourth dimension to reduce signal overlapping. The magnetization originating at aliphatic ^{1}H is transferred to the directly attached ^{13}C via the one-bond scalar coupling ($^1J_{CH} \approx 140$ Hz) after the evolution of ^{1}H chemical shift during t_1. The ^{13}C chemical shift evolves during t_2 before the ^{13}C magnetization is transferred further to the neighboring carbons via one-bond $^1J_{CC}$ (30–40 Hz) during the isotropic mixing period. The ^{13}C magnetization is transferred to other ^{13}C spins within the spin system (amino acid residue) by isotropic mixing of ^{13}C spins via $^1J_{CC}$, which is more efficient than the ^{1}H isotropic mixing in the 2D TOCSY experiment via $^3J_{HH}$. Finally, the magnetization dispersed along the carbon side-chain is transferred to proton spins for detection. The transfer pathway for the pulse sequence is:

3D H(C)CH–TOCSY:

$$H_i(t_1) \xrightarrow{^1J_{HC} \approx 140 \text{ Hz}} C_i \xrightarrow{^1J_{C_iC_j} 30-40 \text{ Hz (TOCSY)}} C_j(t_2) \xrightarrow{^1J_{HC} \approx 140 \text{ Hz}} H(t_3)$$

(5.89)

4D HCCH–TOCSY:

$$H_i(t_1) \xrightarrow{^1J_{HC} \approx 140 \text{ Hz}} C_i(t_2) \xrightarrow{^1J_{C_iC_j} 30-40 \text{ Hz (TOCSY)}} C_j(t_3) \xrightarrow{^1J_{HC} \approx 140 \text{ Hz}} H(t_4)$$

(5.90)

Aliphatic proton TOCSY-type correlations of a given spin system are located at the ^{13}C 2D slices through all carbon frequencies within the spin system (same residue), whereas a ^{1}H 2D slice gives the total correlations between the ^{1}H and ^{13}C involved in the same spin system. As a result, the 3D H(C)CH–TOCSY (Fig. 5.13; Sattler et al. 1995a, b) provides aliphatic ^{1}H and ^{13}C chemical shifts for the complete assignment of side-chain ^{1}H and ^{13}C resonances. Because the coherence

a 3D H(C)CH–TOCSY

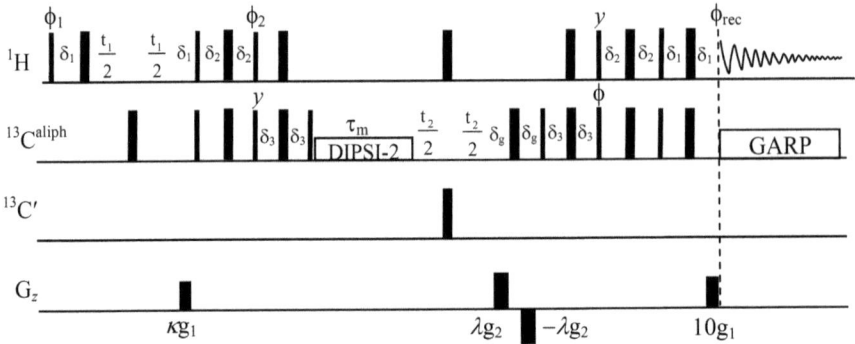

b

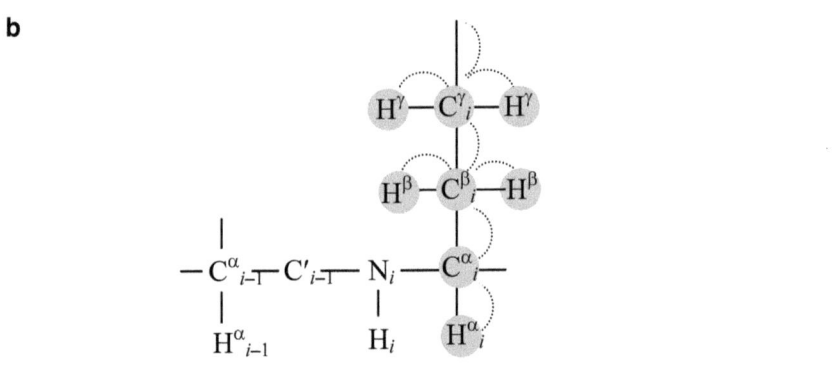

H(C)CH–TOCSY

Fig. 5.13 Pulse sequence for the 3D H(C)CH–TOCSY (**a**) and the resonance connectivity (**b**). All pulses are x phase including DIPSI-2, except $\phi_1 = x, -x$, $\phi_2 = y$, $\phi = y$, and $\phi_{rec} = x, -x$. Four FIDs are acquired with $\kappa = 1, -1, -1, 1$, $\lambda = 1, -1, 1, -1$, $\phi_2 = y, y, -y, -y$, and $\phi = y, -y, y, -y$, and stored in different memory locations for the double sensitivity enhancement using both t_1 and t_2 coherences. Gradients are used as $g_2 = 18g_1$. $\delta_1 = 1.8$ ms, $\delta_2 = 0.8$ ms, $\delta_3 = 1.1$ ms, and δ_g equals the G_2 gradient pulse length. (**b**) The observed correlations are between shaded nuclei

is transferred via large coupling constants, the experiment provides excellent sensitivity that can further be improved by a maximum factor of $2(\sqrt{2}\sqrt{2})$ utilizing double sensitivity enhancement after the t_1 and t_2 evolution periods. The sensitivity-enhanced gradient pulse sequence also provides superior water suppression, which makes it possible to obtain the spectra for side-chain ^{1}H and ^{13}C assignment with the same aqueous samples used for the backbone assignment so that the ^{2}H effect on the chemical shifts of ^{1}H and ^{13}C can be avoided.

The description of the coherence transfer by product operators for the building blocks in the H(C)CH–TOCSY experiment (Fig. 5.13) is given as follows, without considering the second sensitivity enhancement sequence. The initial magnetization of H^{aliph} is transferred to C^{aliph} by an HMQC-type sequence, during which the J_{CH}

5.5 Experiments for Side-Chain Assignment

coupling evolves for a period of $2\delta_1$ ($\delta_1 - \frac{t_1}{2} + \frac{t_1}{2} + \delta_1$), while the H^{aliph} chemical shift evolves for a period of t_1 (both 180° ^{1}H and ^{13}C pulses reverse the sign of delays for the J_{CH} coupling, whereas only 180° ^{1}H pulse reverses the sign of the delays for ^{1}H chemical shift-free precession):

$$H_z \xrightarrow{\left(\frac{\pi}{2}\right)H_x} -H_y \quad (5.91)$$

$$-H_y \xrightarrow{\delta_1 \to \pi H_x \to \frac{t_1}{2} \to \pi C_x \to \frac{t_1}{2} \to \delta_1}$$

$$2H_xC_z \sin(2\pi J_{\text{CH}}\delta_1)\cos(\Omega_H t_1) + 2H_yC_z \sin(2\pi J_{\text{CH}}\delta_1)\sin(\Omega_H t_1) \quad (5.92)$$

in which two terms will yield two components for the first sensitivity enhancement. The delay δ_1 is optimized for J_{CH}, which is set to $1/(4J_{\text{CH}}) = 1.8$ ms in the experiment. For simplicity, the terms of t_1 evolution are omitted temporarily from the equation and will be retrieved later. For the first term of t_1, the 90° ^{1}H and ^{13}C pulses transfer the coherence to C^{aliph} followed by sensitivity enhancement:

$$2H_xC_z \xrightarrow{\left(\frac{\pi}{2}\right)(H_x + C_x)} -2H_xC_y$$

$$\xrightarrow{\delta_2 \to \pi(H_x + C_x) \to \delta_2} -2H_xC_y \xrightarrow{\left(\frac{\pi}{2}\right)(H_y + C_y)} 2H_zC_y \sin(2\pi J_{\text{CH}}\delta_2)$$

$$\xrightarrow{\delta_3 \to \pi(H_x + C_x) \to \delta_3 \to \left(\frac{\pi}{2}\right)C_x} -C_x \sin(2\pi J_{\text{CH}}\delta_2)\sin(2\pi n J_{\text{CH}}\delta_3) \quad (5.93)$$

in which n is for different carbon multiplicities, CH$_n$. The MQ coherence (2H$_x$C$_y$) does not evolve under the influence of scalar coupling, J_{CH} during δ_2. The coefficients are different for different carbon multiplicities. Delay δ_2 is chosen as 0.8 ms approximated for both methylene and methyl carbon (the maximum transfer at $1/(8J_{\text{CH}})$ for methylene and $0.196/(2J_{\text{CH}})$ for methyl carbons) and δ_3 is set to 1.1 ms to optimize the coherence transfer simultaneously for all carbon multiplicities.

After the DIPSI-2 isotropic mixing pulse sequence, a portion of the C$_x$ coherence is transferred to neighboring carbons throughout the spin system:

$$-C_x \xrightarrow{\tau_m} -\sum_{k}^{K} C_x^k \quad (5.94)$$

During mixing time τ_m, the IP C$_x$ magnetization is transferred to neighbor carbons C$_x^k$ within residues via J_{CC} coupling. After retrieving the t_1 and coefficient terms, the above coherence becomes:

$$-\sum_{k}^{K} C_x^k \cos(\Omega_H t_1) \xrightarrow{\frac{t_2}{2} \to \pi(C_x) \to \frac{t_2}{2}}$$

$$-\sum_{k}^{K} C_x^k \cos(\Omega_H t_1)\cos(\Omega_{C^k} t_2) + \sum_{k}^{K} C_y^k \cos(\Omega_H t_1)\sin(\Omega_{C^k} t_2) \quad (5.95)$$

The 180° C_x^{aliph} pulse inverts the coherence and the gradient echo dephases any coherence generated due to imperfect refocusing by the 180° pulse. The isotropic mixing period is usually set to 25–30 ms with a field strength of approximately 8 kHz ($pw_{90} = \sim 30$ μs).

The two terms produced by t_2 evolution are used to obtain an additional set of FIDs by the second PEP sequence, resulting in a set of four FIDs to be recorded and stored in separated memory locations. The evolution of the first t_2 term throughout the remaining sequence is given by:

$$-C_x^k \xrightarrow{\left(\frac{\pi}{2}\right)C_x \to \delta_3 \to \pi(H_x+C_x) \to \delta_3} -2H_zC_y^k \xrightarrow{\left(\frac{\pi}{2}\right)(H_y+C_y)} -2H_xC_y^k$$

$$\xrightarrow{\delta_2 \to \pi(H_x+C_x) \to \delta_2} -2H_xC_y^k \xrightarrow{\left(\frac{\pi}{2}\right)(H_x+C_x)} -2H_xC_z^k$$

$$\xrightarrow{\delta \to \pi(H_x+C_x) \to \delta} H_y \tag{5.96}$$

After bringing in the coefficients of t_1 and t_2, the observable magnetization from the first t_1 term with first t_2 term is given by:

$$\sum_k^K H_y \cos(\Omega_H t_1) \cos(\Omega_{C^k} t_2) \tag{5.97}$$

The second t_1 term in (5.92) throughout the pulse sequence can be described as:

$$2H_x C_z \cos(\Omega_H t_1) + 2H_y C_z \sin(\Omega_H t_1) \tag{5.98}$$

$$2H_y C_z \xrightarrow{\left(\frac{\pi}{2}\right)(H_x+C_x)} -2H_z C_y \xrightarrow{\delta_2 \to \pi(H_x+C_x) \to \delta_2} C_x \xrightarrow{\left(\frac{\pi}{2}\right)(H_y+C_y)} -C_z$$

$$\xrightarrow{\delta_3 \to \pi(H_x+C_x) \to \delta_3} C_z \xrightarrow{\left(\frac{\pi}{2}\right)C_x} -C_y \xrightarrow{\tau_m} -\sum_k^K C_y^k \tag{5.99}$$

$$-\sum_k^K C_y^k \cos(\Omega_H t_1) \xrightarrow{\frac{t_2}{2} \to \pi(C_x+N_x) \to \frac{t_2}{2}}$$

$$-\sum_k^K C_y^k \sin(\Omega_H t_1)\cos(\Omega_{C^k} t_2) + \sum_k^K C_x^k \sin(\Omega_H t_1)\sin(\Omega_{C^k} t_2) \tag{5.100}$$

From the first t_2 term of the above equation, the coherence is transferred as:

$$-C_y^k \xrightarrow{\left(\frac{\pi}{2}\right)C_x} -C_z^k \xrightarrow{\delta_3 \to \pi(H_x+C_x) \to \delta_3} C_z^k \xrightarrow{\left(\frac{\pi}{2}\right)(H_y+C_y)} C_x^k$$

$$\xrightarrow{\delta_2 \to \pi(H_x+C_x) \to \delta_2} 2H_zC_y^k \xrightarrow{\left(\frac{\pi}{2}\right)(H_x+C_x)} -2H_yC_z^k$$

$$\xrightarrow{\delta \to \pi(H_x+C_x) \to \delta} H_x \tag{5.101}$$

5.5 Experiments for Side-Chain Assignment

After bringing in the coefficients of t_1 and t_2, the observable magnetization from the second t_1 term with first t_2 term is given by:

$$\sum_{k}^{K} H_x \sin(\Omega_H t_1) \cos(\Omega_{C^k} t_2) \tag{5.102}$$

Using the result for the second t_1 and first t_2 ((5.98)–(5.101)) during the PEP sequence:

$$-C_y^k \xrightarrow{\text{PEP}} H_x \tag{5.103}$$

the term from first t_1 and second t_2 in (5.95) gives:

$$C_y^k \xrightarrow{\text{PEP}} -\sum_{k}^{K} H_x \cos(\Omega_H t_1) \sin(\Omega_{C^k} t_2) \tag{5.104}$$

Similarly, the term of second t_1 and second t_2 in (5.100) after the PEP sequence is given by:

$$C_x^k \xrightarrow{\text{PEP}} -\sum_{k}^{K} H_y \sin(\Omega_H t_1) \sin(\Omega_{C^k} t_2) \tag{5.105}$$

The coherence transfers for the four terms generated after t_2 evolution throughout the PEP sequence can be summarized as:

$$t_1 \begin{cases} 2H_xC_z \xrightarrow{t_2} \begin{cases} -C_x^k \to \sum_{k}^{K} H_y \cos(\Omega_H t_1) \cos(\Omega_{C^k} t_2) \\ C_y^k \to -\sum_{k}^{K} H_x \cos(\Omega_H t_1) \sin(\Omega_{C^k} t_2) \end{cases} \\ 2H_yC_z \xrightarrow{t_2} \begin{cases} -C_y^k \to \sum_{k}^{K} H_x \sin(\Omega_H t_1) \cos(\Omega_{C^k} t_2) \\ C_x^k \to -\sum_{k}^{K} H_y \sin(\Omega_H t_1) \sin(\Omega_{C^k} t_2) \end{cases} \end{cases} \tag{5.106}$$

Therefore, the product operators for the first FID are given by:

$$\sum_{k}^{K} [H_y \cos(\Omega_H t_1) \cos(\Omega_{C^k} t_2) - H_x \cos(\Omega_H t_1) \sin(\Omega_{C^k} t_2) \\ + H_x \sin(\Omega_H t_1) \cos(\Omega_{C^k} t_2) - H_y \sin(\Omega_H t_1) \sin(\Omega_{C^k} t_2)] \tag{5.107}$$

The second FID is obtained by inverting the phase ϕ and the first three gradients. The C_x^k coherence obtained after t_2 evolution is not affected by the phase inversion, while C_y^k coherence changes sign. Consequently, the second FID is given by:

$$\sum_{k}^{K} [H_y \cos(\Omega_H t_1) \cos(\Omega_{C^k} t_2) + H_x \cos(\Omega_H t_1) \sin(\Omega_{C^k} t_2)$$
$$- H_x \sin(\Omega_H t_1) \cos(\Omega_{C^k} t_2) - H_y \sin(\Omega_H t_1) \sin(\Omega_{C^k} t_2)] \quad (5.108)$$

The third FID is acquired by inverting the phase of ϕ_1, which changes the sign of the first t_1 term:

$$\sum_{k}^{K} [-H_y \cos(\Omega_H t_1) \cos(\Omega_{C^k} t_2) + H_x \cos(\Omega_H t_1) \sin(\Omega_{C^k} t_2)$$
$$+ H_x \sin(\Omega_H t_1) \cos(\Omega_{C^k} t_2) - H_y \sin(\Omega_H t_1) \sin(\Omega_{C^k} t_2)] \quad (5.109)$$

The last FID is acquired with inverting both phases of ϕ_1 and ϕ, which has a form of:

$$\sum_{k}^{K} [-H_y \cos(\Omega_H t_1) \cos(\Omega_{C^k} t_2) - H_x \cos(\Omega_H t_1) \sin(\Omega_{C^k} t_2)$$
$$- H_x \sin(\Omega_H t_1) \cos(\Omega_{C^k} t_2) - H_y \sin(\Omega_H t_1) \sin(\Omega_{C^k} t_2)] \quad (5.110)$$

The four FIDs are recorded and stored separately. The combinations of the data yield four data sets which can be transformed to four spectra and combined to a single spectrum with pure phase in all three dimensions. A maximum sensitivity enhancement by a factor of 2 can be obtained in the combined spectrum.

$$FID_1 + FID_2 - FID_3 - FID_4 = 4H_y \cos(\Omega_H t_1) \cos(\Omega_{C^k} t_2)$$
$$-FID_1 + FID_2 + FID_3 - FID_4 = 4H_x \sin(\Omega_H t_1) \cos(\Omega_{C^k} t_2)$$
$$FID_1 - FID_2 + FID_3 - FID_4 = 4H_x \cos(\Omega_H t_1) \sin(\Omega_{C^k} t_2)$$
$$-FID_1 - FID_2 - FID_3 - FID_4 = 4H_y \sin(\Omega_H t_1) \sin(\Omega_{C^k} t_2) \quad (5.111)$$

5.6 3D Isotope-Edited Experiments

5.6.1 ^{15}N-HSQC–NOESY

The signal overlapping of ^{1}H spins becomes more severe with an increase in molecular size. The 3D heteronuclear-edited experiments resolve the overlapped ^{1}H resonances over the chemical shift frequencies of the directly attached heteronuclei (^{15}N and/or ^{13}C), resulting in a significant increase in resolution in the ^{1}H dimensions. The simplest way to form a 3D pulse sequence is to combine

5.6 3D Isotope-Edited Experiments

two 2D pulse sequences. The 3D NOESY–HSQC is formed by combining an HSQC with an NOESY after removing the acquisition period of the NOESY and the preparation period of the HSQC (Fig. 6.7b; Marion et al. 1989). After the evolution of ^{1}H chemical shifts, the magnetization is transferred to vicinal protons by cross relaxation during the NOESY mixing period, τ_m. The scalar coupling of ^{15}N to ^{1}H is refocused by the 180° ^{15}N pulses in the middle of the t_1 evolution of the ^{1}H chemical shifts. In the following step, the magnetization of the amide proton spins is transferred to ^{15}N and then back to amide protons after the evolution of ^{15}N chemical shifts in the same pathway as in the HSQC experiment. The PEP scheme is used for sensitivity enhancement.

$$H(t_1) \xrightarrow{\text{NOE}} H \xrightarrow{J_{\text{NH}}} N(t_2) \xrightarrow{J_{\text{NH}}} H(t_3) \quad (5.112)$$

The description of the coherence transfer by the product operators is given by:

$$H_z \xrightarrow{\left(\frac{\pi}{2}\right)H_x \to t_1} -H_y \cos(\Omega_H t_1) + H_x \sin(\Omega_H t_1)$$
$$\xrightarrow{\left(\frac{\pi}{2}\right)H_x} -H_z \cos(\Omega_H t_1) + H_x \sin(\Omega_H t_1) \quad (5.113)$$
$$\xrightarrow{\tau_m} -H_z \cos(\Omega_H t_1)$$

The transverse magnetization H_x is removed by the gradient pulse during the mixing period. The ^{1}H magnetization is transferred to ^{15}N:

$$-H_z \cos(\Omega_H t_1) \xrightarrow{\left(\frac{\pi}{2}\right)H_x} H_y \cos(\Omega_H t_1)$$
$$\xrightarrow{\tau \to \pi(H_x + N_x) \to \tau \to \left(\frac{\pi}{2}\right)H_y} 2H_z N_z \cos(\Omega_H t_1) \sin(2\pi J_{\text{NH}} \tau) \quad (5.114)$$

The delay t is optimized to $1/(4J_{\text{NH}})$, 2.7 ms, so that the sine coefficient equals 1. The water magnetization is brought to the transverse plane by the flip-back selective pulse and consequently destroyed by the gradient pulse. After the 90° ^{15}N pulse converts the IP coherence into AP $-2H_z N_y$, the ^{15}N chemical shifts evolve during t_2 evolution period:

$$2H_z N_z \cos(\Omega_H t_1) \xrightarrow{\left(\frac{\pi}{2}\right)N_x} -2H_z N_y \cos(\Omega_H t_1)$$
$$\xrightarrow{t_2} -2H_z N_y \cos(\Omega_H t_1) \cos(\Omega_N t_2) + 2H_z N_x \cos(\Omega_H t_1) \sin(\Omega_N t_2) \quad (5.115)$$

The two terms will be used to achieve sensitivity enhancement by the PEP sequence. The gradient echo scheme before the beginning of PEP sensitivity enhancement is used to select the desired coherence pathway and to invert the coherence order for the PEP sequence. From the results of PEP in the HNCO sequence, the two FIDs after PEP can be obtained as:

$$H_y \cos(\Omega_H t_1) \cos(\Omega_N t_2) - H_x \cos(\Omega_H t_1) \sin(\Omega_N t_2) \quad (5.116)$$

$$H_y \cos(\Omega_H t_1) \cos(\Omega_N t_2) + H_x \cos(\Omega_H t_1) \sin(\Omega_N t_2) \quad (5.117)$$

The two FIDs are recorded separately and stored in different memory locations. The combinations of the two FIDs yield two sets of data:

$$2H_y \cos(\Omega_H t_1) \cos(\Omega_N t_2) \quad (5.118)$$

$$2H_x \cos(\Omega_H t_1) \sin(\Omega_N t_2) \quad (5.119)$$

After Fourier transformations, the two spectra are added to form a single spectrum with enhanced sensitivity.

5.7 Sequence-Specific Resonance Assignments of Proteins

5.7.1 Assignments Using ^{15}N-Labeled Proteins

Through-bond correlation via ^{1}H–^{1}H scalar couplings can be observed only for protons separated by two or three bonds since long-range ^{1}H–^{1}H scalar couplings are usually negligibly weak. Therefore, correlation spectra via ^{1}H–^{1}H scalar couplings do not give correlations between H_i^N and H_{i+1}^α (four bonds away). The observed cross peaks are the correlations between protons within the same amino acid residue, or spin system. Each ^{1}H spin system corresponds to an amino acid residue. Nuclei with different chemical environments have different chemical shift ranges. Chemical shifts for H^N, H^α, and aliphatic side-chain protons have characteristic ranges, with H^N resonances between 7 and 10 ppm, H^α 3.5–6 ppm, and H^{aliph} 1–3.5 ppm. For the majority of residues, the type of protons (H^N, H^α, H^{aliph}) are readily identified when their spin systems are located in the spectrum. However, it is troublesome to obtain complete ^{1}H assignments for a protein with more than 50 residues because the resonance overlap becomes severe.

To reduce resonance overlap, the magnetization of protons with resonances in the crowded chemical shift regions (H^α, H^{aliph}) is observed in (or transferred to) the less crowded H^N region using the TOCSY experiment. In the TOCSY spectrum, each ^{1}H spin system is observed along the resonance of the backbone H^N in the fingerprint region defined by H^N resonances in the F_2 dimension and those of H^α and H^{aliph} in the F_1 dimension.

The connectivity between adjacent residues is established by the NOE cross peaks of H_i^N to H_{i+1}^α in an NOESY or ROESY spectrum. The cross peaks from H_i^N to

H_{i+1}^N can also yield sequential connectivities. To improve the spectral resolution and reduce the resonance overlap, ^1H correlations in the fingerprint region of the TOCSY or NOESY can be expanded along the ^{15}N frequency using an ^{15}N-edited TOCSY or NOESY (Marion et al. 1989; Muhandiram et al. 1993). Correlations between amide ^1H–^{15}N are observed by HSQC. The HSQC sequence can be incorporated into the experiment either before or after TOCSY or NOESY sequence. The advantage of ^{15}N-edited experiments is that they retain sufficient resolution in ^1H–^1H correlation plane. A typical assignment protocol involves identifying spin systems at each ^{15}N frequency slice, assigning the resonances of the spin systems, and categorizing the spin systems to classes of amino acid residues based on the spin coupling topology. For example, Ile and Leu will belong to the same category because their spin systems have a similar coupling topology. Several amino acid residues have unique spin systems such as Gly, Ala, Val, and Thr. It is always helpful to know the chemical shifts and spin topology of the above unique residues when identifying spin systems. It should be noted that proline does not have a spin system staring at an amide proton because of the absence of H^N. The total number of spin systems should match or be close to the number of amino acid residues minus the number of prolines. The next stage is to extend the sequential assignment from the starting amino acid residues to both directions in the sequence via NOE connectivities of H_i^N to H_{i+1}^α and/or to H_{i+1}^β. Suitable starting residues may be selected at Gly, Ala, or Val because they have unique sets of chemical shifts and are easily identified in the TOCSY spectrum. The ambiguous area is where long-range H^N–H^α NOEs exist such as in the case of β sheets. The observed NOE may arise from nonsequential residues due to secondary or tertiary structure rather than the sequential connectivity. Usually, an HSQC experiment is carried out first to examine the dispersion of amide ^1H and ^{15}N chemical shifts. If the degeneracy of $^1H/^{15}N$ pair is observed, it may be helpful to acquire data at two temperatures since H^N chemical shifts are sensitive to the change of temperature. When the temperature is changed, for example by 10 °C, the chemical shifts of degenerate H^N protons will move away from each other. As a result, the overlapped spin systems are resolved in the TOCSY spectrum.

5.7.2 Sequence-Specific Assignment Using Doubly Labeled Proteins

Backbone H^N, N, C^α, C' resonances are assigned by analyzing 3D HNCO, HNCA, HN(CA)CO, and HN(CO)CA experiments. HNCO and HN(CO)CA experiments provide single sets of correlations between $(H, N)_i$ and C'_{i-1}, and between $(H, N)_i$ and C^α_{i-1}, respectively, whereas HNCA and HN(CA)CO give rise to correlations both within the residues and to the preceding residues: $(H, N, C^\alpha)_i$, and $(H, N)_i, C^\alpha_{i-1}$ for HNCA, $(H, N, C')_i$ and $(H, N)_i, C'_{i-1}$ for HN(CA)CO experiments, because of the compatible size of scalar couplings $^1J_{NC\alpha}$ and $^2J_{NC\alpha}$. In principle, the four experiments will provide complete information for backbone assignments. However, since the experiments rely on the correlations of $(H, N)_i$, the degeneracy of the pair will cause ambiguities in the chemical shift assignment of C^α and C'. Hence, the degeneracy of H, N is usually inspected in an HSQC experiment before setting up the 3D experiments. If it exits, the degeneracy of H, N can be bypassed by the correlation between C'_{i-1} and C^α_i in 4D $HNCO_{i-1}CA_i$ or 3D $H(N)CO_{i-1}CA_i$ (Konrat et al. 1999) so that the sequential assignment can be extended. An alternative is to move the H^N chemical shifts by lowering the temperature if the solubility of the sample is allowed.

Once the complete assignment of backbone nuclei is achieved, they can be extended to C^β by a pair of 3D experiments, CBCA(CO)NH and CBCANH, which correlate the chemical shifts of H_i^N, N_i with $C^\alpha_{i-1}, C^\beta_{i-1}$ and with $C^\alpha_i, C^\beta_i, C^\alpha_{i-1}$, and C^β_{i-1}, respectively. Since resonances of the backbone nuclei have been assigned previously, the assignment of C^β is relatively straightforward. The final stage of the assignment focuses on the aliphatic side-chain carbons and protons using a 3D or 4D HCCH–TOCSY and a 3D ^{15}N-edited TOCSY. The HCCH–TOCSY is much more sensitive than the ^{15}N-edited TOCSY because it utilizes the large $^1J_{CC}$ scalar coupling to transfer magnetization along the side-chain rather than relying on the $^3J_{HH}$ coupling as in the case of ^{15}N-edited experiments. Each cross peak of a 3D HCCH–TOCSY correlates the chemical shifts of H_i, H_j, and C_j for a spin system of $H_iC_i - \cdots - H_jC_j$. The chemical shifts of aliphatic side-chain protons originating from $C^\alpha H$ protons can be located at all resonances of side-chain carbons involved in the same spin system. The chemical shifts of the side-chain carbons are obtained by tracing their correlations with the proton spin system. The cross peaks of a 4D HCCH–TOCSY correlate all four protons and carbons, H_i, C_i, H_j, and C_j in the spin system. Therefore, each cross peak provides two proton and two carbon frequencies. By now, the resonance assignments are complete and are ready to be used to identify NOE cross peaks for structural calculations.

5.8 Assignment of NOE Cross Peaks

After the complete assignment of backbone and side-chain resonances, the connectivities among the protons are readily assigned using a 4D ^{13}C–^{13}C NOESY. In the initial stage of the assignment, usually only a fraction of the total NOESY cross peaks can be assigned unambiguously, because of chemical shift degeneracy and inconsistency in some extent of the NOESY cross-peak positions compared to those obtained by resonance assignment. Additional NOESY cross peaks are assigned during the iterative steps of the structure calculation (see Chap. 7).

Questions

1. What is PEP sequence used for? And how is it achieved?
2. Why is it necessary to obtain the resonance assignments?
3. What is the IPAP–HSQC experiment used for and how are the IP and AP doublets generated by the pulse sequence?
4. What is an "out and back" experiment and how is this type of experiments named?
5. Why are the correlations of H_i^N and N_i with both C_i^α and C_{i-1}^α observed in HNCA experiment but the correlations of H_i^N and N_i with C_i^α are not present in HNCO experiment?
6. The ^1H sequence element "2τ – DIPSI-2" is present in many 3D triple-resonance experiments. What is this sequence used for and how is the delay τ set?
7. What are the coefficients used to optimize the delay (shown in Fig. 5.10) for different carbon multiplicities?
8. Why do C^α cross peaks in CBCANH experiment have an opposite sign to C^β cross peaks?
9. Which nucleus is observed in the 3D experiments? and explain why.
10. How can the degeneracy of N, H resonances in the 3D experiments be resolved?

References

Akke M, Carr PA, Palmer AG III (1994) Heteronuclear-correlation NMR spectroscopy with simultaneous isotope filtration, quadrature detection, and sensitivity enhancement using z rotations. J Magn Reson B104:298–302

Bax A, Clore GM, Gronenborn AM (1990a) 1H-1H correlation via isotropic mixing of 13C magnetization, a new three-dimensional approach for assigning 1H and 13C spectra of 13C-enriched proteins. J Magn Reson 88:425–431

Bax A, Ikura M, Kay LE, Torchia DA, Tschudin R (1990b) Comparison of different modes of two-dimensional reverse-correlation NMR for the study of proteins. J Magn Reson 86:304–318

Bodenhausen G, Ruben DJ (1980) Natural abundance nitrogen-15 NMR by enhanced heteronuclear spectroscopy. Chem Phys Lett 69:185–189

Cavanagh J, Palmer AG III, Wright PE, Rance M (1991a) Sensitivity improvement in proton-detected two-dimensional heteronuclear relay spectroscopy. J Magn Reson 91:429–436

Cavanagh J, Rance M (1993) Sensitivity-enhanced NMR techniques for the study of biomolecules. Annu Rep NMR Spectros 27:1–58

Clubb RT, Thanabal V, Wagner G (1992) A constant-time three-dimensional triple-resonance pulse scheme to correlate intraresidue 1HN, 15N, and 13C' chemical shifts in 15N—13C-labelled proteins. J Magn Reson 97:213–217

Coy MA, Mueller L (1993) Selective decoupling. J Magn Reson A101:122–130

Delaglio F, Grzesiek S, Vuister GW, Zhu G, Pfeifer J, Bax A (1995) NMRPipe: a multidimensional spectral processing system based on UNIX pipes. J Biomol NMR 6:277–293

Delaglio F, Grzesiek S, Vuister GW, Zhu G, Pfeifer J, Bax A (2004) http://spin.niddk.nih.gov/bax/software/NMRPipe/

Emsley L, Bodenhausen G (1990) Gaussian pulse cascades: new analytical functions for rectangular selective inversion and in-phase excitation in NMR. Chem Phys Lett 165:469–476

Engelke J, Rüterjans H (1995) Sequential protein backbone resonance assignments using an improved 3D-HN(CA)CO pulse scheme. J Magn Reson B109:318–322

Ernst RR, Bodenhausen G, Wokaun A (1987) The principles of nuclear magnetic resonance in one and two dimensions. Oxford University Press, New York

Fesik SW, Eaton HL, Olejniczak ET, Zuiderweg ERP, McIntosh LP, Dahlquist FW (1990) 2D and 3D NMR spectroscopy employing carbon-13/carbon-13 magnetization transfer by isotropic mixing. Spin system identification in large proteins. J Am Chem Soc 112:886–888

Geen H, Freeman R (1991) Band-selective radiofrequency pulses. J Magn Reson 93:93–141

Gerald RE, Bernhard T, Haegberlen U, Rendell J, Opella S (1993) Chemical shift and electric field gradient tensors for the amide and carboxyl hydrogens in the model peptide N-acetyl-D, L-valine. Single-crystal deuterium NMR study. J Am Chem Soc 115:777–782

Grzesiek S, Bax A (1992a) An efficient experiment for sequential backbone assignment of medium-sized isotopically enriched proteins. J Magn Reson 99:201–207

Grzesiek S, Bax A (1992b) Correlating backbone amide and side chain resonances in larger proteins by multiple relayed triple resonance NMR. J Am Chem Soc 114:6291–6293

Grzesiek S, Bax A (1993) The importance of not saturating water in protein NMR. Application to sensitivity enhancement and NOE measurements. J Am Chem Soc 115:12593–12594

Grzesiek S, Anglister J, Bax A (1993) Correlation of backbone amide and aliphatic side-chain resonances in 13C/15N-enriched proteins by isotropic mixing of 13C magnetization. J Magn Reson B101:114–119

Hurd RE, John BK (1991) Gradient-enhanced proton-detected heteronuclear multiple-quantum coherence spectroscopy. J Magn Reson 91:648–653

Ikura M, Kay LE, Bax A (1990a) A novel approach for sequential assignment of proton, carbon-13, and nitrogen-15 spectra of larger proteins: heteronuclear triple-resonance three-dimensional NMR spectroscopy. Application to calmodulin. Biochemistry 29:4659–4667

Ikura M, Marion D, Kay LE, Shih H, Krinks M, Klee CB, Bax A (1990b) Heteronuclear 3D NMR and isotopic labeling of calmodulin: towards the complete assignment of the 1H NMR spectrum. Biochem Pharmacol 40:153–160

Johnson BA (2004) http://onemoonscientific.com/nmrview/

Kay LE, Ikura M, Tschudin R, Bax A (1990) Three-dimensional triple-resonance NMR spectroscopy of isotopically enriched proteins. J Magn Reson 89:496–514

Kay LE, Ikura M, Zhu G, Bax A (1991) Four-dimensional heteronuclear triple-resonance NMR of isotopically enriched proteins for sequential assignment of backbone atoms. J Magn Reson 91:422–428

Kay LE, Keifer P, Saarinen T (1992) Pure absorption gradient enhanced heteronuclear single quantum correlation spectroscopy with improved sensitivity. J Am Chem Soc 114:10663–10665

Kay LE, Xu GY, Yamazaki T (1994) Enhanced-sensitivity triple-resonance spectroscopy with minimal H2O saturation. J Magn Reson A109:129–133

Konrat R, Yang D, Kay LE (1999) A 4D TROSY-based pulse scheme for correlating 1HNi,15Ni,13Cαi,13C'i−1 chemical shifts in high molecular weight, 15N,13C, 2H labeled proteins. J Biomol NMR 15:309–313

Logan TM, Olejniczak ET, Xu RX, Fesik SW (1993) A general method for assigning NMR spectra of denatured proteins using 3D HC(CO)NH-TOCSY triple resonance experiments. J Biomol NMR 3:225–231

Lyons BA, Montelione GT (1993) An HCCNH triple-resonance experiment using carbon-13 isotropic mixing for correlating backbone amide and side-chain aliphatic resonances in isotopically enriched proteins. J Magn Reson B101:206–209

Majumdar A, Wang H, Morshauser RC, Zuiderweg ERP (1993) Sensitivity improvement in 2D and 3D HCCH spectroscopy using heteronuclear cross-polarization. J Biomol NMR 4:387–397

Marion D, Kay LE, Sparks SW, Torchia DA, Bax A (1989) Three-dimensional heteronuclear NMR of 15N labeled proteins. J Am Chem Soc 111:1515–1517

Matsuo H, Kupce E, Li H, Wagner G (1996) Increased sensitivity in HNCA and HN(CO)CA experiments by selective Cβ Decoupling. J Magn Reson B113:91–96

Montelione GT, Lyons BA, Emerson SD, Tashiro M (1992) An efficient triple resonance experiment using carbon-13 isotropic mixing for determining sequence-specific resonance assignments of isotopically-enriched proteins. J Am Chem Soc 114:10974–10975

Mueller L (1979) Sensitivity enhanced detection of weak nuclei using heteronuclear multiple quantum coherence. J Am Chem Soc 101:4481–4484

Muhandiram DR, Xu GY, Kay LE (1993) An enhanced-sensitivity pure absorption gradient 4D 15N, 13C-edited NOESY experiment. J Biomol NMR 3:463–470

Muhandiram DR, Kay LE (1994) Gradient-enhanced triple-resonance three-dimensional NMR experiments with improved sensitivity. J Magn Reson B103:203–216

Olejniczak ET, Xu RX, Fesik SW (1992) A 4D HCCH-TOCSY experiment for assigning the side chain 1H and 13C resonances of proteins. J Biomol NMR 2:655–659

Ottiger M, Delaglio F, Bax A (1998) Measurement of J and dipolar couplings from simplified two-dimensional NMR spectra. J Magn Reson 131:373–378

Cavanagh J, Palmer AG III, Wright PE, Rance M (1991b) Sensitivity improvement in proton-detected two-dimensional heteronuclear relay spectroscopy. J Magn Reson 91:429–436

Pervushin K, Riek R, Wider G, Wüthrich K (1997) Attenuated T2 relaxation by mutual cancellation of dipole–dipole coupling and chemical shift anisotropy indicates an avenue to NMR structures of very large biological macromolecules in solution. Proc Natl Acad Sci U S A 94:12366–12371

Pervushin K, Wider G, Riek R, Wüthrich K (1999) The 3D NOESY-[1H,15N,1H]-ZQ-TROSY NMR experiment with diagonal peak suppression. Proc Natl Acad Sci U S A 96:9607–9612

Sattler M, Schmidt P, Scleucher J, Schedletzky O, Glaser SJ, Griesinger C (1995a) Novel pulse sequences with sensitivity enhancement for in-phase coherence transfer employing pulsed field gradients. J Magn Reson B108:235–242

Sattler M, Schwedinger M, Schleucher J, Griesinger C (1995b) Novel strategies for sensitivity enhancement in heteronuclear multi—dimensional NMR experiments employing pulsed field gradients. J Biomol NMR 6:11–22

Schleucher J, Schwendinger MG, Sattler M, Schmidt P, Glaser SJ, Sørensen OW, Griesinger C (1994) A general enhancement scheme in heteronuclear multidimensional NMR employing pulsed field gradients. J Biomol NMR 4:301–306

Sørensen OW, Eich GW, Levitt MH, Bodenhausen G, Ernst RR (1983) Product operator formalism for the description of NMR pulse experiments. Prog Nucl Magn Reson Spectrosc 16:163–192

Teng Q, Cross TA (1989) The in situ determination of the 15N chemical-shift tensor orientation in a polypeptide. J Magn Reson 85:439–447

Chapter 6
Studies of Small Biological Molecules

In this chapter, the interactions of small molecules with proteins are discussed in terms of different experimental methods with examples. The last section describes examples to study metabolic pathways using NMR experiments. Key questions to be answered include the following:

1. What are the experiments available to study the interactions of small molecules with proteins?
2. What kind of information can the experiments provide?
3. What types of systems are suitable for a particular experiment?
4. Can simple NMR experiments be used to study metabolic pathways?
5. What are the advantages of NMR experiments compared to other technique in such kind of research?

6.1 Ligand–Protein Complexes

As NMR spectroscopy has been widely used to determine structures and dynamics of molecules ranging from synthetic compounds to macro biomolecules, it has become a powerful approach for studying the interactions between proteins (and/or nucleic acids) and ligands. The interactions can be studied by observing a change in NMR phenomena (signal) that is induced by the binding. For this purpose, a variety of pulse sequences have been implemented to observe changes in chemical shifts, mobility, relaxation properties and NOEs, etc. Some of the methods make full use of the difference in the mass between protein and ligands, such as methods measuring the diffusion and relaxation of ligands, whereas others observe binding-induced changes such as chemical shifts, NOE, and ^{1}H exchange rate.

6.1.1 SAR-by-NMR Method

SAR by NMR (structure–activity relationship) measures the chemical shift changes of ^{1}H and ^{15}N spins at binding sites of target proteins upon the binding of small molecule ligands. The binding affinity of ligands can be improved by directly linking together two weak-binding ligands to obtain a high-affinity binding ligand. The SAR-by-NMR method can also be used to locate binding sites on the target protein (Shuker et al. 1996).

The main point of SAR by NMR can be understood by looking at the dissociation constants of the complexes. For each binding site, an equilibrium is established by three species: the target protein, ligand, and the complex. For a binding equilibrium:

$$P + L \underset{K_A}{\overset{K_D}{\rightleftharpoons}} PL \tag{6.1}$$

the dissociation constant K_D is given by ΔG:

$$RT \ln(K_D) = \Delta G \tag{6.2}$$

in which R is the ideal gas constant, T is the temperature in Kelvin, and ΔG is the free energy difference. For the individual binding sites A and B, the dissociation constants K_D^A and K_D^B are given by:

$$RT \ln(K_D^A) = \Delta G^A \quad \text{and} \quad RT \ln(K_D^B) = \Delta G^B \tag{6.3}$$

When a single ligand occupies two sites simultaneously,

$$L + P \rightleftharpoons C \tag{6.4}$$

in which P, L, and C stand for protein, ligand, and complex, respectively. Then

$$RT \ln(K_D^{AB}) = \Delta G^{AB} = RT \ln(K_D^A) + RT \ln(K_D^B) \tag{6.5}$$

$$\Delta G^{AB} = \Delta G^A + \Delta G^B \tag{6.6}$$

Therefore, $K_D^{AB} = K_D^A \times K_D^B$. If the binding dissociation constant K_D^A is 3×10^{-4} and K_D^B 1×10^{-3}, the dissociation constant K_D^{AB} of the structurally linked ligand is close to 3×10^{-7}, which is much lower than that of each for the individual ligands.

The dissociation constant K_D can be estimated from the observed chemical shift changes in the complex. For a single-site binding, K_D is given by:

$$K_D = \frac{[L][P]}{[C]} \tag{6.7}$$

6.1 Ligand–Protein Complexes

in which [L], [P], and [C] are the equilibrium concentrations of the free ligand, free protein, and the complex, respectively. Assuming that the complex is formed by a 1:1 ratio and the ligand concentration $[L]_0$ is in high excess to that of the protein $[P]_0$, the equilibrium concentrations can be expressed by:

$$[C] = \frac{\delta - \delta_f}{\delta_s - \delta_f}[P]_0 \qquad (6.8)$$

$$[L] = [L]_0 - [C] \qquad (6.9)$$

$$[P] = [P]_0 - [C] \qquad (6.10)$$

in which δ and δ_f are the chemical shifts in the presence and absence of ligand, respectively, and δ_s is the chemical shift at saturation level of the ligand, that is, the target protein is completely bound. Therefore, K_D can be estimated according to:

$$K_D = \frac{([L]_0 - [C])([P]_0 - [C])}{[C]} \qquad (6.11)$$

In some situations, the affinity of the binding ligand is described in terms of its concentration IC_{50}, which is the concentration of the inhibitor (or ligand) required for 50 % inhibition (or binding) of the target protein. The K_D of a ligand with an IC_{50} can also be derived from the known dissociation constant K_I of the inhibitor for a given concentration [I] of the inhibitor:

$$K_D = \frac{IC_{50}K_I}{[I]} \qquad (6.12)$$

The equation is obtained using the previously stated assumptions, and [I] is much higher than K_I (Cheng and Prusoff 1973).

A ^1H–^{15}N HSQC experiment is utilized to observe the changes in chemical shifts of an ^{15}N-labeled target protein with and without small molecules. The ^1H chemical shifts of an unlabeled small molecule will not interfere with the observed signals of the protein because only ^1H–^{15}N correlations can be observed. The binding of the ligand to the protein is determined by comparing the HSQC spectrum of the target protein along with the one in the presence of the small molecule. If there are significant cross-peak shifts in the mixture compared to the free protein, the small molecule compound is determined to bind to the protein and is considered a lead compound. A library of small molecule compounds is used for the screening. Once a weak-binding lead compound is identified based on the chemical shift change, the value of the dissociation constant K_D is determined. The binding affinity to this site is optimized using derivative analogues of the lead compound, which leads to a relatively strong-binding ligand. The binding of a new ligand on a second site is located by observing, in the presence of the optimized ligand, the chemical shift

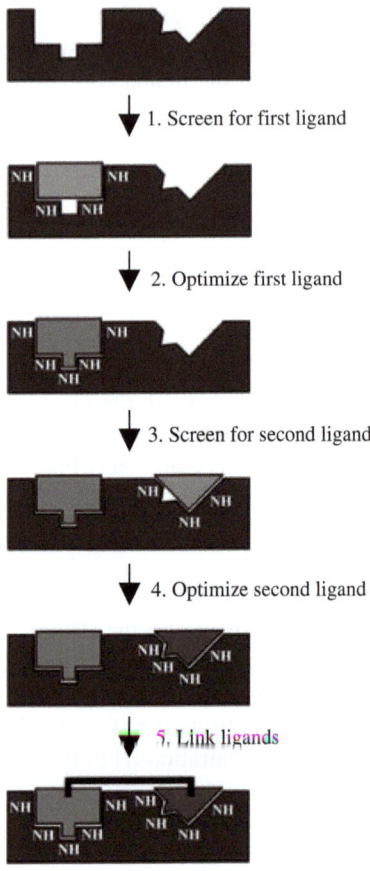

Fig. 6.1 Drug screening by SAR-by-NMR method (reproduced with permission from Shuker et al. (1996), Copyright © 1996 AAAS)

changes of a different set of amide ^{1}H–^{15}N cross peaks that come from different residues than the first site. Then, the second ligand is optimized in the same way as the first one. The two ligands are structurally linked together to form the final ligand that binds to the two sites of the protein simultaneously (Fig. 6.1). The location of the linkage and the stereo orientation of the two ligands play an important role in obtaining a high-affinity ligand, and are determined based on the information of protein structure as well as the information on the binding geometry of the two individual ligands with respect to the protein.

High-affinity ligands binding to a number of proteins have been discovered utilizing the SAR-by-NMR method (Shuker et al. 1996; Hajduk et al. 1997b). For instance, the method has been utilized to identify new ligands for leukocyte function-associated antigen-1 (LFA-1) that is a cell-surface adhesion receptor involved in the inflammatory and certain T cell immune responses (Gahmberg 1997) when complexed with intracellular adhesion molecules (ICAM-1). Inhibitors to the interaction between LFA-1 and ICAM-1 may have therapeutic uses in treating inflammatory diseases (Sligh et al. 1993). Although certain compounds

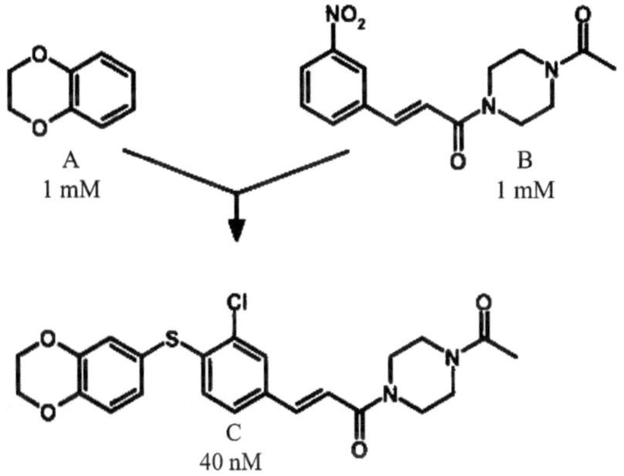

6.1 Ligand–Protein Complexes

Fig. 6.2 Constructing the ligand for leukocyte function-associated antigen-1 (LFA-1) utilizing the SAR-by-NMR method. Compound A was identified first by NMR screening, and was used at saturating concentration to identify compound B binding to a different region of the protein. Based on the structural information of the two compounds and the target protein, compound C was synthesized and binds to the target protein with improved affinity and other pharmaceutical properties

have been identified to prevent the binding of ICAM-1 to LFA-1, poor solubility and side effects make them undesirable (Liu et al. 2000). The NMR method was used to screen for new compounds with improved pharmaceutical properties. Compound A (Fig. 6.2) was identified to bind to LFA-1 with a K_D of approximately 1 mM by observing the chemical shift changes of amide ^{1}H and ^{15}N in an HSQC spectrum. Subsequently, the second ligand (compound B) was found to bind to a different region of LFA-1 in the presence of compound A at saturating concentration with a similar K_D value. Based on the structural information of two ligands and LFA-1, a number of compounds were synthesized, of which compound C has an IC_{50} value of 40 nM with increased solubility (Liu et al. 2001).

6.1.2 Diffusion Method

While SAR by NMR measures the change in chemical shifts of the target protein, the diffusion method makes use of the change in the translational diffusion rate of a ligand upon binding to the target protein. Because the observed signals are from the ligand, it is not necessary to use isotope-labeled protein. The translational diffusion coefficient D for a spherical molecule with radius r in a solvent of viscosity η has a dependence of $1/r$, according to the Stockes–Einstein equation (Stilbs 1987):

$$D = \frac{KT}{6\pi\eta r} \quad (6.13)$$

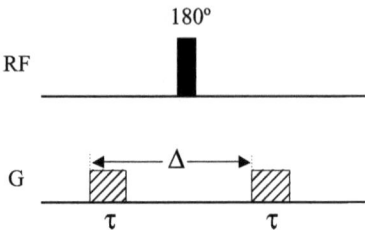

Fig. 6.3 Gradient spin-echo sequence for measuring diffusion constant

in which K is the Boltzmann constant and T is the temperature in Kelvin. Therefore, the D of a ligand has a smaller value when a complex is formed by the ligand with a protein.

There are many versions of NMR experiments available to measure the diffusion coefficient, based on the pioneering work by Stejskal and Tanner using pulsed field gradient NMR methods (Stejskal and Tanner 1965; Johnson 1999). The LED sequence (longitudinal eddy-current delay) was proposed to reduce artifacts caused by eddy current as well as to avoid extensive loss from T_2 relaxation by placing the magnetization along the z-axis for most of the experiment time. The signal intensities of a ligand are attenuated in a series of spectra as a function of the gradient strength G and duration τ of a rectangular gradient and the diffusion time Δ between the two gradient echo pulses (Fig. 6.3) according to:

$$\ln\frac{I(G)}{I(0)} = -(\gamma\tau G)^2(\Delta - \frac{\tau}{3})D \qquad (6.14)$$

in which $I(G)$ and $I(0)$ are the signal intensities observed with and without gradient G, respectively, and γ is the proton gyromagnetic ratio (Price and Kuchel 1991). The signal attenuation for large molecules requires increasing the amplitude of the quantity $(\gamma\delta G)^2(\Delta - \tau/3)$. The value of D is obtained from the slope by plotting $I(G)/I(0)$ vs. G^2. For large proteins, diffusion coefficients are in order of 10^{-6} cm^2 s^{-1}, whereas small molecules have diffusion coefficients of approximately 10^{-5} cm^2 s^{-1}. For example, lysozyme, a globular protein of 14.5 kDa, has a D value of 1.06×10^{-6} cm^2 s^{-1} and hemoglobin, a protein of 65 kDa, has a diffusion coefficient of 0.69×10^{-6} cm^2 s^{-1} compared to sucrose and alanine which have D values of 0.52×10^{-5} cm^2 s^{-1} and 0.86×10^{-5} cm^2 s^{-1}, respectively (Stilbs 1987; Dalvit and Böhlen 1997). The delay Δ is typically set in the range of 100–500 ms, whereas the delay δ is selected within several milliseconds. The spectra are obtained with variable gradient strengths, usually less than 10 G cm^{-1} for small free ligands and up to 50 G cm^{-1} for ligands bound to proteins.

Although the diffusion coefficient can be determined by the LED sequence with acceptable accuracy, a qualitative analysis of the diffusion behaviors of free and bound ligands provides useful information for studying protein–ligand binding. Two spectra are acquired for the ligand sample in the presence and absence of the protein.

6.1 Ligand–Protein Complexes

Because the translational motion of a bound ligand will be slower than that of a free ligand, the gradient strength required to decrease the signals of ligands in a mixture with protein is higher if the ligand forms a complex with the protein than if the ligand stays free in solution. For example, if a free ligand has a diffusion constant of five times larger than the protein, the gradient strength, G_f, used to reduce the signal intensity of the free ligand is half the value of gradient, G_b, needed to decrease the intensity of the bound ligand by the same amount:

$$G_f = \frac{G_b}{\sqrt{5}} \approx \frac{G_b}{2} \qquad (6.15)$$

or the intensity of the bound ligand will be higher than the nonbound one at the same gradient strength:

$$\frac{I_f}{I_0} = \left(\frac{I_b}{I_0}\right)^{5 \times 2.303} \approx \left(\frac{I_b}{I_0}\right)^{11.5} \qquad (6.16)$$

Since I_b/I_0 is always less than one, the intensity of the free ligand is much smaller than the complexed ligand when applying the same gradient strength. Therefore, the bound state can be recognized by comparing the intensity change of the two sets of spectra acquired by varying the gradient in the same steps. The observable change in the diffusion property of a ligand is dependent on both the diffusion coefficient and the concentration of the bound ligand, [C]. If [C] is low due to a low concentration of protein, the change caused by the binding may be too small to be observed. To prevent the binding-induced diffusion change from being below the detection limit, the concentration of ligand should not exceed twice that of the protein. Figure 6.4 shows an example of using the diffusion experiments to identify ligands.

6.1.3 Transferred NOE

The cross-relaxation rate σ is dependent on the distance between the two spins and the rotational correlation time. For large molecules such as proteins which have a large correlation time τ_c, the cross-relaxation rate has an opposite sign to that of small molecules and the rate is significantly higher in magnitude than the small molecules. In bound state, the ligand will have a correlation time determined by the protein of the complex. Therefore, a ligand exchanging between the bound and unbound states will have alternating cross-relaxation rates with opposite signs and different magnitudes. When the chemical exchange of a ligand between equilibrated free and bound states is much faster than the cross-relaxation rate, the change in magnetization arising from the NOE between the protein and bound

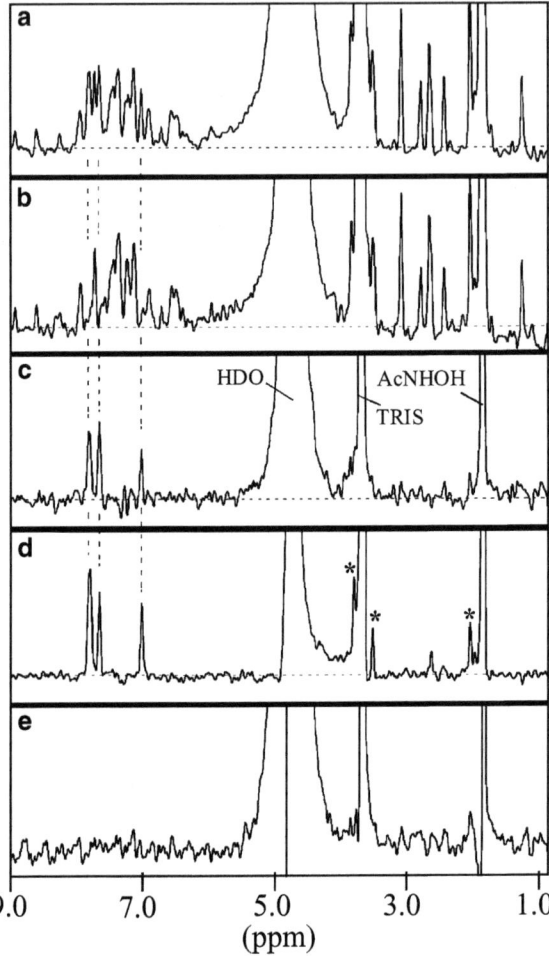

Fig. 6.4 Identification of 4-cyano-4-hydroxybiphenyl as a ligand for stromelysin from a mixture containing eight other nonbinding compounds using diffusion editing. (**a**) Diffusion-edited ^{1}H NMR spectrum of the nine compounds in the absence of stromelysin recorded with low gradient strength. (**b**) Diffusion-edited ^{1}H NMR spectrum of the nine compounds in the presence of stromelysin recorded with low gradient strength after subtracting a similar spectrum recorded with high gradient strength (to remove the protein signals). (**c**) Difference spectrum of (**a**) minus (**b**), which shows the resonances of the bound ligand. (**d**) Reference spectrum of the ligand. (**e**) Difference spectrum similar to spectrum (**c**) recorded on solutions containing only the eight nonbinding compounds. The *dashed lines* in (**c, d**) correspond to the ligand resonances. Buffer impurities are denoted by *asterisks* (reproduced with permission from Hajduk et al. (1997a), Copyright © 1997 American Chemical Society)

ligand protons is transferred to the proton spins of the free ligand. Therefore, for small ligands, the cross-relaxation rate and hence the observed NOE changes sign upon binding to proteins. This transferred NOE has been used to characterize the binding of ligands to proteins (Clore and Gronenborn 1982; Ni 1994).

6.1 Ligand–Protein Complexes

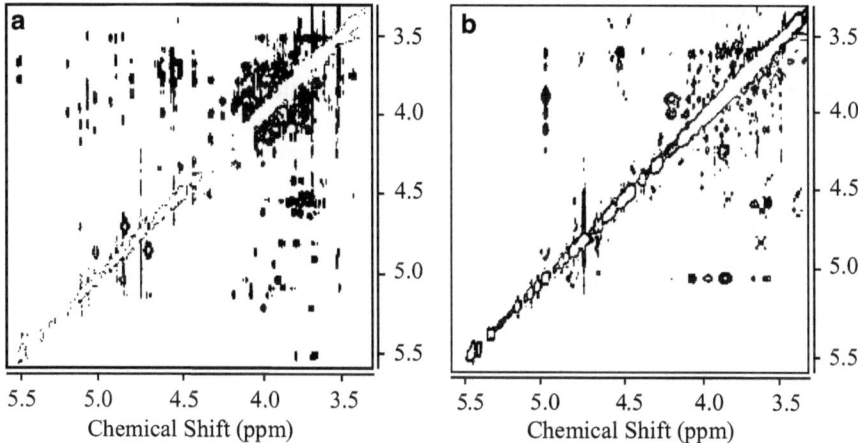

Fig. 6.5 Two-dimensional ^{1}H NOESY spectra of a 15-member oligosaccharide library in the (**a**) absence and (**b**) presence of agglutinin. Observed NOEs are positive in (**a**) and negative in (**b**). In (**b**), a spin-lock filter was used to remove protein resonances, and transfer NOE correlations are observed only for the oligosaccharide α-L-Fuc-(1→6)-β-D-GlcNAc-OMe (reproduced with permission from Meyer et al. (1997), Copyright © 1997 Blackwell Publishing)

NOESY and ROESY experiments are used to observe the transferred NOE cross peaks of the ligand. The observed NOE of a ligand in a NOESY will change sign upon the formation of a protein–ligand complex. Since both intermolecular (between ligand and protein) and intramolecular NOEs are observed simultaneously, the opposite sign of the cross peaks of the free ligand may scale down the intensities of bound ligand in NOESY spectra. A standard NOESY pulse sequence can be utilized with minor modifications to remove the protein resonances. The crucial modification is to insert a relaxation-filter sequence, such as a spin echo in the relaxation period or sometimes at the end of the NOESY sequence. When using the spin echo at the end of NOESY, it also serves the purpose to obtain solvent suppression by suppressing both the protein resonances and the water signal simultaneously.

An alternative way to suppress the protein resonances is to place two orthogonal composite-pulse spin-lock trains (e.g., DIPSI-2) at the beginning of the relaxation period of the NOESY sequence. However, the composite-pulse spin-lock sequence requires a longer relaxation filtering time owing to the fact that the effective relaxation during the composite pulses is determined by the trajectory average of the T_1 and T_2 relaxation rates. Figure 6.5 shows the application of the transferred NOESY experiment in identifying the ligand binding to agglutinin from a mixture of small compounds (Meyer et al. 1997). For the mixture sample in the absence of agglutinin (Fig. 6.5a), the compounds give rise to NOE cross peaks with positive intensities. In the presence of the protein, the cross peaks of the compounds remain the same sign as the diagonal peaks, except for one ligand showing transferred NOE correlations observed with negative intensities (Fig. 6.5b).

6.1.4 Saturation Transfer Difference

For large proteins, the cross relaxation directly proportional to the correlation time τ_c dominates the relaxation process, causing extremely rapid magnetization transfer throughout the protein. Selective saturation of any protein resonances results in saturation of all protein protons as a consequence of the rapid magnetization transfer via the efficient cross relaxation within the protein. Therefore, saturation can be achieved by a long irradiation on any resonance of the protein (i.e., any spectral region) in the 1D NOE difference experiment (Mayer and Meyer 1999; Klein et al. 1999).

In practice, a spectral region away from ligand resonances is selected for the saturation, typically the upfield aliphatic resonances. The 1D saturation transfer difference (STD) experiment subtracts the spectrum observed by the selective saturation of protein resonances with the one recorded by selectively saturating an off-resonance region away from protein and ligand resonances. The subtraction is normally achieved during the experiment by inverting the receiver phase for alternating on- and off-resonance acquisitions in an interleaved manner to avoid the introduction of artifacts induced by the subtraction.

When a ligand binds to a protein, the saturation of the protein resonances will also saturate ligand resonances owing to cross relaxation. The 1D STD method consists of two experiments collected with interleaved acquisition. The first experiment is collected with on-resonance irradiation selective at an aliphatic resonance of the protein. The intensity of the bound ligand will decrease as a consequence of the saturation transfer via cross relaxation. The second experiment is carried out with the irradiation selected at an off-resonance empty region with inverted receiver phase. In this experiment, resonances of neither the protein nor ligand are saturated, hence, no saturation transfer occurs. Subsequently, the two datasets are added and stored (the net effect is the subtraction of the data because of the inversion in the receiver phase for the off-resonance experiment). The difference spectrum contains only signals of the ligand bound to protein, whose intensity is decreased by the saturation.

As in the NOE-based experiment, a T_2-relaxation filter can be inserted into the 1D STD pulse sequence to suppress the unwanted resonances from the protein. Hahn spin-echo and spin-lock sequences yield superior suppression of signals arising from protein. The STD experiment is a significantly sensitive method since a large ratio of ligand to protein can be used to observe the saturation transfer.

Figure 6.6 illustrates an example for the application of the STD method for studying the binding of RCA120 lectin protein to a ligand (Mayer and Meyer 2001). The peaks from the ligand binding to the protein appear in the STD spectrum. It is demonstrated that the background signals from protein resonances have been significantly suppressed in the STD (Fig. 6.6f) and protein reference (c) spectra recorded with the spin-lock transverse relaxation filter, compared to the one acquired using the standard STD sequence (e).

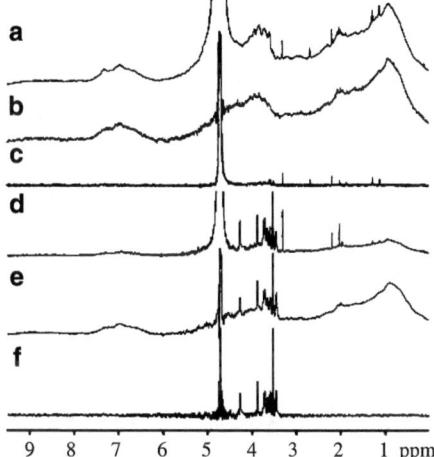

Fig. 6.6 STD ^{1}H experiments to study the binding of methyl-β-D-galactopyranoside as a ligand to the RCA120 lectin. (**a**) Reference spectrum of RCA120 lectin. (**b**) STD spectrum of RCA120 lectin. (**c**) Reference spectrum of RCA120 lectin recorded with a spin-lock filter. (**d**) Reference spectrum of RCA120 lectin in the presence of a 30-fold excess of methyl-β-D-galactopyranoside. (**e**) STD spectrum of RCA120 lectin plus methyl-β-D-galactopyranoside. (**f**) STD spectrum as in (**e**) but with the addition of a spin-lock filter (reproduced with permission from Mayer and Meyer (2001), Copyright © 2001 American Chemical Society)

6.1.5 Isotope-Editing Spectroscopy

Isotope enrichment has made it possible to observe different partners of a complex individually in such a way that only the magnetization originating from the desired part of the complex and then transferred to other part of the complex via cross relaxation is observed. Either the ligand or protein can be isotope labeled with ^{15}N, ^{13}C, and/or ^{2}H. Isotope-edited (also known as isotope-selected) experiments can be used to study a ligand/protein complex with isotope-enriched ligand. For ligands labeled with ^{13}C, the 2D version of the 3D NOESY-^{1}H, ^{13}C-HMQC experiment (Fig. 6.7a) can be applied to observe the intermolecular NOE cross peaks of ^{13}C-labeled ligand to the complexed protein and ^{1}H signals from unlabeled protein are not observable. As a result, these cross peaks are readily identified due to the fact that they only appear on one side of the diagonal. The delay between the two ^{13}C 90° pulses (where the t_2 evolution was in the 3D sequence) is set as short as possible. The sample of the labeled ligand with unlabeled protein is usually dissolved in ^{2}H$_2$O solution. Thus, the HMQC block in the sequence provides sufficient suppression of residual water in ^{2}H$_2$O.

Although this experiment provides a convenient and reliable means to determine the binding of the complex, isotope-labeling synthetic compounds requires tremendous efforts, if it is even possible. However, isotope-labeled peptides serving as ligands are commercially available and are easier to prepare. For a protein–peptide complex, ^{15}N-edited 2D or 3D NOESY experiments can also be applied to obtain

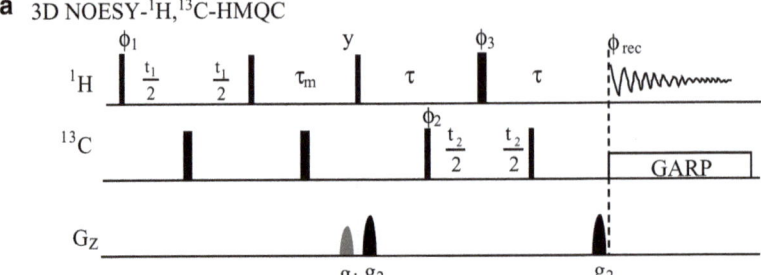

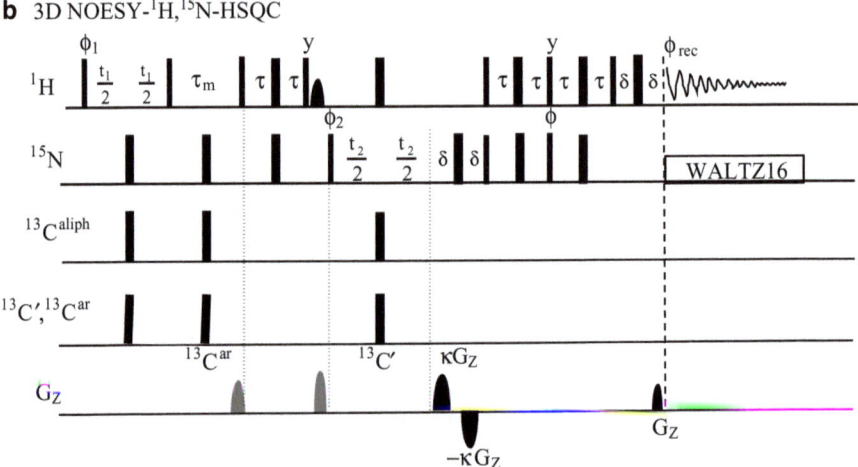

Fig. 6.7 Pulse sequences for 3D ^{13}C-edited NOESY (NOESY-^{1}H, ^{13}C-HMQC) and 3D ^{15}N-edited NOESY (NOESY-^{1}H, ^{15}N-HSQC) experiments. (**a**) In the 3D ^{13}C-edited NOESY, all pulses are x phased, except that $\phi_1 = x, -x +$ States-TPPI (t_1), $\phi_2 = x, x, -x, -x +$ States-TPPI (t_2), and $\phi_{rec} = x, -x, -x, x$. The delay $\tau = 3.8$ ms. The *shaded* gradient g_1 is used to destroy the residual transverse magnetization due to the imperfect 180° pulses, whereas g_2 gradients for the coherence selection. (**b**) In the 3D ^{15}N-edited NOESY experiment with sensitivity enhancement, the delay $\tau = 2.7$ ms, δ equals to G_z gradient pulse length. All pulses are x phased, except that $\phi_1 = 45°, 225° +$ States-TPPI (t_1), $\phi_2 = x, x, -x, -x +$ States-TPPI (t_2), and $\phi_{rec} = x, -x, -x, x$. For PEP sensitivity enhancement, $k = \pm 10, \phi = \pm y$

information on the binding site. The PEP sensitivity-enhanced version of the 3D NOESY-^{1}H,^{15}N-HSQC pulse sequence shown in Fig. 6.7b utilizes a gradient echo to select the heteronuclear ^{15}N coherence of the peptide whose amide protons have NOEs to the bound protein. For a large protein, it is necessary to introduce the heteronuclear frequency dimension in order to overcome difficulty in extracting the NOE intensity caused by the overlap of ^{1}H cross peaks. Resolving the NOESY cross peaks into the additional heteronuclear dimension significantly improves the resolution.

Shown in Fig. 6.8 is an example of the application of the ^{13}C-edited NOESY experiment to study the complex formed by unlabeled cyclophilin with cyclosporin

6.1 Ligand–Protein Complexes

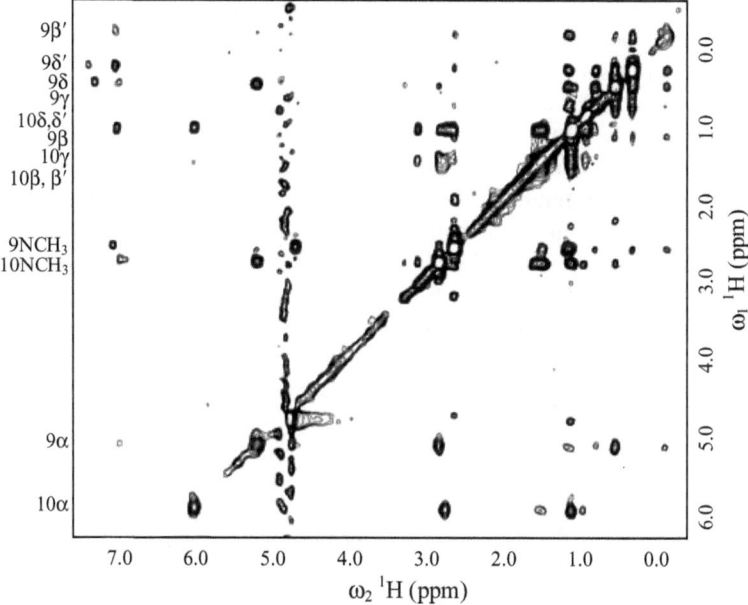

Fig. 6.8 ^{13}C-isotope-edited NOESY spectrum of [U-^{13}C-MeLeu^{9,10}] cyclosporin A (CsA) bound to human cyclophilin. Assignments for the MeLeu^9 and MeLeu^{10} protons of CsA are given on the *left* of the spectrum (reproduced with permission from Fesik et al. (1990), Copyright © 1990 AAAS)

A (ligand) uniformly ^{13}C labeled at MeLeu9 and MeLeu10 (Fesik et al. 1990). In the ω_1 (F_1) dimension, only ^{13}C attached protons of the ligand are observed. In the ω_2 (F_2) dimension, the NOE cross peaks between the ligand and the protein are observed. Since the cross peaks of protein–ligand only appear in the ω_2 dimension, they are readily identified by comparing the both dimensions.

6.1.6 Isotope-Filtering Spectroscopy

When isotope-labeled protein is available, structural information on the protein–ligand complex can be obtained by ^{15}N, ^{13}C isotope-filtered experiment, in which only the intermolecular NOEs of the unlabeled ligand with the labeled protein are observed by suppressing the intramolecular NOEs among the protein resonances. The term "^{13}C-filtered" means that the signals from the ^{13}C-attached protons are suppressed in the experiment, whereas "^{13}C-edited" or "^{13}C-selected" denotes that the signals from the ^{13}C-attached protons are selected in the experiment (Breeze 2000). The application of the heteronuclear filter on one dimension of the 2D experiment is also called half X-filter and, thus, a 2D NOESY-^{1}H, ^{13}C-HMQC pulse sequence may be termed as a ^{13}C-half-filtered (ω_1) NOESY if the filter is on t_1 dimension, or as a ^{13}C-half-filtered (ω_2) NOESY if the filter is on t_2

dimension. Alternatively, the X-filtering can also be applied to both dimensions. However, additional half X-filtering will lengthen the pulse sequence, which makes the experiment less sensitive due to relaxation effects.

Because of a wide range of ^{1}H–^{13}C couplings, a ^{13}C half-filtered experiment cannot completely filter out the magnetization of protons directly bound to heteronucleus. As a consequence, the spectrum contains residual signals from the labeled proteins. To minimize the residual magnetization of the labeled protein, several heteronuclear filtering building blocks have been utilized. The double-isotope filter (also known as double-tuned filter) is a sequential combination of two single filters, as shown in Fig. 6.9a (Gemmecker et al. 1992). The delays in the two filters are selected for different couplings. At the end of the first isotope-filter, the anti-phase sine component after the echo is transformed by a spin-S 90° pulse to double-quantum coherence $2I_xS_y$, which is not transformed into observable magnetization throughout the rest of the pulse sequence:

Spin I-S:

$$-I_y \xrightarrow{\tau' \to \pi(I_x+S_x) \to \tau'} -I_y\cos(\pi J_{IS}2\tau') - 2I_xS_z\sin(\pi J_{IS}2\tau')$$

$$\xrightarrow{\left(\frac{\pi}{2}\right)S_x} -I_y\cos(\pi J_{IS}2\tau') + 2I_xS_y\sin(\pi J_{IS}2\tau') \quad (6.17)$$

After the second echo, the anti-phase sine component ($-2I_xS_y$) is transformed by the spin-S 90° pulse into unobservable zero-quantum coherence $2I_xS_x$:

$$\xrightarrow{\tau'' \to \pi(I_x+S_x) \to \tau''} I_y\cos(\pi J_{IS}2\tau')\cos(\pi J_{IS}2\tau'') - 2I_xS_z\cos(\pi J_{IS}2\tau')\sin(\pi J_{IS}2\tau'')$$

$$\xrightarrow{\left(\frac{\pi}{2}\right)S_y} -I_y\cos(\pi J_{IS}2\tau')\cos(\pi J_{IS}2\tau'') - 2I_xS_x\cos(\pi J_{IS}2\tau')\sin(\pi J_{IS}2\tau'')$$

$$(6.18)$$

Therefore, the undesired magnetization of the labeled protein is scaled according to the factor of $\cos(\pi J_{IS}2\tau')\cos(\pi J_{IS}2\tau'')$. The magnetization of the uncoupled protons is not changed at the end of the double filtering:

$$-I_y \xrightarrow{\tau' \to \pi(I_x+S_x) \to \tau' \to \left(\frac{\pi}{2}\right)S_x} -I_y$$

$$\xrightarrow{\tau'' \to \pi(I_x+S_x) \to \tau'' \to \left(\frac{\pi}{2}\right)S_y} -I_y \quad (6.19)$$

The two delays are set to $1/(4J_{IS})$, typically to 2.0 and 1.785 ms, to optimize the $^1J_{CH}$ couplings of 125 Hz and 140 Hz, respectively, which yields an excellent suppression for aliphatic ^{1}H–^{13}C signals, but moderate reduction of aromatic magnetization. Application of a spin-lock pulse on the y-axis with different lengths (1.0–2.0 ms) after each filter suppresses the anti-phase coherence and, thus, improves the suppression of the unwanted heteronucleus-coupled magnetization.

Shown in Fig. 6.9b is another example sequence for suppressing the magnetization of heteronucleus-attached proton. A "second order J filter" sequence uses two different

6.1 Ligand–Protein Complexes

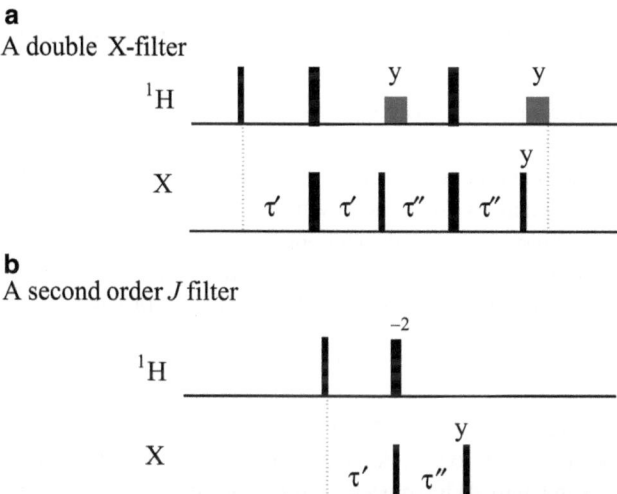

Fig. 6.9 Isotopic-filter schemes for suppressing the signals of the heteronucleus-attached protons. All pulses are x phased except as indicated. The *shaded* pulses are spin-lock pulses. (**a**) A double-isotope filter (or double-tuned filter) uses two consecutive X-half filters with different delays. The magnetization is suppressed by a factor of cos $(2\pi J_{IS}\tau')$ cos $(2\pi J_{IS}\tau'')$. (**b**) A second order J filter uses two different delays that can be optimized for different values of the heteronuclear couplings, which provides a suppression by a factor of cos $(\pi J_{IS}\tau')$ cos $(\pi J_{IS}\tau'')$. Both filters can be implemented in 2D or 3D pulse sequences

delays that can be optimized for different values of the heteronuclear couplings, which provides efficient suppression of the heteronuclear coherence. The final heteronuclear magnetization at the end of the building block is scaled by the same product of two coefficients as described in the double-filtering scheme (Fig. 6.9a). The heteronuclear coupled and uncoupled magnetization can be explained by the product operator as follows. After the mixing period of the NOESY, the magnetization transfers for uncoupled and for coupled spin I are given by:

Spin I (uncoupled):

$$-I_y \xrightarrow{\tau'} -I_y \xrightarrow{\pi I_x + \left(\frac{\pi}{2}\right)S_x} I_y \xrightarrow{\tau''} I_y \xrightarrow{\left(\frac{\pi}{2}\right)S_y} I_y \quad (6.20)$$

Because the delays τ' and τ'' are short compared to the chemical shift, the magnetization of the I spin is almost unchanged after the evolutions of τ' and τ'', whereas the coupled spin I undergoes the following coherence transfer process:

Spin I-S (coupled):

$$-I_y \xrightarrow{\tau'} -I_y \cos(\pi J_{IS}\tau') + 2I_xS_z \sin(\pi J_{IS}\tau')$$
$$\xrightarrow{\pi I_x + \left(\frac{\pi}{2}\right)S_x} I_y \cos(\pi J_{IS}\tau') - 2I_xS_y \sin(\pi J_{IS}\tau') \quad (6.21)$$

$$\xrightarrow{\tau''} I_y \cos(\pi J_{IS}\tau')\cos(\pi J_{IS}\tau'') - 2I_xS_z\cos(\pi J_{IS}\tau')\sin(\pi J_{IS}\tau'')$$

$$\xrightarrow{\left(\frac{\pi}{2}\right)S_y} I_y\cos(\pi J_{IS}\tau')\cos(\pi J_{IS}\tau'') - 2I_xS_x\cos(\pi J_{IS}\tau')\sin(\pi J_{IS}\tau'') \quad (6.22)$$

The anti-phase sine component $2I_xS_z$ after the τ' and τ'' periods is converted into double- and zero-quantum coherence by the spin-S 90° pulses, and are not transferred into observable coherence throughout the isotope-filtered experiment. Therefore, in the end of the sequence the proton magnetization is unchanged, whereas the heteronulcear coherence is attenuated by the coefficient product:

$$\cos(\pi J_{IS}\tau')\cos(\pi J_{IS}\tau'') \quad (6.23)$$

The two delays are set to $1/(2J_{IS})$. Note that τ' and τ'' are twice as long as those in double-isotope filter (Fig. 6.9a). Adiabatic inversion pulses (WURST pulses, see adiabatic pulses in Chap. 4) have been used in isotope-filter experiments to provide more efficient suppression of the heteronuclear coupling with variable sizes (Zwahlen et al. 1997).

6.2 Study of Metabolic Pathways by NMR

The study of metabolic pathways in biological systems has been given new direction by applications of NMR spectroscopy. Traditionally, ^{14}C carbon tracers have been extensively used for these studies. However, ^{14}C tracers have many practical disadvantages due to the radiation precautions and laborious sample handling that limit their applications for studies of the pathways in animals and humans. The use of ^1H and ^{13}C NMR to study the metabolic pathways by tracking the ^{13}C-enriched metabolic substances provides significant insight into how different metabolic processes are regulated in vitro and in vivo. Metabolic pathways are a series of consecutive chemical reactions to degrade specific simple molecules such as glucose and amino acids and produce specific complex molecules in cells. Their reactants, intermediates, and products are referred to as metabolites. Each reaction is catalyzed by a distinct enzyme produced by the expression of a gene. Simple 1D ^1H and ^{13}C NMR experiments in combination with a ^{13}C-isotope labeling approach provide a means of studying the metabolites of specific pathways (more details are discussed in Chap. 9).

The first example is chosen from the study of the initial steps of the common aromatic amino acid pathway in *Methanococcus maripaludis* (Tumbula et al. 1997). The pentose phosphate pathway produces pentoses for nucleosides and erythrose-4-phosphate (E4P) for the biosynthesis of aromatic amino acids (AroAAs). In most methanogens, pentoses are produced by the oxidative pentose pathway via oxidative decarboxylation of hexoses (Fig. 6.10b; Choquet et al. 1994). It was proposed that *M. maripaludis* makes pentoses by a nonoxidative pentose phosphate (NOPP) pathway (Fig. 6.10a, Yu et al. 1994). Some studies of several

6.2 Study of Metabolic Pathways by NMR

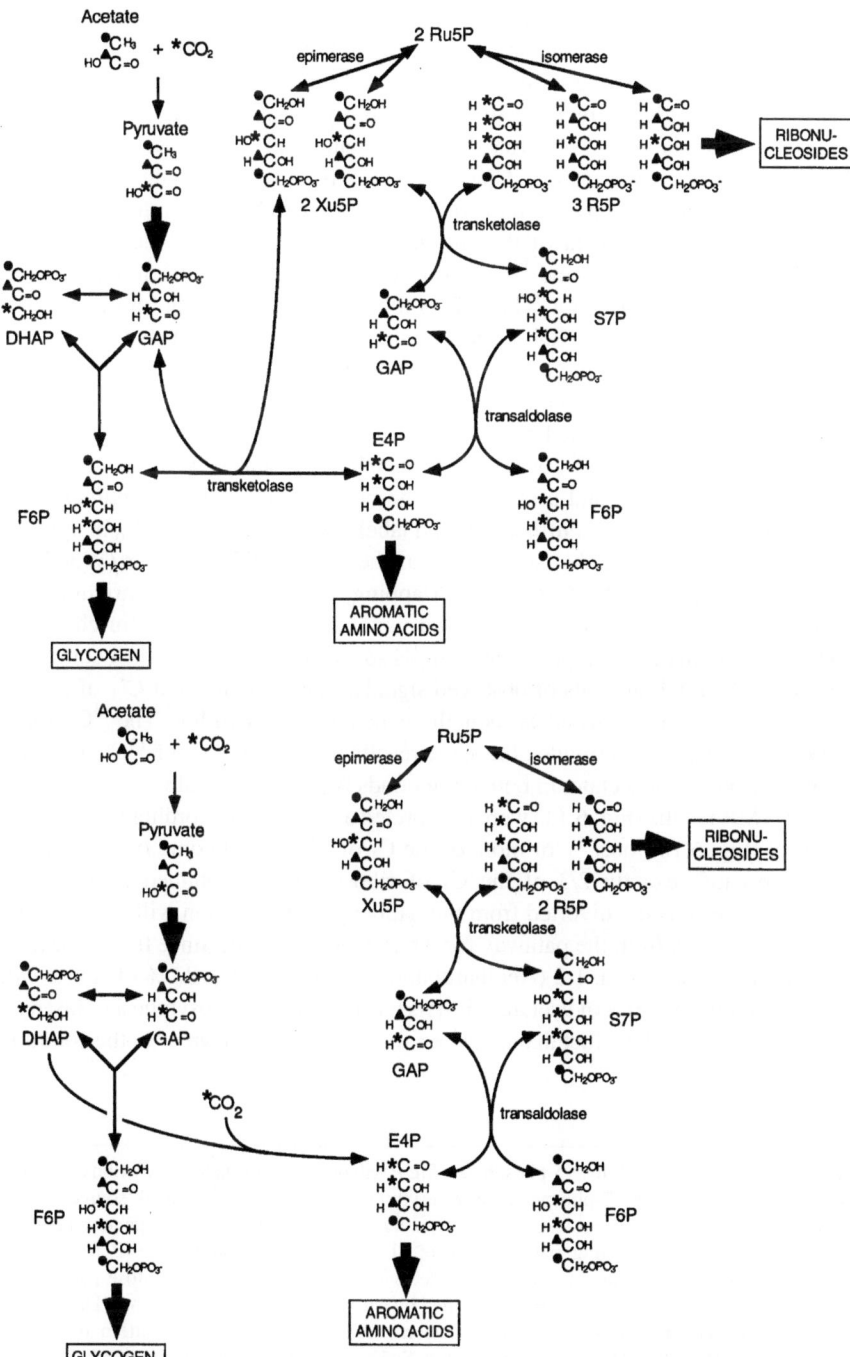

Fig. 6.10 Pentose pathways in methanococci. *Thicker arrows* indicate multiple steps. *DHAP* dihydroxyacetone phosphate; *GAP* glyceraldehyde-3-phosphate; *F6P* fructose-6-phosphate; *Xu5P*

organisms confirmed certain aspects of the proposed pathway, whereas others provided the evidence in contrast to the formation of AroAAs by E4P through the NOPP pathway. The alternative explanation of the results has been studied, that E4P may not be a precursor of AroAAs biosynthesis. A study was carried to determine the labeling patterns by NMR analysis of ribose in cells of *M. maripaludis* grown on [2-^{13}C]acetate since isotope enrichment of ribose is expected to be affected by the removal of E4P for AroAA biosynthesis. Because a quantitative measurement of the enrichment ratio depends on the accuracy of signal integrations, the ^{1}H and ^{13}C data were collected with sufficient long relaxation delays that were 10 s for ^{1}H and 20 s for ^{13}C, and decoupling for either ^{1}H or ^{13}C was not used to avoid NOE effects on the observed signal intensities.

The study found that the extent of ^{13}C enrichment at the C'$_1$ of cytidine and uridine was not consistent with the oxidative pentose pathway (Fig. 6.10b) which should yield 50 % of labeled C'$_1$ derived from ^{13}C$_2$ of acetate. The enrichment determined by the ^{1}H NMR was 66.6 % at the C'$_1$ of cytidine (Fig. 6.11a, ^{1}H spectra of ^{13}C-labeled cytidine and uridine from the biosynthesis of *M. maripaludis* cells grown on [2-^{13}C]acetate). Furthermore, no label at C'$_2$, C'$_3$, C'$_4$, and C$_6$ of cytidine was detected, indicating that scrambling of the isotope did not occur (Wood and Katz 1958). Because of ^{13}C signal overlapping, the extent of ^{13}C enrichment of uridine was determined using the resolved signal of ^{1}H attached to unenriched ^{12}C$_5$ and was based on the assumption that both ^{1}H spins attached to C'$_1$ and C$_5$ contribute equally to the total integrals of observed signals. The enrichment at C'$_1$ of uridine was observed as 66.3 %, which is about the same level as in cytidine. The ^{13}C spectra of the two compounds provided the same results that the amount of ^{13}C'$_1$ carbons, obtained from C$_2$ of acetate, in both compounds is 2/3 of the total C'$_1$ carbons.

The NMR results (Fig. 6.11) were interpreted as follows. According to the NOPP pathway (Fig. 6.10), 66.7 % or more of the C$_1$ of ribose will come from the C$_2$ of acetate because exactly 2/3 of the C$_1$ of ribose will be obtained from the C$_2$ of acetate if E4P is not diverted from the pathway or the fraction will be increased if E4P is removed from the pathway for AroAA biosynthesis. Since the molar ratio of AroAA to ribose is ca. 1:2 (Neidhardt and Umbarger 1996), 83 % of ribose will be produced from the C$_2$ of acetate if E4P is used for AroAA biosynthesis. Based on the NMR results and the fact that the genes for the first two enzymes in the common

---

Fig. 6.10 (continued) xylulose-5-phosphate; *Ru5P* ribulose-5-phosphate; *R5P* ribose-5-phosphate; *S7P* sedoheptulose-7-phosphate; *E4P* erythrose-4-phosphate. ^{13}C label sources: *filled circle*, C$_2$ of acetate; *filled triangle*, C$_1$ of acetate; *asterisk*, CO$_2$. (**a**) Nonoxidative pentose pathway (NOPP) and expected labeling patterns (Yu et al. 1994). Consumption of E4P for AroAA biosynthesis increases the amount of R5P formed via Xu5P. (**b**) Oxidative pentose pathway as proposed by Choquet et al. (1994). E4P is formed by carboxylation of a triose such as DHAP, and the F6P-dependent transketolase reaction is absent. Although the labeling pattern of E4P is unchanged, 50 % of the R5P is now formed via Xu5P, and the labeling pattern of ribose is not affected by the consumption of E4P for AroAA biosynthesis (reproduced with permission from Tumbula et al. (1997), Copyright © 1997 American Association for Microbiology)

6.2 Study of Metabolic Pathways by NMR

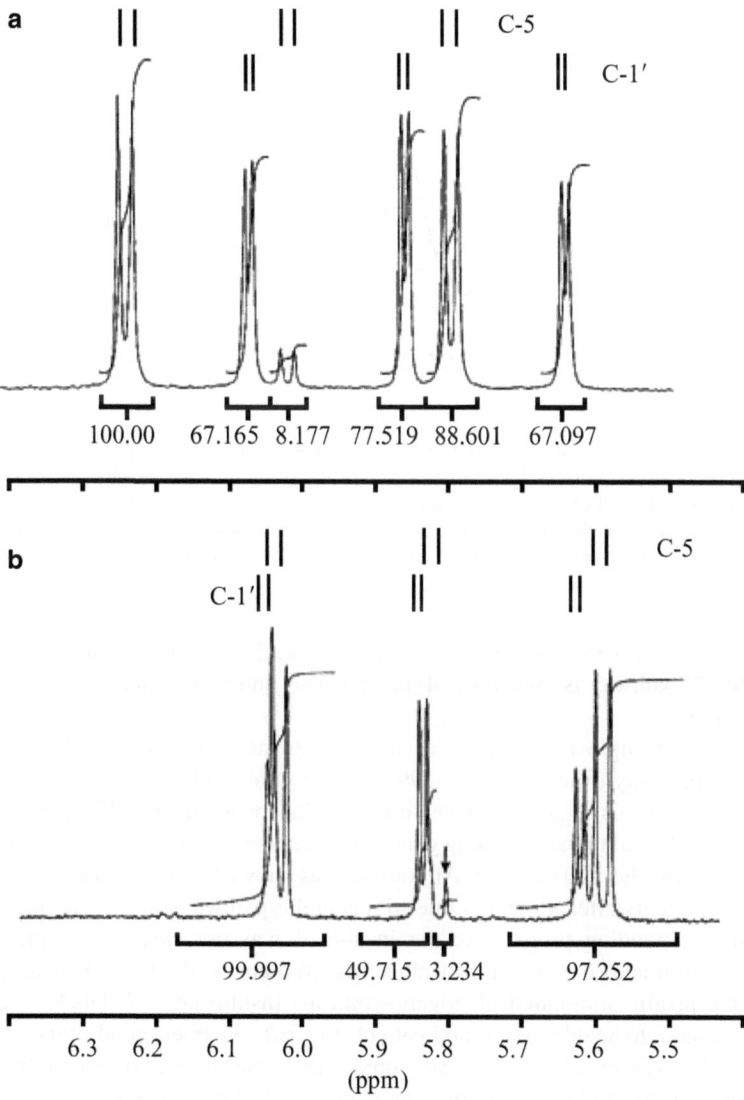

Fig. 6.11 Proton NMR spectra for protons at C'_1 and C_5 of cytidine and uridine from *M. maripaludis* following growth on [2-^{13}C]acetate. The integrals of the peaks are given below the spectra. (a) Cytidine. The $^1J_{CH}$ coupling constants for the C'_1 and C_5 protons were 168 Hz and 172 Hz, respectively. For calculation of the enrichment of cytidine: $C'_1\% = (67.165 + 67.097)/(67.165 + 67.097 + 77.519) = 63.8\%$, $C_5\% = (88.6 + 100)/(88.6 + 100 + 8.17) = 95.8\%$. (b) Uridine. The $^1J_{CH}$ coupling constants for the C'_1 and C_5 protons were 170 Hz and 176 Hz, respectively. The *arrow* indicates one of the ^{12}C peaks for C_5, whereas the other ^{12}C peak overlaps the ^{12}C peaks for C'_1. For calculation of the ^{13}C enrichment, it was assumed that the total signal $(99.997 + 49.715 + 3.234 + 97.252 = 250.198)$ could be divided equally between C'_1 and C_5 (125.099). The ^{13}C enrichment of C_5 was calculated as $125.099 - (3.234 \times 2)/125.099 = 94.8\%$. The ^{13}C enrichment of C'_1, uncorrected for the maximal enrichment, was calculated as: $(125.099 - 49.715 + 3.234)/125.099 = 62.8\%$ (reproduced with permission from Tumbula et al. (1997), Copyright © 1997 American Association for Microbiology)

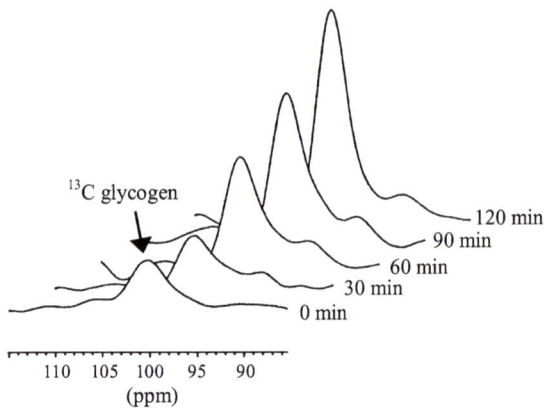

Fig. 6.12 ^{13}C spectra of muscle glycogen as a function of time. The samples were obtained at different times from healthy subjects during an infusion of insulin and ^{13}C-labeled glucose. The resonance frequency of the ^{13}C isotope in the glycogen molecule was resolved from background signals of ^{13}C-labeled glucose and other biological metabolites in the muscle (reproduced with permission from Shulman et al. (1990), Copyright © 1990 Massachusetts Medical Society)

AroAA pathway were not identified in the genome, the conclusion was reached that only NOPP pathway is used for E4P biosynthesis and E4P is not used for AroAA biosynthesis.

Another example is the study of insulin regulation by the muscle glycogen synthesis pathway (Shulman et al. 1990; Taylor et al. 1992; Gruetter et al. 1994). Malfunction of this regulation leads to insulin-independent (type II) diabetes. The disease is believed to have a strong genetic component in addition to environmental factors, such as diet and exercise. Although it was known that the increase of glucose levels in patients after a meal is due to metabolic pathways in muscle and/or the liver not responding properly to the insulin, it was not clear which metabolic pathway dominates the insulin resistance. One-dimensional ^{13}C NMR was used to study the insulin-stimulated glycogen synthesis. Insulin and ^{13}C-labeled glucose were infused into healthy adults and patients to create postmeal conditions. The ^{13}C signal of ^{13}C-labeled glucose was measured at different times during the infusion to monitor the flow of glucose into muscle glycogen as a function of time. The muscle glycogen is increased as showed by the increase of the ^{13}C signal with time (Fig. 6.12). In the ^{13}C spectra, the resonance frequency of the ^{13}C spins in glycogen biosynthesized from the glucose is well resolved from the signals of ^{13}C-labeled glucose and other metabolites in the muscle. The rate of muscle glycogen synthesis in the patients is twofold slower than that obtained from the healthy group (Fig. 6.13), which quantitatively explains the patients' lower insulin-stimulated glucose uptake. Therefore, insulin-stimulated glycogen synthesis in muscle is the major metabolic pathway for consuming excess glucose in healthy adults. A defect in muscle glycogen synthesis is a major cause for the decreased insulin sensitivity in the insulin-independent diabetes patients.

6.2 Study of Metabolic Pathways by NMR

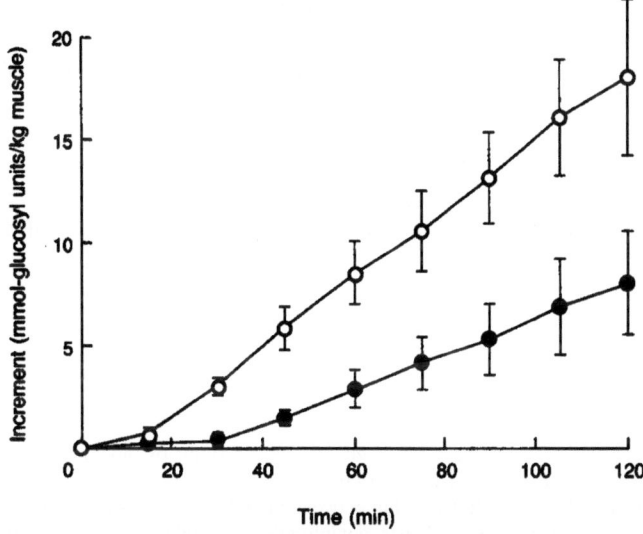

Fig. 6.13 Muscle glycogen concentration calculated from the ^{13}C NMR spectra during an insulin and glucose infusion for patients and healthy controls. The diabetics (*filled circles*) synthesize glycogen more slowly than control subjects (*open circles*). Quantitative features of this study showed that insulin-stimulated muscle glycogen synthesis is the major metabolic pathway of glucose disposal in both groups, and that impairments in this pathway are responsible for the chronic hyperglycemia in patients (reproduced with permission from Shulman et al. (1990), Copyright © 1990 Massachusetts Medical Society)

Questions

1. What is the maximum concentration of ligand, compared to the concentration of protein, used in the diffusion method in the study of ligand–protein binding? And explain why.
2. What is the primary principle underlying the SAR-by-NMR method? And why can NMR be used for that purpose?
3. If compound A has a dissociation constant K_D of 2×10^{-3} M and compound B has a K_D of 2×10^{-6} M, which are binding to two different sites of a protein, what is the dissociation constant K_D likely for compound C that is made from the structurally linked A and B? And explain why.
4. Both transfer NOE and saturation transfer difference (STD) experiments make use of dipolar cross relaxation to determine the binding complex of ligand to protein. What are the differences between these two methods? And what are the advantages and limitations of the two?
5. If a free ligand has a diffusion constant eight times larger than the protein, what is the ratio of the intensities of peaks from the free ligand to those from bound ligand likely to be when using the same parameters (gradient strength, delays, etc.) in the gradient diffusion measurements for the samples with and without protein?

6. From a series of ^{1}H-diffusion experiments, the slope from the plot fitting of ln (I/I_0) vs. G^2 is 28.68. The gradient duration of 5 ms and diffusion time Δ 300 ms were used for all experiments. What is the determined diffusion constant for the solute?
7. Why can STD experiments only be used for studying large proteins? What would happen if the method is used for smaller proteins?
8. In addition to a library of compounds, what kind of NMR sample do you need to use the SAR-by-NMR method to screen the binding affinity of ligand to a protein?

References

Breeze A (2000) Isotope-filtered NMR methods for the study of biomolecular structure and interactions. Prog Nucl Magn Reson Spectrosc 36:323–372

Cheng Y, Prusoff WH (1973) Relationship between the inhibition constant (KI) and the concentration of inhibitor which causes 50 per cent inhibition (I50) of an enzymatic reaction. Biochem Pharmacol 22:3099–3108

Choquet CG, Richards JC, Patel GB, Sprott GD (1994) Ribose biosynthesis in methanogenic bacteria. Arch Microbiol 161:481–488

Clore GM, Gronenborn AM (1982) Theory and applications of the transferred nuclear overhauser effect to the study of the conformations of small ligands bound to proteins. J Magn Reson 48:402–417

Dalvit C, Böhlen JM (1997) Analysis of biofluids and chemical mixtures in non-deuterated solvents with 1H diffusion-weighted PFG phase-sensitive double-quantum NMR spectroscopy. NMR Biomed 10:285–291

Fesik SW, Gampe RT Jr, Holzman TF, Egan DA, Edalji R, Luly JR, Simmer R, Helfrich R, Kishore V, Rich DH (1990) Isotope-edited NMR of cyclosporin A bound to cyclophilin: evidence for a trans 9,10 amide bond. Science 250:1406–1409

Gahmberg CG (1997) Leukocyte adhesion: CD11/CD18 integrins and intercellular adhesion molecules. Curr Opin Cell Biol 9:643–650

Gemmecker G, Olejniczak ET, Fesik S (1992) An improved method for selectively observing protons attached to 12C in the presence of 1H-13C spin pairs. J Magn Reson 96:199–204

Gruetter R, Magnusson I, Rothman DL, Avison MJ, Shulman RG, Shulman GI (1994) Validation of 13C NMR measurements of liver glycogen in vivo. Magn Reson Med 31:583–588

Hajduk PJ, Sheppard G, Nettesheim DG, Olejniczak ET, Shuker SB, Meadows RP, Steinman DH, Carrera GM Jr, Marocotte PA, Severin J, Walter K, Smith H, Gubbins E, Simmer R, Holzman TF, Morgan DW, Davidsen SK, Summers JB, Fesik SW (1997a) Discovery of potent nonpeptide inhibitors of stromelysin using SAR by NMR. J Am Chem Soc 119:5818–5827

Hajduk PJ, Dinges J, Miknis GF, Merlock M, Middleton T, Kempf DJ, Egan DA, Walter KA, Robins TS, Shuker SB, Holzman TF, Fesik SW (1997b) NMR-based discovery of lead inhibitors that block DNA binding of the human papillomavirus E2 protein. J Med Chem 40:3144–3150

Johnson CS Jr (1999) Diffusion ordered nuclear magnetic resonance spectroscopy: principles and applications. Prog Nucl Magn Reson Spectrosc 34:203–256

Klein J, Meinecke R, Mayer M, Meyer B (1999) Detecting binding affinity to immobilized receptor proteins in compound libraries by HR-MAS STD NMR. J Am Chem Soc 121:5336–5337

Liu G, Huth JR, Olejniczak ET, Mendoza R, DeVries P, Leitza S, Reilly EB, Okasinski GF, Fesik SW, von Geldern TW (2001) Novel p-arylthio cinnamides as antagonists of leukocyte function-associated antigen-1/intracellular adhesion molecule-1 interaction. 2. Mechanism of inhibition and structure-based improvement of pharmaceutical properties. J Med Chem 44:1202–1210

Liu G, Link JT, Pei Z, Reilly EB, Leitza S, Ngygen B, Marsh KC, Okasinski GF, von Geldern TW, Ormes M, Fowler K, Gallatin M (2000) Discovery of novel p-arylthio cinnamides as antagonists of leukocyte function-associated antigen-1/intracellular adhesion molecule-1 interaction. 1. Identification of an additional binding pocket based on an anilino diaryl sulfide lead. J Med Chem 43:4025–4040

Mayer M, Meyer B (1999) Characterization of ligand binding by saturation transfer difference NMR spectroscopy. Angew Chem Int Ed 38:1784–1788

Mayer M, Meyer B (2001) Group epitope mapping by saturation transfer difference NMR to identify segments of a ligand in direct contact with a protein receptor. J Am Chem Soc 123:6108–6117

Meyer B, Weimar T, Peters T (1997) Screening mixtures for biological activity by NMR. Eur J Biochem 246:705–709

Neidhardt FC, Umbarger HE (1996) Chemical composition of Escherichia coli. In: Neidhardt FC, Curtiss III R, Ingraham JL, Lin ECC, Low KB, Magasanik B, Reznikoff WS, Riley M, Schaechter M, Umbarger HE (eds) Escherichia coli and Salmonella: cellular and molecular biology, 2nd edn. American Society for Microbiology, Washington, DC, p 13

Ni F (1994) Recent developments in transferred NOE methods. Prog Nucl Magn Reson Spectrosc 26:517–606

Price WS, Kuchel PW (1991) Effect of nonrectangular field gradient pulses in the stejskal and tanner (diffusion) pulse sequence. J Magn Reson 94:133–139

Shuker SB, Hajduk PJ, Meadows RP, Fesik SW (1996) Discovering high-affinity ligands for proteins: SAR by NMR. Science 274:1531–1534

Shulman GI, Rothman DL, Jue T, Stein P, DeFronzo RA, Shulman RG (1990) Quantitation of muscle glycogen synthesis in normal subjects and subjects with non-insulin-dependent diabetes by 13C nuclear magnetic resonance spectroscopy. N Engl J Med 322:223–228

Sligh JE Jr, Ballantyne CM, Rich SS, Hawkins HK, Smith CW, Bradley A, Beaudet AL (1993) Inflammatory and immune responses are impaired in mice deficient in intercellular adhesion molecule 1. Proc Natl Acd Sci USA 90:8529–8533

Stejskal EO, Tanner JE (1965) Spin diffusion measurements: spin echoes in the presence of a time-dependent field gradient. J Chem Phys 42:288–292

Stilbs P (1987) Fourier transform pulsed-gradient spin-echo studies of molecular diffusion. Prog Nucl Magn Reson Spectrosc 19:1–45

Taylor R, Price TB, Rothman DL, Shulman RG, Shulman GI (1992) Validation of 13C NMR measurement of human skeletal muscle glycogen by direct biochemical assay of needle biopsy samples. Magn Reson Med 27:13–20

Tumbula DL, Teng Q, Bartlett MG, Whitman WB (1997) Ribose biosynthesis and evidence for an alternative first step in the common aromatic amino acid pathway in Methanococcus maripaludis. J Bacteriol 179:6010–6013

Wood HG, Katz J (1958) The distribution of C14 in the hexose phosphates and the effect of recycling in the pentose cycle. J Biol Chem 233:1279–1282

Yu JP, Ladapo J, Whitman WB (1994) Pathway of glycogen metabolism in Methanococcus maripaludis. J Bacteriol 176:325–332

Zwahlen C, Legault P, Vincent SJF, Greenblatt J, Konrat R, Kay LE (1997) Methods for measurement of intermolecular NOEs by multinuclear NMR spectroscopy: application to a bacteriophage λ N-peptide/boxB RNA complex. J Am Chem Soc 119:6711–6721

Chapter 7
Protein Structure Determination from NMR Data

7.1 Introduction and Historical Overview

Although NMR was discovered in 1946, its application to biological systems only started in the late 1960s and early 1970s. The application was very limited due to the poor sensitivity and very low resolution offered by the one-dimensional techniques used at that time. Two major breakthroughs in the 1970s revolutionized the field: Fourier transformation (FT) NMR that allowed rapid recording of NMR signals and two-dimensional NMR spectroscopy that dramatically increased the spectral resolution. These advances in combination with the development of stable magnets at higher fields led to explosive investigations using NMR spectroscopy in the late 1970s and early 1980s, which centered on exploring its potential in determining the 3D structures of macromolecules. Even though X-ray crystallography was already a method of choice for structure determination during that period, it was believed that NMR may provide complementary structural information in the more physiologically relevant solution environment. Moreover, since some biomolecules are difficult to crystallize, NMR could be used as an alternative method for obtaining 3D structures.

In the mid-1980s, several groups reported the first generation of solution structures of proteins and oligonucleotides using 2D NMR methods. The protocols used in these NMR structure calculations proved to be valid when the same structure of the α-amylase inhibitor Tendamistat was determined in 1986 independently by NMR and crystallographic groups. After that, the field witnessed an exponential growth with the excitement over NMR being another powerful method for macromolecule structure determination. However, it was soon realized that signal degeneracy and the intrinsic relaxation behavior of macromolecules limited 2D NMR application within the range of small proteins and nucleic acids (< 10 kDa).

In late 1980s and early 1990s, another quantum jump came when multidimensional heteronuclear NMR methods were developed thereby pushing the molecular size limit of NMR structures up to 35 kDa. Advances in molecular biology that lead to the overexpression and isotope labeling of proteins also played an important role.

Hence, multidimensional heteronuclear NMR has opened the door for studying a wide variety of proteins and protein domains. The developments of TROSY and residual dipolar coupling (RDC) allow NMR to study even larger proteins and protein complexes. Hence, although NMR is still in its developmental stage and lags behind macromolecular crystallography by almost 30 years (the first crystal structure of a protein was published in 1957, whereas the first NMR structure came to the world in the mid 1980s), it has certainly become one of the most powerful players in molecular/structural biology. Today, about one fifth of the macromolecular structures deposited in the PDB (Protein Databank) were derived from NMR spectroscopy. Despite its size limitation for macromolecular structure determination, NMR has the following unique features: (a) it allows structural studies in a physiologically relevant solution environment, which avoids experimental artifacts such as crystal packing seen in crystal structures; (b) it allows structural studies of some molecules that are difficult to crystallize such as flexible protein domains, weakly bound protein complexes, etc.; (c) it can provide information about protein dynamics, flexibility, folding/unfolding transitions, etc. (see Chap. 8). With the completion of the human genome, NMR will also play a major role in the post-genome era in areas such as structural genomics, proteomics, and metabolomics.

The outline for the NMR-based structure determination (Wüthrich 1986) shown in Fig. 7.1 includes three stages: (1) sample preparation, NMR experiments, data processing; (2) sequence specific assignment, NOESY assignment, assignments of other conformational restraints such as J coupling, hydrogen bonding, dipolar coupling; and (3) structure calculation and structure refinement. One starts with a well-behaved sample (protein, nucleic acid, etc.) and performs a suite of NMR experiments designed for resonance assignment and structural analysis (Chap. 5). Four important parameters are generated for structure calculations: (a) Chemical shifts that provide mostly secondary structural information for proteins; (b) J coupling constants that provide geometric angles within molecules; (c) Nuclear Overhauser effects (NOEs) that provide ^1H–^1H distances within 5 Å. The NOE data are considered to be the most important and are rich in providing especially tertiary structural information. (d) The RDCs can provide valuable structural restraints. This fourth parameter is complementary to NOE data since it provides long-range distance information (>5 Å), whereas the NOE data is only restricted to <5 Å. Each of the parameters will first briefly be described and then a protocol will be used to describe in detail how a structure is calculated and how each parameter is implemented during the structure calculation.

Key questions to be addressed in the current chapter include the following:

1. What types of structural information can be obtained based on the results of the resonance assignments?
2. What are the methods currently used for structure calculation from NMR data?
3. How are J coupling constants used in the structure determination of a protein?
4. What other nuclear interactions can be utilized for structure determination?
5. How are the NMR parameters utilized in structure calculation?

7.2 NMR Structure Calculation Methods

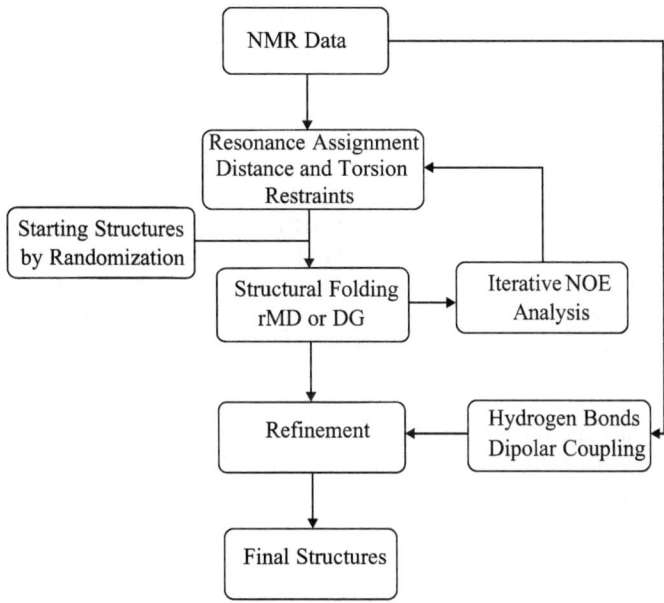

Fig. 7.1 Strategy for NMR-based structure determination

6. What are the strategies to carry out the calculation?
7. How is the structural quality analyzed?
8. How are the precision and accuracy of the calculated structures determined?
9. What is the role of iterative NOE analysis during the structure calculation?
10. Step-by-step illustration of structural calculation using a typical XPLOR protocol.

7.2 NMR Structure Calculation Methods

To date, the majority of structures characterized by NMR spectroscopy are obtained using distance and orientational restraints derived from the measurements of NOEs, J coupling constants, RDCs, and chemical shifts. The calculation of a 3D structure is usually formulated as a minimization problem for a target function that measures the agreement between a conformation and the given set of restraints. There are several algorithms developed over the past three decades, of which four are widely used (Table 7.1): (a) metric matrix distance geometry, represented by the DG-II protocol; (b) variable target function method, represented by the DIANA protocol; (c) Cartesian space or torsion angle space restrained molecular dynamics (rMD), represented by the XPLOR (or CNS) protocol; (d) Torsion angle dynamics, represented by the DYANA protocol. These methods generate and refine biomolecular structures by searching globally to get an ensemble of molecular structures

Table 7.1 Structure calculation methods and programs

Method	Program[a]	Reference
Metric matrix distance geometry	DIG-II	Havel (1991)
Variable target function method	DIANA	Guüntert et al. (1991)
Cartesian space or torsion angle space rMD	AMBER	Pearlman et al. (1991)
	CHARM	Brooks et al. (1983)
	GROMOS	van Gunsteren et al. (1996)
	XPLOR	Brünger (1992a, 1992b)
	CNS	Brünger et al. (1998)
Torsion angle dynamics	DYANA	Güntert et al. (1997)

[a] These are the programs used for structure calculation

that fit with the experimentally measured restraints within the range of experimental error. Distance geometry and rMD (simulated annealing) calculations are discussed in this section.

7.2.1 Distance Geometry

One approach is to generate structures with the distance and orientational restraints derived from NOEs and J coupling constants using a metric matrix or dihedral angle space distance algorithm. The metric matrix method utilizes all possible interatomic distances as restraints including the known distances from covalent bonds and experimentally estimated distances from NOE data to generate an n-dimensional matrix for a molecule with n atoms (Crippen 1981; Braun 1987). For the remaining distances, upper and lower bounds are chosen and altered until no further alterations can be made according to the triangle inequalities:

$$u_{ij} \leq u_{ik} + u_{kj}$$
$$l_{ij} \leq l_{ik} + l_{kj} \tag{7.1}$$

in which i, j, k are the three atoms defining a triangle, and u and l are the upper and lower bounds between any two given points of the triangle. An ensemble of structures is generated from randomly selected distances within the boundary conditions.

The initial structures obtained by the randomization are further refined by minimizing a penalty function V (or potential function) such as:

$$V = \sum_{i>j} \begin{cases} k(r_{ij}^2 - l_{ij}^2)^2 & \text{if } r_{ij} < l_{ij} \\ k(r_{ij}^2 - u_{ij}^2)^2 & \text{if } r_{ij} < u_{ij} \\ 0 & \text{if } l_{ij} \leq r_{ij} \leq u_{ij} \end{cases} \tag{7.2}$$

7.2 NMR Structure Calculation Methods

in which r_{ij} is the distance between atom i and j and k is a weighting factor, u, l are defined as in (7.1). The true global minimum is found only if $V = 0$, meaning that all restraints are satisfied. However, in practice, V is always greater than zero because of the insufficient number of restraints available. Function V is also called a target function. The minimization is achieved by comparing the interproton distances of the calculated structures with the distances within the chosen boundaries. The improved distance geometry uses dihedral angles rather than Cartesian coordinates to fold protein structures based on the short-range restraints and then expands the calculation to eventually include all restraints. These distance geometry methods (e.g., the DIANA program) have played an important role in the determination of protein structures by solution NMR spectroscopy.

7.2.2 Restrained Molecular Dynamics

Restrained molecular dynamics methods (e.g., the programs CNS and XPLOR) calculate the structures using NMR experimental restraints and energy minimization with potential energy functions similar to the above restrained potential energy function. The potential energy (or target function) is calculated for an array of initial atomic coordinates based on a series of potential energy functions (van Gunsteren 1993):

$$V_{tot} = V_{classic} + V_{NOE} + V_{J\ coupling} + V_{H\ bond} + V_{dipolar} + V_{cs} + V_{other} \quad (7.3)$$

in which term $V_{classic}$ is the potential function from the classic energy of the molecule, which contains $V_{bond} + V_{angle} + V_{dihedral} + V_{Van\ der\ Waals} + V_{electrostatic}$. The rest of (7.3) takes the NMR data in terms of distances derived from NOE, torsion angles from J coupling constants, molecular bond orientation restraints from residual dipolar couplings, chemical shifts and other restraints such as disulfide bridges, hydrogen bonding, and planarity. Although torsion angles and residual dipolar coupling restraints are sometimes not used, NOE distance restraints are always used in the rMD calculations. Among several different functions used to characterize the potential energy, a flat-well potential is commonly used, which consists of the energy contributions of NOE distance violations relative to the lower and upper distance bounds (Clore et al. 1986):

$$V_{NOE} = \sum_{i}^{all\ NOEs} V_{NOEi}$$

$$= \sum_{i}^{all\ NOEs} \begin{cases} k_{NOE}(r_i - r_u)^2 & \text{if } r_i > r_u \\ k_{NOE}(r_i - r_l)^2 & \text{if } r_i < r_l \\ 0 & \text{if } r_l < r_i < r_u \end{cases} \quad (7.4)$$

in which k_{NOE} is the force constant of the NOE potential function (also called the target function), r_l and r_u are lower and upper bounds for individual NOE intensities, respectively, and r_i is the interproton distance for each proton pair from the generated structure. If the precise interproton distances (instead of a range of distances) are obtained from NOESY spectra with different mixing times, the NOE potential function can be described using a biharmonic potential:

$$V_{NOE} = \sum_{i}^{\text{all NOEs}} \begin{cases} C_1(r_i - r_0)^2 & \text{if } r_i > r_0 \\ C_2(r_i - r_l)^2 & \text{if } r_i < r_0 \end{cases} \quad (7.5)$$

in which C_1 and C_2 are the force constants which are temperature-dependent and r_i and r_0 are the calculated and experimental distance, respectively.

An alternative method to calculate the NOE potential energy is to compare the calculated NOE intensities from the structure to the experimental NOE intensities at any given step during an rMD calculation using the NOE potential function (Brünger 1992a, 1992b):

$$V_{NOE} = k(a_{exp} - a_{cal})^2 \quad (7.6)$$

in which a_{exp} and a_{cal} are, respectively, matrices of experimental NOE cross-peak intensities used for the calculation and calculated intensities from the structure obtained by the rMD simulation.

The energy barriers between local minima are more easily overcome in rMD because molecular dynamics is used in the energy minimization, which makes the method less sensitive to the initial structures (Allen and Tildesley 1987). A molecular dynamics simulation is performed by solving Newton's equations of motion using the forces generated by varying the potential energies of the macromolecular structures. A minimum energy structure is obtained by solving the first derivative of the potential energy with respect to the coordinates of each atom using the condition that the derivative is zero. From Newton's law, the force for an individual atom can be written as:

$$F = ma = -\frac{dV}{dr} \quad (7.7)$$

or

$$-\frac{dV}{dr} = m\frac{d^2r}{dt^2} \quad (7.8)$$

in which m is mass of the atom, a is the acceleration, V is the potential energy, t is time, and r the coordinates of the atom. The equation of motion is solved by numerical integration algorithms, and the trajectory for each atom as a function of time is calculated.

In order to maintain an accurate and stable simulation, the time step should be kept sufficiently smaller than the fastest local motion of the molecule. Typically, the time step size chosen in the simulation is in the range of femtoseconds (10^{-15} s) for simulations in a picosecond (10^{-12} s) timescale. During the simulation, energy barriers of the system (whose amplitude approximately equals kT) are passed by raising the temperature of the system high enough to increase the kinetic energy during the simulation so that a global energy minimum is located by balancing both the classical energy terms and the energy terms fitted with the experimental restraints. In the first stage of the calculation, an ensemble of initial structures is selected by randomization. The initial structures must gain kinetic energy, which is commonly provided by increasing temperature of the simulated system, to move away from their local energy minima and then pass over higher energy barriers. With the higher energy, the system contains a greater range of structural space. The system is then slowly cooled down to room temperature. During the cooling process, the system energy is minimized over the surface of potential energy to search for stable structures at low temperature. The cycles of heating and cooling are repeated until an ensemble of stable structures with an acceptable penalty (or violations) is eventually determined.

A typical procedure (Güntert 1998, 2003; de Alba and Tjandra 2002; Lipstitz and Tjandra 2004) for structure calculation includes (a) a stage of randomization in which a set of initial structures is generated with the idealized covalent geometry restraints such as bond length, bond angles, dihedral angles, and improper torsions; (b) global folding in which a variety of energy terms with both the geometry restraints and experimentally obtained distance and torsion restraints are used to obtain folded structures; and (c) refinement which utilizes the same energy terms as in the previous stage but in a smaller step size (typical 0.5 fs) for ps rMD processes to refine the structures generated in the previous stage. The refinement can also involve RDCs in which the structures are refined using observed RDCs by increasing dipolar force constants slowly and simultaneously refining the principal components of the alignment tensor.

7.3 NMR Parameters for Structure Calculation

7.3.1 Chemical Shifts

In principle, chemical shifts of NMR-active nuclei such as ^{1}H, ^{15}N, and ^{13}C are dictated by the structural and chemical environment of the atoms (see Chap. 1). Vigorous efforts have been made to deduce protein structure from chemical shifts. Chemical shift-based secondary structural prediction has met with some success. In particular, the deviations of ^{13}C$^\alpha$ (and, to some extent, ^{13}C$^\beta$) chemical shifts from their random coil values can be well correlated with the α-helix or β-sheet conformations: ^{13}C$^\alpha$ chemical shifts larger than the random coil values tend to occur for helical residues, whereas the opposite is observed for β-sheet residues (Spera and Bax 1991; de Dios et al. 1993; Wishart et al. 1991, 1992). A good correlation was also observed for proton shifts with secondary structures: $^1H^\alpha$ shifts smaller than the random coil values tend to occur for helical residues, whereas the opposite is observed for β-sheet residues. Although the information is useful for tertiary structure calculations (Luginbühl et al. 1995; Kuszewski et al. 1995a, b), it is much more valuable for the initial secondary structural analysis in combination with NOE data (see below).

Methods have been developed to predict the dihedral angles using backbone chemical shifts, such as the programs TALOS (torsion angle likelihood obtained from the shift and sequence similarity, Cormilescu et al. 1999) and SHIFTOR (Zhang 2001). The prediction is based on the observation that similar amino acid sequences with similar backbone chemical shifts have similar backbone torsion angles. First, TALOS breaks the sequence of a target protein into overlapping amino acid triplets. Then, for each triplet, the program searches its database which contains proteins with known chemical shifts ($^1H^\alpha$, $^{13}C^\alpha$, $^{13}C^\beta$, $^{13}C'$, and ^{15}N) and high-resolution x-ray crystal structures to compare the chemical shift and sequence similarity. The torsion angles for the central residue from the best ten matches are chosen as the predicted torsion angles for the residue, which are used as backbone dihedral angles in the structure calculation. Typically, TALOS can predict the dihedral angles for ~70 % of the residues within ±20°. The incorrect predictions can be removed by the inconsistency with other types of constraints during the structure calculation.

7.3.2 J Coupling Constants

J coupling constants are derived from the scalar interactions between atoms. They provide geometric information between atoms in a molecule. The most useful and obtainable coupling constants are vicinal scalar coupling constants, ^3J, between atoms separated from each other by three covalent bonds. Its relation with dihedral angle θ is defined as follows (Karplus equation, Karplus 1959, 1963):

$$^3J(\theta) = A \cos^2\theta + B \cos\theta + C \qquad (7.9)$$

in which A, B, C are coefficients for various types of couplings, and θ is the dihedral angle. Using the Karplus relationship, one can convert the J coupling constants into the dihedral angles, commonly, φ, ψ, and χ^1. The dihedral angles can be determined by best fitting the measured J values to the corresponding values calculated with (7.9) for known protein structures (see Fig. 7.2). These dihedral angles can be used as structural restraints later during calculations (see below). Table 7.2 lists ^3J commonly used for deducing various dihedral angles for proteins.

7.3 NMR Parameters for Structure Calculation

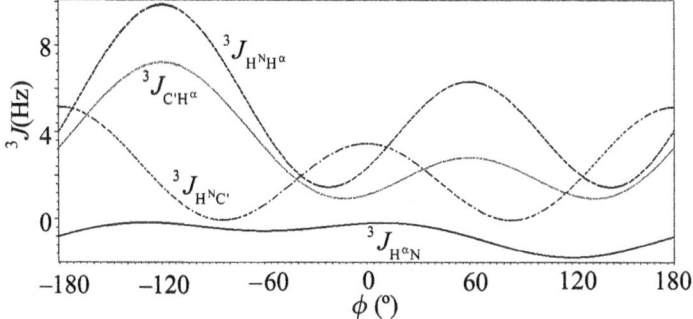

Fig. 7.2 Karplus curves describing the relationships of vicinal J coupling constants and torsion angle ϕ using the constants listed in Table 7.2. The *solid line* is for the coupling between protons, *dotted lines* for the heteronuclear couplings. The angle $\theta = \phi$ + offset

Table 7.2 Karplus constants

Coupling constant	Torsion angle	Dihedral	A	B	C	Offset (°)	Reference
$^3J_{H^NH^\alpha}$	ϕ	H^N–N–C^α–H^α	6.51	−1.76	1.6	−60	Vuister and Bax (1993)
			6.98	−1.38	1.72	−60	Wang and Bax (1996)
$^3J_{H^NC'}$		H^N–N–C^α–C'	4.32	0.84	0.00	180	Wang and Bax (1996)
$^3J_{C'_{i-1}H^\alpha}$		C'–N–C^α–H^α	3.75	2.19	1.28	120	Wang and Bax (1996)
$^3J_{H^\alpha N_{i+1}}$	ψ	H^α–C^α–C'–N	−0.88	−0.61	−0.27	−120	Wang and Bax (1995)
$^3J_{H^\alpha H^\beta}$	χ^1	H^α–C^α–C^β–H^β	9.50	−1.60	1.80	−120/0	de Marco et al. (1978a)
$^3J_{NH^\beta}$		N–C^α–C^β–H^β	−4.40	1.20	0.10	120/−120	de Marco et al. (1978b)
$^3J_{C'H^\beta}$		C'–C^α–C^β–H^β	7.20	−2.04	0.60	0/120	Fischman et al. (1980)

7.3.3 Nuclear Overhauser Effect

NOEs are the most important NMR parameters for structure determination because they provide short-range as well as long-range distance information between pairs of hydrogen atoms separated by less than 5Å. Whereas the short-range NOEs are valuable for defining secondary structure elements such as α-helix or β-sheet, the long-range NOEs provide crucial tertiary structural information (Wüthrich 1986). The spectral intensity of an NOE (I) is related to the distance r between the proton pair, $I = f(\tau_c)<r^{-6}>$ in which $f(\tau_c)$ is a function of the rotational correlation time τ_c of the molecule. Because of many technical factors such as highly variable τ_c for different molecules at different temperatures and solvent conditions, it is common to use intensity I (or cross-peak volume) to obtain qualitative distance information. The information is usually grouped into three different distance categories: 1.8–2.5 Å (strong), 1.8–3.5 Å (medium), and 1.8–5.0 Å (weak). Note that the lower bound for all three categories is 1.8Å, corresponding to the Van der Waals repulsion range. This treatment is due to the consideration that weak NOEs may not be related longer distances such as >4 Å. Instead, they may be related to the chemical exchange or protein motions that diminish the NOE intensities.

When performing a NOESY experiment to obtain NOE information, it is important to choose a proper mixing time that is in principle proportional to the distance (NOE intensity). Short mixing times may lead to the loss of weak NOEs that may contain important tertiary structural information. However, long mixing times may induce so-called "spin-diffusion," that is, NOEs indirectly generated by spins in the vicinity >5 Å. The mixing time can be accurately determined by analyzing an NOE build-up curve (Neuhaus and Williamson 1989). However, the build-up curves vary considerably among different spins in the same molecule such as between methylene protons and methyl protons. A compromise is usually given to suppress spin diffusion and to maintain sufficient NOE intensities based on NOE build-up curves. As a rule of thumb, 80–120 ms is usually used for small–medium-sized proteins. Sometimes, NOESY experiments with several different mixing times are performed to make sure that the "spin-diffusion" peaks are not picked.

Several programs are available for NOE analyses such as nmrview and PIPP. These programs can store all the assigned NOEs in a table that can be converted into distance format for structure calculations (see Appendix A). The assigned NOEs can also be plotted as a function of protein sequence to gain information about the protein secondary structural information and topology of tertiary fold in conjunction with chemical shifts and J coupling information (Fig. 7.3).

7.3.4 Residual Dipolar Couplings

Between the early 1980s and late 1990s, all NMR structures were determined based primarily on NOE data supplemented by J coupling constants and chemical shifts. In the late 1990s, a new class of conformational restraints emerged, which originate from internuclear RDC in weakly aligned media such as bicelles (Tjandra and Bax 1997; Prestegard et al. 2000; de Alba and Tjandra 2002; Lipstitz and Tjandra 2004). The RDC gives information on angles between covalent bonds and on long-range order. The addition of this structural parameter has proven to greatly improve the precision as well as the accuracy of NMR structures.

Although internuclear DD (dipole–dipole) couplings are typically averaged out due to molecular tumbling, RDC occurs when there is a small degree of molecular alignment with the external magnetic field (see Chap. 1). The RDCs are manifested

7.3 NMR Parameters for Structure Calculation

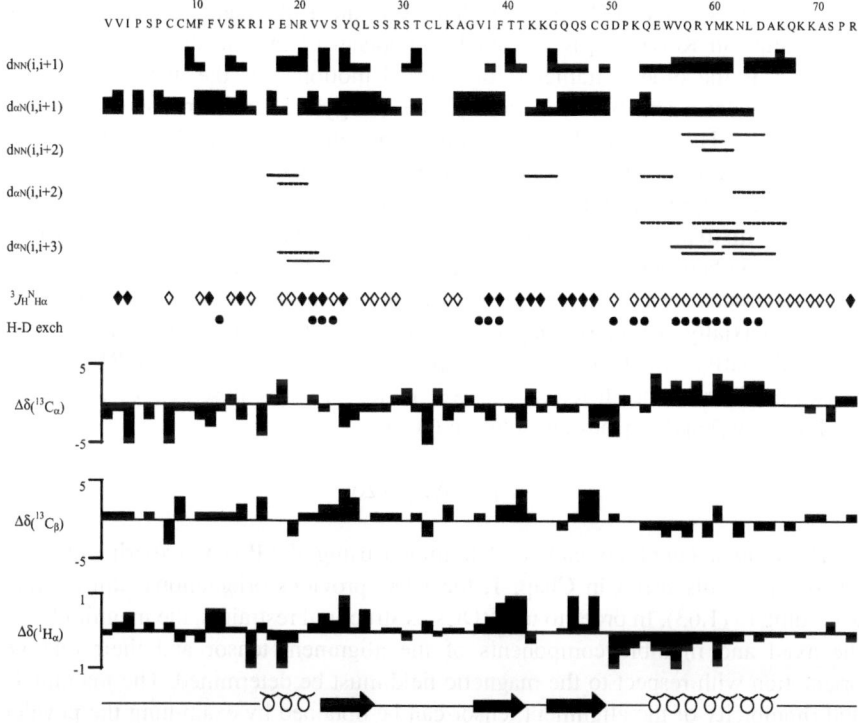

Fig. 7.3 Experimental restraints for eotaxin-2, including NH exchange $^3J_{H^NH^\alpha}$ coupling constants, sequential, short- and medium-range NOEs and H$^\alpha$, C$^\alpha$, and C$^\beta$ secondary chemical shifts, along with the secondary structure deduced from the data (reproduced with permission from Mayer and Stone (2000), Copyright © 2000 American Chemical Society). The amino acid sequence and numbering are shown at top. Sequential N–N and α–N NOEs are indicated by *black bars*; the thickness of the bar represents the strength of the observed NOE. The presence of medium-range N–N and α–N NOEs is indicated by solid lines. *Gray bars* and *dashed lines* represent ambiguous assignments. $^3J_{H^NH^\alpha}$ coupling constants are represented by diamonds corresponding to values of <6 Hz (*open*), 6–8 Hz (*gray*), and >8 Hz (*black*). Residues whose amide protons show protection from exchange with solvent are indicated with filled circles. The chemical shift indices shown for C$_\alpha$, C$_\beta$, and H$_\alpha$ were calculated according to the method developed by Wishart et al. (1992). The locations of the secondary structure elements identified in the calculated family of structures are shown at the bottom

as small, field-dependent changes of the splitting normally caused by one-bond J couplings between directly bound nuclei. With the assumption of an axially symmetric magnetic susceptibility tensor and neglecting the contribution from "dynamic frequency shifts," the frequency difference Δv^{obs} between the apparent J values at two different magnetic field strengths, B_0^1 and B_0^2, is given by (Tjandra et al. 1997):

$$\Delta v^{obs} = \frac{\hbar \gamma_a \gamma_b \chi_a S}{30\pi r k T}(B_0^2 - B_0^1)(3\cos^2\theta - 1) \qquad (7.10)$$

in which $\hbar$ is reduced Plank's constant, k is the Boltzmann constant, T is the temperature in Kelvin, χ_a is the axial component of the magnetic susceptibility tensor, S is the order parameter for internal motion, r is the distance between coupled nuclei a and b, and γ_a and γ_b are the gyromagnetic ratios of a and b, respectively. The structural information is contained in the angle θ between the covalent bond formed by two scalar coupled atoms a and b and the main axis of the magnetic susceptibility tensor. It is then straightforward to add an orientational restraint term to the target function of a structure calculation program that measures the deviation between the experimental and calculated values of θ.

An alternative way to obtain RDC orientational restraints without the assumption of an axially symmetric magnetic susceptibility tensor is to obtain the magnitude and relative orientation of the alignment tensor. The value for the RDC (Δv_D) is extracted from the difference between the splittings observed in alignment medium (Δv_A) and in isotropic solution (Δv_J):

$$\Delta v_D = \Delta v_A - \Delta v_J \qquad (7.11)$$

The dipolar couplings can be determined using IPAP type experiments (see Chap. 5). As discussed in Chap. 1, the RDC provides orientational information according to (1.63). In order to use RDCs as structural restraints, the magnitudes of the axial and rhombic components of the alignment tensor and their relative orientation with respect to the magnetic field must be determined. The magnitude and rhombicity of the alignment tensor can be obtained by examining the powder pattern distribution of all normalized observed dipolar couplings for the molecule. When structures are calculated, all of the variables can be obtained by fitting the equation with a large number of RDCs. If the structures are accurate, the calculated dipolar couplings of the structures will be in good agreement with the observed RDCs within the experimental error range. Such orientational restraints have been shown to improve the quality of the structures. They are also extremely valuable when calculating protein complex structures.

7.4 Preliminary Secondary Structural Analysis

Prior to structure calculations, it is useful to determine the secondary structure using a combination of chemical shifts, J coupling constants, and NOE data. As mentioned above, ^{13}C chemical shifts are particularly indicative of α-helices and β-sheets. Three-bond $J_{NH-H\alpha}$ coupling constants are often small for helical residues (<5 Hz), but large for β-sheet residues (>8 Hz). Regular secondary structure elements can also be easily identified from sequential NOEs, as each type of secondary structure element is characterized by a particular pattern of short-range NOEs ($|r_i - r_j| < 5$ Å). For instance, α-helices are characterized by a stretch of strong and medium $NH_i\_NH_{i+1}$ NOEs, and medium or weak $C^\alpha H_i-NH_{i+3}$, $C^\alpha H_i-C^\beta H_{i+3}$ NOEs, and $C^\alpha H_i-NH_{i+1}$ NOEs, sometimes supplemented by

NH$_i$–NH$_{i+2}$ and C$^\alpha$H$_i$–NH$_{i+4}$ NOEs. β-Strands, on the other hand, are characterized by very strong C$^\alpha$H$_i$–NH$_{i+1}$ NOEs and by the absence of other short-range NOEs involving the NH and C$^\alpha$H protons. β-Sheets can be identified and aligned from interstrand NOEs involving the NH, C$^\alpha$H, and C$^\beta$H protons. Hydrogen-exchange experiments are also often performed to extract information for slowly exchanging amides that are normally involved in helices or β-sheets. This adds great confidence later when dealing with H-bonds of the backbone amides involved in α-helices or β-sheets.

A computer program written in a shell script can convert all the NOE, J coupling, exchange data, and chemical shifts into a figure for analyzing the secondary structures of proteins. This is illustrated in Fig. 7.3. Note that this approach tends to perform poorly in ill-defined secondary structural regions such as loops. In addition, the exact start and end of helices tends to be less accurate since the pattern of these parameters is similar to that present in turns. Thus, a turn at the end of a helix could be misinterpreted as still being part of the helix. In the case of β-sheets, the definition of the start and end is more accurate as the alignment is accomplished from interstrand NOEs involving NH and C$^\alpha$H protons.

Another preliminary structural analysis is the stereospecific assignment of diasterotopic protons. There are two major types of diastereotopic protons in amino acids: (a) methylene protons in Lys, Arg, etc.; (b) Methyl groups of Val, Leu. If the signals are well resolved, stereospecific assignments of *β*-methylene protons can be assigned by a combination of ^{15}N-edited TOCSY and ^{15}N-edited NOESY. Some methylene protons can also be stereospecifically assigned during the course of structure calculations. Stereospecific assignments of Val and Leu methyls can be made experimentally by the partial ^{13}C labeling or fractional deuteration method (Neri et al. 1989; Senn et al. 1989), provided the signals are well resolved. If signals are degenerate or weak, which prevents the stereospecific assignments, diastereotopic protons are referred to as pseudoatoms, which may result in less well-defined structures. Stereospecific assignments not only provide more accurate distance information, but also provide dihedral angle information including χ^1 and χ^2 (Powers et al. 1991). Hence, it is important to have as many stereospecific assignments as possible in order to obtain a high quality structure.

7.5 Tertiary Structure Determination

7.5.1 Computational Strategies

Because proteins typically consist of more than a thousand atoms that are restrained by thousands of experimentally determined NOE restraints in conjunction with stereochemical and steric conditions, it is in general neither feasible to do an exhaustive search of allowed conformations nor to find solutions by interactive model building. In practice, as mentioned in the previous section, the calculation of a molecular structure is performed by minimizing the target function that represents the agreement between a conformation and a set of experimentally derived restraints. In the following section, a step-by-step description of structure calculation is provided using the most widely used XPLOR protocol.

7.5.2 Illustration of Step-by-Step Structure Calculations Using a Typical XPLOR Protocol

General guidance for the rMD protocol is given in Table 7.3. A complete protocol for protein structural calculations using simulated annealing XPLOR program (sa.inp) is provided in Appendix B. The file names in bold require modifications for specific protein structure determination and generally include the different input files such as distance restraints, PDB coordinates, etc. In the protocol, readers are referred to specific remarks on important lines such as "read the PSF file and initial structure," which will help in understanding the protocol.

7.5.2.1 Preparation of Input Files

1. *Example of NOE table*. All the assigned NOEs in a peak-pick table generated by programs such as PIPP or nmrPipe can be converted into a distance restraint table using a shell script. An example of an XPLOR distance restraint table can be found in Appendix C.
2. *Example of dihedral angle restraint table*. Dihedral angles derived from J coupling constants can be assembled into the format for the XPLOR program (Appendix D).
3. *Example of chemical shift restraint table*. The carbon chemical shifts of C^α and C^β can be formatted for the XPLOR program (Appendix E).
4. *Example of H-bond table*. Although NMR experiments have been developed to directly measure the H-bonds, most of the H-bond restraints are still derived indirectly from amide exchange experiments. These H-bond restraints are normally used for structure refinement after the initial structure is calculated. The H-bond input table is as shown in Appendix F.

7.5.2.2 Preparation of Initial Random-Coil Coordinates and Geometric File

1. Input file to generate random-coil coordinates based on the protein sequence (Appendix G).
2. Input file to generate geometric PSF file (Appendix H). This file contains information on the molecular bonds, angles, peptide planes, etc. present in the structure (i.e., how the atoms are connected together).

7.5 Tertiary Structure Determination

Table 7.3 Structure calculation protocol using rMD

Randomization
10 ps Restrained molecular dynamics
  Energy terms: bonds, angles, improper torsions
  Temperature: 1,000 K
  Number of calculated structures: 100
500 Cycles of Powell energy minimization

Global folding
5 ps Restrained molecular dynamics
  Energy terms: bonds, angles, improper torsions, NOE (soft-square NOE potential), Van der Waals (Lennard-Jones). Van der Waals radii are scaled by 0.9
  $k_{NOE} = 30$ kcal mol^{-1} Å^{-2}, $k_{dih} = 10$ kcal mol^{-1} rad^{-2}
  Step size: 5 fs
  Number of steps: 1,000
  Temperature: 2,000 K
15 ps Restrained molecular dynamics while cooling to 300 K
  34 Cycles of 0.44 ps each
  Energy terms: bonds, angles, improper torsions, NOE (soft-square NOE potential), van der Waals (Lennard–Jones). Van der Waals radii are gradually reduced by the factor from 0.9 to 0.8
  k_{NOE} is gradually increased from 2 to 30 kcal mol^{-1} Å^{-2}, $k_{dih} = 200$ kcal mol^{-1} rad^{-2}
  Step size: 5 fs
  Total number of steps: 3,000
  Temperature decrement 50 K per step
500 Cycles of Powell energy minimization

Refinement
500 Cycles of energy minimization
  Energy terms: bonds, angles, improper torsions, soft-square NOE potential, van der Waals (Lennard–Jones)
  $k_{NOE} = 50$ kcal mol^{-1} Å^{-2}, $k_{dih} = 5$ kcal mol^{-1} rad^{-2}
2.5 ps Restrained molecular dynamics while increasing force constants of all torsion angles
  Energy terms: bonds, angles, improper torsions, soft-square NOE potential
  20 Cycles of 0.125 ps each
  k_{NOE}: from 2.2 to 30 kcal mol^{-1} Å^{-2}, increased by a factor of 1.14 per cycle
  k_{dih}: from 1 to 200 kcal mol^{-1} rad^{-2}, increased by a factor of 1.304 per cycle
  Step size: 0.5 fs
  Temperature: 2,000 K
200 Cycles of energy minimization
  Energy terms: bonds, angles, improper torsions, soft-square NOE potential, van der Waals (Lennard–Jones)
Restrained molecular dynamics while cooling to 300 K
  34 Cycles of 0.44 ps each
  Step size: 0.5 fs
  Temperature decrement 50 K per step
500 Cycles of Powell energy minimization

Refinement with dipolar couplings
  Energy terms: bonds, angles, improper torsions, NOE (soft-square NOE potential), van der Waals (Lennard–Jones), electrostatic, dipolar

(continued)

Table 7.3 (continued)

500 Cycles of energy minimization
  $k_{dip} = 0.001$ cal mol^{-1} Å^{-2}
15 ps Restrained molecular dynamics
  50 Cycles of 0.3 ps each
  k_{NOE}: from 0.001 to 0.2 kcal mol^{-1} Å^{-2}, increased by a factor of 1.11 per cycle
  Step size: 0.5 fs
Temperature: 300 K
500 Cycles of energy minimization

7.5.2.3 Randomization

In the initial stage of the calculation, an array of random (or semi-random) initial structures is generated based on covalent geometry restraints including bond length, bond angles, dihedral angles, and improper torsions. After 10 ps of rMD is carried out at a temperature of 1,000 K, a total of 50–100 initial structures are obtained, which will be used for the structure calculation using experimental restraints in next step of the calculation. The energy of the randomized structures is minimized by 500 cycles of Powell minimization (Brooks et al. 1983) against the force of bond length, bond angles, dihedral angles, and improper torsions.

7.5.2.4 First-Round Structure Calculation: Global Folding

After the starting structures are obtained and all other files in bold in sa.inp are prepared, one can start the first-round structure calculations. On a UNIX-based SGI workstation on which XPLOR or CNS is installed, simply type "XPLOR<sa.inp>sa.out &" to initiate the calculation process. The detailed process and output parameters are all contained in the sa.out file. Calculation is often terminated in the beginning due to errors in the input files, nomenclature, metal coordination, etc. These errors are usually reflected in the sa.out file and the readers are referred to the XPLOR or CNS manual for instructions of specific file format. PDB coordinates of a set of calculated structures are stored in the directory during the calculation for visualization and analysis.

The starting structures are first calculated by 5 ps of rMD at 2,000 K with a step size of 5 fs and forces of the covalent geometry restraints such as bond length, bond angles, dihedral angles, improper torsions, and Van der Waals (Lennard–Jones bonds), and experimental restraints of NOE and J coupling. The NOE restraints are used with a force constant of $k_{NOE} = 30$–50 kcal mol^{-1} Å^{-2}, whereas the torsion angle restraints are applied usually with a relatively weak constant, $k_{dih} = 5$–10 kcal mol^{-1} rad^{-2}. The soft repulsive Van der Waals radii (Lennard–Jones) are scaled by a factor of 0.9. In the next step, the temperature of the system is decreased by 50 K per cycle during the slow cooling down to 300 K by 34 cycles (0.44 ps for each cycle; 0.44 ps × 34 = 15 ps) of rMD calculation with a step size of 5 fs. During the cooling, the Van der Waals radii are reduced from 90 to 80 % of their true values, k_{NOE} is gradually increased from 2 to 30 kcal mol^{-1} Å^{-2}, and $k_{dih} = 200$ kcal mol^{-1} rad^{-2}. The last step of the second-round of calculation consists of 500 cycles of Powell energy minimizations. The above procedure is looped for 100 cycles.

Although one will use as many NOEs as possible for the structure calculation, the interresidual distance restraints play a more important role in calculation. In order to obtain a high quality structure, more than 10 distance restraints should be used for each residue.

7.5.2.5 NOE Violations and Removal of Incorrect Distance Restraints

Once the first-round structures are calculated using the experimental restraints, mainly NOE data, it is necessary to analyze and validate the derived structures. Because of experimental errors and imperfect restraints, the calculated structures always contain violations of distance and torsion angles. A distance violation is the difference between the interproton distance in the structure and the closest distance bound (upper or lower) defined by the observed NOE intensity. NOE violations are output into a file after the calculation. At this stage, one should carefully examine the violations that appear in a large number of the structures. These consistent violations are likely caused by misassignment of NOEs or incorrect NOE volume integration. Frequently, finding the consistent violations is not always straightforward because the structure calculation is done by minimizing the potential function over all restraints. As a result, the large violations caused by the incorrect restraints are spread to neighboring regions, leading to a region being consistently violated with lesser scale, or sometimes to violations too small to be recognized after being distributed over a large number of minor violated restraints. Removal of the incorrect restraints after the first-round calculation will improve the quality of structures.

7.5.2.6 Iterative Steps for NOE Analysis and Structure Calculations

Because it is not possible to assign all NOESY cross peaks after sequence specific assignment due to chemical shift degeneracy and inconsistency in some extent of NOESY cross-peak positions compared to those obtained by resonance assignment, only a fraction of NOESY cross peaks are assigned unambiguously and used for the structure calculation at the initial stage. Even with only 30 % of the final number of NOEs, the generated structures are usually well defined although the resolution is relatively low. These structures are then used to resolve the ambiguous NOESY cross-peaks based on the spatial information of the first-round structures. In order to utilize the NOEs, criteria must be set such as chemical shift tolerance range (usually <0.02 ppm) and corresponding distance between the proton pair in the structures. The newly assigned NOEs are then used as restraints for the next round of structure calculation. It is necessary to carry out several rounds of NOE assignment and structure calculation to assign a majority of ambiguous NOE cross-peaks.

The above process of assignment/calculation can also be performed automatically. First, the NOEs are listed with possible assignments for a given chemical shift tolerance. After the first-round calculation, the program such as ARIA (Nilges et al. 1997; Nilges and O'Donoghue 1998; Linge et al. 2001, 2003) uses output structures to reduce the assignment possibilities by comparing the interproton distance from the structure to that from the ambiguous NOEs for all assignment possibilities. The new distance restraints are tested during the next round of calculation. Usually, multiple restraints are given to the reassigned ambiguous NOE, of which only one will be correct. Therefore, these restraints are taken to be more flexible range during the structure calculation.

An alternative approach for iterative NOE Analysis is back-calculation of NOESY cross-peaks based on the generated structure. Once the folded structure is generated by rMD, the intensities (in terms of volume) of NOESY diagonal and cross-peaks can be calculated for the structures using a relaxation matrix approach with the consideration of spin diffusion. During the back-calculation, the relaxation matrix is first defined with the assumption of isotropic motion in absence of the cross correlation contribution to the relaxation. The relaxation matrix is then used to calculate theoretical NOESY spectra from the calculated structures. The theoretical NOESY spectrum is compared with the experimental data either manually or automatically to assign additional NOESY cross-peaks which are used as distance restraints for further structure calculations as described above.

7.5.3 Criteria of Structural Quality

The quality of structures is usually reported in terms of several statistic characters including rmsd (root mean square deviation) of the distance and dihedral restraints, rsmd of idealized covalent geometry, rsmd of backbone atoms, rsmd of heavy atoms, the total number of distance violations, and dihedral restraint violations. These criteria provide insights about how consistent the structures are with the experimental restraints. However, these statistical criteria are imperfect when describing the accuracy of the structural calculations. The quality factor is a preferable parameter to describe the consistency of the derived structures with the experimentally determined restraints. The quality factor or Q factor is defined for a type of restraint A as follows (Yip and Case 1989; Nilges et al. 1991; Baleja et al. 1990; Gochin and James 1990):

$$Q = \frac{\text{rms}(A^{\text{obs}} - A^{\text{calc}})}{\text{rms}(A^{\text{obs}})} = \frac{\sqrt{\sum_i (a_i^{\text{obs}} - a_i^{\text{calc}})^2}}{\sqrt{\sum_i (a_i^{\text{obs}})^2}} \quad (7.12)$$

in which rms($A^{obs} - A^{calc}$) is the root mean square of the difference between the observed values and calculated value of restraint, and rms(A^{obs}) is a normalization factor. The restraint can be the NOE intensity (defined by peak volume), J coupling constant, or RDC. As the equation depicts, the Q factor can be used as an indicator for how close the calculated structures are to the actual one for a given set of NMR data.

7.5.4 Second-Round Structure Calculation: Structure Refinement

The second-round calculation involves structural refinement by optimizing the calculated structures after the folding stage with small time step rMD processes while simultaneously minimizing the Q factor of the generated structures. Usually, hydrogen bonding restraints observed for slowly exchangeable amide protons of secondary structural elements are included in the target function during the final stage of the refinement since the hydrogen bonding distance cannot be longer than 2.4 Å and the bonding angle must be within ±35° of linearity. If RDC data are available, the refinement is also carried out using an empirical energy term containing dipolar couplings for the target function. Since dipolar couplings provide long-range restraints, they can be used to correct misassigned NOEs, and hence reduce the number of violations and increase the accuracy of the structures.

7.5.5 Presentation of the NMR Structure

Once the structures are determined from NMR data, it is necessary to display them. In a ribbon representation of the average structure, the secondary structural elements are easily recognized, whereas the deviation of the determined structures is visualized by the backbone superposition of a set of final structures. The flexible and rigid regions of the structures are clearly indicated by the superposition representation. The structures can also be represented by molecular surface or electrostatic potential, which is helpful in studying the binding sites of a complex or the overall shape of a molecule. The detailed molecular structures can also be displayed.

There are many software packages available for displaying molecular structures in both schematic and detailed representations, of which MOLMOL (MOLecule analysis and MOLecule display) and MOLSCRIPT are widely used. The software takes the coordinates of the atoms in a structure file to generate 3D structures in the above representations. In addition to displaying superimposed structures, MOLMOL (http://www.mol.biol.ethz.ch/groups/wuthrich_group/software) can be used to display hydrogen bonds, electrostatic surfaces, and Ramamchandran plots. A unique feature of MOLSCRIPT is that the output image can be saved in various formats such as PNG, JPEG, GIF, and many other image formats (http://www.avatar.se/molscript). Midasplus is another program for structural display, which also calculates molecular surfaces, electrostatic potentials, and draw distances between protons (http://www.cgl.ucsf.edu/Outreach/midasplus). Figure 7.4 shows a sample display of a set of structures in ribbon and superimposition representations using MOLMOL.

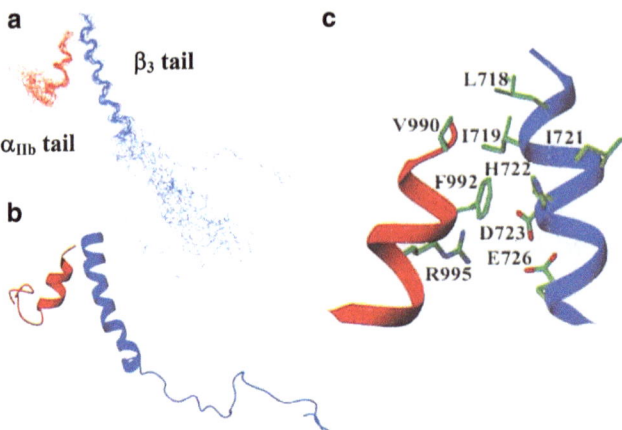

Fig. 7.4 Solution structure of the cytoplasmic domain of a prototypic integrin $\alpha_{IIb}\beta_3$ and binding interface (Vinogradova et al. 2002). (**a**) A backbone superposition of 20 best structures of $\alpha_{IIb}\beta_3$ tail complex. (**b**) Backbone ribbon display of the average structure of the $\alpha_{IIb}\beta_3$ tail complex same as in (**a**). (**c**) Zoom-in detailed view of the $\alpha_{IIb}\beta_3$ binding interface showing the hydrophobic and electrostatic contacts (reproduced from Vinogradova et al. (2002). Copyright © 2002 Elsevier)

7.5.6 Precision of NMR Structures

The precision of NMR structures is related to the precision of the experimental data. Errors in measurements of NOE, J coupling, and dipolar coupling will affect the precision in the estimation of distance and orientational restraints derived from the data. The precision of the calculated structures is usually presented in terms of the atomic rmsd such as the rmsd of backbone atoms and the rmsd of all heavy atoms. A low rmsd value of the structures indicates that the calculated structures are close to the average structure, which represents a high precision of the structure calculation. A smaller range of the errors in the restraints will produce structures with improved rmsd values. Several factors will contribute errors in the measurements such as low digital resolution in multidimensional experiments, noise level, resonance overlapping, etc. The rmsd of the calculated NOE intensity, J coupling, and chemical shift for the structures compared to the experimental data will also validate the quality of NMR structures as mentioned previously.

In general, an increase in the number of experimental restraints will improve the precision of the calculated structures. However, the precision of the structure determination does not guarantee the accuracy of the NMR structures. For example, if the distances derived from NOE are scaled by a factor due to incorrect NOE volume measurement, the calculated structures will be significantly different from the structures obtained with the correct distance restraints. Therefore, the accuracy of NMR structures is required to be examined with additional criteria.

7.5.7 Accuracy of NMR Structures

It is thought that an accurate structure should not have substantial violations in Ramachandran diagrams and covalent bond geometry. Programs have been developed such as PROCHECK (Laskowski et al. 1996) and WHAT_CHECK (Hooft et al. 1996) for checking the values of bond lengths and angles, the appearance of Ramachandran diagrams, the number and scale of violations of experimental restraints, potential energy, etc. Structures with poor scores do not necessarily indicate errors in the structure, but they require attention to locate possible misassigned experimental data. On the other hand, structures with high scores also do not assure the accuracy of the calculation.

As mentioned earlier in this chapter, the quality factor is frequently used to describe the consistency of the generated structures with the experimentally obtained restraints. Actual NMR structures must possess a small Q factor. Consequently, the Q factor is often minimized during the refinement stage of the structural calculation. Although an ensemble of structures can be obtained with small rmsd and Q factor values, the accuracy of the structures cannot be validated using the restraints which are used to generate the structures. The accuracy of the structures requires cross validation with other criteria, of which a free R factor has been used for this purpose (Brünger 1992a, 1992b). The idea is to set aside some portion of experimental data which will be used for validation of the accuracy of NMR structures. The prerequisite to do this is that there must be sufficient restraints to generate the structures after excluding those to be set aside. For example, NMR structures can be calculated and refined using restraints from the measurements of NOE, J coupling, chemical shift, and hydrogen bonding. RDCs can then be used for validation of the accuracy and to further refine the calculated structure. Good agreement of the validated structures with the refined structures using the dipolar couplings will confirm the accuracy of the NMR structures. NOE back-calculation is also a valuable indicator of the accuracy of structures determined using NMR data.

7.6 Protein Complexes

7.6.1 Protein–Protein Complexes

Protein–protein interactions play an essential role at various levels in information flow associated with various biological process such as gene transcription and translation, cell growth and differentiation, the neurotransmission, and immune responses. The interactions lead to changes in shape and dynamics as well as in chemical or physical properties of the proteins involved. Solution NMR spectroscopy provides a powerful tool to characterize these interactions at the atomic level and at near-physiological conditions. With the use of isotopic labeling, structures of many protein complexes in the 40 kDa total molecular mass regime have been determined (Clore and Gronenborn 1998). The development of novel NMR techniques and sample preparation has been increasing the mass size further for the structural determination of protein complexes. Furthermore, NMR has been utilized to quickly identify the binding sites of the complexes based on the results of chemical shift mapping or hydrogen bonding experiments. Because it is particularly difficult and sometimes impossible to crystallize weakly bound protein complexes ($K_d > 10^{-6}$), the chemical mapping method is uniquely suitable to characterize such complexes. The binding surfaces of proteins with molecular mass less than 30 kDa to large target proteins (unlabeled, up to 100 kDa) can be identified by solution NMR in combination with isotopic labeling (Mastsuo et al. 1999; Takahashi et al. 2000). As discussed in Chap. 6, structures of small ligands weakly bound to the proteins can be determined by transferred NOE experiments. The structures of peptides or small protein domains of weakly bound protein complexes can also be characterized by NMR techniques, which may be beneficial to the discovery and design of new drugs with high affinity. In addition to the structural investigation of protein complexes, NMR is a unique and powerful technique to study the molecular dynamics involved in protein–protein reorganization.

7.6.2 Protein–Peptide Complexes

The contact surface contributing to the interactions of high affinity and specificity generally involves 30 or less amino acid residues from each protein of the complex (de Vos et al. 1992; Song and Ni 1998). Frequently, this contact surface is located in a single continuous fragment of one of the proteins, which can be identified by mutation and deletion experiments. Therefore, fragments can be chemically synthesized in large amount and studied by ^{1}H NMR experiments owing to their small molecular size (Wüthrich 1986). In the study of protein–peptide complexes, samples prepared according to the procedure discussed in Chap. 3 for isotopic-labeled protein and unlabeled peptide are most commonly used since the availability of labeled peptide is prohibited by the expense of chemical synthesis from labeled amino acids and the difficulty of biosynthesis due to peptide stability problems during expression and purification.

Data collection and resonance assignment for the complex can be carried out in three stages: for labeled protein, for unlabeled peptide, and for the complex. In the first two stages, the protein and peptide are treated as two independent entities. Standard multidimensional heteronuclear experiments can be carried out using the complex sample for resonance assignment, including J coupling measurement, NOE analysis, and RDC measurement. Backbone H^N, N, C^α, C' resonances are assigned using HNCO, HNCA, HN(CA)CO, and HN(CO)CA and aliphatic side-chain resonances using CBCA(CO)NH, CBCANH, 3D or 4D HCCH-TOCSY, and ^{15}N edited TOCSY as described previously. The distance restraints are obtained from 3D ^{13}C or 4D ^{13}C–^{13}C NOESY based on the resonance assignments (Qin et al. 2001).

Questions

1. Why are small rmsd values of the calculated structure insufficient to describe the accuracy of the structure?
2. Why is the temperature increased to 2,000 K and then cooled down to 300 K during rMD calculation?
3. How are the chemical shift indexes used for identifying secondary structural elements?
4. Why is the iterative NOE analysis important to the structure calculation?
5. What is the Q factor of the structure calculation? What does it describe?
6. What are the parameters used for protein structure calculation and how are they obtained?
7. What kind of restraints is used for secondary structure determination and what for tertiary structure determination?
8. Both NOEs and RDCs are used as restraints for structure calculation. What kinds of structure information are they provide?

Appendix B: sa.inp—Xplor Protocol for Protein Structure Calculation

REMARKS This protocol has very slow cooling with increase of vdw
evaluate ($seed=287346589)

set seed $seed end
!————————————————————————————
! read in the PSF file and initial structure
param @**parallhdg_ILK.pro** end
structure @**ILK_new.psf** end
coor @**ILK_aves_min.pdb**

coor copy end
!————————————————————————————
! set the weights for the experimental energy terms
evaluate ($knoe=25.0) ! noes
evaluate ($asym=0.1) ! slope of NOE potential
evaluate ($kcdi=10.0) ! torsion angles
!————————————————————————————
! The next statement makes sure the experimental energies are used in the
! calculation, and switches off the unwanted energy terms.
! note that the NMR torsions are only switched on in the cooling stage
! we include the noncrystallographic symmetry right from the start
!————————————————————————————

```
! Read experimental restraints
noe
    reset
    nrestraints=6000           ! allocate space for NOEs
      ceiling 100
      set echo off message off end
      class           all
      set message off echo off end
      @ILK_mod1.tbl
      @hbond.tbl
      set echo on message on end
      averaging all center
      potential all square
      scale all $knoe

      sqconstant all 1.0
      sqexponent all 2
! soexponent all 1
! rswitch all 1.0
! sqoffset all 0.0
! asymptote all 2.0
end
couplings
potential harmonic
    class phi
    force 1.0
    nres 300
    degeneracy 1
    coefficients 6.98 -1.38 1.72 -60.0
@dihed_talos.tbl
end

carbon
    nres=200
    class all
    force 0.5
    potential harmonic
    @rcoil_c13.tbl
    @expected_edited_c13.tbl
    @shift_qm.tbl
end
evaluate ($rcon=0.003)

parameters
  nbonds
```

Appendix B: sa.inp—Xplor Protocol for Protein Structure Calculation

```
    atom
    nbxmod 3
    wmin=0.01 ! warning off
    cutnb=4.5 ! nonbonded cutoff
        tolerance 0.5
    repel=0.9 ! scale factor for vdW radii=1 (L-J radii)
    rexp=2 ! exponents in (r^irex - R0^irex)^rexp
    irex=2
    rcon=$rcon ! actually set the vdW weight
  end
end

set message off echo off end
restraints dihed
scale $kcdi
@dihed_talos.tbl
end
set message on echo on end

flags
exclude * include bonds angle impr vdw noe cdih coup carb end
evaluate ($cool_steps=3000)
evaluate ($init_t=2000.01)
vector do (mass=100.0) (all)          ! uniform mass for all atoms
vector do (fbeta=10.0) (all)          ! coupling to heat bath
eval ($endcount=100)
coor copy end

eval ($count=0)
while ($count<$endcount) loop main
evaluate ($count=$count+1)

coor swap end
coor copy end

vector do (x=xcomp) (all)
vector do (y=ycomp) (all)
vector do (z=zcomp) (all)

evaluate ($ini_rad=0.9)      evaluate ($fin_rad=0.80)
evaluate ($ini_con=0.004)       evaluate ($fin_con=4.0)
evaluate ($ini_ang=0.4)      evaluate ($fin_ang=1.0)
evaluate ($ini_imp=0.1)       evaluate ($fin_imp=1.0)
evaluate ($ini_noe=2.0)       evaluate ($fin_noe=30.0)
```

```
evaluate ($knoe=$ini_noe)      ! slope of NOE potential
evaluate ($kcdi=10.0)     ! torsion angles

noe
  averaging all center
  potential all square
  scale all $knoe
  sqconstant all 1.0
  sqexponent all 2
end

restraints dihed
  scale $kcdi
end
evaluate ($rcon=1.0)

parameters
  nbonds
    atom
    nbxmod 3
    wmin=0.01      ! warning off
    cutnb=100      ! nonbonded cutoff
        tolerance 45
    repel=1.2 ! scale factor for vdW radii=1 (L-J radii)
    rexp=2 ! exponents in (r^irex - R0^irex)^rexp
    irex=2
    rcon=$rcon ! actually set the vdW weight
    end
end

constraints
    interaction (not name ca) (all)
    weights * 1 angl 0.4 impr 0.1 vdw 0 elec 0 end
    interaction (name ca) (name ca)
    weights * 1 angl 0.4 impr 0.1 vdw 1.0 end
end

dynamics verlet
    nstep=1000       !
    timestep=0.005       !
    iasvel=maxwell      firsttemp=$init_t
    tcoupling=true
    tbath=$init_t
    nprint=50
```

Appendix B: sa.inp—Xplor Protocol for Protein Structure Calculation

```
    iprfrq=0
    ntrfr=99999999
end

parameters
  nbonds
    atom
    nbxmod 3
    wmin=0.01 ! warning off
    cutnb=4.5 ! nonbonded cutoff
    tolerance 0.5
    repel=0.9 ! scale factor for vdW radii=1 (L-J radii)
    rexp=2 ! exponents in (r^irex - R0^irex)^rexp
    irex=2
    rcon=1.0 ! actually set the vdW weight
  end
end

evaluate ($kcdi=200)
restraints dihed
  scale $kcdi
end

evaluate ($final_t=100) {K}
evaluate ($tempstep=50) {K}
evaluate ($ncycle=($init_t-$final_t)/$tempstep)
evaluate ($nstep=int($cool_steps/$ncycle))
evaluate ($bath=$init_t)
evaluate ($k_vdw=$ini_con)
evaluate ($k_vdwfact=($fin_con/$ini_con)^(1/$ncycle))
evaluate ($radius=$ini_rad)
evaluate ($radfact=($fin_rad/$ini_rad)^(1/$ncycle))
evaluate ($k_ang=$ini_ang)
evaluate ($ang_fac=($fin_ang/$ini_ang)^(1/$ncycle))
evaluate ($k_imp=$ini_imp)
evaluate ($imp_fac=($fin_imp/$ini_imp)^(1/$ncycle))
evaluate ($noe_fac=($fin_noe/$ini_noe)^(1/$ncycle))
evaluate ($knoe=$ini_noe)

vector do (vx=maxwell($bath)) (all)
vector do (vy=maxwell($bath)) (all)
vector do (vz=maxwell($bath)) (all)
```

```
evaluate ($i_cool=0)
while ($i_cool<$ncycle) loop cool
       evaluate ($i_cool=$i_cool+1)
       evaluate ($bath=$bath - $tempstep)
       evaluate ($k_vdw=min($fin_con,$k_vdw*$k_vdwfact))
       evaluate ($radius=max($fin_rad,$radius*$radfact))
       evaluate ($k_ang=$k_ang*$ang_fac)
       evaluate ($k_imp=$k_imp*$imp_fac)
       evaluate ($knoe=$knoe*$noe_fac)
       constraints interaction (all) (all) weights
          * 1 angles $k_ang improper $k_imp
       end end
          parameter nbonds
          cutnb=4.5 rcon=$k_vdw nbxmod=3 repel=$radius
       end       end
       noe scale all $knoe end

       dynamics verlet
       nstep=$nstep timestep=0.005 iasvel=current firsttemp=$bath
       tcoupling=true tbath=$bath nprint=$nstep iprfrq=0
       ntrfr=99999999
    end
end loop cool

mini powell nstep=500 nprint=50 end

{* NOE Data Analysis *}
   print threshold=0.5 noe
   evaluate ($noe5=$violations)
   print threshold=0.0 noe
   evaluate ($noe0=$violations)
   evaluate ($rms_noe=$result)

{* CDIH Data Analysis *}
  print threshold=5.0 cdih
  evaluate ($cdih5=$violations)
  print threshold=0.0 cdih
  evaluate ($cdih0=$violations)
  evaluate ($rms_cdih=$result)

{* BOND Data Analysis *}
  print thres=0.05 bond
  evaluate ($bond5=$violations)
  evaluate ($rms_bond=$result)
```

Appendix B: sa.inp—Xplor Protocol for Protein Structure Calculation

```
{* ANGLE Data Analysis *}
  print thres=5.0 angle
  evaluate ($angle5=$violations)
  evaluate ($rms_angle=$result)

{* IMPROPER Data Analysis *}
  print thres=5.0 improper
  evaluate ($improper5=$violations)
  evaluate ($rms_improper=$result)

{* ENERGY Data Analysis *}
  energy end

{* J-coupling constant analysis *}
  couplings print threshold 1.0 all end
  evaluate ($rms_coup=$result)
  evaluate ($viol_coup=$violations)

{* Carbon chemical shift analysis *}
  carbon print threshold=1.0 end
  evaluate ($rms_carbashift=$rmsca)
  evaluate ($rms_carbbshift=$rmscb)
  evaluate ($viol_carb=$violations)

remarks==================================
remarks   noe, cdih, bonds, angles, improp
remarks violations.: $noe5[I5], $cdih5[I5], $bond5[I5], $angle5[I5], $improper5[I5]
remarks RMSD .: $rms_noe[F6.3], $rms_cdih[F6.3], $rms_bond[F6.3], $rms_angle[F6.3], $rms_improper[F6.3]
remarks 0-viol .: $noe0[I5], $cdih0[I5]
remarks coup, carb-a, carb-b
remarks violations: $viol_coup[I5], $viol_carb[I5], \
remarks RMSD: $rms_coup[F6.3], $rms_carbashift[F6.3], $rms_carbbshift[F6.3], \
remarks ==================================
remarks overall=$ener
remarks noe=$NOE
remarks dih=$CDIH
remarks vdw=$VDW
remarks bon=$BOND
remarks ang=$ANGL
remarks imp=$IMPR
remarks coup=$COUP
remarks carb=$CARB
remarks prot=$PROT
```

```
remarks ========================================
evaluate ($file="ILK_tal_"+encode($count)+".pdb")
write coordinates output=$file end
end loop main
write coordinates output=$filename end
stop
```

Appendix C: Example of NOE Table

!K1
assign (resid 1 and name HG#) (resid 1 and name HD#) 2.5 0.7 0.2   !#A 526 9.18e+05
assign (resid 1 and name HG#) (resid 1 and name HB#) 3.0 1.2 0.3   !#A 521 2.31e+05
assign (resid 1 and name HD#) (resid 1 and name HE#) 2.5 0.7 0.2   !#A 518 5.72e+05
assign (resid 1 and name HG#) (resid 1 and name HE#) 3.0 1.2 0.3   !m#A 516 4.30e+05
assign (resid 1 and name HG#) (resid 1 and name HA) 4.0 2.2 2.0   !#A 510 2.25e+05
assign (resid 1 and name HD#) (resid 1 and name HA) 4.0 2.2 1.0   !#A 509 1.45e+05
assign (resid 1 and name HB#) (resid 1 and name HA) 3.0 1.2 0.3   !#A 508 3.20e+05
assign (resid 1 and name HE#) (resid 1 and name HA) 4.0 2.2 1.0   !#A 500 8.04e+04
assign (resid 1 and name HA) (resid 2 and name HB) 4.0 2.2 1.0   !#A 512 1.24e+05
assign (resid 1 and name HB#) (resid 2 and name HA) 4.0 2.2 2.0   !added
!assign (resid 1 and name HG#) (resid 2 and name HA) 4.0 2.2 2.0   !added
!assign (resid 1 and name HG#) (resid 3 and name HA#) 4.0 2.2 1.0   !#A 513 9.34e+04
!assign (resid 1 and name HG#) (resid 3 and name HN) 4.0 2.2 1.0   !added
assign (resid 1 and name HB#) (resid 3 and name HN) 4.0 2.2 1.0   !added
assign (resid 1 and name HB#) (resid 3 and name HA#) 4.0 2.2 2.0   !added
!assign (resid 1 and name HD#) (resid 3 and name HA#) 4.0 2.2 2.0   !added
assign (resid 1 and name HA) (resid 4 and name HB#) 4.0 2.2 1.0   !jun
assign (resid 1 and name HB#) (resid 4 and name HN) 4.0 2.2 2.0   !m#A 498 8.90e+04

!V2
assign (resid 2 and name HG1#) (resid 2 and name HB) 2.5 0.7 0.2   !#A 314 6.73e+05
assign (resid 2 and name HG1#) (resid 2 and name HA) 3.0 1.2 0.3   !m#A 267 5.15e+05
assign (resid 2 and name HG1#) (resid 3 and name HN) 4.0 2.2 1.0   !#A 601 1.48e+05
assign (resid 2 and name HB) (resid 3 and name HA#) 4.0 2.2 1.0   !#A 293 1.42e+05
assign (resid 2 and name HG1#) (resid 3 and name HA#) 4.0 2.2 1.0   !m#A 269 2.73e+05
assign (resid 2 and name HA) (resid 3 and name HA#) 4.0 2.2 1.0   !m#A 254 2.15e+05
assign (resid 2 and name HG1#) (resid 3 and name HN) 4.0 2.2 2.0   !#A 173 2.73e+05
assign (resid 2 and name HB) (resid 3 and name HN) 4.0 2.2 1.0   !#A 166 1.49e+05
assign (resid 2 and name HA) (resid 3 and name HN) 3.0 1.2 0.3   !m#A 57 5.58e+05
assign (resid 2 and name HB) (resid 4 and name HN) 4.0 2.2 1.0   !#A 230 5.67e+04
assign (resid 2 and name HG1#) (resid 4 and name HN) 4.0 2.2 2.0   !#A 229 9.28e+04
assign (resid 2 and name HG2#) (resid 4 and name HN) 4.0 2.2 2.0   !#A 603 4.72e+04
!assign (resid 2 and name HG2#) (resid 4 and name HE#) 4.0 2.2 2.0   !#A 611 1.30e+05

Appendix C: Example of NOE Table

!assign (resid 2 and name HB) (resid 4 and name HD#) 4.0 2.2 2.0   !#A 224 7.79e+04
assign (resid 2 and name HA) (resid 4 and name HD#) 4.0 2.2 2.0   !#A 187 5.78e+04
assign (resid 2 and name HA) (resid 4 and name HN) 4.0 2.2 1.0   !#A 99 1.15e+05
assign (resid 2 and name HB) (resid 5 and name HE#) 4.0 2.2 2.0   !#A 440 1.13e+05
assign (resid 2 and name HG1#) (resid 5 and name HD#) 4.0 2.2 2.0   !#A 212 1.60e+05
assign (resid 2 and name HG1#) (resid 5 and name HE#) 4.0 2.2 2.0   !#A 433 2.33e+05
assign (resid 2 and name HA) (resid 5 and name HD#) 4.0 2.2 2.0   !#A 406 9.65e+04
assign (resid 2 and name HB) (resid 5 and name HD#) 4.0 2.2 2.0   !#A 220 1.00e+05
assign (resid 2 and name HA) (resid 5 and name HE#) 4.0 2.2 2.0   !#A 186 8.51e+04
assign (resid 2 and name HA) (resid 5 and name HB#) 4.0 2.2 1.0   !#Jun

!G3
assign (resid 3 and name HA#) (resid 3 and name HN) 2.5 0.7 0.2   !#A 58 4.42e+05
assign (resid 3 and name HA#) (resid 4 and name HA) 4.0 2.2 1.0   !m#A 257 1.69e+05
assign (resid 3 and name HA#) (resid 4 and name HN) 3.0 1.2 0.3   !#A 59 6.34e+05
assign (resid 3 and name HN) (resid 4 and name HE#) 4.0 2.2 1.0   !#A 28 3.96e+04
assign (resid 3 and name HN) (resid 4 and name HN) 3.0 1.2 0.3   !#MA 4 6.69e+04
assign (resid 3 and name HA#) (resid 5 and name HB1) 4.0 2.2 2.0   !m#A 503 6.23e+04
assign (resid 3 and name HA#) (resid 5 and name HE#) 4.0 2.2 2.0   !olga
assign (resid 3 and name HA#) (resid 5 and name HN) 4.0 2.2 1.0   !#A 78 1.39e+05
assign (resid 3 and name HN) (resid 5 and name HD#) 4.0 2.2 2.0   !#A 50 3.75e+04
assign (resid 3 and name HN) (resid 6 and name HN) 4.0 2.2 1.0   !#A 20 4.57e+04
assign (resid 3 and name HA#) (resid 6 and name HN) 4.0 2.2 1.0   !#A 486 7.65e+04
assign (resid 3 and name HA#) (resid 6 and name HB#) 4.0 2.2 2.0   !added
assign (resid 3 and name HA#) (resid 6 and name HG#) 4.0 2.2 2.0   !added

!F4
assign (resid 4 and name HA) (resid 4 and name HB2) 2.5 0.7 0.2   !#A 242 3.84e+05
assign (resid 4 and name HA) (resid 4 and name HB1) 3.0 1.2 0.3   !#A 241 3.46e+05
assign (resid 4 and name HB1) (resid 4 and name HD#) 3.0 1.2 0.3   !m#A 204 4.66e+05
assign (resid 4 and name HB2) (resid 4 and name HD#) 3.0 1.2 0.3   !m#A 203 4.42e+05
assign (resid 4 and name HA) (resid 4 and name HD#) 4.0 2.2 1.0   !m#A 175 4.05e+05
assign (resid 4 and name HB1) (resid 4 and name HN) 3.0 1.2 0.3   !#A 102 2.49e+05
assign (resid 4 and name HB2) (resid 4 and name HN) 3.0 1.2 0.3   !#A 101 2.48e+05
assign (resid 4 and name HA) (resid 4 and name HN) 3.0 1.2 0.3   !#A 60 2.23e+05
assign (resid 4 and name HD#) (resid 4 and name HE#) 2.5 0.7 0.2   !#A 46 2.48e+06
assign (resid 4 and name HE#) (resid 4 and name HN) 4.0 2.2 1.0   !#A 40 1.05e+05
assign (resid 4 and name HD#) (resid 4 and name HN) 4.0 2.2 1.0   !m#A 38 1.77e+05
assign (resid 4 and name HN) (resid 5 and name HB1) 4.0 2.2 1.0   !#A 428 6.11e+04
assign (resid 4 and name HD#) (resid 5 and name HB1) 4.0 2.2 1.0   !#A 206 9.60e+04
assign (resid 4 and name HA) (resid 5 and name HN) 3.0 1.2 0.3   !m#A 61 3.98e+05
assign (resid 4 and name HN) (resid 5 and name HD#) 4.0 2.2 2.0   !#A 39 7.34e+04
assign (resid 4 and name HE#) (resid 5 and name HN) 4.0 2.2 1.0   !#A 36 1.29e+05
assign (resid 4 and name HD#) (resid 5 and name HN) 4.0 2.2 1.0   !#A 35 1.07e+05
assign (resid 4 and name HN) (resid 5 and name HN) 3.0 1.2 0.3   !#MA 8 1.34e+05

```
assign (resid 4 and name HA) (resid 7 and name HN) 4.0 2.2 1.0      !#A 392 2.71e+05
assign (resid 4 and name HB1) (resid 7 and name HN) 4.0 2.2 2.0     !#A 116 1.09e+05
!assign (resid 4 and name HD#) (resid 7 and name HD#) 4.0 2.2 2.0   !added 1.09e+05
assign (resid 4 and name HD#) (resid 8 and name HB2) 4.0 2.2 2.0    !added 1.09e+05
assign (resid 4 and name HD#) (resid 17 and name HG#) 4.0 2.2 2.0

!F5
assign (resid 5 and name HA) (resid 5 and name HB1) 3.0 1.2 0.3     !#A 240 2.24e+05
assign (resid 5 and name HB2) (resid 5 and name HB1) 2.5 0.7 0.2    !#A 232 5.40e+05
assign (resid 5 and name HB2) (resid 5 and name HD#) 3.0 1.2 0.3    !m#A 202 5.11e+05
assign (resid 5 and name HB1) (resid 5 and name HD#) 2.5 0.7 0.2    !#A 201 4.51e+05
assign (resid 5 and name HA) (resid 5 and name HD#) 3.0 1.2 0.3     !#A 176 4.58e+05
assign (resid 5 and name HB1) (resid 5 and name HN) 3.0 1.2 0.3     !#A 104 1.68e+05
assign (resid 5 and name HB2) (resid 5 and name HN) 4.0 2.2 1.0     !m#A 103 2.72e+05
assign (resid 5 and name HA) (resid 5 and name HN) 3.0 1.2 0.3      !#A 64 3.98e+05
assign (resid 5 and name HD#) (resid 5 and name HE#) 2.5 0.7 0.2    !#A 45 3.13e+06
assign (resid 5 and name HD#) (resid 5 and name HN) 3.0 1.2 0.3     !#A 34 1.97e+05
assign (resid 5 and name HE#) (resid 6 and name HB#) 4.0 2.2 1.0    !#A 447 1.20e+05
assign (resid 5 and name HD#) (resid 6 and name HB#) 4.0 2.2 1.0    !#A 446 1.23e+05
assign (resid 5 and name HE#) (resid 6 and name HD#) 4.0 2.2 2.0    !#A 441 2.27e+05
!assign (resid 5 and name HD#) (resid 6 and name HG#) 4.0 2.2 1.0   !#A 221 1.05e+05
assign (resid 5 and name HE#) (resid 6 and name HG#) 4.0 2.2 1.0    !#A 214 1.02e+05
assign (resid 5 and name HD#) (resid 6 and name HA) 4.0 2.2 1.0     !#A 183 1.27e+05
assign (resid 5 and name HB2) (resid 6 and name HN) 4.0 2.2 1.0     !m#A 112 1.86e+05
assign (resid 5 and name HB1) (resid 6 and name HN) 4.0 2.2 1.0     !#A 111 1.39e+05
assign (resid 5 and name HA) (resid 6 and name HN) 3.0 1.2 0.3      !#A 62 3.22e+05
assign (resid 5 and name HD#) (resid 6 and name HN) 4.0 2.2 1.0     !#A 37 1.27e+05
assign (resid 5 and name HE#) (resid 6 and name HN) 4.0 2.2 1.0     !#A 33 8.23e+04
```

INCOMPLETED

Appendix D: Example of Dihedral Angle Restraint Table

```
!remark phi angle constraints
!!  r22
assign (resid 21 and name c and segid b) (resid 22 and name n and segid b)
       (resid 22 and name ca and segid b) (resid 22 and name c and segid b)      1.0 -64.0 20.0 2
!!  23
assign (resid 22 and name c and segid b) (resid 23 and name n and segid b)
       (resid 23 and name ca and segid b) (resid 23 and name c and segid b)      1.0 -67.0 20.0 2
!!  24
assign (resid 23 and name c and segid b) (resid 24 and name n and segid b)
       (resid 24 and name ca and segid b) (resid 24 and name c and segid b)      1.0 -73.0 20.0 2
```

Appendix D: Example of Dihedral Angle Restraint Table

```
!! 25
assign (resid 24 and name c and segid b) (resid 25 and name n and segid b)
      (resid 25 and name ca and segid b) (resid 25 and name c and segid b)      1.0 -79.0 20.0 2
!! 26
assign (resid 25 and name c and segid b) (resid 26 and name n and segid b)
      (resid 26 and name ca and segid b) (resid 26 and name c and segid b)      1.0 -68.0 20.0 2
!! 27
assign (resid 26 and name c and segid b) (resid 27 and name n and segid b)
      (resid 27 and name ca and segid b) (resid 27 and name c and segid b)      1.0 -66.0 20.0 2
!! 28
assign (resid 27 and name c and segid b) (resid 28 and name n and segid b)
      (resid 28 and name ca and segid b) (resid 28 and name c and segid b)      1.0 -94.0 20.0 2
!! 29
assign (resid 28 and name c and segid b) (resid 29 and name n and segid b)
      (resid 29 and name ca and segid b) (resid 29 and name c and segid b)      1.0 -62.0 20.0 2
!! 30
assign (resid 29 and name c and segid b) (resid 30 and name n and segid b)
      (resid 30 and name ca and segid b) (resid 30 and name c and segid b)      1.0 -65.0 20.0 2
!! 31
assign (resid 30 and name c and segid b) (resid 31 and name n and segid b)
      (resid 31 and name ca and segid b) (resid 31 and name c and segid b)      1.0 -63.0 20.0 2
!! 32
assign (resid 31 and name c and segid b) (resid 32 and name n and segid b)
      (resid 32 and name ca and segid b) (resid 32 and name c and segid b)      1.0 -63.0 20.0 2
!! 33
assign (resid 32 and name c and segid b) (resid 33 and name n and segid b)
      (resid 33 and name ca and segid b) (resid 33 and name c and segid b)      1.0 -63.0 20.0 2
!! 34
assign (resid 33 and name c and segid b) (resid 34 and name n and segid b)
      (resid 34 and name ca and segid b) (resid 34 and name c and segid b)      1.0 -64.0 20.0 2
!! 35
assign (resid 34 and name c and segid b) (resid 35 and name n and segid b)
      (resid 35 and name ca and segid b) (resid 35 and name c and segid b)      1.0 -67.0 20.0 2
!! 36
assign (resid 35 and name c and segid b) (resid 36 and name n and segid b)
      (resid 36 and name ca and segid b) (resid 36 and name c and segid b)      1.0 -63.0 20.0 2
!! 37
assign (resid 36 and name c and segid b) (resid 37 and name n and segid b)
      (resid 37 and name ca and segid b) (resid 37 and name c and segid b)      1.0 -64.0 20.0 2
!! 38
assign (resid 37 and name c and segid b) (resid 38 and name n and segid b)
      (resid 38 and name ca and segid b) (resid 38 and name c and segid b)      1.0 -64.0 20.0 2
```

!! 39
assign (resid 38 and name c and segid b) (resid 39 and name n and segid b)
    (resid 39 and name ca and segid b) (resid 39 and name c and segid b)   1.0 -63.0 20.0 2
!! 40
INCOMPLETED

!remark psi angles constraints
!! 22
assign (resid 22 and name n and segid b) (resid 22 and name ca and segid b)
    (resid 22 and name c and segid b) (resid 23 and name n and segid b)   1.0 -41.0 20.0 2
!! 23
assign (resid 23 and name n and segid b) (resid 23 and name ca and segid b)
    (resid 23 and name c and segid b) (resid 24 and name n and segid b)   1.0 -39.0 20.0 2
!! 24
assign (resid 24 and name n and segid b) (resid 24 and name ca and segid b)
    (resid 24 and name c and segid b) (resid 25 and name n and segid b)   1.0 -30.0 20.0 2
!! 25
assign (resid 25 and name n and segid b) (resid 25 and name ca and segid b)
    (resid 25 and name c and segid b) (resid 26 and name n and segid b)   1.0 -33.0 20.0 2
!! 26
assign (resid 26 and name n and segid b) (resid 26 and name ca and segid b)
    (resid 26 and name c and segid b) (resid 27 and name n and segid b)   1.0 -36.0 20.0 2
!! 27
assign (resid 27 and name n and segid b) (resid 27 and name ca and segid b)
    (resid 27 and name c and segid b) (resid 28 and name n and segid b)   1.0 -34.0 20.0 2
!! 28
assign (resid 28 and name n and segid b) (resid 28 and name ca and segid b)
    (resid 28 and name c and segid b) (resid 29 and name n and segid b)   1.0 -9.0 20.0 2
!! 29
assign (resid 29 and name n and segid b) (resid 29 and name ca and segid b)
    (resid 29 and name c and segid b) (resid 30 and name n and segid b)   1.0 -36.0 20.0 2
!! 30
assign (resid 30 and name n and segid b) (resid 30 and name ca and segid b)
    (resid 30 and name c and segid b) (resid 31 and name n and segid b)   1.0 -40.0 20.0 2
!! 31
assign (resid 31 and name n and segid b) (resid 31 and name ca and segid b)
    (resid 31 and name c and segid b) (resid 32 and name n and segid b)   1.0 -42.0 20.0 2
!! 32
assign (resid 32 and name n and segid b) (resid 32 and name ca and segid b)
    (resid 32 and name c and segid b) (resid 33 and name n and segid b)   1.0 -40.0 20.0 2
!! 33
assign (resid 33 and name n and segid b) (resid 33 and name ca and segid b)
    (resid 33 and name c and segid b) (resid 34 and name n and segid b)   1.0 -44.0 20.0 2

!! 34
assign (resid 34 and name n and segid b) (resid 34 and name ca and segid b)
   (resid 34 and name c and segid b) (resid 35 and name n and segid b)   1.0 -42.0 20.0 2
!! 35
assign (resid 35 and name n and segid b) (resid 35 and name ca and segid b)
   (resid 35 and name c and segid b) (resid 36 and name n and segid b)   1.0 -35.0 20.0 2
!! 36
assign (resid 36 and name n and segid b) (resid 36 and name ca and segid b)
   (resid 36 and name c and segid b) (resid 37 and name n and segid b)   1.0 -42.0 20.0 2
!! 37
assign (resid 37 and name n and segid b) (resid 37 and name ca and segid b)
   (resid 37 and name c and segid b) (resid 38 and name n and segid b)   1.0 -40.0 20.0 2
!! 38
INCOMPLETED

Appendix E: Example of Chemical Shift Table for Talos

REMARK AlfaIIb fused to MBP in complex with beta3, input for TALOS

DATA SEQUENCE KVGFFKRNRP PLEEDDEEGE

VARS RESID RESNAME ATOMNAME SHIFT

FORMAT %4d %1s %4s %8.3f

1K	N	120.93
1K	HA	4.08
1K	C	176.52
1K	CA	56.26
1K	CB	32.81
2V	N	117.94
2V	HA	4.15
2V	C	176.47
2V	CA	62.48
2V	CB	32.55
3G	N	109.71
3G	HA	3.88
3G	C	173.63
3G	CA	45.15
4F	N	118.02
4F	HA	4.56
4F	C	175.36
4F	CA	57.93

4F	CB	39.55
5F	N	118.99
5F	HA	4.58
5F	C	175.12
5F	CA	57.51
5F	CB	39.60
6K	N	120.89
6K	HA	4.21
6K	C	175.94
6K	CA	56.20
6K	CB	33.00
7 R	N	119.97
7 R	HA	4.28
7 R	C	175.84
7 R	CA	55.98
7 R	CB	30.90
8N	N	118.01
8N	HA	4.66
8N	C	175.17
8N	CA	53.14
8N	CB	38.78
9 R	N	119.81
9 R	HA	4.62
9 R	CA	55.37
9 R	CB	30.19
10 P	HA	4.67
11 P	HA	4.39
11 P	C	176.81
11 P	CA	63.10
11 P	CB	31.84
12L	N	119.06
12L	HA	4.33
12L	C	174.45
12L	CA	55.18
12L	CB	42.19
13 E	N	119.81
13 E	HA	4.29
13 E	C	176.23
13 E	CA	56.30
13 E	CB	29.93
14 E	N	119.17
14 E	HA	4.30
14 E	C	177.34
14 E	CA	56.30
14 E	CB	29.93

15 D	N	119.13	
15 D	HA	4.63	
15 D	C	175.70	
15 D	CA	53.91	
15 D	CB	40.64	
16 D	N	119.14	
16 D	HA	4.61	
16 D	C	175.99	
16 D	CA	53.92	
16 D	CB	40.67	
17 E	N	119.02	
17 E	HA	4.30	
17 E	C	176.37	
17 E	CA	56.29	
17 E	CB	29.86	
18 E	N	119.40	
18 E	HA	4.30	
18 E	C	176.81	
18 E	CA	56.38	
18 E	CB	29.80	
19 G	N	108.49	
19 G	HA	3.96	
19 G	C	173.15	
19 G	CA	45.20	
20 E	N	123.55	
20 E	HA	4.16	
20 E	CA	57.45	
20 E	CB	30.62	

Appendix F: Example of Hydrogen Bond Table

assign (resid 2 and name o) (resid 6 and name n) 3.0 0.7 0.5
assign (resid 2 and name o) (resid 6 and name hn) 2.5 0.7 0.5
assign (resid 3 and name o) (resid 7 and name n) 3.0 0.7 0.5
assign (resid 3 and name o) (resid 7 and name hn) 2.5 0.7 0.5
assign (resid 4 and name o) (resid 8 and name n) 3.0 0.7 0.5
assign (resid 4 and name o) (resid 8 and name hn) 2.5 0.7 0.5
assign (resid 5 and name o) (resid 9 and name n) 3.0 0.7 0.5
assign (resid 5 and name o) (resid 9 and name hn) 2.5 0.7 0.5

Appendix G: Example of Input File To Generate A Random-Coil Coordinates

remarks file nmr/generate_template.inp
remarks Generates a "template" coordinate set. This produces
remarks an arbitrary extended conformation with ideal geometry.
remarks
remarks Author: Axel T. Brunger

topology reset @topallhdg_new.pro end
parameter reset @parallhdg_new.pro end

{====>}
structure @alfa_RQ.psf end   {*Read structure file.*}

vector ident (x) (all)
vector do (x=x/10.) (all)
vector do (y=random(0.5)) (all)
vector do (z=random(0.5)) (all)

vector do (fbeta=50) (all)   {*Friction coefficient, in 1/ps.*}
vector do (mass=100) (all)   {*Heavy masses, in amus.*}

parameter
   nbonds
      cutnb=5.5 rcon=20. nbxmod=−2 repel=0.9 wmin=0.1 tolerance=1.
      rexp=2 irexp=2 inhibit=0.25
   end
end

flags exclude * include bond angle vdw end

minimize powell nstep=50 nprint=10 end

flags include impr end

minimize powell nstep=50 nprint=10 end

dynamics verlet
   nstep=50 timestep=0.001 iasvel=maxwell firsttemp=300.
   tcoupling=true tbath=300. nprint=50 iprfrq=0
end

```
parameter
   nbonds
      rcon=2. nbxmod=-3 repel=0.75
   end
end

minimize powell nstep=100 nprint=25 end

dynamics verlet
   nstep=500 timestep=0.005 iasvel=maxwell firsttemp=300.
   tcoupling=true tbath=300. nprint=100 iprfrq=0
end

flags exclude vdw elec end
vector do (mass=1.) (name h*)
hbuild selection=(name h*) phistep=360 end
flags include vdw elec end

minimize powell nstep=200 nprint=50 end
            {*Write coordinates.*}
remarks produced by nmr/generate_template.inp
write coordinates output=alfa_RQ_00.pdb end

stop
```

Appendix H: Example of Input File to Generate a Geometric PSF File

```
remarks file nmr/generate.inp
remarks Generate structure file for a protein
remarks using the SA parameter and topology files.

topology
   @../topallhdg_new.pro
   end         {*Read topology file *}

segment             {*Generate protein *}
   name=" "         {*This name has to match the *}
                    {*four characters in columns 73 *}
                    {*through 76 in the coordinate *}
                    {*file, in XPLOR this name is *}
                    {*name is referred to as SEGId. *}
```

```
            chain
                @TOPPAR:toph19.pep {*Read peptide bond file *}
                    sequence LYS VAL GLY PHE PHE LYS GLN ASN ARG PRO
                        PRO LEU GLU GLU ASP ASP GLU GLU GLY GLU
                end
            end
        end

        write structure output=alfa_RQ.psf end

        stop
```

References

Allen MP, Tildesley DJ (1987) Computer simulation of liquids. Clarendon Press, Oxford

Baleja JD, Pon RT, Sykes BD (1990) Solution structure of phage.lambda. half-operator DNA by use of NMR, restrained molecular dynamics, and NOE-based refinement. Biochemistry 29:4828–4839

Braun W (1987) Distance geometry and related methods for protein structure determination from NMR data. Quart Rev Biophys 19:115–157

Brooks BR, Bruccoleri R, Olafson B, States D, Swaninathan S, Karplus M (1983) CHARMM: a program for macromolecular energy, minimization, and dynamics calculations. J Comp Chem 4:187–217

Brünger A (1992a) Free R value: a novel statistical quantity for assessing the accuracy of crystal structures. Nature 355:472–475

Brünger AT (1992b) X-PLOR, version 3.1. A system for X-ray crystallography and NMR. Yale University Press, New Haven

Brünger AT, Adams PD, Clore GM, DeLano WL, Gros P, Grosse-Kunstleve RW, Jiang JS, Kuszewski J, Nilges M, Pannu NS, Read RJ, Rice LM, Simonson T, Warren GL (1998) Crystallography & NMR system: a new software suite for macromolecular structure determination. Acta Cryst D54:905–921, http://cns.csb.yale.edu/

Clore GM, Gronenborn AM (1998) Determining the structures of large proteins and protein complexes by NMR. Trends Biotech 16:22–34

Clore GM, Brünger AT, Karplus M, Gronenborn AM (1986) Application of molecular dynamics with interproton distance restraints to three-dimensional protein structure determination: a model study of crambin. J Mol Biol 191:523–551

Crippen GM (1981) Distance geometry and conformational calculations. Research Studies Press, Chichester, New York

de Alba E, Tjandra N (2002) NMR dipolar couplings for the structure determination of biopolymers in solution. Prog Nucl Magn Reson Spectrosc 40:175–197

de Dios AC, Pearson JG, Oldfield E (1993) Secondary and tertiary structural effects on protein NMR chemical shifts: an ab initio approach. Science 260:1491–1496

Demarco A, Llinás M, Wüthrich K (1978a) Analysis of the 1H-NMR spectra of ferrichrome peptides. I. The non-amide protons. Biopolymers 17:617–636

Demarco A, Llinás M, Wüthrich K (1978b) 1H-15N Spin–spin couplings in alumichrome. Biopolymers 17:2727–2742

de Vos AM, Ultsch M, Kossiakoff AA (1992) Human growth hormone and extracellular domain of its receptor: crystal structure of the complex. Science 255:306–312

Fischman AJ, Live DH, Wyssbrod HR, Agosta WC, Cowburn D (1980) Torsion angles in the cystine bridge of oxytocin in aqueous solution. Measurements of circumjacent vicinal couplings between proton, carbon-13, and nitrogen-15. J Am Chem Soc 102:2533–2539

Cornilescu G, Delagio F, Bax A (1999) Protein backbone angle restraints from searching a database for chemical shift and sequence homology. J Biomol NMR 13:289–302, http://spin.niddk.nih.gov/bax/software/TOLAS/info.html

Gochin M, James TL (1990) Solution structure studies of d(AC)4.cntdot.d(GT)4 via restrained molecular dynamics simulations with NMR constraints derived from two-dimensional NOE and double-quantum-filtered COSY experiments. Biochemistry 29:11172–11180

Güntert P, Mumenthaler C, Wüthrich K (1997) Torsion angle dynamics for NMR structure calculation with the new program Dyana. J Mol Biol 273:283–298

Güntert P (2003) Automated NMR protein structure calculation. Prog Nucl Magn Reson Spectrosc 43:105–125

Güntert P (1998) Structure calculation of biological macromolecules from NMR data. Quart Rev Biophys 31:145–237

Guüntert P, Braun W, Wüthrich K (1991) Efficient computation of three-dimensional protein structures in solution from nuclear magnetic resonance data using the program DIANA and the supporting programs CALIBA, HABAS and GLOMSA. J Mol Biol 217:517–530, http://www.mol.biol.ethz.ch/groups/wuthrich_group/software/

Havel TF (1991) An evaluation of computational strategies for use in the determination of protein structure from distance constraints obtained by nuclear magnetic resonance. Prog Biophys Mol Biol 56:43–78

Hooft RW, Vriend G, Sander C, Abola EE (1996) Errors in protein structures. Nature 381:272

Karplus K (1963) Vicinal proton coupling in nuclear magnetic resonance. J Am Chem Soc 85:2870–2871

Karplus K (1959) Contact electron-spin interactions of nuclear magnetic moments. J Phys Chem 30:11–15

Kuszewski J, Gronenborn AM, Clore GM (1995a) The impact of direct refinement against proton chemical shifts on protein structure determination by NMR. J Magn Reson B107:293–297

Kuszewski J, Qin J, Gronenborn AM, Clore GM (1995b) The impact on direct refinement against 13Cα and 13Cβ chemical shifts on protein structure determination by NMR. J Magn Reson B106:92–96

Laskowski RA, Rullmann JAC, MacArthur MW, Kaptein R, Thornton JM (1996) AQUA and PROCHECK-NMR: programs for checking the quality of protein structures solved by NMR. J Biomol NMR 8:477–486

Linge JP, Habeck M, Rieping W, Nilges M (2003) ARIA: automated NOE assignment and NMR structure calculation. Bioinformatic 19:315–316

Linge JP, O'Donoghue SI, Nilges M (2001) Automated assignment of ambiguous nuclear overhauser effects with ARIA. Methods Enzymol 339:71–90

Lipstitz RS, Tjandra N (2004) Residual dipolar couplings in NMR structure analysis. Annu Rev Biophys Biomol Struct 33:387–413

Luginbühl P, Szyperski T, Wuüthrich K (1995) Statistical basis for the use of 13Cα chemical shifts in protein structure determination. J Magn Reson B109:229–233

Mastsuo H, Walters KJ, Teruya K, Tanaka T, Gassner GT, Lippard SJ, Kyogoku Y, Wagner G (1999) Identification by NMR spectroscopy of residues at contact surfaces in large, slowly exchanging macromolecular complexes. J Am Chem Soc 121:9903–9904

Mayer KL, Stone MJ (2000) NMR solution structure and receptor peptide binding of the CC chemokine eotaxin-2. Biochemistry 39:8382–8395

Neri D, Szyperski T, Otting G, Senn H, Wüthrich K (1989) Stereospecific nuclear magnetic resonance assignments of the methyl groups of valine and leucine in the DNA-binding domain of the 434 repressor by biosynthetically directed fractional carbon-13 labeling. Biochemistry 28:7510–7516

Neuhaus D, Williamson MP (1989) The nuclear overhauser effect in structural and conformational analysis. VCH Publishers, New York

Nilges M, O'Donoghue SI (1998) Ambiguous NOEs and automated NOE assignment. Prog NMR Spectrosc 32:107–139

Nilges M, Habazettl J, Brünger A, Holak TA (1991) Relaxation matrix refinement of the solution structure of squash trypsin inhibitor. J Mol Biol 219:499–510

Nilges M, Macias MJ, O'Donoghue SI, Oschkinat H (1997) Automated NOESY interpretation with ambiguous distance restraints: the refined NMR solution structure of the pleckstrin homology domain from β-spectrin. J Mol Biol 269:408–422

Pearlman DA, Case DA, Caldwell JC, Seibel GL, Singh UC, Weiner P, Kollman PA (1991) Amber 4.0. University of California, San Francisco, http://www.amber.ucsf.edu/amber/amber.html

Powers R, Gronenborn AM, Clore GM, Bax A (1991) Three-dimensional triple-resonance NMR of 13C/15N-enriched proteins using constant-time evolution. J Magn Reson 94:209–213

Prestegard JH, Al-Hashimi HM, Tolman JR (2000) NMR structures of biomolecules using field oriented media and residual dipolar couplings. Quart Rev Biophys 33:371–424

Qin J, Vinogradova O, Gronenborn AM (2001) Protein-protein interactions probed by nuclear magnetic resonance spectroscopy. Meth Enzmol 339:377–389

Senn H, Werner B, Messerle BA, Weber C, Traber R, Wüthrich K (1989) Stereospecific assignment of the methyl 1H NMR lines of valine and leucine in polypeptides by nonrandom 13C labelling. FEBS Lett 249:113–118

Song J, Ni F (1998) NMR for the design of functional mimetics of protein-protein interactions: one key is in the building of bridges. Biochem Cell Biol 76:177–188

Spera S, Bax A (1991) Empirical correlation between protein backbone conformation and C.alpha. and C.beta. 13C nuclear magnetic resonance chemical shifts. J Am Chem Soc 113:5490–5492

Takahashi H, Nahanishi T, Kami K, Arata Y, Shimada I (2000) A novel NMR method for determining the interfaces of large protein–protein complexes. Nat Struct Biol 7:220–223

Tjandra N, Bax A (1997) Direct measurement of distances and angles in biomolecules by NMR in a dilute liquid crystalline medium. Science 278:1111–1114

Tjandra N, Omichinski JG, Gronenborn AM, Clore GM, Bax A (1997) Use of dipolar 1H–15N and 1H–13C couplings in the structure determination of magnetically oriented macromolecules in solution. Nature Struct Biol 4:732–738

van Gunsteren WF, Billeter SR, Eising AA, Hünenberger PH, Krüger P, Mark AE, Scott WRP, Tironi IG (1996) Biomolecular simulation: the GROMOS96 manual and user guide. vdf Hochschulverlag, Zürich

van Gunsteren WF (1993) Molecular dynamics studies of proteins. Curr Opin Struct Biol 3:277–281

Vinogradova O, Velyvis A, Velyvience A, Hu B, Haas AT, Plow EF, Qin J (2002) A structural mechanism of integrin αIIbβ3 'inside-out' activation as regulated by its cytoplasmic face. Cell 110:587–597

Vuister GW, Bax A (1993) Quantitative J correlation: a new approach for measuring homonuclear three-bond J(HNH.alpha.) coupling constants in 15N-enriched proteins. J Am Chem Soc 115:7772–7777

Wang AC, Bax A (1995) Reparametrization of the karplus relation for 3J(H.alpha.-N) and 3J(HN-C') in peptides from uniformly 13C/15N-enriched human ubiquitin. J Am Chem Soc 117:1810–1813

Wang AC, Bax A (1996) Determination of the backbone dihedral angles φ in human ubiquitin from reparametrized empirical Karplus equations. J Am Chem Soc 118:2483–2494

Wishart DS, Sykes BD, Richards FM (1992) The chemical shift index: a fast and simple method for the assignment of protein secondary structure through NMR spectroscopy. Biochemistry 31:1647–1651

Wishart DS, Sykes BD, Richards FM (1991) Relationship between nuclear magnetic resonance chemical shift and protein secondary structure. J Mol Biol 222:311–333

Wüthrich K (1986) NMR of proteins and nucleic acids. Wiley, New York

Yip P, Case DA (1989) A new method for refinement of macro molecular structures based on nuclear overhauser effect spectra. J Magn Reson 83:643–648

Zhang H (2001) M.Sc. Thesis, University of Alberta http://redpoll.pharmacy.ualberta.ca

Chapter 8
Protein Dynamics

High-resolution NMR spectroscopy has become a unique and powerful approach with atomic resolution not only for determining structures of biological macromolecules but also for characterizing the overall and internal rotational motions in proteins. The dynamic behavior of proteins at different timescales can be monitored experimentally by different methods because it is difficult, if not impossible, to completely characterize all motional processes by a single approach. Nuclear spin relaxation measurement provides information on fast motions on the timescales of picosecond to nanosecond (laboratory frame nuclear spin relaxation experiments), and slow motions on the timescales of microsecond to millisecond (rotating frame nuclear spin relaxation measurements), whereas magnetization exchange spectroscopy deals with motions on the timescales of millisecond to second. This chapter focuses on the experiments and data analysis for heteronuclear spin relaxation approaches used to characterize the dynamic processes of proteins in solution.

Key questions to be addressed include the following:

1. What is spectral density?
2. What is correlation time?
3. How can they be interpreted to describe protein dynamics?
4. What types of nuclear interactions can be used for dynamics study?
5. How can these interactions be used to obtained spectral density and correlation time?
6. How are T_1, T_2, and NOE measured?
7. How are the relaxation parameters derived from the experimental data?
8. What timescales of protein internal motions can NMR be used to study?
9. How are the results of relaxation parameters presented to illustrate the protein motions?
10. How are the experiments for protein dynamics measurements set up?

8.1 Theory of Spin Relaxation in Proteins

The relaxation rates of proteins are affected primarily by dipolar interactions and chemical shift anisotropy (CSA). For ^{2}H labeled proteins, the ^{2}H quadrupolar interaction also contributes to the relaxation rates. The overall relaxation rates are the linear combination of all rates of the interactions. The relaxation rates can be expressed in terms of the combination of spectral density functions (Abragam 1961; Kay et al. 1989). For an isolated XH spin system, the relaxation rate constants of the X spin (^{15}N or ^{13}C) caused by the dipolar interaction of the X spin with the ^{1}H spin and by the magnetic shielding arising from the CSA interaction of the X spin are given by:

$$R_1 = R_1^D + R_1^{CSA}$$
$$= \frac{d^2}{4}[6J(\omega_H + \omega_X) + J(\omega_H - \omega_X) + 3J(\omega_X)] + c^2 J(\omega_X) \tag{8.1}$$

$$R_2 = R_2^D + R_2^{CSA}$$
$$= \frac{d^2}{8}[6J(\omega_H + \omega_X) + 6J(\omega_H) + J(\omega_H - \omega_X) + 3J(\omega_X) + 4J(0)]$$
$$+ \frac{c^2}{6}[3J(\omega_X) + 4J(0)] \tag{8.2}$$

$$\sigma_{XH} = \frac{d^2}{4}[6J(\omega_H + \omega_X) - J(\omega_H - \omega_X)] \tag{8.3}$$

in which R_1, R_2, and σ_{XH} are the rate constants of spin–lattice relaxation, spin–spin relaxation, and cross-relaxation, respectively, which are dependent on the spectral density functions evaluated at five frequencies ($\omega_H = \omega_X$, ω_H, $\omega_H - \omega_X$, ω_X, and 0); $d = \mu_0 \hbar \gamma_X \gamma_H \langle r_{XH}^{-3} \rangle / 8\pi$; μ_0 is the permeability of vacuum ($4\pi \times 10^{-7}$ T mA); $\hbar$ is reduced Plank's constant; r_{XH} is the XH bond length; γ_X and γ_H are the gyromagnetic ratios; $c = \Delta\sigma\omega_X/\sqrt{3}$; $\Delta\sigma$ is the CSA of the X spin with the assumption that the chemical shift tensor is axially symmetrical, which has been demonstrated to be valid for peptide bond ^{15}N with $\Delta\sigma = -160$ to -170 ppm (Hiyama et al. 1988), $\Delta\sigma = 25 - 35$ ppm for peptide carbonyl ^{13}C and $\Delta\sigma = 30$ ppm for ^{13}C$^\alpha$ (Ye et al. 1993). The R_1 and R_2 rate constants can be directly determined experimentally ($R_1 = 1/T_1$ and $R_2 = 1/T_2$), whereas σ_{XH} is determined from stead-state {^{1}H}-X NOE via the relationship (Kay et al. 1989; Yamazaki et al. 1994):

$$\sigma_{XH} = \frac{d^2}{4}[6J(\omega_H + \omega_X) - J(\omega_H - \omega_X)] = \frac{\gamma_X}{\gamma_H} R_1 (\text{NOE} - 1) \tag{8.4}$$

which can be recast to:

$$\text{NOE} = 1 + \frac{\sigma_{XH}}{R_1} \frac{\gamma_X}{\gamma_H} = 1 + \frac{d^2}{4R_1} \frac{\gamma_X}{\gamma_H} [6J(\omega_H + \omega_X) - J(\omega_H - \omega_X)] \tag{8.5}$$

8.1 Theory of Spin Relaxation in Proteins

Without any assumptions, the spectral density functions at the five frequencies cannot be determined from the three experimentally determined relaxation rate constants by measuring T_1, T_2, and NOE. Assumptions must be made so that only three unknowns need to be determined from the three known values.

There are various mathematical models for mapping the spectral density functions (Wittebort et al. 1978; London 1980), of which the model-free analysis is widely used to obtain information about site-specific internal motions of proteins (Lipari and Szabo 1982a, b; Clore et al. 1990a, b; Dayie et al. 1996; Palmer 2001). Rather than fitting the experimental data to any specific physical models, the method depends on the mathematical analysis of spectral density functions by assuming two types of motions contributing to the dynamic process for isotropically tumbling proteins: overall tumbling of the protein as a whole and internal dynamics for the heteronuclear bonds. Therefore, the analysis characterizes the amplitude and rate of internal dynamics for individual chemical bond vectors (e.g., peptide NH bond) via model-free order parameters S (or the generalized order parameter), the overall rotational correlation time τ_m (or global correlation time) and effective correlation time τ_e according to the relationship:

$$J(\omega) = \frac{2}{5}\left[\frac{S^2 \tau_m}{1+(\omega \tau_m)^2} + \frac{(1-S^2)\tau}{1+(\omega \tau)^2}\right] \quad (8.6)$$

in which τ is given by

$$\tau = \left(\frac{1}{\tau_m} + \frac{1}{\tau_e}\right)^{-1} \quad (8.7)$$

The effective correlation time τ_e is the internal correlation time for motions of a bond vector in a molecular frame. The squared generalized order parameter S^2 measures the degree of spatial restriction of the bond vector in a molecular frame, which provides information about the angular amplitude of the internal motions of bond vectors. If the bond vector diffuses in a cone with an angle θ defined by the diffusion tensor and the equilibrium orientation of the bond vector, S^2 is highly sensitive to the cone angle in the range from 0° to 75°, and decreases dramatically as the cone angle increases (Ishima and Torchia 2000; Fig. 8.1). The value of θ may vary from 1 when the bond is rigid to 0 when the internal motion is completely isotropic. The overall rotational correlation time τ_m characterizes the molecular tumbling whereas internal correlation time τ_e describes the internal dynamics.

The model-free formalism has been extended to include internal motions both on a fast timescale and slow timescale (Clore et al. 1990a, b; Farrow et al. 1994). The spectral density function described by the extended model-free formalism is given by:

$$J(\omega) = \frac{2}{5}\left[\frac{S^2 \tau_m}{1+(\omega \tau_m)^2} + \frac{(S_f^2 - S^2)\tau}{1+(\omega \tau)^2}\right] \quad (8.8)$$

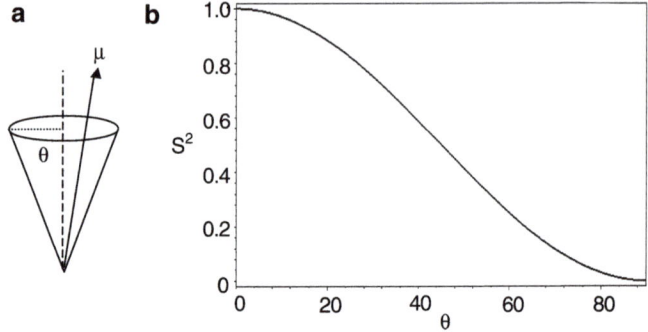

Fig. 8.1 Model-free parameters for characterizing dynamics of proteins. (**a**) Relationship of bond vector μ and the angular amplitude of the internal motion of the bond with respect to its equilibrium position, defined by θ. (**b**) The value of the squared generalized order parameter S^2 changes as a function of θ described by (8.26) for diffusion in a cone as shown in (**a**) (reproduced with permission from Ishima and Torchia (2000), Copyright © 2000 Nature Publishing Group)

in which

$$\tau = \left(\frac{1}{\tau_m} + \frac{1}{\tau_s}\right)^{-1} \tag{8.9}$$

and τ_s is the internal correlation time for slow motion, S^2, the squared generalized order parameter $= S_f^2 S_s^2$, and S_f^2 and S_s^2 are the squared generalized order parameters characterizing the fast and slow internal motions, respectively.

The squared generalized order parameter S^2 and correlation times τ_m, τ_e, τ_f, and τ_s can be determined by two different types of approaches. The first type of approach relies on valid assumptions to simply the equations for spectral density functions, including the R_2/R_1 ratio method and the constant high-frequency spectral density method. The second type is based on the optimization of fitting the data to obtain the dynamic parameters, which tends to generate quantitative analysis. From the expressions for R_2 and R_1, the ratio R_2/R_1 is given by (Kay et al. 1989; Mayo et al. 2000):

$$\frac{R_2}{R_1} = \frac{d^2[6J(\omega_H + \omega_X) + 6J(\omega_H) + J(\omega_H - \omega_X)] + (d^2 + (4c^2/3))[3J(\omega_X) + 4J(0)]}{2d^2[6J(\omega_H + \omega_X) + J(\omega_H - \omega_X) + 3J(\omega_X)] + c^2 J(\omega_X)} \tag{8.10}$$

R_1 and R_2 are the relaxation rates for each backbone ^{15}N spin. The R_2/R_1 ratio method assumes that internal motions of bond vectors are sufficiently faster than overall tumbling ($\tau_e \leq 200$ ps) and have low amplitude ($S^2 \geq 0.5$) so that the ratio of ^{15}N R_2 and R_1 relaxation rate constants is essentially independent of the internal correlation time τ_e. Since τ_e is relatively small based on the assumption, the spectral

8.1 Theory of Spin Relaxation in Proteins

density function $J(\omega)$ primarily relies only on single correlation time—overall correlation time τ_m, which simplifies the expression to the form of:

$$J(\omega) = \frac{2}{5} \frac{S^2 \tau_m}{1 + (\omega \tau_m)^2} \tag{8.11}$$

By replacing $J(\omega)$ in (8.10) with (8.11), the R_2/R_1 ratio is independent of S^2 and the overall correlation time τ_m can be determined via computer minimization of the deviation of the following equation using all observed values of the R_2/R_1 ratio at different static magnetic field strengths for each backbone ^{15}N site:

$$\tau_m = \frac{1}{\omega_N} \sqrt{\frac{6R_2}{R_1} - 17} \tag{8.12}$$

The R_2 and R_1 constants are sensitive to motions on different timescales. R_1 is sensitive to the dynamics on the timescale of picosecond to microsecond, whereas R_2 is sensitive to the motions on both the picosecond to microsecond and microsecond to millisecond timescales. For the ^{15}N spins that have a R_2/R_1 ratio below the average value by a difference larger than the standard deviation, local conformational averaging at a rate comparable to the chemical shift difference between the conformational forms is assumed to be responsible for the shortening of the T_2 relaxation. For the ^{15}N sites at which the R_2/R_1 ratio is above the average value by a difference larger than the standard deviation, the prolongation of T_1 is caused by a motion on a timescale comparable to τ_m. The squared generalized order parameter S^2 for an individual site can in turn be obtained using the expression either for R_1 or R_2 ((8.1) or (8.2)) with the average value of τ_m. In practice, S^2 is obtained using R_1 and NOE (8.1) without T_2 relaxation data because the measured T_2 may contain contributions from other mechanisms, such as slow motions, scalar relaxation, chemical exchange, antiphase magnetization, pulse imperfection, off-resonance effect of the CPMG pulse train, cross-correlation of dipolar/CSA interactions, etc.

Another simplification method is to approximate the spectral density functions in (8.1)–(8.3) to the first order as a single term $\alpha J(\beta \omega)$, in which α and β are constants (Farrow et al. 1995; Ishima and Nagayama 1995a, b), by assuming that it can be described by a linear combination of the contributions from overall rotation and internal motions. The spectral density function is given by:

$$J(\omega) = \frac{\lambda_1}{\omega^2} + \lambda_2 \tag{8.13}$$

in which the first term and second term are the contributions from overall rotation and internal motions, respectively. The rate constants consisting of the linear combination of the five spectral densities may then be simplified as:

$$R_1 = \frac{d^2}{4}[7J(\beta_1 \omega_H) + 3J(\omega_X)] + c^2 J(\omega_X) \tag{8.14}$$

$$R_2 = \frac{d^2}{8}[13J(\beta_2\omega_H) + 3J(\omega_X) + 4J(0)] + \frac{c^2}{6}[3J(\omega_X) + 4J(0)] \quad (8.15)$$

$$\text{NOE} = 1 + \frac{5d^2}{4R_1}\frac{\gamma_X}{\gamma_H}J(\beta_3\omega_H) \quad (8.16)$$

in which β_1, β_2, and β_3 can be obtained using the relationships (Farrow et al. 1995):

$$\frac{6}{(\omega_H + \omega_X)^2} + \frac{1}{(\omega_H - \omega_X)^2} = \frac{7}{(\beta_1\omega_X)^2} \quad (8.17)$$

$$\frac{6}{(\omega_H + \omega_X)^2} + \frac{6}{\omega_H^2} + \frac{1}{(\omega_H - \omega_X)^2} = \frac{13}{(\beta_2\omega_X)^2} \quad (8.18)$$

$$\frac{6}{(\omega_H + \omega_X)^2} - \frac{1}{(\omega_H - \omega_X)^2} = \frac{5}{(\beta_3\omega_H)^2} \quad (8.19)$$

The equations yield $\beta_1 = 0.921$, $\beta_2 = 0.955$, and $\beta_3 = 0.87$ for ^{15}N spin relaxation, and $\beta_1 = 1.12$, $\beta_2 = 1.06$, and $\beta_3 = 1.56$ for ^{13}C spin relaxation. The method does not assume that the molecular tumbling is isotropic. The spectral density functions are first obtained from experimental data of T_1, T_2, and NOE, and then used to determine the squared generalized order parameter and correlation times. For backbone ^{15}N spins, the value of $J(0.78\omega_H)$ is calculated directly from the observed values of T_1 and NOE using the equation:

$$J(0.87\omega_H) = \frac{4\sigma_{NH}}{5d^2} \quad (8.20)$$

in which d is defined as in (8.1) and σ_{NH} is given in (8.4).

The spectral density functions for the other four frequencies are extracted either by assuming $J(\omega) \propto 1/\omega^2$ in the range of $\omega_H \pm \omega_N$, or from the values of $J(0.78\omega)$ obtained at different field strengths. When $J(\omega_H) \propto 1/\omega_H^2$, $J(\beta_i\omega_H)$ can be estimated according to (Farrow et al. 1995)

$$J(\beta_i\omega_H) \approx \left(\frac{0.87}{\beta_i}\right)^2 J(0.87\omega_H) \quad (8.21)$$

in which $i = 1, 2$ or 3. Therefore, $J(\omega)$ at 0 and ω_N are given by:

$$J(0) = \frac{6R_2 - 3R_1 - 2.72\sigma_{NH}}{3d^2 + 4c^2} \quad (8.22)$$

$$J(\omega_N) = \frac{4R_1 - 5\sigma_{NH}}{3d^2 + 4c^2} \quad (8.23)$$

8.1 Theory of Spin Relaxation in Proteins

If values of $J(0.78\omega_H)$ obtained at different field strengths are employed, $J(\beta_i\omega_H)$ can be estimated from:

$$J(\beta_i\omega_H) \approx J(0.87\omega_H) - (\beta_i - 0.87)\omega_H \frac{J(0.87\omega_H) - J(0.87\omega'_H)}{0.87(\omega_H - \omega'_H)} \quad (8.24)$$

in which ω'_H is the proton Larmor frequency at which $J(0.78\omega'_H)$ is obtained and ω'_H is different than ω_H.

The second type of method utilizes extensive optimization of data fitting globally or locally for all peptide residues by minimizing an error function to obtain the overall correlation time τ_m (Dellwo and Wand 1989; Palmer et al. 1991; Mandel et al. 1995). For global fitting, the global error function may have a form of:

$$\chi^2 = \sum_{j=1}^{M} \left[\left(\frac{R_{1j}^{obs} - R_{1j}^{cal}}{\lambda_{R_{1j}}} \right)^2 + \left(\frac{R_{2j}^{obs} - R_{2j}^{cal}}{\lambda_{R_{2j}}} \right)^2 + \left(\frac{NOE_j^{obs} - NOE_j^{cal}}{\lambda_{NOE_j}} \right)^2 \right] \quad (8.25)$$

in which M is the number of residues for which the relaxation parameters have been measured; λ is the standard deviation in R_1, R_2 or NOE for residue j; the superscripts obs and cal denote the observed and calculated relaxation parameters, respectively. The minimization can also be done for a local error function with $M = 1$ for an individual residue. In either case, an array of presumed values of correlation time is selected for fitting the parameters (S^2 and τ_e) for internal motion. The correlation time τ_m is identified when the sum of the deviations between the observed and calculated relaxation parameters has reached a minimum. Using the data observed at different field strengths reduces the fitting error and improves the quality of the extracted dynamic parameters.

The squared generalized order parameter, S^2, is the measure of the orientational distribution of internal motions by the bond vector in the molecular frame. For the model describing the internal motion of bond vector as a restricted diffusion in a cone, the quantity S^2 is given by (Lipari and Szabo 1982a, b):

$$S^2 = \left[\frac{\cos\theta(1 + \cos\theta)}{2} \right]^2 \quad (8.26)$$

in which θ is the angle between the bond vector and the diffusion cone as defined in Fig. 8.1a, which characterizes the angular amplitude of the internal motion. When θ is equal to zero, the motion of the vector is restricted to the fixed orientations and S^2 is unity, the maximum value. As θ increases, S^2 decreases rapidly and the motion of the bond vector becomes less restricted. The motion becomes completely isotropic when θ is 75°, leading to an S^2 of almost zero.

In addition to S^2, the internal correlation time τ_e also is an important quantity for characterizing the internal motions. Although quantitative interpretation of τ_e relies on how realistic the model is to describe the motion, the determined value provides a qualitative insight about the rate of internal motion. However, as (8.11) ($J(\omega) \propto S^2$, τ_m, τ_e) indicates, τ_e can precisely be characterized only over a very narrow frequency range. In general, the accurate determination of τ_e based on the equation is limited in the range of >30 ps (slower than 30 ps) and $\ll \tau_m$ (much faster than τ_m; Palmer 2001). The accuracy of τ_e determination can be significantly increased by applying the relaxation data obtained for additional nuclei besides ^{15}N, such as ^{13}C and/or ^{2}H.

For quadrupolar interaction of spin-1 nuclei such as ^{2}H bound to ^{13}C, the quadrupolar relaxation is much more efficient than the dipolar interaction and hence the relaxation rate constants contain only quadrupolar relaxation and are expressed by (Wittebort and Szabo 1978):

$$R_1 = \frac{3C_Q^2}{16}[2J(2\omega_D) + J(\omega_D)] \qquad (8.27)$$

$$R_2 = \frac{3}{32}C_Q^2[2J(2\omega_D) + 5J(\omega_D) + 3J(0)] \qquad (8.28)$$

in which $C_Q = e^2qQ/\hbar$ is the quadrupolar coupling constant, e is the charge of an electron, q is the principal value of the electric field gradient tensor, Q is the nuclear quadrupolar moment, ω_D is the ^{2}H frequency, and $J(0)$, $J(\omega_D)$ and $J(2\omega_D)$ correspond to the spectral density function at the zero-, single- and double-quantum ^{2}H frequencies. Generally, the assumption that the principal axis of the electric field gradient tensor is collinear with the C–^{2}H bond vector is valid except for methyl groups.

Cross-correlation between different nuclear interactions may potentially complicate the determination of molecular dynamics based on spin relaxation measurement. For an isolated HX spin system, the relaxation rates including cross-correlation (or cross correlated relaxation) of dipolar and CSA interactions are different for the downfield and upfield lines of an ^{15}N or ^{13}C doublet (Goldman 1984; Bull 1992):

$$R^+ = R + R^c \quad \text{and} \quad R^- = R - R^c \qquad (8.29)$$

in which $R^\pm$ are the rate constants for the upfield (right) and downfield (left) line, respectively, R is defined as in (8.1) and (8.2), and R^c is the rate from the cross-correlation between the two interactions. The influence of the cross-correlation on the spin relaxation is usually removed during spin relaxation rate measurements by applying a continuous inversion of the proton resonance during the relaxation periods. However, this interaction provides useful information in such cases as when the relative orientation of the CSA tensor with respect to the dipolar interaction or bond vector is known. For an axially symmetric CSA tensor, the rate constants from cross-correlation with the dipolar interaction can be described in terms of:

$$R_1^c = -\sqrt{3}dcP_2(\cos\theta)J(\omega_X) \qquad (8.30)$$

8.1 Theory of Spin Relaxation in Proteins

$$R_2^C = -\frac{\sqrt{3}}{6} dc P_2(\cos\theta)[3J(\omega_X) + 4J(0)] \quad (8.31)$$

in which R_1^C and R_2^C are the longitudinal and transverse relaxation rates of the cross-correlation, respectively, c and d are the constants defined in (8.1), $P_2(x) = (3x^2 - 1)/2$ is the second Legendre polynomial, and θ is the angle between the HX bond vector and the symmetric axis of the CSA tensor. By measuring R_1 and R_2 in the absence and presence of ^1H decoupling during the relaxation period T, the contribution from the cross-correlation of the heteronuclear dipole and CSA interactions can be obtained according to the expression for R^+ or R^-.

The spin–spin relaxation can also be affected by additional internal motions induced by chemical exchange processes such as those arising from microsecond to millisecond exchange of spins between magnetic environments during the δ delays of a CPMG sequence. Consequently, the $J(0)$ determined according to (8.30) and (8.31) or (8.22) may not be accurate. The exchange rate constant R_{ex} was proposed to add to the expression of transverse relaxation rate R_2:

$$R_2 = R_2^D + R_2^{CSA} + R_{ex} \quad (8.32)$$

The exchange rate can be determined by its magnetic field dependence with the assumption $R_{ex} = \lambda_{ex} B_0^2$, in which λ_{ex} is a constant (Peng and Wagner 1995; Phan et al. 1996; Kroenke et al. 1999):

$$R_2 - \frac{1}{2} R_1 = \frac{d^2}{4} [3J(\omega_H) + 2J(0)] + \left[\frac{2}{9} \gamma_X^2 \Delta\sigma^2 J(0) + \lambda_{ex}\right] B_0^2 \quad (8.33)$$

By fitting the relaxation data obtained at different field strengths vs. the squared static field strength B_0^2, R_{ex} as well as $J(0)$ can be calculated from the intercept since $J(\omega_H)$ is determined from τ_m and τ_e by the methods described previously. Alternatively, the transverse rates of the dipole/CSA cross-correlation, which are not affected by chemical exchange processes, can be used to directly identify the contribution to R_2 arising from chemical exchange effects. The spectral density function $J(0)$ can be represented in terms of R_1^C and R_2^C:

$$J(0) = -\frac{2\sqrt{3}}{dc P_2(\cos\theta)} \left(R_2^C - \frac{1}{2} R_1^C\right) \quad (8.34)$$

Once $J(0)$ is obtained, R_{ex} can be determined from the slope of fitting $(R_2 - \frac{1}{2} R_1)$ vs. B_0^2 in (8.33).

For anisotropically tumbling proteins with axially symmetric rotational diffusion tensors, the model-free spectral density functions are given by (Brüschweiler et al. 1995):

$$J(\omega) = \frac{2}{5} \sum_{j=0}^{2} A_j \left[\frac{S^2 \tau_j}{1 + (\omega\tau_j)^2} + \frac{(1-S^2)\tau'_j}{1 + (\omega\tau'_j)^2}\right] \quad (8.35)$$

in which $1/\tau'_j = (1/\tau_j) + (1/\tau_e)$; $1/\tau_j = 6D_\perp - j^2(D_\perp - D_\|)$; $D_\perp$ and $D_\|$ are the perpendicular and parallel components of the axially symmetric diffusion tensor; $A_0 = P_2(\cos\theta)/2$; $A_1 = 3\cos^2\theta\sin^2\theta$; $A_2 = (3\sin^4\theta)/4$; and θ is the angle between the average orientation of the bond vector and the parallel component of the axially symmetric diffusion tensor (Fig. 8.1a), which is obtained from the known structure of the protein. For isotropic rotational motions, $D_\perp = D_\|$. Then, $\tau_j = \tau_m$ and $\tau'_j = \tau$, and $\sum A_j = 1$. Equation (8.35) reduces to the model-free expression for isotropically tumbling proteins.

8.2 Experiments for Measurement of Relaxation Parameters

The measurement of spin relaxation rates is achieved by carrying out a series of 2D heteronuclear HSQC- or HMQC-type experiments. The pulse sequences for T_1 and T_2 relaxation measurement include a relaxation period inserted either before or after the t_1 evolution period, whereas for heteronuclear NOE measurement, the cross-relaxation period (saturation period) is incorporated into the preparation period in the 2D steady-state NOE sequence.

8.2.1 T_1 Measurement

8.2.1.1 Water Flip-Back Sensitivity-Enhanced T_1 HSQC

The inversion technique described in Chap. 1 is widely used to measure the longitudinal relaxation time T_1. The scheme (Fig. 8.2a, Farrow et al. 1994; Kay et al. 1992) consists of a 180° inversion pulse that inverts the heteronuclear magnetization S_z to $-S_z$, a relaxation period T during which the magnetization relaxes along the z axis, and a 90° pulse to create observable transverse magnetization for detection. For T_1 measurement, this relaxation block is inserted to the seHSQC sequence before the t_1 evolution time. The HSQC is slightly different than the ^{15}N seHSQC, in which double refocused INEPT-type sequences are utilized to transfer the magnetization from the directly bound proton to the heteronucleus and back to the proton for observation via the reversed refocused-INEPT pathway. In-phase magnetization is generated before the relaxation period rather than the antiphase coherence created in the conventional seHSQC sequence. The selective water flip-back pulse is used to ensure that the water magnetization remains along the z axis during the experiment so that saturation transfer is minimized. ^{1}H decoupling consisting of 180° pulses is used during the relaxation period T to eliminate effects of the cross-correlation between dipolar and CSA interactions. The decoupling pulse train should not perturb the longitudinal water magnetization, and consists of shaped selective 180° pulses such as cosine-modulated 180° pulses.

8.2 Experiments for Measurement of Relaxation Parameters

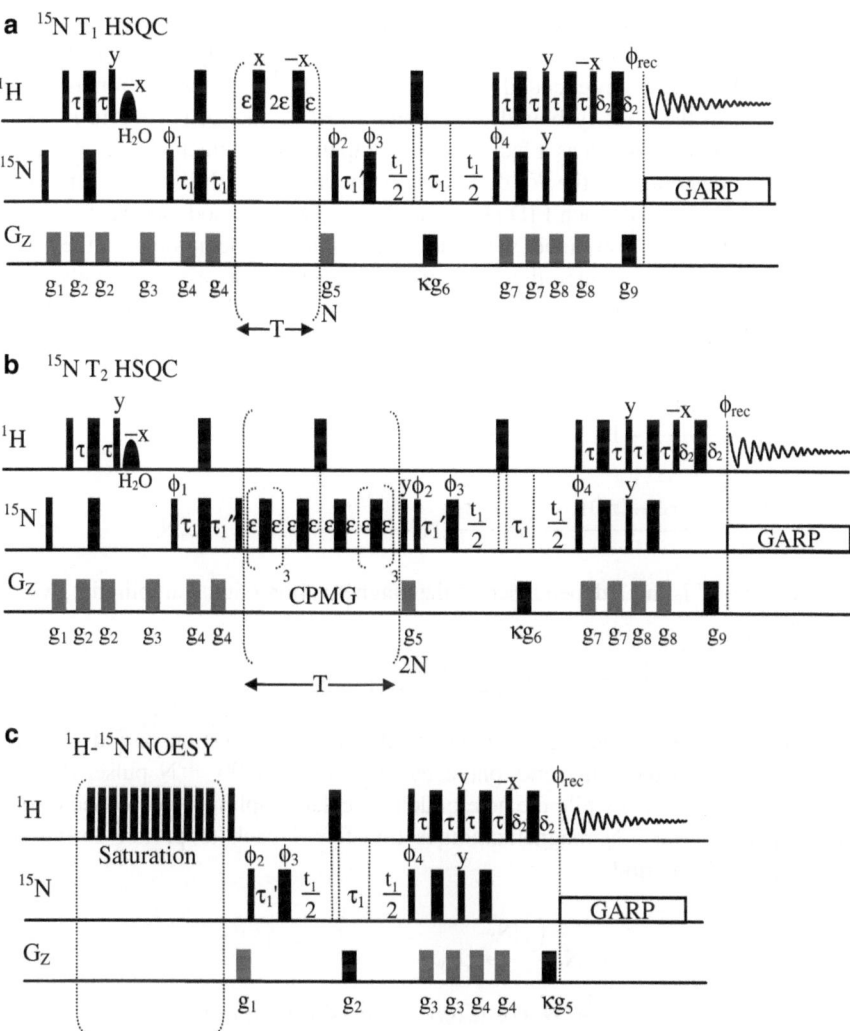

Fig. 8.2 Pulse sequences for spin relaxation measurements of ^{15}N (**a**) T_1, (**b**) T_2 relaxation times and (**c**) NOE value with sensitivity enhancement and gradient coherence selection. $R_1 = 1/T_1$ and $R_2 = 1/T_2$. In all experiments, narrow and *wide bars* represent 90° and 180° pulses, respectively; water flip-back selective pulses are shown in *rounded small bars* which are 1.8-ms *rectangular pulses*. Unless otherwise specified, all pulses have *x* phase. The coherence selection is achieved by the black gradients and $\kappa = \pm 1$ and stored in different memory locations. For PEP sensitivity enhancement two FIDs are recorded by inverting the phase ϕ_4 and the sign of κ in second experiment for every t_1. The T_1 and T_2 relaxation data are recorded by a series of experiments with different relaxation periods T. In all experiments, the delay τ is set to 2.25 ms, τ_1 to 2.75 ms, $\tau'_1 = \tau_1 + 2pw$ in which pw is ^{1}H 90° pulse length, $\tau''_1 = \tau_1 + (2/\pi)pw_N$ in which pw_N is ^{15}N 90° pulse length, and δ_2 to 0.5 ms. Recycle delays are 1.5 s for T_1 and T_2 experiments and 5 s for NOE and NONOE experiments. For the T_1 experiment (**a**) the gradient pulses are applied as $g_1 = 1$ ms, 5 G cm^{-1}, $g_2 = 0.5$ ms, 4 G cm^{-1}, $g_3 = 2$ ms, 10 G cm^{-1}, $g_4 = 0.5$ ms, 8 G cm^{-1}, $g_5 = 1$ ms, 10 G cm^{-1}, $g_6 = 1.25$ ms, 30 G cm^{-1}, $g_7 = g_8 = 0.5$ ms, 4 G cm^{-1}, $g_9 = 0.125$, 27.8 G cm^{-1}. The phase cycle is $\phi_1 = x, -x, \phi_2 = y$, +States–TPPI, $\phi_3 = 2(x), 2(y), 2(-x), 2(-y), \phi_4 = x$, and

The relaxation-encoded ^{15}N magnetization consequently evolves with the scalar coupling being refocused during the t_1 evolution period. During the "back" transfer pathway, the relaxation-encoded frequency-labeled ^{15}N magnetization is transferred back to the directly bound proton and a PEP sequence is used to increase the sensitivity by as much as a factor of $\sqrt{2}$ compared to the unenhanced spectrum. Quadrature detection in the F_1 dimension is obtained by shifting the phases of ϕ_2 and the receiver for each FID using the States–TPPI method (see Sect. 4.10.2). The ^{15}N 90° pulse combined with the gradient at the beginning of the pulse sequence is used to ensure that the initial magnetization originates only from amide proton spins. The magnetization transfer during the experiment can be described in terms of product operators:

$$H_z \xrightarrow{\left(\frac{\pi}{2}\right)H_x} -H_y \xrightarrow{\tau \to \pi(H_x+N_x) \to \tau \to \left(\frac{\pi}{2}\right)(H_y+N_x)} -2H_zN_y \quad (8.36)$$

$$\xrightarrow{\tau_1 \to \pi(H_x+N_x) \to \tau_1} N_x \xrightarrow{\left(\frac{\pi}{2}\right)N_y} -N_z \xrightarrow{T} -\zeta N_z \quad (8.37)$$

The factor ζ is the T dependence of the magnetization (signal amplitude), which is given by:

$$\zeta = 1 - 2e^{-T/T_1} = 1 - 2e^{-TR_1} \quad (8.38)$$

The ^{1}H–^{15}N scalar coupling is decoupled during the relaxation period T. After being brought to the transverse plane by the following 90° ^{15}N pulse, the ^{15}N magnetization evolves while the heteronuclear scalar coupling is decoupled during the t_1 period. The ^{15}N magnetization is transferred back to proton during the evolution of $2\tau_1$ period.

$$-\zeta N_z \xrightarrow{\left(\frac{\pi}{2}\right)N_x} \zeta N_y$$

$$\xrightarrow{t_1} \zeta N_y \cos(\Omega_N t_1) - \zeta N_x \sin(\Omega_N t_1) \quad (8.39)$$

Fig. 8.2 (continued) receiver phase $\phi_{rec} = x, -x, -x, x$, +States–TPPI. The T_2 experiment (**b**) uses the same levels and durations of the gradient pulses as used in the T_1 experiment (**a**), and the phase cycle is the same as in (**a**). During CPMG pulse trains, the ^{15}N 180° pulses are applied every 0.9 ms and ^{1}H 180° pulses are applied every 4 ms. The ^{1}H 180° pulses are calibrated to ~40 μs corresponding to a field of 3.4 kHz to avoid sample heating problems. (**c**) The saturation in both NOE and NONOE experiments is achieved by applying ^{1}H 120° pulses every 5 ms for 3 s. The gradient pulses are used as $g_1 = 3$ ms, -20 G cm^{-1}, $g_2 = 1.25$ ms, 30 G cm^{-1}, $g_3 = g_4 = 0.5$ ms, 4 G cm^{-1}, $g_5 = 0.125$ ms, 27.8 G cm^{-1}. The phase cycle is $\phi_2 = y$ + States–TPPI, $\phi_3 = x, y, -x, -y$, $\phi_4 = x$, and receiver phase $\phi_{rec} = x, -x$, +States–TPPI. The saturation frequency is placed off-resonance for NONOE and switched back to on-resonance before the first ^{1}H 180° pulse. The phase of the last ^{1}H 90° pulse is used to ensure the water magnetization is along the z axis (not the $-z$ axis) immediately before acquisition (Farrow et al. 1994)

8.2 Experiments for Measurement of Relaxation Parameters

$$\xrightarrow{2\tau_1} -\zeta 2H_zN_x \cos(\Omega_N t_1) - \zeta 2H_zN_y \sin(\Omega_N t_1)$$

Both components are retained to generate two time domain data sets by the PEP sequence:

$$\xrightarrow{\left(\frac{\pi}{2}\right)(H_x + N_x)} \zeta 2H_yN_x \cos(\Omega_N t_1) + \zeta 2H_yN_z \sin(\Omega_N t_1)$$

$$\xrightarrow{\tau \to \pi(H_x + N_x) \to \tau} \zeta 2H_yN_x \cos(\Omega_N t_1) - \zeta H_x \sin(\Omega_N t_1)$$

$$\xrightarrow{\left(\frac{\pi}{2}\right)(H_y + N_y)} -\zeta 2H_yN_z \cos(\Omega_N t_1) - \zeta H_z \sin(\Omega_N t_1)$$

$$\xrightarrow{\tau \to \pi(H_x + N_x) \to \tau \to \left(\frac{\pi}{2}\right)H_x} \zeta H_x \cos(\Omega_N t_1) + \zeta H_y \sin(\Omega_N t_1) \quad (8.40)$$

The second FID is obtained by inverting both phase ϕ and the sign of gradient factor κ:

$$\zeta H_x \cos(\Omega_N t_1) - \zeta H_y \sin(\Omega_N t_1) \quad (8.41)$$

The two FIDs are recorded for a given t_1 value and stored in separate memory locations. The data are manipulated as described in Chap. 5 for the seHSQC to obtain pure phase data, which are processed using the States–TPPI method.

8.2.1.2 Experiment Setup and Data Processing

In addition to the setup procedure common to 2D heteronuclear experiments such as 90° pulse calibrations for transmitter and decoupler and spectral window selection, the typical setup for heteronuclear T_1 relaxation measurement includes a recycle delay set to 1.5–2.0 s; an array of 8–12 T delays ranging from 5 ms to 1.5 s; for ^{15}N, the delay τ for the INEPT sequence set to 2.25 ms [$<1/(4J_{XH}) = 2.75$ ms], τ_1 set to 2.75 ms [$=1/(4J_{XH})$], τ'_1 and τ''_1 are set according to the figure legend (Fig. 8.2) and the delay δ_2 is usually set to 0.5 ms. During period T, 180° shaped pulses selected for amide protons (or 120° hard ^{1}H pulses) are spaced at 5-ms intervals to eliminate the effects of dipole/CSA cross-correlation and cross-relaxation. The gradient pulses are set to 2 ms with ~10 G cm^{-1} for residual water suppression (g_3), 1 and 0.1 ms with ~30 G cm^{-1} for coherence dephase (g_6) and refocus (g_9), respectively, and 0.5 ms with 4–8 G cm^{-1} for all other gradients. The carrier frequency is set to the water resonance for ^{1}H and 118 ppm for the ^{15}N dimension. The data are acquired with 128 complex t_1 (^{15}N) increments and 1,024 complex t_2 (^{1}H) points, with the same spectral windows in ppm for ^{1}H (~15 ppm) and ^{15}N (~35 ppm) in all data sets.

Prior to Fourier transformation, it is necessary to rearrange the data according to the procedure described in Sect. 5.1.3 (PEP seHSQC). The arranged PEP data can be processed separately as States–TPPI data and then the resulting in spectra combined together, or the data sets can be combined first and then processed in the conventional manner for States–TPPI. Prior to Fourier transformation, the FIDs are first multiplied by a 90°-shifted squared sine-bell or Gaussian window function and zero-filled twice to yield a digital resolution better than 2 Hz/point. A 90°-shifted squared sine-bell window function is applied to the indirect ^{15}N dimension. The data are zero-filled twice before the second Fourier transformation is applied. In general, linear prediction to improve digital resolution of the indirect dimension is not used since it may introduce a deviation from the real value of the T_1 relaxation rate. Each of the series of spectra is phase-corrected after being Fourier transformed: phase of ^{1}H dimension is adjusted according to the phase of the first FID of the shuffled data, whereas the phase of the ^{15}N dimension is corrected using two F_1 slices. The amplitudes of the cross-peaks are measured using either peak volume integrals or intensities if signal overlapping becomes severe. The intensity or volume integral $I_j(T)$ of the cross-peak for residue j is measured for all spectra with different values of T. The longitudinal relaxation time constant is calculated by fitting (1.83b) in Chap. 1 for all $I_j(T)$ values with the approximation that $I_{0j} = I_j(1.5\ \mathrm{s})$.

8.2.2 T_2 and $T_{1\rho}$ Measurements

8.2.2.1 Sensitivity-Enhanced HSQC for T_2 and $T_{1\rho}$ Measurements

The sequence for T_2 measurement (Fig. 8.2b, Farrow et al. 1994; Messerlie et al. 1989) is identical to the sequence used for T_1 measurement if the inversion scheme is replaced with a CPMG or spin lock sequence (Carr and Purcell 1954; Meiboom and Gill 1958). Rather than decaying along the longitudinal direction, the heteronuclear magnetization relaxes on the transverse plane during the T period of the T_2 pulse sequence. In addition to spin–spin interactions, the inhomogeneity of the magnetic field also contributes to the transverse relaxation. To remove the effect of field inhomogeneity, a CPMG spin echo sequence, which was developed by Carr and Purcell, and by Meiboom and Gill, is frequently applied in the measurement of transverse relaxation. In the CPMG scheme, the heteronuclear magnetization S_x evolves during a period ε under the interaction of the chemical shift and field inhomogeneity. After the 180° ^{15}N pulse reverses the direction of precession of the nuclear spins, the evolution due to the chemical shift and field inhomogeneity is refocused during the second ε period, provided that the spins being refocused remain in the identical magnetic field during both ε periods. The resulting transverse magnetization at the end of an even number of echoes in a CPMG pulse train has an amplitude decayed according to:

$$I = I_0 e^{-TR_2} \qquad (8.42)$$

8.2 Experiments for Measurement of Relaxation Parameters

in which $T = 2n(2\varepsilon + \text{pw}_{180°})$, n is an integer and $\text{pw}_{180°}$ is the pulse length of a 180° ^{15}N pulse in the CPMG pulse train.

In the $R_{1\rho}$ experiment, the transverse magnetization is locked in the rotating frame by applying a spin lock train or continuous radio frequency field (Peng et al. 1991; Desvaux and Berthault 1999). The relaxation rate constant of the magnetization along the effective field direction in the rotating frame is called $R_{1\rho}$. The $R_{1\rho}$ measurement depends on such experimental parameters as the amplitudes of the applied B_1 field, ω_1, and the effective field in the rotating frame, ω_e, and the offset frequency Ω, $\omega_e^2 = \Omega^2 + \omega_1^2$. In the rotating frame, the tilt angle of the effective field from the RF field is given by:

$$\tan\theta = \frac{\omega_1}{\Omega} \tag{8.43}$$

The measured $R_{1\rho}$ is the combination of R_1 and R_2 via the dependence of the tilt angle:

$$R_{1\rho} = R_2\sin^2\theta + R_1\cos^2\theta \tag{8.44}$$

For an on-resonance spin lock field, θ is close to 90° for all resonances and the effective field is along the applied B_1 field (Meiboom 1961; Szyperski et al. 1993). The measured $R_{1\rho}$ represents the transverse relaxation rate constant R_2. For off-resonance $R_{1\rho}$ experiment (Akke and Palmer 1996; Zinn-Justin et al. 1997; Mulder et al. 1998), the RF transmitter frequency is placed far enough off-resonance so that θ in (8.43) is less than 70°. A pair of adiabatic ramp pulses is used to align the magnetization along the spin lock axis and rotate it back to the original direction. The magnetization at the beginning and the end of spin lock period is along the z axis. A continuous and small increase in the amplitude of the RF field causes the magnetization to follow the effective field in an adiabatic manner, resulting in a rotation, instead of projection, of the magnetization. At the end of the spin lock, the magnetization follows the effective field, by decreasing the amplitude of the B_1 field, back to the z axis.

In both the R_2 and $R_{1\rho}$ experiments, the two PEP FIDs obtained in the same manner to those obtained in the T_1 seHSQC sequence are given by:

$$\xi H_x \cos(\Omega_N t_1) + \xi H_y \sin(\Omega_N t_1)$$

$$\xi H_x \cos(\Omega_N t_1) - \xi H_y \sin(\Omega_N t_1) \tag{8.45}$$

in which $\xi = e^{-TR_2}$. The two FIDs are recorded for each given t_1 value and stored in separate memory locations. The data are treated as described in Sect. 5.1.3 for the PEP seHSQC to obtain pure phase data, which are processed using the States–TPPI method.

Analogous to the T_1 measurement, the contribution to the T_2 from cross-correlation of dipole/CSA is required to be minimized during the CPMG spin echo. The cross-correlation effect can be effectively removed by the combination of applying 180° ^{15}N pulses every 0.9 ms ($\varepsilon = 0.45$ ms) during the entire CPMG pulse train and applying 180° ^{1}H pulse centered in the CPMG pulse train. An alternative way is to apply 180° ^{15}N pulses every 0.9 ms and 180° ^{1}H pulses every 4 ms during the CPMG pulse train.

8.2.2.2 Experiment Setup and Data Processing

The setup procedure for the heteronuclear T_2 relaxation measurement is primarily identical to the procedure for T_1 relaxation measurement except for the setup of parameters for the relaxation period T. The number of echo cycles must be chosen as $2n$, in which n is an integer to ensure that the magnetization has same sign after the echo period. An array of 8–12 relaxation delays ranging from 5 to 150 ms is typically used. The time of the CPMG pulse train is set to be shorter than 150 ms to avoid sample heating problems caused by the pulse train. For the same reason, the ^{15}N RF field strength is set to less than 6 kHz, corresponding to a 90° ^{15}N pulse length of longer than 40 µs. As described previously, in order to remove the cross-correlation effect, a combination of 180° ^{15}N pulses every 0.9 ms during the entire CPMG pulse train and 180° ^{1}H pulses centered in the CPMG pulse train is applied. A different approach is also often used by applying 180° ^{15}N pulses every 0.9 ms and applying 180° (or 120°) ^{1}H pulses every 4–5 ms during the CPMG pulse train. The gradient pulse length and amplitude are the same as used in the T_1 experiment.

In the $R_{1\rho}$ experiment, the ^{15}N spin lock replaces the CPMG spin echo in the T_2 experiment, and is applied with a continuous RF ^{15}N pulse with a field strength less than 3.5 kHz to minimize sample heating problems. The sample heating during the spin lock can also be minimized by using a predelay time longer than 3 s. During the spin lock, 180° ^{1}H pulses spaced at 5-ms intervals are applied to eliminate the effects of dipole/CSA cross-correlation and cross-relaxation.

The data are processed in the same procedure as described for T_1 measurement after the rearrangement of the data according the procedure for PEP FIDs. Fitting for $T_{1\rho}$ values is required for the correction of the resonance offset effect (Peng and Wagner 1994).

8.2.3 Heteronuclear NOE Measurement

8.2.3.1 Heteronuclear NOE Experiment

The heteronuclear NOE is determined from the change in intensity of the NMR signal of heteronucleus X when the equilibrium magnetization of protons in the vicinity is perturbed by saturation in experiments such as the transient NOE or steady-state NOE. The pulse sequence of a steady-state heteronuclear NOE experiment shown in Fig. 8.2c utilizes a pulse train to saturate proton equilibrium magnetization prior to the heteronuclear magnetization being excited. The first 90° ^{1}H pulse combined with the gradient is used to ensure that the ^{15}N magnetization is the only initial magnetization of the experiment. After the first 90° X pulse, the chemical shift of the heteronucleus X evolves during t_1 and the heteronuclear magnetization is transferred to proton with decoupling of the scalar coupling J_{XH}. In the final stage of the pulse sequence, the two orthogonal transverse magnetization components generated during t_1 are refocused by the PEP sequence for simultaneous detection by inverting the phase of ϕ_4 and sign of κ. The two FIDs are recorded and stored in separated memory locations. Quadrature detection in the F_1 dimension is obtained by shifting the phases of ϕ_2 and the receiver for each FID in a States–TPPI manner (see Sect. 4.10.2). To measure the NOE, a pair of experiments is recorded with the saturation (NOE) and without the saturation (NONOE) of the protons bound to the heteronuclei. The intensity of the heteronuclear NOE can be obtained from the longitudinal magnetization and relaxation rates of the heteronucleus X and proton H, which is the ratio of signal intensities between the NOE and NONOE (unsaturation) spectra (Goldman 1998: Farrow et al. 1994):

$$I_{sat} = \langle X_z \rangle_{eq} + \frac{\sigma_{XH}}{R_1} \langle H_z \rangle_{eq} = I_{unsat} \left(1 + \frac{\sigma_{XH}}{R_1} \frac{\gamma_H}{\gamma_X}\right) \quad (8.46)$$

in which I_{sat} and I_{unsat} represent the measured intensities of a resonance in the presence and absence of proton saturation, respectively; σ_{XH} is the rate constant of cross-relaxation; γ_X and γ_H are the gyromagnetic ratios. The values of NOE in (8.4) are obtained by the steady-state NOE values which are determined by the ratios of the peak intensities in the NOE and NONOE spectra:

$$\text{NOE} = \frac{I_{sat}}{I_{unsat}} = 1 + \frac{\sigma_{XH}}{R_1} \frac{\gamma_H}{\gamma_X} \quad (8.47)$$

The standard deviation of NOE value, $\overline{\delta}_{NOE}$ can be determined using the measured background noise levels:

$$\frac{\overline{\delta}_{NOE}}{\text{NOE}} = \sqrt{\frac{\overline{\delta}^2_{sat}}{I^2_{sat}} + \frac{\overline{\delta}^2_{unsat}}{I^2_{unsat}}} \quad (8.48)$$

in which $\overline{\delta}_{sat}$ and $\overline{\delta}^2_{unsat}$ represent the standard deviations of I_{sat} and I_{unsat}, respectively, calculated from the root-mean-squared noise of background regions (Nicholson et al. 1992). In the condition of the extreme narrowing limit ($\omega \tau_m \ll 1$) in which τ_m is short, $\sigma_{XH} = \frac{1}{2} R_1$ which yields a maximum magnitude of NOE. Therefore, ^{15}N spins have NOE values between 1 and −4 because of its negative gyromagnetic ratio, whereas ^{13}C NOE values are in the range of 1–5.

The magnetization transfer in the experiment can be described by product operators. The transverse ^{15}N magnetization is frequency-labeled and transferred to ^{1}H spins during the first period:

$$N_z \xrightarrow{\left(\frac{\pi}{2}\right)N_x} N_y \xrightarrow{t_1} N_y \cos(\Omega_N t_1) - N_x \sin(\Omega_N t_1)$$

$$\xrightarrow{2\tau_1} -2H_z N_x \cos(\Omega_N t_1) - 2H_z N_y \sin(\Omega_N t_1) \tag{8.49}$$

The scalar coupling J_{XH} is refocused during the t_1 evolution time if τ_1 is set to $1/(4J_{XH})$. Both orthogonal components of the ^{1}H magnetization are refocused and observed by the PEP sequence (8.40), resulting in the two FIDs:

$$H_x \cos(\Omega_N t_1) + H_y \sin(\Omega_N t_1)$$

$$H_x \cos(\Omega_N t_1) - H_y \sin(\Omega_N t_1) \tag{8.50}$$

8.2.3.2 Experiment Setup and Data Processing

The saturation of the ^{1}H magnetization is obtained by applying either 120° ^{1}H pulses every 5 ms for 3 s or a WALTZ16 pulse train for 3 s. The 90° ^{1}H pulse length for the WALTZ16 is calibrated to about 30 μs. A total recycle time of at least 5 s is used for ^{15}N measurement to allow the longitudinal magnetization to relax back to equilibrium. Usually, the NOE and NONOE spectra are recorded in an interleaved manner to reduce artifacts. The gradient amplitude and durations are selected as $g_1 = 3$ ms, -20 G cm^{-1}; $g_2 = 1.25$ ms, 30 G cm^{-1}; $g_3 = g_4 = 0.5$ ms, 4 G cm^{-1}; $g_5 = 0.125$, 27.8 G cm^{-1}. The delays are set to $\tau = 2.25$ ms, $\tau_1 = 2.75$ ms, and $\delta_2 = 0.5$ ms. A total of 128 complex t_1 points is usually recorded.

After rearrangement of the PEP data, the FIDs are processed into a 512 × 1,024 matrix with 90°-shifted squared sine-bell window functions in both dimensions. The ^{15}N NOE values are calculated from the ratio I_{sat}/I_{unsat} of the cross-peak intensities (8.47) in the NOE and NONOE spectra.

8.3 Relaxation Data Analysis

The R_1 and R_2 values are obtained by fitting the intensities of individual cross-peaks with a series of values of relaxation times T using (1.83b) and (8.42), respectively. The NOE values are extracted from the intensity ratios of individual cross-peaks in the NOE and NONOE experiments using (8.47) for the data recorded at different static magnetic field strengths. Once the values of the relaxation rates and NOE are

8.3 Relaxation Data Analysis

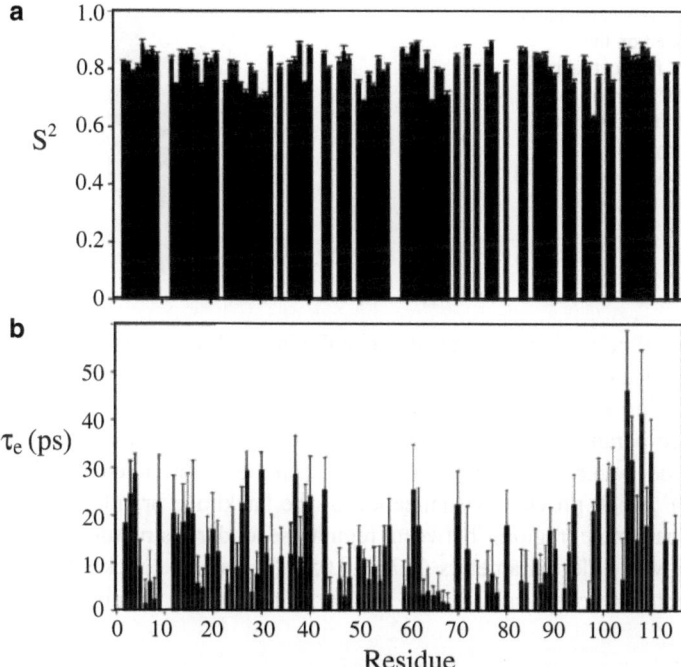

Fig. 8.3 Simple model-free parameters for local motion of the backbone amide N–H of ferrocytochrome $c2$ derived from ^{15}N relaxation data recorded at 30 °C and analyzed using an axially symmetric diffusion tensor. (**a**) Squared generalized order parameters (S^2). (**b**) Effective correlation time constants (reproduced with permission from Flynn et al. (2001), Copyright © 2001American Chemical Society)

calculated, the overall correlation time τ_m is usually determined from the 10 % trimmed mean of the R_2/R_1 ratio (Mandel et al. 1995). In the next step, dynamic parameters (squared generalized order parameter S^2, and τ_c) are obtained via such methods described in the theory section as a grid search by minimizing the global error function or local error functions using the estimated τ_m value (Mandel et al. 1995; Dellow and Wand 1989). The program CurveFit is available for determining R_1 and R_2 from experimental data, and the program "Modelfree" can be used to fit the R_1 and R_2 and NOE data to heteronuclear relaxation data to obtain model-free parameters according to the extended model-free formalism using minimization of the error function (http://cpmcnet.columbia.edu/dept/gsas/biochem/labs/palmer/).

Once S^2 and τ_e are obtained, interpretation of the results is straightforward. The dynamic parameters can be plotted for each residue as shown in Fig. 8.3. Backbone and side-chain dynamics information can be obtained based on the distribution of S^2 and τ_e over the residues. As mentioned earlier, S^2 with higher amplitude indicates that the motion of the bond vector is restricted to the rigid orientations. As S^2 decreases, the motion of the bond vector becomes less restricted. The motion becomes completely isotropic as S^2 approaches zero. In addition to S^2, the internal

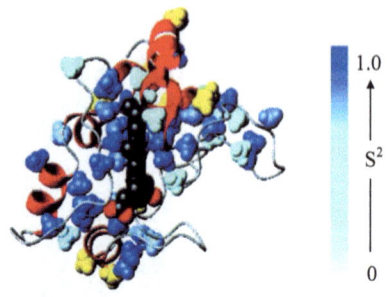

Fig. 8.4 S^2 axis parameters colored-coded on the structure of ferrocytochrome $c2$ (reproduced with permission from Flynn et al. (2001), Copyright © 2001 American Chemical Society)

correlation time τ_e characterizes how fast the internal motion is. As the example in Fig. 8.3 indicates, the majority of the amide NH S^2 are distributed near 0.8 with the corresponding effective correlation times in the range of 1–50 ps. The obtained model-free parameters of the protein are basically consistent with a well-ordered polypeptide backbone. The values of S^2 can be color-coded on the structure (Fig. 8.4), which provides visualization of the backbone or side-chain dynamics. Several backbone regions between regular secondary structure elements have slightly lower order parameters (0.7 ± 0.05). Overall, the backbone dynamics of ferrocytochrome $c2$ reveals that the interior of the protein is unusually rigid.

Questions

1. What are the experimental parameters measured for the study of protein dynamics?
2. What are the two methods for calculating the spectral density functions from experimental data using model-free analysis?
3. What is the assumption of model-free analysis?
4. What is the physical meaning of the squared generalized order parameters S^2? What is the range of S^2 value? What kind of motion does an S^2 with a value near 0.8 describe?
5. What is the relationship of R_1, R_2, and $R_{1\rho}$?
6. Derive (8.22) and (8.23) from (8.14) and (8.15) using (8.20) and (8.21), and $\beta_1 = 0.921$ and $\beta_2 = 0.955$ for ^{15}N. Hint: derive (8.23) first. $J(\omega_X)$ in (8.14) and (8.15) equals $J(\omega_N)$. To derive (8.22) for ^{15}N, substitute $J(\omega_N)$ with (8.23).

References

Abragam A (1961) Principles of nuclear magnetism. Clarendon Press, Oxford
Akke M, Palmer G (1996) Monitoring macromolecular motions on microsecond to millisecond time scales by R1ρ–R1 constant relaxation time NMR spectroscopy. J Am Chem Soc 118:911–912

References

Brüschweiler R, Liao X, Wright PE (1995) Long-range motional restrictions in a multidomain zinc-finger protein from anisotropic tumbling. Science 268:886–889

Bull TE (1992) Relaxation in the rotating frame in liquids. Prog Nucl Magn Reson Spectrosc 24:377–410

Carr HY, Purcell EM (1954) Effects of diffusion on free precession in nuclear magnetic resonance experiments. Phys Rev 94:630–638

Clore GM, Szabo A, Bax A, Kay LE, Driscoll PC, Gronenborn AM (1990a) Deviations from the simple two-parameter model-free approach to the interpretation of nitrogen-15 nuclear magnetic relaxation of proteins. J Am Chem Soc 112:4989–4991

Clore GM, Driscoll PC, Wingfield PT, Gronenborn AM (1990b) Analysis of the backbone dynamics of interleukin-1.beta. using two-dimensional inverse detected heteronuclear nitrogen-15-proton NMR spectroscopy. Biochemistry 29:7387–7401

Dayie KT, Wagner G, Lefevre JF (1996) Theory and practice of nuclear spin relaxation in proteins. Annu Rev Phys Chem 47:243–282

Dellwo MJ, Wand AJ (1989) Model-independent and model-dependent analysis of the global and internal dynamics of cyclosporin A. J Am Chem Soc 111:4571–4578

Desvaux H, Berthault P (1999) Study of dynamic processes in liquids using off-resonance rf irradiation. Prog Nucl Magn Reson Spectrosc 35:295–340

Farrow NA, Zhang O, Szabo A, Torchoia DA, Kay LE (1995) Spectral density function mapping using 15N relaxation data exclusively. J Biomol NMR 6:153–162

Farrow NA, Muhandiram R, Singer AU, Pascal SM, Kay CM, Gish G, Shoelson SE, Pawson T, Forman-Kay JD, Kay LE (1994) Backbone dynamics of a free and a phosphopeptide-complexed Src homology 2 domain studied by 15N NMR relaxation. Biochemistry 33:5984–6003

Flynn PF, Bieber Urbauer RJ, Zhang H, Lee AL, Wand AJ (2001) Main chain and side chain dynamics of a heme protein: 15N and 2H NMR relaxation studies of R. capsulatus ferrocytochrome c2. Biochemistry 40:6559–6569

Goldman M (1984) Interference effects in the relaxation of a pair of unlike spin-1/2 nuclei. J Magn Reson 60:437–452

Hiyama Y, Niu C, Silverton JV, Bavoso A, Torchia DA (1988) Determination of 15N chemical shift tensor via 15N-2H dipolar coupling in Boc-glycylglycyl[15N glycine]benzyl ester. J Am Chem Soc 110:2378–2383

Ishima R, Torchia DA (2000) Protein dynamics from NMR. Nat Struct Biol 7:740–743

Ishima R, Nagayama K (1995a) Protein backbone dynamics revealed by quasi spectral density function analysis of amide N-15 nuclei. Biochemistry 34:3162–3171

Ishima R, Nagayama K (1995b) Quasi-spectral-density function analysis for nitrogen-15 nuclei in proteins. J Magn Reson B108:73–76

Kay LE, Torchia DA, Bax A (1989) Backbone dynamics of proteins as studied by 15N inverse detected heteronuclear NMR spectroscopy: application to staphylococcal nuclease. Biochemistry 28:8972–8979

Kay LE, Keifer P, Saarinen T (1992) Pure absorption gradient enhanced heteronuclear single quantum correlation spectroscopy with improved sensitivity. J Am Chem Soc 114:10663–10665

Kroenke CD, Rance M, Palmer AG III (1999) Variability of the 15N chemical shift anisotropy in Escherichia coli Ribonuclease H in solution. J Am Chem Soc 121:10119–10125

Lipari G, Szabo A (1982a) Model-free approach to the interpretation of nuclear magnetic resonance relaxation in macromolecules. 1. Theory and range of validity. J Am Chem Soc 104:4546–4559

Lipari G, Szabo A (1982b) Model-free approach to the interpretation of nuclear magnetic resonance relaxation in macromolecules. 2. Analysis of experimental results. J Am Chem Soc 104:4559–4570

London RE (1980) Intramolecular dynamics of proteins and peptides as monitored by nuclear magnetic relaxation experiments. in Magnetic Resonance in Biology, J.S. Cohen (ed.), Wiley, New York, pp. 1–69

Mandel AM, Akke M, Palmer AG III (1995) Backbone dynamics of Escherichia coli ribonuclease HI: correlations with structure and function in an active enzyme. J Mol Biol 246:144–163

Mayo KH, Daragan VA, Idiyatullin D, Nesmelova I (2000) Peptide internal motions on nanosecond time scale derived from direct fitting of 13C and 15N NMR spectral density functions. J Magn Reson 146:188–195

Meiboom S, Gill D (1958) Modified spin-echo method for measuring nuclear relaxation times. Rev Sci Instrum 29:688–691

Meiboom S (1961) Nuclear magnetic resonance study of the proton transfer in water. J Chem Phys 34:375–388

Messerlie BA, Wider G, Otting G, Weber C, Wuthrich K (1989) Solvent suppression using a spin lock in 2D and 3D NMR spectroscopy with H2O solutions. J Magn Reson 85:608–613

Mulder FAA, de Graaf RA, Kaptein R, Boelens R (1998) An off-resonance rotating frame relaxation experiment for the investigation of macromolecular dynamics using adiabatic rotations. J Magn Reson 131:351–357

Nicholson LK, Kay LE, Baldisseri DM, Arango J, Young PE, Torchia DA (1992) Dynamics of methyl groups in proteins as studied by proton-detected carbon-13 NMR spectroscopy. Application to the leucine residues of staphylococcal nuclease. Biochemistry 31:5253–5263

Palmer AG III (2001) NMR probes of molecular dynamics: overview and comparison with other techniques. Annu Rev Biophys Biomol Struct 30:129–155

Palmer AG III, Rance M, Wright PE (1991) Intramolecular motions of a zinc finger DNA-binding domain from Xfin characterized by proton-detected natural abundance carbon-13 heteronuclear NMR spectroscopy. J Am Chem Soc 113:4371–4380

Peng JW, Wagner G (1994) Protein Mobility from Multiple ^{15}N relaxation Parameters in Tycko R (ed.) Nuclear Magnetic Resonance Probes of Molecular Dynamics. Kluwer, Dordrecht, pp. 373–454

Peng JW, Wagner G (1995) Frequency spectrum of NH bonds in eglin c from spectral density mapping at multiple fields. Biochemistry 34:16733–16752

Peng JW, Thanabal V, Wagner G (1991) 2D heteronuclear NMR measurements of spin–lattice relaxation times in the rotating frame of X nuclei in heteronuclear HX spin systems. J Magn Reson 94:82–100

Phan IQ, Boyd J, Campbell ID (1996) Dynamic studies of a fibronectin type I module pair at three frequencies: anisotropic modelling and direct determination of conformational exchange. J Biomol NMR 8:369–378

Szyperski T, Luginbul P, Otting G, Guntert P, Wuthrich K (1993) Protein dynamics studied by rotating frame 15N spin relaxation times. J Biomol NMR 3:151–164

Wittebort R, Szabo A (1978) Theory of NMR relaxation in macromolecules: restricted diffusion and jump models for multiple internal rotations in amino acid side chains. J Chem Phys 69:1722–1736

Yamazaki T, Muhandiram R, Kay LE (1994) NMR experiments for the measurement of carbon relaxation properties in highly enriched, uniformly 13C, 15N-labeled proteins: application to 13Cα carbons. J Am Chem Soc 116:8266–8278

Ye C, Fu R, Hu J, Hou L, Ding S (1993) Carbon-13 chemical shift anisotropies of solid amino acids. Magn Reson Chem 31:699–704

Zinn-Justin S, Berthault P, Guenneuges M, Desvaux H (1997) Off-resonance rf fields in heteronuclear NMR: application to the study of slow motions. J Biomol NMR 10:363–372

Chapter 9
NMR-Based Metabolomics

9.1 Introduction

Metabolomics is a relatively new and emerging field of the omics research compared to other well-established platforms (genomics, transcriptomics, and proteomics) and is rapidly growing as evidenced by the increasing number of publications in this field (Fig. 9.1). It studies the global profiles of metabolites in a biological system (cell, tissue, or organism) under a given set of conditions (Goodacre et al. 2004). Arguably, its history can be traced back to ancient times (1500–2000 BC) when traditional Chinese doctors used ants to detect high concentrations of glucose in patient's urine for diagnosing diabetes (Van der Greef and Smilde 2005). The concept that individuals might have different "metabolic patterns" that can be detected in their biological fluids was first introduced by Roger Williams in the late 1940s (Williams 1956; Gates and Sweeley 1978). Since then, several terms (or definitions) have been proposed to describe the field of metabolomics. "Metabolic profile" was introduced by Horning and Horning (1971) to describe the quantitative measurement of metabolite concentrations in urine. "Metabolome" was proposed by Oliver et al. (1998) as referring to the complete set of small-molecule (<1 kDa) endogenous metabolites in an organism, and "metabonomics" by Nicholson et al. as "the quantitative measurement of the dynamic multiparametric response to living systems to pathophysiological stimuli or genetic modification" (Nicholson et al. 1999). Fiehn subsequently extended "metabolome" terminology to metabolomics as the comprehensive and quantitative analysis of all metabolites of an organism (Fiehn 2001). Although these terms are frequently used interchangeably, there is a growing consensus that the field is named as metabolomics, as reflected by the establishment of the Metabolomics Society (an international society) in 2004 and its official journal *Metabolomics* in 2005.

Metabolomics is considered as a complementary tool to other omics platforms because it more directly reflects cellular physiological states as being the most downstream in the omics family (Fig. 9.2). While a genome contains all genetic information of a living organism, the information of genes is expressed to produce

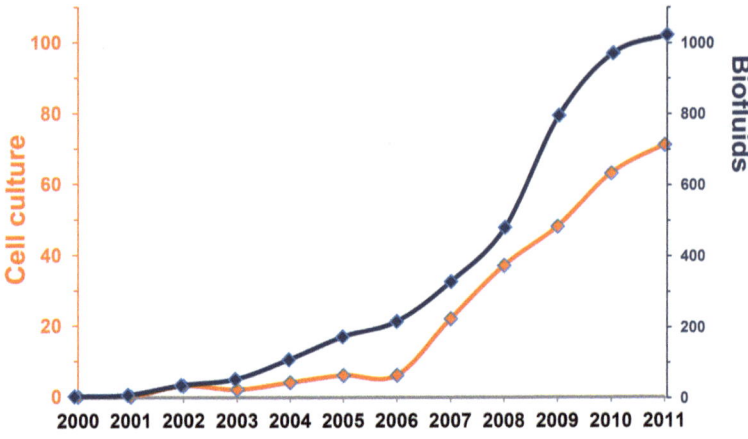

Fig. 9.1 Annual numbers of publications in the field of metabolomics from search results of PubMed database. The numbers present the rapidly growing rate of metabolomics research

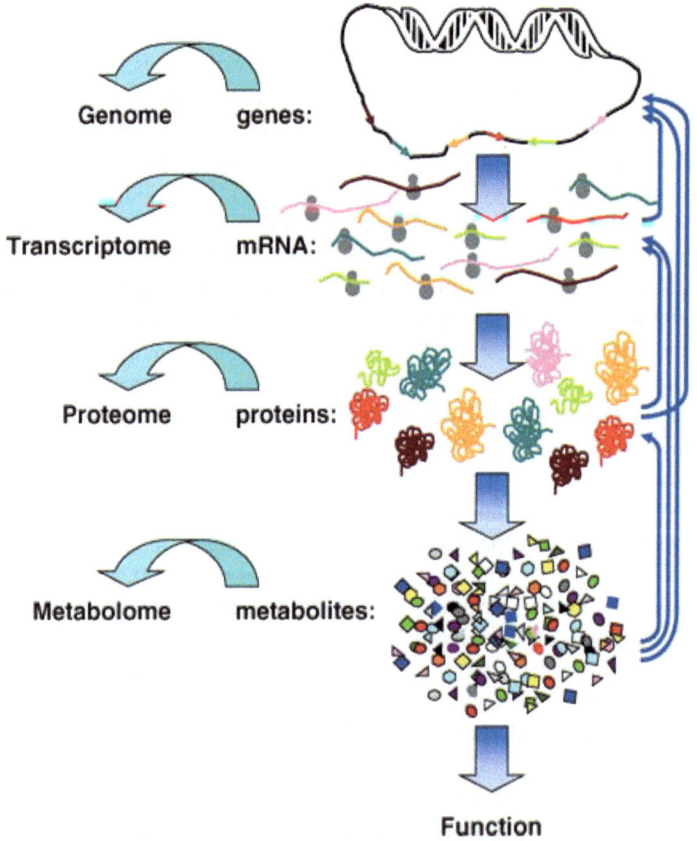

Fig. 9.2 Schematic overview of omics technologies. The information flows from genes to transcripts to proteins to metabolites to function—phenotype. *Blue arrows* indicate interactions regulating the respective translations (reproduced with permission from Goodacre (2005). Copyright © 2005 Springer)

transcriptome—a complete set of RNA transcripts of the genes (transcriptomics) and levels of mRNA determine the concentration of expressed proteins (proteomics). The translations of genes and mRNA are complex processes and sometimes errors are introduced during the translations or chemical modifications occur after the translations. Moreover, many proteins are not enzymes and only enzymes regulate the concentrations of metabolites. Therefore, it is necessary to conduct a comprehensive and quantitative analysis of the metabolic profile of a living organism in order to fully study its systemic responses to environmental or genetic perturbations (Fiehn 2001; Khoo and Al-Rubeai 2007). Metabolomics can thus provide a snapshot of the cellular biochemistry and physiology taking place at any instant. Metabolomics combined with system biology can finally bridge the gap between genotype and phenotype (Goodacre 2005).

In this chapter, methodologies of multivariate statistical analysis used in metabolomics and several examples of metabolomics applications will be described. These examples constitute only a small portion of enormous current and potential applications of metabolomics.

9.2 Fundamentals of Multivariate Statistical Analysis for Metabolomics

Multivariate data analysis (MVDA; Rencher 2002) has been applied to the analysis of complicated chemical and biochemical data with efficient and robust statistical methods and has provided meaningful and reliable models to analyze the metabolomic data obtained by NMR spectroscopy, mass spectrometry, Raman spectroscopy, FTIR, etc. The MVDA methods routinely used in metabolomics include principal component analysis (PCA; Wold et al. 1987), partial least squares discriminant analysis (PLS-DA; Wold et al. 1984), and orthogonal PLS-DA (OPLS-DA; Wold et al. 1998).

In this section, the following key questions will be addressed:

1. What is PCA?
2. What is data scaling and why do we need it?
3. What is preprocessing and how is it used to improve models?
4. How is PCA applied to metabolomics data analysis?
5. What do the results of PCA mean?
6. How can the results be interpreted?
7. What are Q^2 and R^2 factors?
8. How are the statistical models validated?

PCA, invented in 1901 by Karl Pearson (Pearson 1901), is the workhorse in multivariate analysis (MVA) for metabolomics and is the first step in the statistical analysis to obtain an overview of metabolomics data. The main objectives of PCA are to identify the variance in the data and transform high-dimensional data into fewer dimensions. The scatter scores plot of the first two components (PC1 and

PC2) is usually examined to see whether the data is homogeneous, any outliers exist, any grouping is formed, and what the trends of the grouping are. An outlier is typically defined as any observation that has fallen outside of the Hotelling's T2 ellipse (Hotelling 1931) at the 95% confidence interval in the scores plots of the first two components.

PCA maximizes the variance of a linear combination of the variables, based on the assumption that the largest variance in a dataset contains most of information. PCA is carried out with no prior knowledge on groupings of observations or partitioning of variables. For a data matrix X with m rows of observations (e.g., samples) and n columns of variables (e.g., NMR chemical shifts), PCA transforms the data into a new coordinate space according to:

$$X = TP' + E \qquad (9.1)$$

where **T** is called the scores matrix, **P** is called the loadings matrix, **P'** is the transposed loadings matrix, and **E** is the residual matrix, which contains remaining variation that is not captured by the PCA. Scores and loadings matrices **T** and **P** can be expressed in the form of:

$$T = \begin{pmatrix} t_{11} & t_{12} & \cdots & t_{1k} \\ t_{21} & t_{22} & \cdots & t_{2k} \\ \vdots & \vdots & \ddots & \vdots \\ t_{m1} & t_{m2} & \cdots & t_{mk} \end{pmatrix} \qquad (9.2)$$

$$P = \begin{pmatrix} p_{11} & p_{12} & \cdots & p_{1k} \\ p_{21} & p_{22} & \cdots & p_{2k} \\ \vdots & \vdots & \ddots & \vdots \\ p_{n1} & p_{n2} & \cdots & p_{nk} \end{pmatrix} \qquad (9.3)$$

in which subscript m represents the number of observations, n represents the number of variables, and k represents the order of principal components, which is smaller than n, and typically no more than 10. For example, $k = 3$ represents the third principal component, PC3. For NMR-based metabolomics, variables are NMR spectral bins, whereas m is the number of NMR spectra. The transposed loadings matrix is defined as:

$$P' = \begin{pmatrix} p_{11} & p_{21} & \cdots & p_{n1} \\ p_{12} & p_{22} & \cdots & p_{n2} \\ \vdots & \vdots & \ddots & \vdots \\ p_{1k} & p_{1k} & \cdots & p_{nk} \end{pmatrix} \qquad (9.4)$$

9.2 Fundamentals of Multivariate Statistical Analysis for Metabolomics

By substituting matrices **T** and **P′** with (9.2) and (9.4), (9.1) becomes:

$$X = \begin{pmatrix} t_{11} & t_{12} & \cdots & t_{1k} \\ t_{21} & t_{22} & \cdots & t_{2k} \\ \vdots & \vdots & \ddots & \vdots \\ t_{m1} & t_{m2} & \cdots & t_{mk} \end{pmatrix} \begin{pmatrix} p_{11} & p_{21} & \cdots & p_{n1} \\ p_{12} & p_{22} & \cdots & p_{n2} \\ \vdots & \vdots & \ddots & \vdots \\ p_{1k} & p_{2k} & \cdots & p_{nk} \end{pmatrix} + E \quad (9.5)$$

Matrices **T** and **P′** can also be rewritten as a series of vectors, t_i and p_i. Then the PCA (9.5) has the form of:

$$X = \begin{pmatrix} t_1 & t_2 & \cdots & t_k \end{pmatrix} \begin{pmatrix} p'_1 \\ p'_2 \\ \vdots \\ p'_k \end{pmatrix} + E \quad (9.6)$$

$$= t_1 p'_1 + t_2 p'_2 + \cdots + t_k p'_k + E \quad (9.7)$$

in which $t_1, t_2, \ldots t_k$ are scores vectors of the first component, second component, ... and k component in scores matrix **T** and $p'_1, p'_2, \ldots p'_k$ are loadings vectors of first component, second component, ... and k component of the transposed loadings matrix **P′**. For $i = 1, 2, \ldots, k$,

$$t_i = \begin{pmatrix} t_{1i} \\ t_{2i} \\ \vdots \\ t_{mi} \end{pmatrix} \quad (9.8)$$

$$p'_i = \begin{pmatrix} p_{1i} & p_{2i} & \cdots & p_{ni} \end{pmatrix} \quad (9.9)$$

Note that t_i and p'_i are vectors while t_{mi} and p_{ni} are individual elements of the vectors. Equations (9.8) and (9.9) state that for each principal component i, there are m scores and n loadings. Each score is the coordinate of the observation (or sample) on the principal component i (the ith axis of the new coordinate system generated by the PCA transformation). For instance, for $k = 3$,

$$X = E + \begin{pmatrix} \text{PC1} & \text{PC2} & \text{PC3} \\ \downarrow & \downarrow & \downarrow \\ t_{11} & t_{12} & t_{13} \\ t_{21} & t_{22} & t_{23} \\ \vdots & \vdots & \vdots \\ t_{m1} & t_{m2} & t_{m3} \end{pmatrix} \begin{pmatrix} p_{11} & p_{21} & \cdots & p_{n1} \\ p_{12} & p_{22} & \cdots & p_{n2} \\ p_{13} & p_{23} & \cdots & p_{n3} \end{pmatrix} \begin{matrix} \leftarrow \text{PC1} \\ \leftarrow \text{PC2} \\ \leftarrow \text{PC3} \end{matrix} \quad (9.10)$$

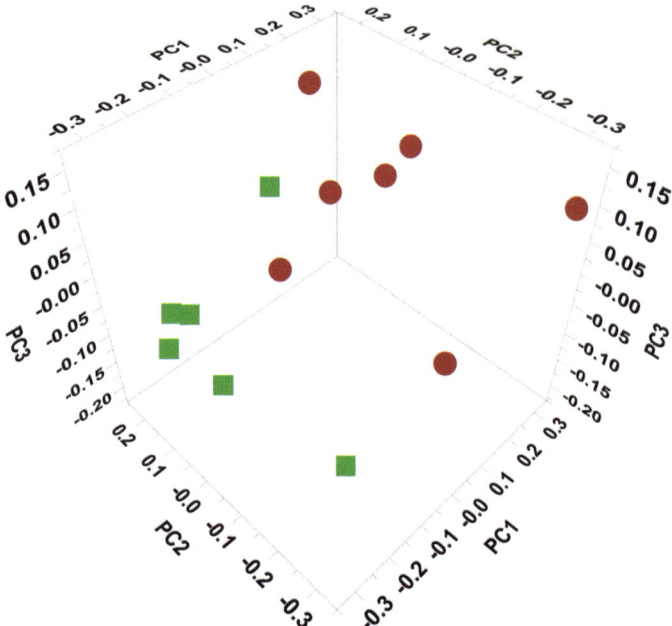

Fig. 9.3 An example of a 3D scores plot of a PCA model. Each score is a linear combination of a complete set of variables according to (9.15). Two classes of scores are represented with two different colors and shapes

Three PCs are three axes of the new coordinate system, and each observation (each sample) has three scores that define the position of the observation in the 3D PCA scores plot (Fig. 9.3). For example, sample m has scores of t_{m1}, t_{m2}, and t_{m3}, corresponding to PC1, PC2, and PC2 components, respectively.

To illustrate the meaning of the loadings matrix $\mathbf{P}$, the matrix is rearranged by multiplying both sides of (9.5) by $\mathbf{p}_i$ from the right:

$$X(p_1 \quad p_2 \quad \cdots \quad p_k) = (t_1 \quad t_2 \quad \cdots \quad t_k) \begin{pmatrix} p'_1 \\ p'_2 \\ \vdots \\ p'_k \end{pmatrix} (p_1 \quad p_2 \quad \cdots \quad p_k)$$

$$+ \mathrm{E}(p_1 \quad p_2 \quad \cdots \quad p_k) \qquad (9.11)$$

Because loadings matrix $\mathbf{P}$ is orthonormal ($\mathbf{p}'_i \mathbf{p}_j = 0$ for $i \neq j$, $\mathbf{p}'_i \mathbf{p}_j = 1$ for $i = j$), scores vector $\mathbf{t}_i$ can be expressed mathematically as

$$t_i = X p_i \qquad (9.12)$$

in which for $i = 1, 2, \ldots k$,

9.2 Fundamentals of Multivariate Statistical Analysis for Metabolomics

$$p_i = \begin{pmatrix} p_{1i} \\ p_{2i} \\ \vdots \\ p_{ni} \end{pmatrix} \qquad (9.13)$$

$$\begin{pmatrix} t_{1i} \\ t_{2i} \\ \vdots \\ t_{mi} \end{pmatrix} = \begin{pmatrix} x_{11} & x_{12} & \cdots & x_{1n} \\ x_{21} & x_{22} & \cdots & x_{2n} \\ \vdots & \vdots & \ddots & \vdots \\ x_{m1} & x_{m2} & \cdots & x_{mn} \end{pmatrix} \begin{pmatrix} p_{1i} \\ p_{2i} \\ \vdots \\ p_{ni} \end{pmatrix} \qquad (9.14)$$

In which subscripts 1, 2, ..., m represent observations or a total of m scores (for each principal component); subscripts 1, 2, ..., n represent n variables or n loadings (for each principal component). A total of n variables are used to describe an observation. In other words, one sample consists of n variables. For a PCA model with the first three principal components, $i = 1, 2, 3$, sample m has three scores and each score has n loadings, whereas one sample has n variables in the original data matrix **X**. This means that PCA reduces n dimensions of each of m samples in the original data matrix X to three dimensions in PCA coordinate system.

From (9.14), the scores of sample m on PC_i axis can be rewritten as:

$$t_{mi} = \sum_{j=1}^{n} x_{mj} p_{ji} \qquad (9.15)$$

The equation tells us that the scores of sample m, t_{mi}, are transformed from the original variables (NMR spectral bins) by the linear combinations of the variables with the corresponding loadings, p_{ji} that serve as coefficients (or contribution) of the transformation. Scores are the coordinates of an observation in the new coordinate system obtained by the PCA, whereas loadings of a score represent how the original variables in the data matrix X contribute to the scores.

Equation (9.15) also implies that a larger value of an original variable, x_{mj}, may contribute more to the model, even if its counterpart loading has a smaller value. It means that large variables weigh more variance in a PCA model. To reduce this drawback of PCA, the original dataset is usually preprocessed by scaling all variables. A scaling approach divides each variable with a scaling factor that is different for each variable:

$$\hat{x}_{ij} = \frac{x_{ij}}{f} \qquad (9.16)$$

In (9.16), subscript i represents the ith observation and j represents the jth variable, x_{ij} is an original variable (the jth variable for the ith observation), $\hat{x}_{ij}$ is a scaled variable, and f is a scaling factor. The scaling is used to reduce the importance of metabolites with high concentrations in a statistical model.

There is a variety of scaling methods used for metabolomics, among which Pareto scaling is a popular one. Pareto method scales each column (variable) of the dataset by the square root of standard deviation (Eriksson et al. 2006):

$$\hat{x}_{ij} = \frac{x_{ij}}{\sqrt{s_j}} \qquad (9.17)$$

in which standard deviation s_j is defined by

$$s_j = \sqrt{\frac{\sum_{i}^{m}(x_{ij} - \bar{x}_j)^2}{m-1}} \qquad (9.18)$$

The mean $\bar{x}_j$ is determined by

$$\bar{x}_j = \frac{1}{m}\sum_{i}^{m} x_{ij} \qquad (9.19)$$

Pareto scaling increases the relative importance of small values (low metabolite concentrations) while it maintains the original data structure relatively intact and keeps the model closer to the original measurement than unit variance (UV) scaling.

UV scales variables of the dataset by standard deviation (Jackson 1991). As a result, each variable (in each column of the data matrix) has a variance with value 1.0:

$$\hat{x}_{ij} = \frac{x_{ij}}{s_j} \qquad (9.20)$$

UV scaling makes all metabolites become equally important, and hence the changes in metabolite concentration are compared based on correlations. However, UV scaling amplifies the experimental errors and noise.

Other available scaling methods include level scaling (van den Berg et al. 2006) that scales the data variables with the mean value:

$$\hat{x}_{ij} = \frac{x_{ij}}{\bar{x}_j} \qquad (9.21)$$

and Poisson scaling (also known as square root mean scaling; Keenan and Kotula 2004) that scales variables with square root of the mean:

$$\hat{x}_{ij} = \frac{x_{ij}}{\sqrt{\bar{x}_j}} \qquad (9.22)$$

9.2 Fundamentals of Multivariate Statistical Analysis for Metabolomics

Log transformation (Baxter 1995) is a scaling method that scales data by log function:

$$\hat{x}_{ij} = \log(x_{ij}) \tag{9.23}$$

It reduces heteroscedasticity of the dataset (random variables), but it has difficulty to scale values with large standard deviations and zeros, and inflates experimental errors.

In order to remove the offset from the dataset, variables need to be mean-centered. Centering converts all metabolite concentration (NMR peak integrations) to the changes with respect to the mean instead of to zero of the concentrations.

$$\dot{x}_{ij} = \hat{x}_{ij} - \bar{x}_j \tag{9.24}$$

Shown in Fig. 9.4 are the influences of different scaling applications on a dataset. Data are centered after scaling is applied. Log transformation, leveling scaling, and UV scaling increase the small values of the data dramatically compared to the Pareto scaling. Pareto scaling keeps the original data structure while it reduces the large values by significant amounts. Hence, for most metabolomics studies, Pareto scaling is a suitable choice to scale the data prior to multivariate statistical analysis.

After a model is obtained, two quantities are used to represent the quality of a model, which are R^2 and Q^2. R^2 describes how well the model explains the dataset—the quality of the model, whereas Q^2 represents how well the model can predict—predictability of the model (Eriksson et al. 2006). For a two-component model, a high R^2 indicates that a large portion of the information generated by a model from the dataset is meaningful and a high Q^2 means that the data predicted by the model will match the original dataset well. By definition, R^2 is the fraction of the sum of squares of all variables explained by the model:

$$R_k^2 = \frac{\sum_{m,n}(t_{mk}p_{kn})^2}{\sum_{m,n}x_{mn}^2} \tag{9.25}$$

$$R^2 = \sum_k R_k^2 \tag{9.26}$$

in which $R^2{}_k$ is R^2 for the kth PC, t_{mk} and p_{kn} are scores and loadings. R^2 has a value from 0 to 1. Q^2 is the fraction defined by

$$Q_k^2 = 1 - \frac{\sum_{m,n}(x_{mn} - t_{mk}p_{kn})^2}{\sum_{m,n}x_{mn}^2} \tag{9.27}$$

and

$$Q^2 = \sum_k Q_k^2 \tag{9.28}$$

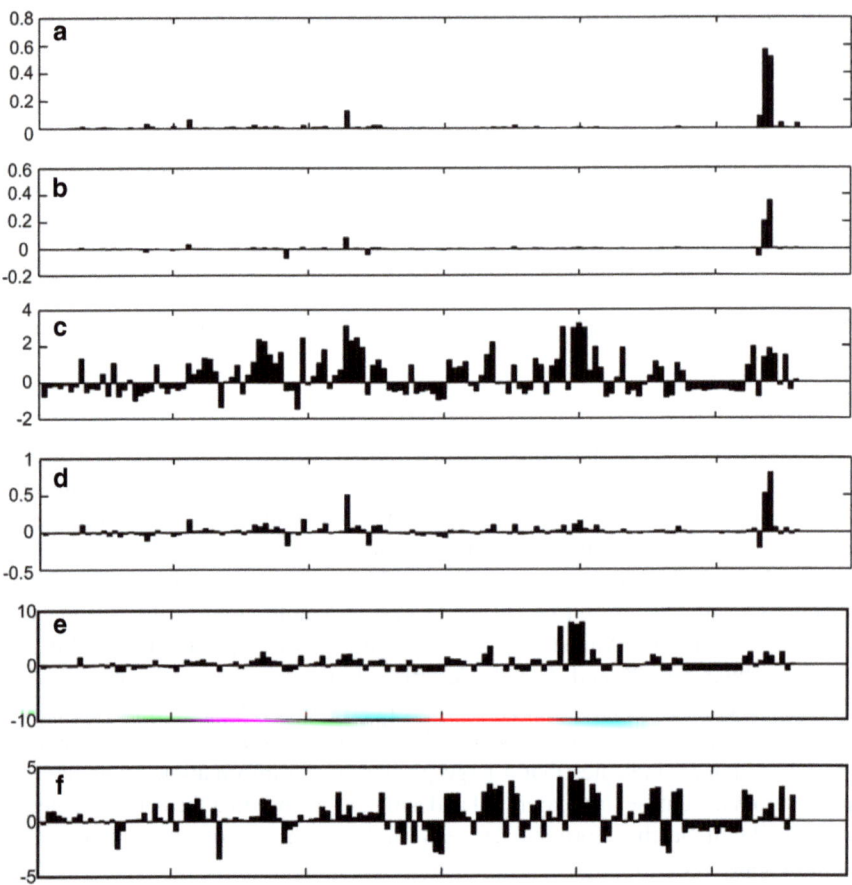

Fig. 9.4 Effects of different scaling techniques on original data. (**a**) Original data, (**b**) after centering, (**c**) UV scaling, (**d**) Pareto scaling, (**e**) level scaling, and (**f**) log transformation (reproduced with permission from van den Berg et al. (2006). Copyright © 2006 BMC Genomics)

in which Q^2_k is Q^2 for the kth PC. A low Q^2 indicates that the PCA model mostly describes noise and does not capture the true data structure.

In summary, PCA is used to identify the variance in the dataset by transforming high-dimensional data into fewer dimensions. The scores are the coordinates of the new "coordinate space" (the model), which are obtained by the linear combination of the original variables factored with the loadings (as coefficients). For a two component PCA model, each sample has two scores (one on PC1 and the other on PC2) and each score has the same number of loadings as that of the original variables. To reduce the influences of large value variables in the model, the dataset is preprocessed with scaling and then centered by the mean. Pareto scaling is best suitable for preprocessing the data because it retains the original data structure. After a model is generated, R^2 is used to describe how well the model explains the dataset—the quality of the model, and Q^2 represents how well the model can predict—predictability of the model. High R^2 and Q^2 represent a good model in both quality and predictability.

9.3 Sample Preparation

Sample preparation is a crucial step for any research project. This section describes detailed procedures on sample preparation, including proper sample handling of different biological samples (biofluids and tissues), extraction of tissue and biofluid samples, sample preparation for different experiments (solution NMR and high-resolution magic angle spinning [HR MAS]). Quenching and extraction of cell culture samples are also discussed.

Questions to be addressed in this section include the following:

1. How are biofluid samples collected?
2. How are the samples stored?
3. What are the procedures of NMR sample preparation for biological samples?
4. How are adherent mammalian cells cultured?
5. What are the procedures for quenching tissue samples and cultured cells?
6. How are cellular metabolites extracted from tissue and cell samples?
7. How are the lipophilic metabolites extracted from biofluids?

9.3.1 Phosphate Buffer for NMR Sample Preparation

Solution NMR samples for metabolomics are usually made of ^2H_2O phosphate buffer (PB). The PB has a final concentration of 100 mM with pH of 7.4. A standard practice is to first make a stock PB solution with higher concentration (e.g., 500 mM). A 500 mM stock PB solution at pH = 7.4 is formulated with 0.1140 g NaH_2PO_4 (monosodium phosphate, 0.95 mM) and 0.5751 g Na_2HPO_4 (disodium phosphate, 4.05 mM) in 10 mL 99.96% ^2H_2O (see Table 9.1). It is recommended that the stock PB is not stored in a refrigerator to avoid precipitation of the phosphates. For urine samples, the buffer solution also contains approximately 0.02% sodium azide (NaN_3, highly toxic) to prevent microbial contamination.

The ^2H_2O before use is required to be degassed to remove as much oxygen as possible, because oxygen can cause line broadening due to its paramagnetic property.

Table 9.1 Formula of phosphate buffer

pH value	Concentration (M)	Compound	Grams/10 mL	mM
7.2	0.5	NaH_2PO_4 (monosodium phosphate)	0.1680	1.40
		Na_2HPO_4 (disodium phosphate)	0.5112	3.60
7.4	0.5	NaH_2PO_4	0.1140	0.95
		Na_2HPO_4	0.5751	4.05

There are a few degas approaches. The simplest one is vacuum degas in which vacuum from a lab vacuum line is applied on a bottle of ^2H_2O for 10 min. The second common one is to place a unsealed, but capped bottle of ^2H_2O in a bath sonicator for 5 min. Caution should be taken to avoid the bottle neck contacting the water. This method is commonly used to remove excess gas (not only O_2) from mobile phase solvent for liquid chromatography (LC) or flow NMR. Thirdly, the traditional way for degassing is a flow-degas procedure that removes O_2 by blowing high-purity argon or nitrogen gas with a 0.1 μm filter into the ^2H_2O for 10 min (see Sect. 3.5.1). Among all these degas procedures, the sonication is most efficient and easiest to use, which requires minimum effort to operate.

9.3.2 Urine Samples

Urine of rodents is collected in wire-bottom cages or metabolic cages (shoebox cages) into tubes with 0.1 mL 1% sodium azide (NaN_3, highly toxic) solution to prevent metabolites in urine from bacterial degradation (Kurien et al. 2004). Note that the final concentration of the NaN_3 solution in the urine should be calculated based on the total volume of the collected urine during the preparation of NMR samples to ensure that the concentration of NaN_3 is not higher than 0.02%. Typically, the addition of NaN_3 will add approximately 10 mM Na^+ to the final NMR samples. The cages are washed after sample collection and at least once daily. There is a waiting period of 48 h before the urine collection is started in order to ease any stress that may be caused by the environment changes.

After collection of urine samples (ideally 250–500 μL), each sample is centrifuged at 3,000 × g for 10 min at 4°C and the supernatants is transferred directly into a cryogenic vial and then frozen in liquid nitrogen or dry ice. Samples are stored at −80°C or colder (Bernini et al. 2011). For preparing NMR samples, a urine sample is mixed with an equal volume of 200 mM PB to obtain a final volume of 220 μL for a 3 mm tube (600 μL for a 5 mm tube, or 150 μL for flow cell), and vortexed until sample is completely mixed (approximately 30 s). The initial concentration of the NMR phosphate buffer (PB) should be 200 mM in 99.8% ^2H_2O to yield the final phosphate concentration of 100 mM.

Because human urine is more diluted than that of rodents, more urine volume may be needed. For a 220 μL NMR sample, 176 μL human urine is mixed with 44 μL 500 mM PB in ^2H_2O in a labeled 2 mL microfuge tube (human urine: PB = 4:1). To achieve better water suppression and flat spectral baseline, 0.5 mL urine sample in a labeled 2 mL microfuge tube is lyophilized (or freeze dry) overnight or dried in a vacuum concentrator for 4 h. Next, 200 μL 100 mM PB is added into the microfuge tube. The solution is vortexed for 1 min and centrifuged briefly. A stock sodium 3-(trimethylsilyl) propionate-2,2,3,3-d_4 (TSP) solution is added to each sample as an internal reference with a final concentration of 0.5 mM. Note that the concentration of TSP may be optimized to yield a TSP NMR signal comparable to those of metabolites in the urine sample.

The final step of the sample preparation includes transferring each sample into an individual NMR tube after centrifuging the sample at 10,000 × g for 10 min at 4°C. For running the samples in flow mode, the samples are filled into a 96-well 0.45 μm filter plate (see Sect. 9.4.2) and centrifuged at 10,000 × g for 2 min at 4°C and transferred to a 96-well plate filled with 500 μL glass inserts.

9.3.3 Blood Plasma and Serum

Blood plasma is the liquid component of blood without blood cells, which is obtained from anticoagulated blood. Blood serum is the blood plasma without the fibrinogens (coagulation factor proteins). Serum is obtained from coagulated blood (clotted blood) and has all the proteins that plasma has, except those used in blood coagulation. Therefore, serum samples can be prepared for NMR experiments in the same way as for plasma after the anticoagulation process.

Blood samples are collected in labeled Vacutainer® (BD, NJ, USA) serum separator tubes without anticoagulant (Vacutainer® SST™ or Vacutainer® SST™ II plus) for serum isolation or Multivette® (Sarstedt™) Lithium-Heparin plasma tubes for plasma preparation (Deprez et al. 2002; Bernini et al. 2011). When a Vacutainer® tube is used for collecting serum samples, the tube is gently inverted five times. The blood collected without anticoagulant is clotted at room temperature for 30–120 min. For plasma preparation, Multivette® tubes containing blood were immediately placed on a roller for up to 35 min. The Vacutainer® or Multivette® tubes are then centrifuged at 1,500–2,000 × g for 15 min at 4°C. It has been suggested that both types of tubes do not introduce any impurity to the sample which generates interfering peaks in the ^{1}H NMR spectrum. Plasma preparations using collection tubes with such anticoagulants as citrate, EDTA (e.g., Vacutainer® SST™ K2E 5.4 mg), or an unspecified clotting activator should be avoided for NMR-based metabolomics because these types of anticoagulants give intense NMR signals that interfere with the resonances of metabolites in NMR spectrum.

After the centrifugation, the plasma (supernatant) or serum is carefully transferred into a cryogenic vial and either analyzed immediately or stored at −80°C. It is recommended that the samples should not be stored at −80°C for longer than 1 year to avoid the degradation of samples. For long-term storage, the cryotubes are stored in the vapor phase of liquid nitrogen in a liquid nitrogen dewar.

To prepare a plasma sample for NMR experiments, 200 μL plasma is mixed with 100 μL 300 mM PB with 0.6 mM DSS (or TSP) by vortexing. After they are centrifuged at 10,000 × g for 10 min at 4°C, the samples are transferred into individual 3 mm NMR tubes. For running the samples in flow mode, the samples are filled into a 96-well 0.45 μm filter plate (see Sect. 9.4.2) after the above centrifugation step and centrifuged at 10,000 × g for 2 min at 4°C, then transferred to a 96-well plate filled with glass inserts.

Similar to urine samples, plasma samples can also be dried to obtain a better water suppression and better spectral baseline. Plasma samples are first

Table 9.2 Biofluid extraction protocol

Preparation
1. Label and add bearing (3.2 mm) to Eppendorf tubes (SafeLock®). Note that bearing is used to facilitate the mix of solvents, not for breaking tissue mass
2. Place tubes in cold racks
3. Add 280 μL sample (serum, plasma, or urine) to each tube
4. Return to the cold rack
5. Mix ice cold solvents: methanol (HPLC grade), chloroform (HPLC grade), and deionized H_2O (dH_2O) in volume ratio of 4:4:2.85

Extraction of biofluid samples
6. Add 1.2 mL of the above solvent mixture into each tube
7. Place in a vortex adapter and vortex for 20 min
8. Centrifuge at 1,000 × g for 15 min at 4 °C. The solutions separate into an upper aqueous phase (polar metabolites) and a lower chloroform phase (lipophilic metabolites) separated by protein debris

Separate and dry samples
9. Using a pipette, carefully remove upper phase leaving a small amount behind (to avoid contamination) and dispense into a labeled 2 mL cylindrical screw-top tube
10. Using a gel loading tip, remove lower phase making sure not to bring any residual upper phase along
11. Dispense into a labeled 2 mL HPLC glass vial (do not use microfuge tube since they are not compatible with chloroform)
12. Dry samples in vacuum concentrator without radiant heat. Four to six hours is usually sufficient

deproteinized by adding 200 μL ice cold methanol (HPLC grade) into 200 μL plasma. After precipitation of proteins for 5 min, the samples are centrifuged at 10,000 × g for 10 min at 4°C. The samples are dried in a vacuum concentrator for 4 h and reconstituted with 200 μL 100 mM PB with 0.2 mM TSP.

Plasma and serum (or urine) samples can also be extracted into polar and lipophilic fractions using a two-phase extraction method. For the extraction, 285 μL biofluid sample is added to an Eppendorf Safelock® 2 mL tube. The samples are extracted with 0.8 mL of an ice cold solvent mixture made with chloroform (HPLC grade) and methanol (1:1 volume ratio). The upper phase is the aqueous phase containing the water soluble metabolites, while lipophilic metabolites are in the lower chloroform phase. Table 9.2 provides a detailed protocol for the extraction. After the extraction, 250 μL $CDCl_3$/Methanol-d_6 (2:1 volume ratio) is added to lipophilic samples. The lipophilic samples are then vortexed for 1 min. and transferred into 3 mm tubes for NMR analysis. Aqueous samples are prepared with 220 μL 100 mM phosphate buffer solution and transferred into 3 mm tubes or 150 μL to 500-μL glass inserts on a 96-well plate for NMR analysis.

9.3.4 Tissue Sample Quench, Storage, and Extraction

For a metabolomics study, it is crucial to instantly stop the metabolic activities of biological samples and to prevent degradation of the samples. The most common step

9.3 Sample Preparation

Table 9.3 Tissue or cell extraction protocol

Preparation and weighing tissue samples
1. Label and add bearing (3.2 mm) to Eppendorf tubes (SafeLock®)
2. Weigh tubes and record (skip steps 2 and 5 for cell extraction)
3. Place tubes in cold racks
4. Add sample to each tube
5. Weigh again and record
6. Return to the cold rack

Extraction of tissue samples
7. Add 400 µL ice cold methanol (HPLC grade) and 85 µL ice cold deionized H_2O (dH_2O) to each tube with 100 mg or less (if weight is >100 mg, adjust volumes accordingly)
8. Place in tissuelyser and run for 10 min at 15 s^{-1} (vortex can be used, instead of tissuelyser in steps 8, 12, and 14)
9. Remove and centrifuge at 1,000 × g briefly (~1 min) at 4 °C to bring solvents down from cap
10. Place tubes back into cold racks
11. Add 200 µL ice cold chloroform (HPLC grade)
12. Place in tissuelyser and run for 20 min at 15 s^{-1}
13. Centrifuge briefly again
14. Add 200 µL ice cold chloroform and 200 µL ice cold dH_2O and run the tissuelyser for 10 min at 15 s^{-1}
15. Centrifuge at 1,000 × g for 15 min at 4 °C. The solutions separate into an upper methanol:water phase (polar metabolites) and a lower chloroform phase (lipophilic metabolites) separated by protein debris

Separate and dry samples
16. Using a pipette, carefully remove upper phase leaving a small amount behind (to avoid protein contamination) and dispense into a labeled 2 mL screw-top cylindrical tube
17. Using a gel loading tip, remove lower phase making sure not to bring any protein debris along
18. Dispense into a labeled 2 mL HPLC glass vial (do not use microfuge tube since they are not compatible with chloroform)
19. Dry samples in vacuum concentrator without radiant heat. Four to six hours is usually sufficient

for quenching samples is to instantly freeze the samples in liquid nitrogen. The quenched samples are usually stored in dry ice or $-80°C$ freezer for short term and in liquid nitrogen vapor or $-80°C$ freezer for long term. It is worth noting that for storage in a liquid nitrogen dewar, samples are stored in the vapor phase, not inside the liquid nitrogen, and cryogenic vials are used. When samples must be stored inside the liquid phase (for instance, due to limited space in the dewar), a cryotube is used to seal the vials to prevent samples from suddenly erupting when taken out of the dewar.

The quenching step starts with collecting approximately 100 mg of tissue sample, immediately placing it in a labeled 2 mL cryogenic vial and flash-freezing it in liquid nitrogen. The samples are then subjected to a dual phase methanol/chloroform extraction for polar and lipophilic metabolites (Table 9.3; Ekman et al. 2008; Viant 2007). Each tissue sample (100 mg or less) is transferred in a labeled Eppendorf Safelock® 2 mL tube with a stainless steel bead (e.g., BioSpec 3.2 mm) in a cold rack. The bead is used for tissue disruption in the next step. The first step of the extraction is to break down the tissue samples using a tissuelyser. After adding 400 µL ice cold methanol (HPLC grade) and 85 µL ice cold deionized water (dH$_2$O, HPLC grade) to each tube with 100 mg or less tissue (if the weight is greater than 100 mg, adjust volumes accordingly), the tubes are placed in the tissuelyser and grinded at 15 s^{-1} for 15 min. After samples are centrifuged briefly at 4°C (1,000 × g or 3,200 RPM for 1 min), 200 μL ice cold chloroform (HPLC grade) is added to each sample. The samples are again grinded at 15 s^{-1} for 20 min. After briefly centrifuging, 200 μL ice cold chloroform and 200 μL ice cold dH$_2$O are added to each sample, and grinded for last time at 15 s^{-1} for 15 min. The last step is to centrifuge the samples at 1,000 × g for 15 min at 4°C.

After the last centrifuge, each sample tube has two phase extracts. The upper phase is the aqueous phase and contains the water soluble metabolites, while lipophilic metabolites are in the lower organic phase. Proteins and other large molecules are precipitated in the solution and trapped as a thick film in the middle layer between the aqueous and organic phases. The upper aqueous phase is carefully removed using a pipette leaving a small amount behind to avoid protein contamination and dispensed into a labeled 2 mL screw-top cylindrical tube (e.g., Sorenson 12980, Sorenson BioScience), while the lower phase is removed using a gel loading tip and dispensed into a labeled 2 mL HPLC glass vial (note: microfuge tubes are not compatible with chloroform). The samples are dried at room temperature in a vacuum concentrator for 4 h. Each polar sample is reconstituted with 220 μL 100 mM PB with 0.1 mM TSP and transferred into each individual 3 mm NMR tubes after centrifuged at 10,000 × g for 10 min at 4°C. For preparing lipophilic samples, each sample is reconstituted with 250 μL methanol-d$_4$ and CDCl$_3$ mixture (volume ration 1:2) with 1.5 mM TMS.

9.3.5 Culture Adherent Cells

Cultured cell lines from human and model animals have, for a long time, been used for omics research. However, not until recently, adherent tissue cultures have been used for metabolomics research. Cell lines can be cultured either in suspension or in an adherent matrix. This section only includes culturing adherent cells. Adherent mammalian cells are available from tissue culture banks (e.g., American Type Culture Collection; http://www.atcc.org) and can be grown on the surface of culture ware (flask or dish) and microcarriers (plastic beads in the size of a few hundreds of micrometers). The plastic surface is usually treated (called tissue culture treated) to increase the binding of cell membrane proteins to the surface.

There is a variety of culture media for growing mammalian cells lines, of which Dulbecco's Modified Eagle Medium (DMEM) is the most common one. DMEM has different formulations, including low or high glucose, with or without pyruvate, glutamine, bicarbonate, or phenol red (phenolsulfonphthalein). For culturing cells, the medium is also supplemented with 100 units/mL penicillin, 100 μg/mL streptomycin, 10% heat-inactivated fetal bovine serum (FBS), and 25 mM HEPES (for buffering pH). For some hard-to-grow cell lines culture medium is also supplemented with 50 μg/L mouse epidermal growth factor (EGF). Usually, the culture medium is changed twice weekly.

Because the mammalian body contains 5% CO_2, the cells are cultured under 5% CO_2 at 37°C. Note that the CO_2 percentage of a cell culture incubator should be calibrated monthly. Since CO_2 is dissoluble in aqueous solution to form bicarbonate, resulting in pH change, the medium usually contains sodium bicarbonate at a concentration of 3.7 g/L to buffer its pH. When CO_2 is not used such as for fish cell lines, culture media contain much less sodium bicarbonate (0.15 g/L). Phenol red is used as a pH indicator for the medium. It becomes yellow when pH is <6.5 and pink at pH approximately 7.4 (Freshney 2005). When cells consume glucose to produce lactate or acetate for the energetic needs, the acidification of medium turns phenol red into yellowish. Additionally, if the medium has not been changed during a period of 3 days, the color of the medium changes to pink, indicating that nitrogenous byproducts are accumulated in the medium and the cells are under oxidative stress.

To illustrate how adherent cells can be cultured, the following examples are described. Details on culturing mammalian cells can be found in several excellent books (Freshney 2010; Davis 2002). In the first example, human brain cancer (LN229) cells were grown as monolayer cultures in a T75 (75 cm^2) tissue culture-treated flask with 10 mL low glucose DMEM supplemented with 100 units/mL penicillin, 100 μg/mL streptomycin, 25 mM HEPES, and 10% FBS at 37°C in a humidified atmosphere of 5% CO_2. The medium was renewed twice weekly.

The cells can be expanded into additional flasks by stripping the cells from growth surface using 0.25% trypsin solution with 0.01% EDTA. Prior to the addition of 2 mL trypsin solution, medium was removed and the flask was washed twice with phosphate-buffered saline (PBS). After reacting for 5 min at room temperature, the enzyme was inactivated by adding 2 mL culture medium into the trypsin solution. The cell solution was transferred into a 15 mL centrifuge tube and centrifuged at 2,500 RPM for 2–3 min. After removing the medium, 5 mL fresh medium was added into the tube and the cell solution was transferred into the T75 flasks. Next, 10–12 mL medium was added into each individual flask. The cells were incubated in a humidified atmosphere of 5% CO_2.

In the second example, zebrafish liver (ZFL) cells were grown at 28°C in a humidified atmosphere. In this case, CO_2 was not used. Because of that, the concentration of sodium bicarbonate (NaH_2CO_3) is reduced to 0.15 g/L. Therefore, regular DMEM (with 3.7 g/L NaH_2CO_3) is not suitable for growing ZFL cells. The medium for ZFL cells is made of high glucose DMEM (no sodium bicarbonate) supplemented with 100 units/mL penicillin, 100 μg/mL streptomycin, 15 mM HEPES, 0.15 g/L NaH_2CO_3, 50 μg/L mouse EGF, 10 mg/mL bovine insulin, and 10% FBS. Note that high glucose is used for ZFL cell culture.

Cells can be grown on microcarriers. For instance, MCF7 human breast cancer cells are first grown in a T75 flask with DMEM in a humidified atmosphere of 5% CO_2. At approximately 80% confluence, the cells are stripped with trypsin and reseeded back to the flask with 1 g sterile microcarriers. The flask is gently shaken every 15 min during the initial 2 h. After growing for 24 h, cells are trypsinized again and reseeded back to the flask with the microcarriers. The cells are allowed to grow for 2 days after shaken every 15 min during the initial 2 h. Note that petri dishes (or non-treated dishes) are not recommended for culturing cells on microcarriers. When petri dishes are used, cells will not attach to the surface of the dishes, which might promote cells to grow on the microcarriers. However, not all of the cells anchor on the microcarriers. The suspension cells will consequently die and produce toxic metabolites. These dead cells are difficult to be separated from the microcarriers.

9.3.6 Quench and Extract Cells

One of the advantages of metabolomics using cell cultures instead of live animals is higher throughput. Although effective extrapolation to whole animal responses is ultimately required, such an approach provides obvious advantages. For example, there is no need to house and sacrifice animals, costs are significantly lower, and cells can be grown and studied rapidly. Also, human cell lines can be employed in order to avoid cross-species extrapolations.

Despite these advantages, owing to some barriers to practicality, conventional cell culture assays for macromolecules such as DNA and proteins are not amendable for cellular metabolomics. Perhaps the most serious of these barriers are related to the conventional method for quenching cell cultures. Preferably, a quenching method would instantly inactivate all cellular metabolism, without first changing the cell environment, since metabolite concentrations are sensitive to any change of the cell environment (de Koning and van Dam 1992; Villas-Boas et al. 2007). Also, a cell quenching method should be rapid, efficient, and reproducible so as to allow an unbiased measurement of cellular metabolite concentrations and to enable direct comparison of a large number of biological samples. Unfortunately, the conventional method for quenching adherent cell cultures does not meet the above requirements.

During the conventional cell quench, first, trypsin is applied to remove cells from their growth surface (Fig. 9.5a). This inevitably changes the profile of cellular metabolites since the enzyme severely alters the physiological state of cells due to its interaction with membrane proteins (Fig. 9.5c). Furthermore, there are no nutrients and supplements available in the trypsin solution, which are present in the culture medium. After the trypsinization, numerous time-consuming steps (e.g., wash and centrifugation) are required in the conventional method before cells are finally quenched or fixed. The total time for the double-wash process is 45–60 min. During this time, a considerable portion of intracellular metabolites are secreted from cells due to their small molecular size, fast turnover rates (Villas-Boas et al. 2007), and the different osmotic strength of the applied solutions (Britten and McClure 1962; Smeaton and Elliott 1967). As a result, the conventional method requires 10^7–10^8 cells in order to obtain suitable concentrations of intracellular metabolites for NMR analysis (Beloueche-Babari et al. 2010; Lane and Fan 2007;

9.3 Sample Preparation

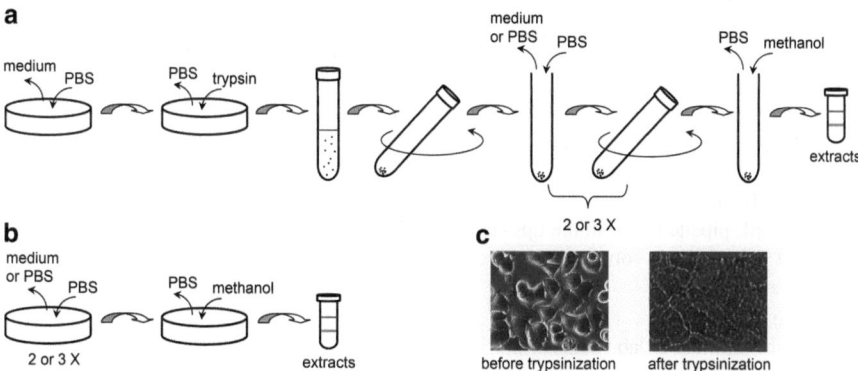

Fig. 9.5 Cell quench procedures. (**a**) In a conventional quenching method, cells are first removed from culture dish (or flask) surfaces by applying trypsin to obtain cell suspensions. Then, cells undergo a series of wash and centrifuge processes. The wash/centrifuge steps usually take approximately 45–60 min. The cells are finally quenched with methanol (or other quench solutions) and extracted. (**b**) In the direct cell quenching method, trypsin is not applied to the cells. After the culture medium is removed, cells are quickly rinsed by spraying with PBS. Residual PBS solution is removed by vacuum. The cells are then immediately quenched by methanol. The procedure takes less than 10 s. Both cells and methanol solution are collected for the cell extraction. (**c**) Images of MCF-7 cells before and after trypsinization. The images of the cells are significantly changed after trypsinization, reflecting that the trypsin severely changes the physiological state of cells due to its interaction with membrane proteins (reproduced with permission from Teng et al. (2009), Copyright © 1996 Springer)

Yang et al. 2007a, b). As a final deterrent, the conventional method is simply too time consuming to process the cell samples for metabolomics studies.

These drawbacks of the conventional method for quenching adherent cell cultures has been overcome by a direct cell quenching method that is efficient and accurate for cellular metabolomics (Teng et al. 2009). Provided in Table 9.4 is a step-by-step protocol for the direct quench (Fig. 9.6). During the direct quench, the culture medium is first removed from the 6 cm tissue culture dish (Fig. 9.5b). If extracellular metabolic profiles will also be used for the metabolomics study, 0.5 mL culture medium is collected before the medium removal. After the cells on the grown surface (culture dish surface) are flush-washed with ice cold PBS, the residual PBS is removed by vacuum. First, a 200 μL pipet tip is connected to a vacuum tubing and the vacuum line is turned on. To vacuum PBS, the dish is tilted at 45° and the residual PBS accumulated at the bottom edge of the dish is vacuumed using the tip. Cells are then quenched with 650 μL 80% methanol (HPLC grade, room temperature). In our hands, the entire quenching process (from the removal of the medium to the addition of methanol) is completed in less than 10 s, compared to approximately 45 min needed in the conventional method. Next, cells are gently detached from the culture dish using a cell lifter without crushing the cells on the dish surface. The methanol solution containing the quenched cells is then pipetted into a labeled Eppendorf safe-lock 2 mL tube in a cold rack, which contains a stainless steel bead for extraction.

Table 9.4 Cell quenching procedure for metabolomics

This procedure requires two people in order to be completed effectively and efficiently

Preparation
1. Obtain the following supplies and put them in a fume hood
   - Disposable 1 mL pipette tips
   - 1 mL pipette
   - Cell lifters
   - 100 µL pipette tips with the tips cut off by a razor
   - 80 % MeOH in a solvent dispenser
   - Waste container
   - Paper towels
   - Tube attached to lab vacuum
   - Ice bucket
   - PBS in a spray bottle
   - Paper towel rolled into a tube to support culture dish
   - Cold racks
2. Place the PBS spray bottle into the ice bucket
3. Label Eppendorf tubes (SafeLock®) with the cell line used, compound used, and sample class (e.g., "control," "low," or "high") and add a bearing (3.2 mm) to each labeled tube
4. Label microfuge tubes, if medium is to be collected
5. Place tubes into a tube rack labeled with the cell line, compound used, and date quenched

Quenching

Quencher's role
6. Note: samples should be quenched in a pseudo random order to reduce systematic error
7. Attach a cut-end 100 µL pipette tip to the vacuum hose attached to lab vaccum and turn the vacuum on
8. Take one dish from the incubator
9. If taking medium
   a. Pipette 500 µL of medium into one labeled microfuge tube
   b. Cap the tube and place it into the cold rack
10. Pour all remaining medium in the dish into the waste container and tap the dish quickly two to three times on the paper towels to remove the remaining medium
11. Holding the dish at an angle over the waste container, and spray PBS onto the dish (Fig. 9.6a)
12. Tap the dish on the paper towel to remove the remaining PBS
13. Use the pipette tip attached to the vacuum to thoroughly remove the remaining PBS from the dish (Fig. 9.6b). Vacuum only from the edges. Do not place the pipette tip directly in the middle of the dish as this will disrupt the cells
14. Dispense 650 µL of 80 % MeOH into the center of the dish using the solvent dispenser and swirl the MeOH quickly around the dish until the dish surface is fully covered by MeOH (Fig. 9.6c). Note: the quench steps (10–14) should be completed within 10 seconds
15. Angle the dish and using the flat end of the cell lifter gently scrape down the bottom of the dish to collect the cells in the pool of MeOH formed from angling the dish (Fig. 9.6d)
16. Rotate the dish 90° and continue to scrape the dish's bottom until the bottom is clear and free of cells
17. Place the dish on an angle on the paper towel tube for the sample collector to collect the cells (repeat the above steps for all samples)

(continued)

Table 9.4 (continued)

Collecting quenched cells

Sample collector's role

18. Gently pick up the dish placed on the paper towel roll by the Quencher
19. Using a 1 mL pipette set on 250 µL, transfer the contents (cells and methanol) of the dish to the labeled Eppendorf tube. This step will take multiple attempts. Make sure to transfer all the cell fragments from the dish to the tube
20. Place the tube on the cold rack
21. After the entire quenching process has completed, extract the samples immediately

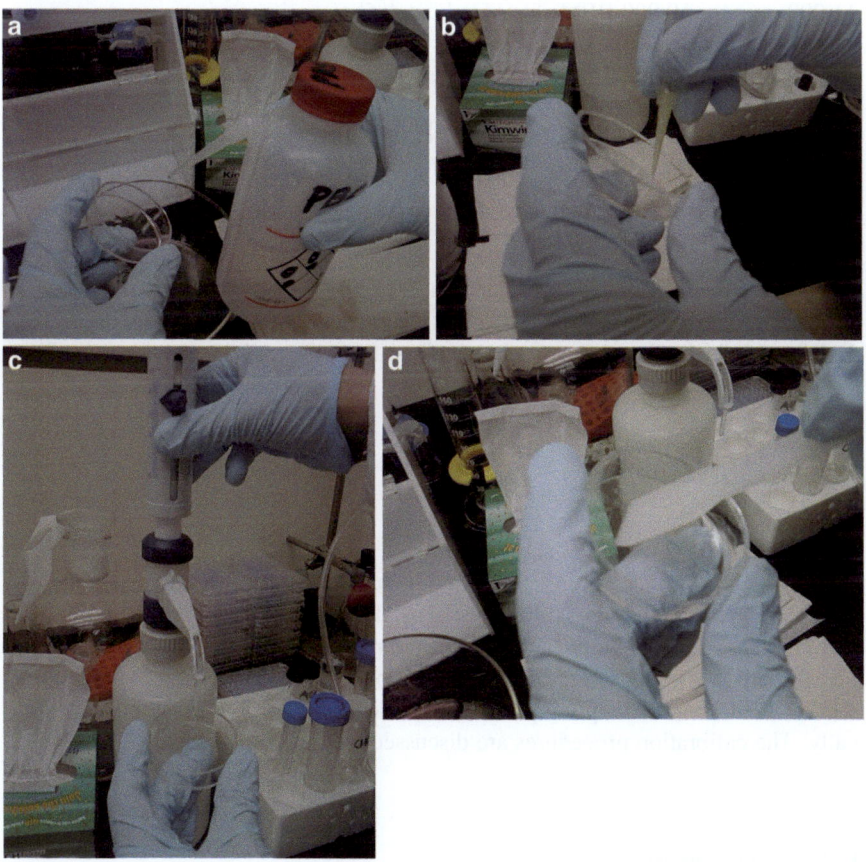

Fig. 9.6 Key steps of the direct cell quench method. (**a**) To flash wash the cells, hold the dish at an angle over the waste container, and spray PBS onto the dish. (**b**) Use the pipette tip attached to the vacuum to thoroughly remove the remaining PBS from the dish. (**c**) After dispensing methanol into the center of the dish using a solvent dispenser, swirl the methanol solution quickly around the dish until the dish surface is fully covered by methanol. (**d**) To collect the cells in the pool of methanol, angle the dish and gently scrape down the bottom of the dish using the flat end of the cell lifter

For an NMR measurement using a cryogenic probe, aqueous extract of 6×10^5 cells will yield a good signal-to-noise ratio with 512 scans. For a conventional NMR probe, it is recommended to grow 2×10^6 cells (in a 6 cm dish) for an NMR sample. The intracellular metabolites are extracted using a two-phase extraction procedure similar to the one described in the Sect. 9.3.4. Briefly, the cells are disrupted in tubes with stainless steel beads by a tissuelyser shaken at 15 s^{-1} for 15 min. After samples are centrifuged briefly at 4°C (1,000 × g or 3,200 RPM for 1 min), 240 μL ice cold chloroform is added into each sample. The samples are grinded again at 15 s^{-1} for 20 min. After briefly centrifuged, each sample is added with 240 μL ice cold chloroform (HPLC grade) and 220 μL ice cold dH$_2$O and grinded at 15 s^{-1} for 15 min. The last step is to centrifuge the samples at 11,000 × g (or 10,000 RPM) for 15 min at 4°C.

9.4 Practical Aspects of NMR Experiments

This section covers the practical aspects of the experiments for metabolomics studies, including the following:

1. How are the instruments calibrated and/or setup?
2. What are the routine NMR experiments for metabolomics and how are they set up?
3. What kind of information can be obtained from the individual experiment?

9.4.1 Calibration

There are a few routine 1D experiments and 2D ^{1}H/^{13}C experiments for metabolomics. Before setting the experiments, the probe should be tuned and then calibrated, including ^{1}H 90° pulse, ^{13}C decoupler 90° and decoupling pulses (for 2D ^{1}H/^{13}C experiments), and variable temperature (VT). A ^{1}H 90° pulse should be calibrated for every sample or a set of samples if samples are run in automation, while ^{13}C 90° and decoupling pulses as well as VT control are calibrated periodically. The calibration procedures are discussed in details in Chap. 4.

9.4.2 Automation

A metabolomics study usually includes a large number of samples. For a certain study, it may consist of more than 1,000 samples. It requires a large amount of time and persistence to run all the samples manually. Manual operations may also cause variations in the data. Therefore, the optimum configuration of the NMR instrument for metabolomics is to have automatic sample handler, either for tubes

9.4 Practical Aspects of NMR Experiments

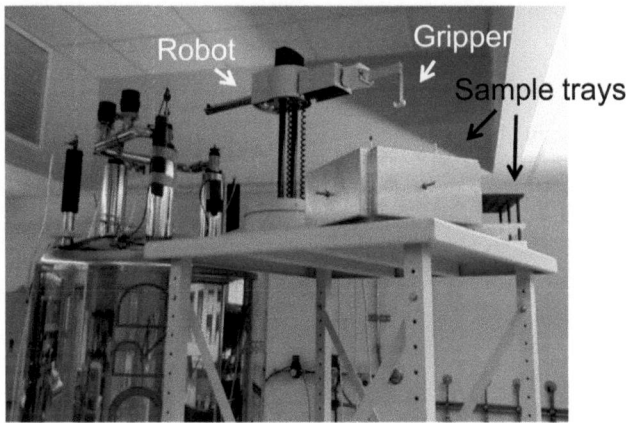

Fig. 9.7 A 100-sample robotic autosampler. The robot grips the sample holder of an NMR tube and moves it from sample trays. It returns the sample back to its original well after NMR analysis

or in flow-through mode. A sample management robot (Fig. 9.7) allows tube submission for up to 100 samples. The left rack has a home-made cooling bath to keep queuing samples at 4°C which requires an FTS precooling system. While the 100-tube robot system does a good job to run samples continuously it has several drawbacks, including that a sample has to be transferred into an NMR tube, which requires a tube washing procedure. It is time and resource consuming to wash hundreds of NMR tubes. Also, the magnetic field homogeneity (shimming) changes from sample to sample. Precaution should also be taken to see if every tube has the same sample volume. Identical sample volume before samples is transferred into NMR tubes may not always yield the identical volume in all tubes due to the sample loss during the sample transfer procedure. There are also chances that sample tubes may be broken or the robot fails to grab samples during tube ejection due to technical issues. The chance of the failures increases with the number of samples. The solution to these drawbacks is flow-through automation.

A typical flow configuration requires a flow cell in the NMR probe and a liquid sample handler to transfer samples from wells of a 96-well plate (usually HPLC glass inserts) to the NMR probe (flow cell) via polyether ether ketone (PEEK) tubing (Teng et al. 2012). The size of flow cells range from 60 to 200 μL. The smaller a flow cell volume, the more difficult the calibration. Shown in Fig. 9.8 is a Gilson® liquid sample hander (LSH) that can hold five racks. Each rack has two 96-well plates and can be cooled at 4°C. With 96-well format, tissue or cell culture samples can be extracted in much higher throughput. Biofluid samples can also be prepared in 96-well plate format. Despite the advantages of the flow NMR, there are a few challenges before flow automation can be used for NMR data collection. For example, the thin HPLC tubing for transferring sample is easy to be clogged. Inevitably, there is sample diffusion at the front end of the sample, causing dilution of sample concentration. The sample diffusion occurs inside the flow cell due to the

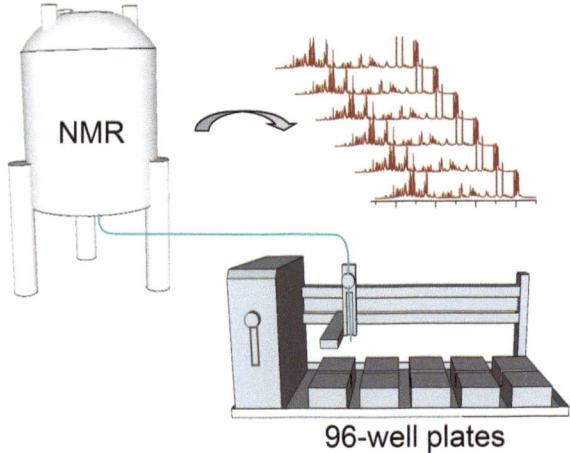

Fig. 9.8 Schematic representation of flow (or direction injection) NMR automation. The Gilson® liquid sample handler takes a sample from a specific well of a cooled 96-well plate and injects it into the probe. After NMR analysis is done, the sample is transferred back to its original vial on the 96-well plate, and the flow cell is completely washed before the next injection (reproduced with permission from Teng et al. (2012), Copyright © The Royal Society of Chemistry, 2012)

push pressure, where the contact surface is between wash solvent (^2H_2O) and sample solution, and hence dilutes sample concentration. Last, the most serious drawback to flow NMR is the well-known carryover contamination that the residual sample is retained in the flow cell during the wash process, which causes cross contamination between samples.

All of the above problems with the flow NMR have been solved recently. Sample filtration is utilized to avoid clogging the sample transfer line. The samples are passing through 0.45 μm filters in 96-well format (e.g., Millipore HTS) while spinning at $10,000 \times g$. Because metabolomics deals with small molecules, the filtration does not alter the metabolic profile observed by NMR spectroscopy. The carryover problem is overcome by a "liquid brush" and push-through procedure integrated into the wash step (Fig. 9.9). During the brush wash, the wash solvent is first pushed through, then pulled back to clean both ends of the flow cell (called delta regions) and double volume wash solvent is then pushed through. The brush step is repeated a couple of times and finally all wash solvent is pushed through the entire tubing into a waste bottle. The use of compressed air before sample injection serves two purposes: push wash solvent into waste without using ^2H_2O (meaning less cost per sample) and eliminate the diffusion since the sample will not easily diffuse into the air phase because of the thin size of the transfer tubing (inner diameter 1/16 in.).

Calibrations of the effective probe volume and volume between the injection needle to the front end of the flow cell (push solvent volume) is the first step for configuring a flow automation. The calibrations should not change as long as the tubing stays the same. The actual sample volume is about twice the size as the flow

9.4 Practical Aspects of NMR Experiments

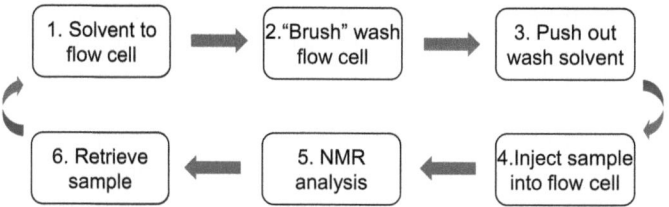

Fig. 9.9 Flow chart of push-through (PT) DI NMR automation. Before a sample in an automation queue is injected into the flow cell, it is necessary to wash the entire flow system thoroughly. The first step is to push wash solvent into the flow cell. Next, the flow cell is washed by a push–pull "brush" routine. The last step is to empty the wash solvent out of the flow line by blowing compressed air. The sample is then injected into the flow cell and the NMR data is collected. After NMR analysis is complete, the sample is retrieved and the above wash routine is repeated

cell volume to fill the space at the ends of the flow cell in order to ensure good shimming results. During the injection procedure, the sample is needed to fully fill all of the flow cell volume. Therefore, the volume for the push solvent must be properly calibrated, which equals to the volume of the tubing from the Gilson® sample syringe to the front end of the flow cell. To calibrate the push volume, first ^2H_2O is manually injected into the flow line to fully fill the flow cell. Then, the probe is properly tuned, the field is shimmed, and the lock level is adjusted to approximately 80%. Next, ^2H_2O with a volume 30 μL more than the volume of the flow cell (90 μL for a 60 μL flow cell) is injected into the probe pushed by H_2O using automation (0.1 μm-filtered and deionized H_2O), while monitoring the lock level change during the injection. If the lock increases to the maximum then decreases, then push volume is larger than the optimum value. If the lock level was increasing during the entire injection but it is not quite close to lock level of the manual injection, then push volume needs to be increased.

The ^2H lock level method discussed above works well for the calibration of push solvent volume. However, it will be convenient if the movement of the sample inside the flow cell can be visualized. For a cryogenic probe, the flow cell can be removed out of the probe. Using a colored sample, the calibration is very straightforward. A red food dye can be used for this purpose. The sample is prepared with a drop of red food dye in 10 mL deionized H_2O and subsequently filtered with a 0.1 μm syringe filter. The sample is injected into the flow cell by a predefined volume of push solvent (filtered H_2O) while observing the movement of the red sample inside the flow cell. A color gradient in either the outlet (front end) or the inlet end of the flow cell indicates that the volume needs to be increased or decreased, respectively. The volume of push solvent is adjusted accordingly until a uniformed color sample in the flow cell is obtained repeatedly.

The probe volume and push volume can be fine-tuned by injecting a sample consisting of 1% H_2O/99% ^2H_2O by push solvent ^2H_2O. Because the signal of H_2O is used for this calibration, the probe is tuned and the sample is shimmed again. After the ^1H 90° pulse is calibrated and gain is properly adjusted, the H_2O signal is observed by a one-pulse experiment. Probe volume and push volume are adjusted to optimize H_2O signal. The Auto-gain setting should not be used because it changes the absolute intensity of H_2O signal and in turn affects the accuracy of the calibration. Once the probe volume (sample volume plus push solvent volume) is calibrated, it does not need be recalibrated if the length and diameter of the tubing from the injection valve to the inlet of the flow cell and the volume of the flow cell are unchanged.

9.4.3 NMR Experiments

There are a set of experiments used routinely for NMR-based metabolomics, including PRESAT, 1D NOESY, PURGE, CPMG, T_1 and T_2 measurements, COSY, TOCSY, 2D J-resolved spectroscopy, gHSQC, gHMBC, and HR MAS. PRESAT, CPMG, T_1 and T_2 measurement, COSY, TOCSY, gHSQC, and gHMBC have been discussed in previous chapters. The setup of these experiments and parameters are similar to those discussed in previous chapters with a few exceptions. Because metabolomics deals containing samples with mixtures of small molecules (endogenous metabolites), it requires high spectral resolution. Consequently, the acquisition of a 1D experiment should be sufficiently long, usually longer than 2 s (2–5 s). For a heteronuclear 2D experiment, the t_1 increments need to be as many as 1,024 to ensure a sufficient resolution required for separating ^{13}C resonances of the metabolites. Other practical considerations include that an internal standard (DSS or TSP for aqueous samples, TMS for lipophilic samples) should be used and its line width should be less than 2 Hz (usually close to 1 Hz).

9.4.3.1 PRESAT

Arguably, PRESAT is the oldest and most popular experiment for solvent suppression. It is simple to setup with fewer parameters that need to be calibrated (see Sect. 4.8.1). The key to success for this experiment is to accurately calibrate the resonance frequency of the solvent (water in most cases) to be suppressed. The saturation frequency must be set to water resonance frequency. Usually, the carrier frequency is set to the same as the saturation frequency, in which case, the center of the spectrum is at the top of the water peak. The selectivity of the saturation is achieved by a long pulse with very low pulse power. A frequently cited drawback with PRESAT is that the sequence is ineffective to suppress the hump regions of the water signal, which originated from the sample outside the probe coil. Additionally, an asymmetric water peak will inevitably deteriorate suppression efficiency of the experiment since the saturation will be only applied to a narrow bandwidth (usually less than 0.1 ppm the spectral range). Therefore, samples should be well shimmed beforehand. Another drawback with PRESAT is the unsatisfactory baseline for samples with low concentrations. All of these problems can be improved by 1D NOESY experiment.

9.4.3.2 1D NOESY

A 1D NOESY (or NOESY1D) experiment is the same as the first increment of a 2D NOESY (Sect. 4.10.7) without an evolution time. Although NOESY is used to obtain the distance information of spins (^{1}H, ^{13}C, ^{15}N or other 1/2 spins), 1D NOESY is used here to achieve a better water suppression with a better spectral baseline compared to PRESAT experiment. NOESY pulse sequence is essentially the same as a WEFT (water eliminated Fourier transform; Patt and Sykes 1972) sequence that suppresses water signal by taking advantage of the difference in T_1 relaxation (spin–lattice relaxation) between solute and solvent nuclei. WEFT sequence is commonly used in paramagnetic NMR spectroscopy for fast data acquisition to suppress signals from diamagnetic nuclei that have longer T_1 relaxation times than paramagnetic nuclei.

Because water has a different T_1 relaxation time (usually longer) than metabolites, 1D NOESY can be used for metabolomics applications to obtain a better solvent suppression and flatter baseline than a PRESAT experiment. During the predelay period, the PRESAT pulse suppresses most intensity of the water signal. The residual water signal is further suppressed by the NOESY sequence (Fig. 4.21). The first two 90° pulses of 1D NOESY invert all magnetizations (solute and solvent) to the −z axis in a vector presentation. When metabolites (solutes) relax toward the +z axis (equilibrium state) during the mixing period, the residual water signal (mostly from the hump regions of the water peak) relaxes slower. Therefore, the mixing time can be optimized to achieve maximal sensitivity for metabolites and suppression for the water peak (when the residual water magnetization is near the origin in a vector frame).

Setup of a 1D NOESY experiment includes tuning, shimming, and calibrations of ^{1}H 90° pulse and water resonance frequency. The data is usually collected with a spectral window of 12 ppm, saturation delay of 2 s, mixing time 100 ms, acquisition time of 2–3 s, saturation frequency on water resonance, 16 steady-state scans (dummy scans), and transients of $n*32$. The saturation power should be weaker than that in a PRESAT experiment. The drawback of 1D NOESY is the reduced sensitivity compared to PRESAT because of the mixing time. A longer mixing time may further improve the spectral baseline, but it will lead to further loss of the spectral intensity. It is worth noting that the loss of spectral intensity is not equal for the metabolites in a sample, which is mixing time-dependent, because of different T_1 relaxation times. Therefore, if 1D NOESY is used in a metabolomics study in which the changes in concentrations of metabolites are measured, it is necessary to use the same mixing time for all samples, including standard samples. The NOESY experiment works very well for samples prepared with "100%" ^{2}H$_2$O. For samples containing relatively a large amount of water (e.g., urine samples), PURGE sequence is one of the options.

9.4.3.3 PURGE

Presaturation utilizing relaxation gradients and echoes (PURGE; Fig. 9.10a) was introduced as an alternative approach for solvent suppression (Simpson and Brown 2005), which is as simple to use as the PRESAT but yields similar results to those by more complex sequences such as Watergate and Water-flip-back experiments. The pulse sequence consists of spin echo block sequences and gradient pulses.

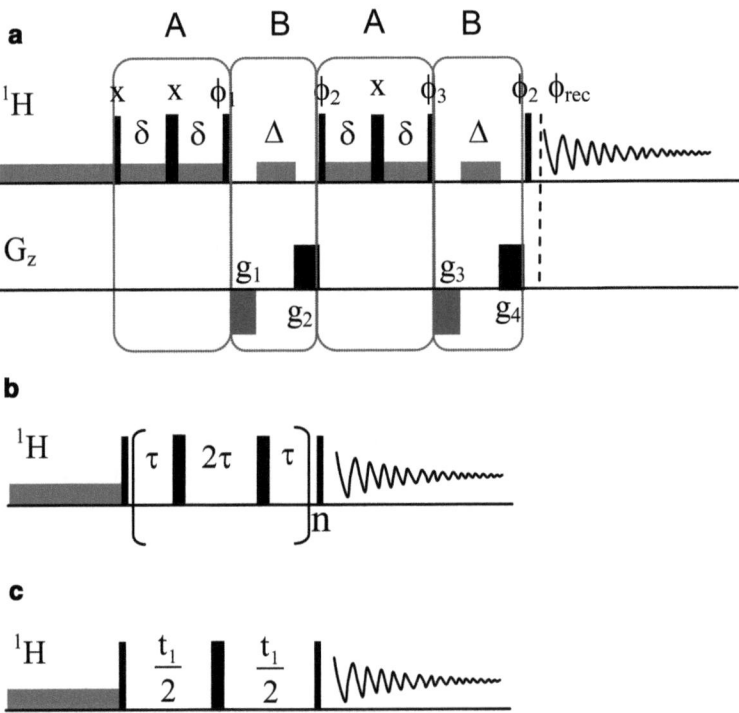

Fig. 9.10 NMR pulse sequences. ^{1}H represents ^{1}H transmitter channel, while G_z represents z gradients. For ^{1}H channel, narrow rectangles represent 90° pulses, wide rectangles are 180° pulses, and flat rectangles are water suppression pulses. Unless stated, ^{1}H pulses are applied along the x axis. (**a**) PURGE sequence (Simpson and Brown 2005). A and B represent the pulse segments. Phase $\phi_1 = x, x, -x, -x$; $\phi_2 = x, x, x, x, -x, -x, -x, -x, y, y, y, y, -y, -y, -y, -y$; $\phi_3 = x, -x, x, -x, -x, x, -x, x, y, -y, y, -y, -y, y, -y, y$; $\phi_{rec} = x, -x, -x, x, -x, x, x, -x, -y, y, y, -y, y, -y, -y, y$. Gradients G_1, G_2, G_3, and G_4 are set to $-7, 30, -9$, and 35 Gcm^{-1} with a duration of 1.0 ms. A typical setup of PURGE includes: calibrating ^{1}H 90° pulse length, setting both ^{1}H transmitter frequency and saturation frequency on water resonance, and setting saturation pulse power in the range of 40–100 Hz. Other parameters include 100 μs for δ, 200 μs for Δ, 2 s presaturation, and 2–3 s acquisition time. (**b**) CPMG sequence. The spin echo sequence is cycled n times to attain a desired T_2 relaxation delay time (or spin echo period, T). The number of cycles (n) must be an even number and is calculated according to spin echo period T. (**c**) J-resolved COSY sequence. The pulse sequence is basically a spin echo sequence, except that the spin echo delay is modulated as evolution time t_1

9.4 Practical Aspects of NMR Experiments

As its name implies, the idea of PURGE sequence is to use selective pulses during a spin echo period to tilt residual water magnetization out of a transverse plane, while the bulk magnetization of solutes is refocused by the spin echo. The 90° pulse at the end of the spin echo rotates the solute magnetization back to the z axis, whereas the residual water magnetization is brought to the transverse plane and dephased by the following z gradient pulse (G_1). Similarly, the bulk water magnetization is tilted away from the z axis by the second short saturation pulse (Δ) and again destroyed by gradient G_2. The sequence is repeated to improve the efficiency of solvent suppression.

Unlike the gradient pulses of Watergate or Water-flip-back sequence which are used to refocus solute and dephase water magnetization, the gradient pulses in PURGE are used exclusively for destroying the water magnetization that was tilted away from solute magnetization by the water selective pulse (long pulse on water resonance with the same pulse power as that of presaturation pulse). The first gradient (G_1 and G_3) after spin echo is used to dephase the magnetization of residual water, whereas the second gradient (G_2 and G_4) for dephasing bulk water magnetization. Therefore, strengths of G_1 and G_3 are much weaker than those of G_2 and G_4. Additionally, these two kinds of z gradients must be applied in different signs in order to avoid water magnetization being refocused. Each gradient pulse is optimized independently, meaning that it is not necessary to have a fixed ratio among the four gradients. Typically, the gradients G_1, G_2, G_3, and G_4 are set to -7, 30, -9, and 35 Gcm^{-1}, respectively for an aqueous sample in 90% H_2O/10% ^2H_2O. Usually, the strength of G_4 is optimized for maximum water suppression (Simpson and Brown 2005). A ^2H_2O sample only needs 20% as much as the gradient strengths used for the aqueous sample. Furthermore, the saturation periods (or saturation pulses) δ and Δ should be optimized so that the solvent magnetization is tilted about 90° out of phase to the solute magnetization.

A typical setup of PURGE includes: calibrate ^1H 90° pulse length, set both ^1H transmitter frequency and saturation frequency on water resonance, set saturation pulse power in the range of 40–100 Hz (see Sect. 4.8.1). Other parameters include 100 μs for δ, 200 μs for Δ, 2 s presaturation, and 2–3 s acquisition time. To achieve the best results of suppression, saturation pulses Δ need to be optimized first and then pulses δ.

9.4.3.4 CPMG

For plasma or serum samples, the broad signals from proteins and lipids distort spectral baseline and partially bury the signals of small metabolites. Carr–Purcell–Meiboom–Gill experiment (CPMG) makes use of the difference in T_2 relaxation between small metabolites and proteins to suppress the signals of proteins. Shown in Fig. 9.10b is a CPMG pulse sequence. The spin echo sequence is cycled n times to attain a desired T_2 relaxation delay time (or spin echo period T), where n must be an even number. During the spin echo period (T), the chemical shift is refocused and the refocused magnetization has decayed by T_2 relaxation. At the end of T period, magnetizations of all ^1H spins are scaled by the relaxation factor e^{-T/T_2}. Therefore,

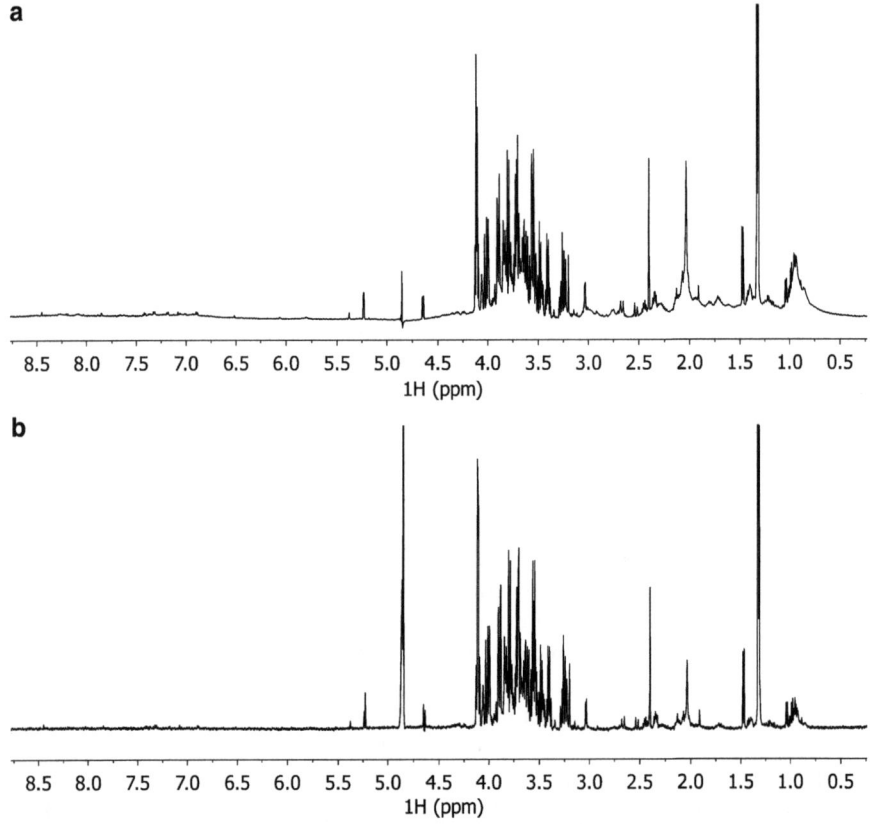

Fig. 9.11 NMR spectra of a serum (heat-inactivated fetal bovine serum). (**a**) 1D NOESY spectrum with a mixing time of 100 ms. The severely distorted baseline was caused by the broad peaks from proteins and lipids in the spectrum. (**b**) CPMG spectrum of the same sample. The data was acquired with 100 ms T_2 relaxation delay and the same number transients as in (**a**). The line shapes of peaks from small molecules and spectral baseline have significantly been improved

for the ^{1}H spins of proteins with T_2 relaxations close to the T value, the signal intensities are significantly suppressed, whereas the signals from small molecules are detected since their T_2 relaxations are much longer. Fig. 9.11 shows ^{1}H spectra of a serum sample acquired by 1D NOESY and CPMG. With a T value of 100 ms, CPMG (Fig. 9.11b) successfully suppressed the protein signals that severely buried the signals from small molecules in the PRESAT spectrum (Fig. 9.11a).

A typical setup for CPMG includes calibration of 90° and 180° ^{1}H pulses. Delay T is usually set to 100 ms or between 50 ms and 200 ms. Note that delay T is used to calculate the even number of spin echo cycles for the CPMG experiment. Delay τ is set to 0.5 ms, which can also be optimized for different samples (usually up to 1.0 ms). The 180° pulse may be fine-tuned during the setup to optimize the quality of CPMG spectrum. Parameters for presaturation are the same as in PRESAT.

9.4.3.5 J-Resolved Spectroscopy

In a J-resolved (JRES) NMR spectrum, the homonuclear spin–spin coupling is separated from ^{1}H chemical shifts and heteronuclear couplings. The ^{1}H–^{1}H couplings appear on F1 dimension after Fourier transformations. The pulse sequence for JRES NMR (Fig. 9.10c) is basically a spin echo sequence, except that the spin echo delay is modulated as evolution time. Because of the spin echo (T_2 editing), the projection of JRES spectrum has a flat baseline. The experiment is mainly used to resolve the overlapped spin couplings. The projection of 2D JRES spectrum (45° tilted) has been used for metabolomics to improve spectral resolution (Viant 2003). The increase in the spectral resolution allows to segment the spectrum in reduced binning size (0.005 ppm compared to 0.01 ppm). It results in the reduced scatter of samples in PCA scores plots. A drawback of JRES is the loss of spectral sensitivity compared to a 1D ^{1}H experiment (e.g., PRESAT).

Setup of a JRES experiment is straightforward, including tuning, shimming, calibrations of 90° and 180° pulses, and finding water resonance. Because it is used as a 1D spectrum (1D projection), the acquisition time is set to 2 s or longer for suitable spectral resolution. The spectral width (SW_1) of F1 is set to 40 Hz while SW is the same as in PRESAT (12 ppm). Other parameters includes saturation delay of 2 s, dummy scans of 16, and t_1 increment same as 1/(2*SW) and 32 increments. The number of transients in the setup is based on the total experimental time, which should be at least double to that of a PRESAT experiment.

The data is processed with unshifted squared-sine-bell or sine-bell functions for both dimensions, linear predicted to 64 points and then zerofilled to 128 points on F1 and zero-filled to 64 k complex points on F2 before Fourier transformations. Shifted squared-sine-bell or sine-bell functions may be applied to both dimensions to increase S/N ratio of the JRES projection spectrum at the price of resolution. The processed 2D spectrum is tilted by 45° and projected along the F2 dimension. Note that JRES spectrum is displayed in magnitude mode. Therefore, spectral phasing is not needed.

9.4.3.6 High-Resolution Magic Angle Spinning

In a solution, rapid molecular tumbling averages out the anisotropic nuclear interactions such as chemical shift anisotropy, dipolar and quadrupolar couplings (see Chap. 1). In intact tissue specimens or cells, however, some of the interactions may not be averaged to zero because of slow or restricted molecular motions. The metabolites in the constrained environments have short transverse relaxation (T_2) times, resulting in reduced spectral resolution (broadened NMR line widths) and detection sensitivity (low intensity). While the J-coupling and CSA is negligible and ^{1}H does not have quadrupolar coupling, the ^{1}H homonuclear dipolar coupling and susceptibility anisotropy are significant in the intact tissue or cell samples, which shorten the T_2 relaxation time and cause the line broadening. Both dipolar coupling and susceptibility anisotropy have an angular dependence on the well-known ($3\cos^2\theta-1$) term, meaning that the interaction can be averaged by spinning the sample at the magical angle ($\theta_m = 54.74°$ or $3\cos^2\theta_m - 1 = 0$). Magical angle spinning (MAS) technique (Moestue et al. 2011) has been used to increase the spectral resolution of the intact tissue and cell samples.

In a tissue or cell sample, metabolites experience different motions. For instance, small metabolites in the cell cytosol move free and rapidly, while those that bind to macromolecules or cell membrane are restricted in mobility. Lipid molecules in the liquid crystal phase of the cell membrane have restricted diffusion motions but rapid axial rotation. For the motion-restricted metabolites or lipids, ^1H–^1H dipolar coupling and susceptibility anisotropy are not canceled by the molecular motion, but averaged in some extent. These reduced couplings can be averaged by slow MAS or HR MAS. Note that HR MAS can only be utilized to study the semisolid samples in which ^1H–^1H dipolar coupling and susceptibility anisotropy are partially averaged by rapid axial molecular motions. Large macromolecules with slow axial motion are not the subject of HR MAS. Solid-state MAS NMR is required to study those large molecules and the ^1H–^1H dipolar coupling can only be suppressed partially even at a spinning rate of 50 kHz or above.

For sample preparation, 5–8 mg intact tissue or 10^6–10^7 cells are placed in the bottom of a rotor. After 30 μL ice cold 100 mM PB/^2H_2O with 0.1 mM TSP is added, the cap is gently screwed in to force out excess PB. The tissue should be small enough to allow self-adjustment of balance during the initial spinning. It is important to note that air bubbles should be completely removed from the sample rotor since the presence of the air bubbles will broaden the peak line width. This can be done by gently screwing in the rotor cap with a slightly excessive amount of the buffer solution.

The general calibration of the probe is done using a liquid sample (e.g., 2 mM sucrose in a glass rotor). The calibration includes ^1H transmitter pulses and ^1H and ^{13}C decoupling pulses. The spin rate is calibrated using an oscilloscope. The first step of the experiment setup is to spin the sample. The pressure of bearing air is first increased to achieve a spin rate of approximately 300 Hz. Next, drive air flow is increased and regulated by the operation software (or manually) to the desired spin rate (2–5 kHz). The fast spin is needed to ensure that all spin sidebands of the peaks lie outside the spectral window (5 kHz for 8.5 ppm on a 600 MHz ^1H frequency). However, the fast spin may cause sample degradation. Since spectral peaks in the aromatic region (downfield) usually have intensities much lower than those in the upfield region, the sidebands of aromatic resonances may not interfere with the statistical analysis of the data. Therefore, spin rate of 2–3 kHz is commonly used for HR MAS when the data in aromatic region is not used for data analysis (spin sidebands of aliphatic resonances overlap with aromatic resonances for MAS of 3 kHz or less).

After the spin rate is regulated, basic setup of an NMR experiment for HR MAS is similar to solution NMR with the following exceptions. For a CPMG experiment, it is crucial to set the τ delays to be synchronized with the MAS spin rate. For instance, the delays are set to 500 μs when the sample is spun at 2 kHz (τ delay in second = 1/spin rate in Hz). Additionally, the sample is cooled at 4°C to prevent sample degradation. Because the sample spinner is surrounded by the bearing air, the probe temperature is regulated using refrigerated bearing air. In practice, in order to achieve 4°C sample temperature, the temperature of bearing air needs to be

9.5 Data Analysis and Model Interpretation

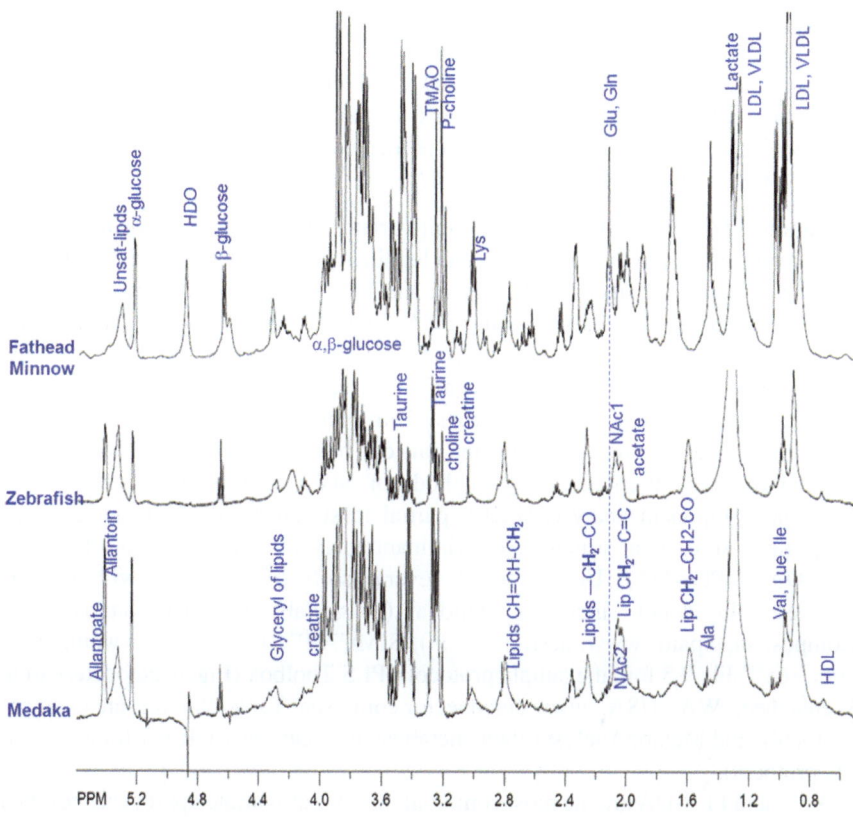

Fig. 9.12 HR MAS spectra of intact fish liver tissues using CPMG sequence. The *top* spectrum is from fathead minnow, *middle* from zebrafish, and *bottom* from medaka. Some identified metabolites are labeled above the spectra. The data were collected with approximately 5 mg tissue spun at 2 kHz

maintained at −25 to −35°C. Shown in Fig. 9.12 are representative MAS CPMG spectra obtained from fish liver tissues using the parameters noted above. Some of the metabolites are labeled on the spectra. Different than the tissue extracts, an HR MAS spectrum has NMR peaks from both polar and lipophilic metabolites.

9.5 Data Analysis and Model Interpretation

In Sect. 9.2, the fundamentals of PCA have been discussed, including scaling functions and quality factors. In the current section, practical aspects of applying multivariate methods to analyze metabolomic data will be covered. Key questions to be addressed including the following:

1. What is a scores plot and how can it be used?
2. What is a loadings plot and how can it be interpreted?

3. What is PLS-DA?
4. How can the models be validated?
5. What is the difference between PLS-DA and OPLS-DA models?
6. What are the advantages and drawbacks of the methods?
7. How is the information obtained by the methods different?
8. How is an OPLS-DA model validated?

Among the results generated by multivariate approaches, four quantities are often analyzed first, which are scores and loadings plots, R^2 and Q^2. As described in Sect. 9.2, scores are the coordinates of the new dimension-reduced coordinate space obtained by the analyses using PCA, PLS-DA, or OPLS-DA, loadings are the contributions of the original variables (NMR spectral bins) to the new coordinates. R^2 indicates how well the model explains the dataset and Q^2 describes what the predictability of the model is. This section mainly discusses how the experimental data is processed; how the data can be utilized for metabolic profiling. Several routinely used modeling methods of MVA will be discussed such as principle component analysis (PCA), partial least squares discriminant analysis (or projection to latent structures discriminant analysis, PLS-DA), and orthogonal PLS-DA (OPLS-DA). Utilization of several software packages will also be discussed in detail, including Mnova (Mestrelab Research, Santiago de Compostela, Spain, www.mestrelab.com), SIMCA-P$^+$ (Umetrics, www.umetrics.com; see Table 9.5 for an example protocol), PLS Toolbox (Eigenvector Research, Wenatchee, WA, USA, www.eigenvector.com; see Table 9.6 for an example protocol), and MetaboAnalyst (www.metaboanalyst.ca; see Table 9.7 for an example protocol).

PCA and PLS-DA are the most commonly used multivariate approaches for data analysis in metabolomics. OPLS-DA is a modification of PLS-DA method by integrating orthogonal signal correction (OSC) into a PLS model. The basic idea of OPLS analysis is to separate the systematic variation in the original dataset (X) into two parts, one that is linearly related to the class information (Y variables, class labels) and the other that is unrelated (orthogonal) to Y. The components that are related to Y are called predictive, whereas the components that are unrelated to Y are called orthogonal. For an OPLS-DA model, there is one predictive component related to Y variables, and several orthogonal components with no or very little relation to Y meaning that only the first component is predictive. Usually, only one orthogonal component is calculated because for every addition of an orthogonal component, the percent of X explained is reduced.

9.5.1 NMR Dada Processing and Normalization

Before the NMR data can be utilized for the multivariate analyses, it is necessary to process the spectra and export binned spectra into a text file or an Excel® file. An example protocol for the preparation of an NMR data table using Mnova is

9.5 Dada Analysis and Model Interpretation

Table 9.5 Multivariate analysis protocol using SIMCA-P+ (Umetrics)

Load a data file
1. Open a data file
   a. Select File → New (or "New File" icon), after starting SIMCA-P+
   b. Go to the folder containing the excel file, select the file, Open, OK
   c. Select "SIMCA-P project," then "Next"
2. Transpose the data arrangement (click on Commands → Transpose)
3. Set primary Variable IDs and primary observation IDs
   a. Click on the first cell
   b. Click on primary under Variable IDs
   c. Click primary for observation IDs, NEXT, Finish

Assign Class IDs and Scale data
4. Open data worksheet (click Workset → Edit → 1)
5. Set Observation IDs
   a. Click Observations
   b. Select Primary ID for Class from Obs ID. Click Set
   c. Change Length to highlight the name for sample class
   d. Click OK
6. Scale variables
   a. Click Scale
   b. Select all variables (Ctrl+a)
   c. Select Par for Scaling type
   d. Click Set, OK

Build a PCA Model
7. Select PCA for modeling: click Analysis → Change Model Type → PCA on X-block
8. Start calculation: click on "Autofit" icon or "Calculate the first two components"
9. Generate a scores plot by clicking "Scores Scatter Plot" icon
10. Set color for classes by clicking on Color, Coloring type: by identifiers
11. Type a number for Length that only shows classes. Click OK
12. Generate a loadings scatter plot by clicking Loadings Scatter Plot icon. p1 for X-Axis, p2 for Series
13. Set label types, var ID (primary) for Point Label, none for Axis Label
14. Set "Save As Default Options," OK
15. Generate a loadings line plot: click Analysis → Loadings → Line plot
16. Use Num for X-axis, p1 for series
17. Label type: Var ID (Primary) for Axis Label, 1 for Start, 4 for Length, 90 for rotation, 100 for Display label interval. Click OK
18. Save the project: click Save icon

Build a PLS-DA or an OPLS-DA model
19. To do PLS-DA, select Analysis → Change Model Type → PLS Discriminant
20. Do steps 8–18
21. For **OPLS-DA**, Edit Excel data spreadsheet to add a row for classes at beginning of the Excel file, e.g., 111111222222333333
22. Do steps 1–2
23. Set Y variable: click the arrow of the first column, then select Y-Variable
24. Select the second cell of first row, then click primary under Variable IDs, and primary for observation IDs
25. Then, click Next, Yes to all for Var-1 (for variable ID), finish

(continued)

Table 9.5 (continued)

26. Answer No to All for exclusion of 0 values. Note that the 0 values are needed for loadings line plot
27. Do steps 4–6
28. Answer No to All for exclusion of 0 values. Click OK
29. Set OPLS-DA for modeling: select Analysis → Change model type → OPLS
30. Calculate first two components
31. Do step 9–1. for generating scores and loadings plots

Edit loading line plot

32. Right-click on the plot, select Plot Settings → Axis...
33. Click X axis, General, Set Sacle minimum: 950. maximum: 1,810. which gives a range of 4.75–0.49 ppm, Apply
34. X axis Title: ppm, Apply
35. Click Y axis, General, set Scale range
36. Title: PC1 (or PLS1. etc.), Apply
37. Axis aspect ratio: type in a ratio
38. Click OK
39. To save a pdf, Ctrl+p, set plot size, OK
40. Right-click on the plot, select Plot Settings → Header and Footer...
41. Right-click on the plot, select Plot Settings → Plot area...

Create Loadings plot with 3. most significant loadings

42. Create a loadings column plot: Analysis → Loadings → Column plot
43. Use p1 for Series, Label types → Axis label: Var ID; start 1. length 4. display interval 1
44. Click OK
45. Sort Ascending: right-click on the plot, select Sorting Ascending, check Sort by values in series: 1
46. Create a list: right-click on the loadings plot, Create → List
47. Save the list: right-click on the list, Save list as. Drag the save txt file into excel. Save the excel file
48. Delete all rows, except 1. rows in the beginning and the end of the file
49. Save the excel file as "top 3. loadings"
50. Select all cells in excel, Data → Sorting the smallest to largest. Note: make sure that the paired cells are sorted together (ppm and the loadings)
51. Right-click the loadings plot → Properties → Item selection
52. Select all Var ID on the selected field (right) and move them to the unselected field (left)
53. Select the Var Id according to the "top 3. loadings" list. Click OK

provided in Table 9.8. The data are first apodized with 0.3 Hz line-broadening and a multiplication of first points by 0.5. The data are then zero-filled to 64 k complex points before Fourier transformation. It is important that the spectra are properly phase- and baseline-corrected because metabolomics deals with small changes in metabolite concentrations (NMR spectral intensities). The next step involves chemical shift calibration and signal alignment to ensure that variables of all samples represent the same physical identities, meaning that the resonance (or resonances) of a metabolite has the same chemical shift values across all samples. Resonance shift occurs across samples partly due to the slight change in pH or concentration of samples. Shown in Fig. 9.13 are representative superimposed NMR spectra

9.5 Data Analysis and Model Interpretation

Table 9.6 Multivariate analysis protocol using solo (or PLS_Toolbox; Eigenvector Research)

Load a data file
1. Open a data file
   a. Click on the Decompose (PCA) icon to open the analysis workspace, after starting Solo
   b. To import the desired .txt (Tab delimited) file, click: File → Import Data → X-block → XY...Delimited Text Files (TXT, XY), then select the desired normalized data file (must be in .txt format)
2. To make sure the data is sorted correctly for labeling, click: Edit → Calibration → X-block Data
   a. When the dialog box opens, click Row Labels. Each row should have a different label that corresponds to each sample. For each class of data (control, dose, etc.), manually change the class so that all samples within a class are matched and all different classes are differentiated (ex: Class 1, Class 2, etc.). Class 0 denotes an unknown, so do not use this class unless there is an unknown involved (certain analysis tools are unavailable for Class 0)
   b. All samples that you wish to include in the analysis (generally every sample) should have a green check in the box next to the row
   c. Click Column Labels. This section corresponds to ppm values. Right-click on the box labeled Axis Scale → Copy. Then right-click on the box that says Label → Paste. Under the label column, you should now see ppm for Name and the ppm values next to each column number. It should look the same as the Axis Scale column
   d. Again, all samples that you wish to include in the analysis (generally every sample) should have a green check in the box next to the row. Do not worry about class for Column Labels
   e. Close the dialog box. Everything has saved automatically
3. Preprocessing: On the right hand side of the screen, click 2. Preprocessing
   a. Highlight Autoscale, then click Remove to remove it from the Selected Methods
   b. From the available methods, first highlight Parato Scale, and click Add to move it to the Selected Methods
   c. Then highlight Mean Center and do the same
   d. In Selected Methods, Mean Center should be listed *above* Parato Scale (this is important). Click OK to close the dialog box

Build PCA model
4. On the right hand side of the screen, click 4. Build Model
5. When a series of values shows up on the screen, highlight the second or third row (PC2 or PC3). The % Variance Cumulative should be >50.0
6. When this is highlighted, again click 4. Build Model

View cores and loadings plots
7. When this is highlighted, again click 4. Build Model
8. Then click 6. Review Scores (the second row should still be highlighted). A graph (**SCORES PLOT**) and a small Plot Controls box will open
   a. For X:, highlight Scores on PC1 from the drop box
   b. For Y:, highlight Scores on PC2 from the drop box
   c. At the bottom, check the box next to Conf. Limits (95 %)
   d. On the top of the box, click View → Labels → Set 1
   e. The graph should now be in four quadrants with a circle around most of the data and labels on each point. Each class of data should be a different color
   f. Without closing the graph or the dialog box, click back onto the analysis workspace
9. With the second row still highlighted, click 7. Review Loadings. Another graph (**LOADINGS PLOT**) and small Plot Controls box will open
   a. For X:, highlight Loadings on PC1 from the drop box
   b. For Y:, highlight Loadings on PC2 from the drop box
   c. On the top of the box, click View → Labels → ppm

(continued)

Table 9.6 (continued)

    d. The graph should now be in four quadrants with the majority of the data densely packed towards the center and labels on each point

       To view both graphs together, at the top of the box, click FigBrowser → Auto-dock User Figures. Then on the top right of the graph, select the image of a box with a single line down the middle. The graphs should now be side-by-side, and you can toggle the Plot Control boxes by clicking on the gray area of each graph

10. For developing **CONFIDENCE ELLIPSES** (Tolerance Ellipses) around the Scores Plot, click on the gray area of the Scores Plot so that its Plot Controls window is displayed
   a. Click View → Classes → Outline Class Groups → Confidence Ellipse
   b. Each class of data should now be surrounded by a corresponding confidence ellipse (solid lines). The previous overall confidence limit circle will still be present as a dashed line, unless the box is unchecked in the Plot Controls window
   c. The default confidence limit is 0.95. In order to adjust this, click View → Classes → Outline Class Groups → Set Confidence Limit. Enter the desired confidence limit and click OK
   d. Note that a confidence ellipse will not be constructed for Class 0 data, which are unknowns (this is the reason we generally do not use Class 0)

11. For developing line graphs for the **LOADINGS OF INDIVIDUAL VARIABLES** on the Loadings Plot, click on the gray area of the Loadings Plot so that its Plot Controls window is displayed
   a. Click Data in the middle section
   b. Then highlight the point you would like to analyze by creating a box around it
   c. A line graph will then pop up, with Sample in the X-range and the loading values for the selected variables in the Y-range
   d. Click View → Labels → Set 1. Each point should now be labeled according to which sample it correlates to

12. For developing **PPM BAR CHARTS FROM LOADINGS**, click on the gray area of the Loadings Plot so that its Plot Controls window is displayed
   a. Click PlotGUI → Duplicate Figure. The same graph will pop up in a new window
   b. In Plot Controls, for X:, select ppm from the dropdown box
   c. In Plot Controls, for Y:, select either Loadings on PC1 or Loadings on PC2, depending on which graph you wish to analyze
   d. There should now be a line graph with ppm on the x-axis and Loadings on the y-axis
   e. In order to create a bar graph, click View → Settings... Under Plot Settings, there should be a box that says plot type. Select bar from the dropdown box. Click OK

obtained from 21 intracellular extracts. Several peaks are shifted in the data (Fig. 9.13a) and aligned well after spectral alignment process (Fig. 9.13b).

Resonance region of water or any other solvent is excluded from the spectra before spectral binning. The binning is a process in which a spectrum is segmented into a number of regions (called bins) by integrating spectral peak area over small ppm ranges (bin size), usually 0.005 ppm or smaller. Integration of a binned segment can be summed to the high limit of the bin or the center point of the bin, either of which produces the same results, provided all data of the dataset are treated in the same way. Binning is used to reduce the effect of minor peak shift and decrease the number of variables used for the statistical analysis. For a spectrum with 10 ppm spectral width and 32 k data points, spectral binning with 0.005 ppm bin size reduces the original 32 k variables to 2,000 bins.

9.5 Data Analysis and Model Interpretation

Table 9.7 Multivariate analysis protocol using web-based MetaboAnalyst (http://www.metaboanalyst.ca)

Upload a data file
1. Go to www.metaboanalyst.ca
2. Click on "click here to start"
3. Make sure your dataset is in the proper format (.csv format)
   a. Open the Excel normalized file you wish to analyze
   b. Where it says ppm in the Excel file, write Sample
   c. Add a row beneath the sample labels. In the first Column, write Label. Then under each sample label, put the sample's class. Make all control samples 1, low dose 2, and high dose 3, etc
   d. Save the file as .csv
4. Upload your data
   a. Under Data type, select spectral bins
   b. Under Format, select Samples in columns (unpaired)
   c. Click Browse, and locate your .csv data file
   d. Click Submit

Preprocessing
5. When the Data Integrity Check screen comes up, click Skip if you are sure your data is in the correct format
6. When Data Filtering comes up click none and then "Submit"
7. Normalize your data
   i. For Row-wise normalization, select Normalization by sum
   ii. For Column-wise normalization, select Pareto Scaling
   iii. Click Process
   iv. Click Next when the normalization figure is shown

Multivariate analysis
8. Select an analysis method
   a. To create a PCA scores plot
      i. Under Multivariate Analysis, click Principle Component Analysis (PCA)
      ii. Click on the tab that says 2D scores Plot
      iii. A scores plot should be displayed with PC1 plotted against PC2 with 95 % confidence ellipses around the classes
   b. To create a PLSDA scores plot
      i. On the left hand menu select PLSDA (directly under the PCA button)
      ii. Click on the tab that says 2D scores Plot
      iii. A scores plot should be displayed with Comp #1 and Comp #2 with 95 % confidence region

Download results
9. In the far left menu click the "Download" button. Click "Download.zip" and save the results to the appropriate folder

Metabolomics uses multivariate analyses (PCA, PLS-DA, etc.) to extract biological information from the changes in metabolite concentrations relevant to biological functions. In order to reduce the errors in the multivariate analyses due to variations in sample concentrations or variable sample dilutions, spectral normalization is applied to the binned spectra. The spectral normalization may not be an ideal method for the data normalization, but it reduces contribution from concentration

Table 9.8 Protocol for spectral alignment and binning using Mnova (Mestrelab)

Load NMR data files
1. Start Mnova
2. Drag all data files one by one in sequential order into Mnova window to load fid files
3. Double check the file order
4. Select all files by clicking on the first data on the pages menu and Ctrl+a. Make sure that all files are selected during following steps (steps 5–12)

Process NMR data
5. Setup apodization: click Processing → Apodization and set 0.3 Hz for line broadening and 0.5 for first point. Click OK
6. Correct spectral phases
    a. Click phase correction icon and click on Biggest
    b. Hold left mouse button (LMB) and move it up or down to adjust 0-order phase (PH0) and RMB to adjust first-order phase (PH1). PH0 should be between 40 and 50° and PH1 $\approx$ 0. Or type in a value for PH0 if it is known
    c. Close the phase window
7. Correct spectral baseline
    a. Click on the arrow next to baseline correction icon
    b. Select Baseline Correction
    c. Select Bernstein Polynomial Fit, and 3 for polynomial order. Click OK

Superimpose all spectra
8. Superimpose all spectra: Select all spectra, then Tools → Superimpose spectra
9. Right-click on the superimposed spectra and left-click the superimposed spectra on the pages window to select the spectra

Calibrate chemical shift
10. Zoom the DSS peak, then click "fit to height"
11. Calibrate 0.00 ppm for DSS peak
    a. Click on TMS icon
    b. Click the cursor on the top of the DSS peak
    c. Type 0 for New Shift, select Auto Tuning, and press enter key

Save the superimposed spectra
12. Save the processed data: click File → Save As and type the sample name, select mnova for file type, change to the data file folder and click Save. Note; the file size should be >10 MB
13. Delete all spectra except the superimposed one and save as new_file.mnova

Align spectra
14. Zoom a region
15. Align peaks
    a. Select "Add region graphically" and click and hold LMB to select a region
    b. Repeat the above step for all peaks
    c. Preview the peak alignment by clicking Preview. Click again to go back to "Align Spectra" window
    d. Accept the alignment (click OK)

Remove water peak and bin the spectra
16. Exclude water resonance: click View → Cuts → Manual Cuts. Enter ppm range (4.7–5.0). Click OK
17. Bin the spectra: click Processing → Binning, uncheck Full Spectrum and type 0.5 (From), press Tab key, type 10.0 (To), press Tab key, type 0.005 for width of each integral region. Click OK

(continued)

9.5 Dada Analysis and Model Interpretation

Table 9.8 (continued)

Normalize and save the spectra

18. Set the total area of individual spectrum to unity: click Processing → Normalize, select total area and value 1.0. Click OK
19. To save the binned spectra into a text file, click File → Save As, select ASCII Text file (*.txt) for file type, use sample name for text file name and click Save. Note that the file size should be ~522 KB
20. Close the data file **without** saving changes. Note Saving changes will overwrite the raw data

Edit the txt file

21. Run excel, and drag the txt file to excel to open the txt file
22. Insert first row and type in sample IDs. Column A should be labeled as ppm. Save the file as xlsx with a new filename
23. Add (or replace with) zeros for the rows containing water
24. Sort the ppm value from largest to smallest: click on the first ppm value (0.5), and click home → sort and filter → largest to smallest
25. Check sum of columns. They should be 1.0. Scroll to the last cell of column B and click on the end blank cell
26. Click Σ and Enter key
27. Save the file as txt file

Process J-resolved COSY data

1. Drag a JRES file into Mnova. The data will be processed automatically and 45° tilted
2. Zero-fill the spectral size to 16 k
3. Rephase the spectrum if necessary. Click on F2
4. Click on Phase Correction and adjust zero-order and first-order phase according to step 6
5. Click F1 and do the phase correction in the same way as step 3
6. Right-click on the spectrum and select Properties
7. Click on 2D Spectrum, and select horizontal Trace and click OK
8. Right-click on the projection and then select Setup
9. Use sum for projection type and click Extract icon. The projection will be extracted to a 1D spectrum

changes unrelated to the biological functions. The last step in the generation of the data table is to export binned and normalized NMR spectra into a text file (or an Excel® file).

9.5.2 Analysis of Metabolomic Data

Once the data table is generated, MVA can be utilized to analyze the data. Routinely used analysis methods include PCA, PLS-DA, and OPLS-DA. After the data table is loaded into an analysis software package (such as SIMCA-P, PLS-Toolbox, and web-based MetaboAnalyst), the dataset is preprocessed with scaling, followed by mean-centering. PCA is first applied to provide an overview on the dataset.

As an unsupervised pattern recognition method, PCA is the first analysis to be conducted for a metabolomics study. Herein, a study on the responses of human

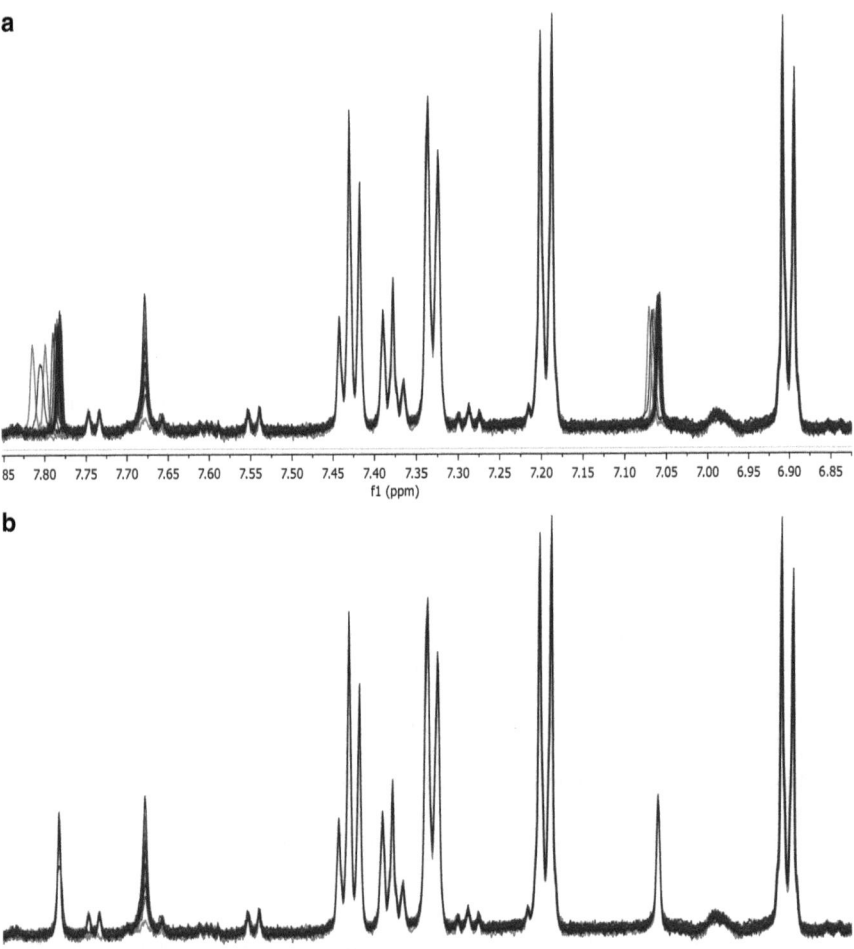

Fig. 9.13 Superimposed spectra of 21cell extracts. (**a**) Peak shifts in the superimposed spectra are clearly visible, which can be problematic during multivariate analyses. (**b**) The spectral alignment was able to correct the peak shift problem (the peaks were aligned with Mnova spectral align routine)

brain cancer cells (cell line LN229) to the treatment of anticancer drug paclitaxel (or Taxol™; PTX) is used as an example to explain the MVA process. Paclitaxel is a mitotic inhibitor used in cancer chemotherapy. In the study, the cells were treated with 1 nM (low dose) and 10 nM (high dose) PTX for 24 h. After 0.4 mL of the culture medium was taken, the treated and untreated (control, treated only with solvent vehicle) cells (7 dishes per class) were quenched and extracted, according to the methods described in Sect. 9.3. The NMR data were processed with spectral alignment, resonance exclusion (spectral region cut), binning, and normalization. Shown in Fig. 9.14 are the superposed spectra before and after resonance exclusion. The intense peaks from water, methanol, and dimethylamine (DMA) were removed before binning.

9.5 Dada Analysis and Model Interpretation

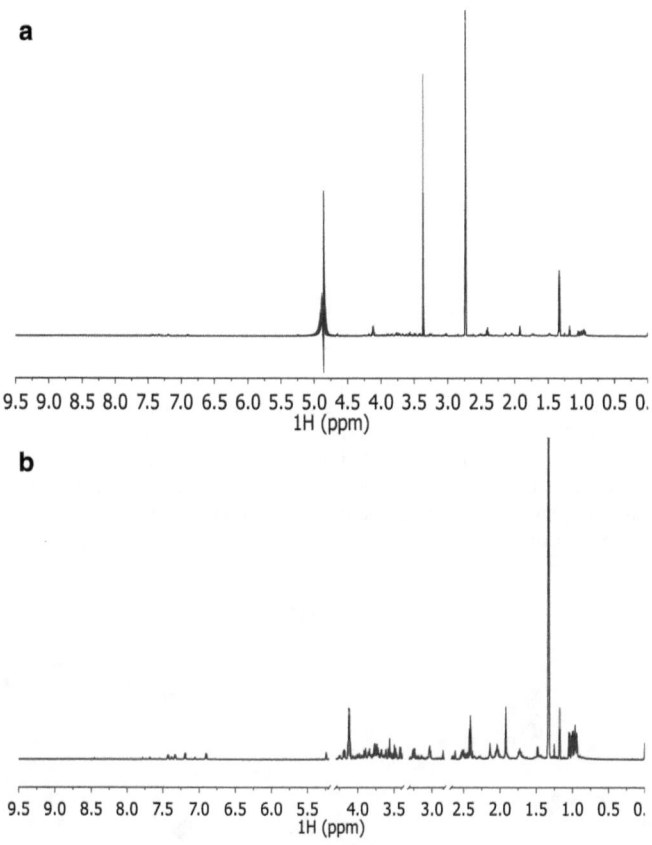

Fig. 9.14 Normalized superimposed spectra of 21cell medium samples from LN229 cells treated with PTX for 24 h as discussed in the text. (**a**) Before and (**b**) after the intense peaks from water, methanol, and dimethylamine (DMA) were removed prior to binning. The removal of the intense peaks clearly improved relative intensities of metabolite resonances

After the data file was loaded into SIMCA-P, (Table 9.5) preprocessed with Pareto scaling, and mean-centering, the PCA model was calculated. Shown in Fig. 9.15 is the summary of R^2 and Q^2 of the first three components of the models for cell culture medium and intracellular polar extracts. Both models have acceptable R^2 and Q^2 values. Next, a scatter two-PC scores plot was generated for the PCA models. In both models (cell culture medium and intracellular extracts), there are clearly class separations between control and high-dose sample classes, and low dose samples spread in between the two classes (Fig. 9.16a, b). The scores can also be visualized by plotting single-component scores vs. sample numbers (Fig. 9.16c, d). For the model of medium samples, most control samples have positive PC1 scores and the high-dose samples have negative scores with a few exceptions. For cellular extracts, however, the segregation is mostly in PC2 as shown in both 2D and 1D scores plots. This means that the variation among sample

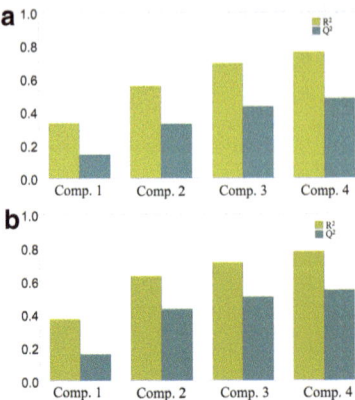

Fig. 9.15 Overview plots highlight the quality and predictability of two PCA models derived from human brain cancer cells (LN229) treated with different dose levels of paclitaxel (PTX). R^2 describes the quality of the model and how well the model explains the dataset, while Q^2 measures predictability of the model, that is how well the model can predict classifications of observations (samples). The models are generated with NMR data collected (**a**) for intracellular extracts and (**b**) for culture media. The PCA model for culture media has a better quality (higher R^2) and predictability (high Q^2)

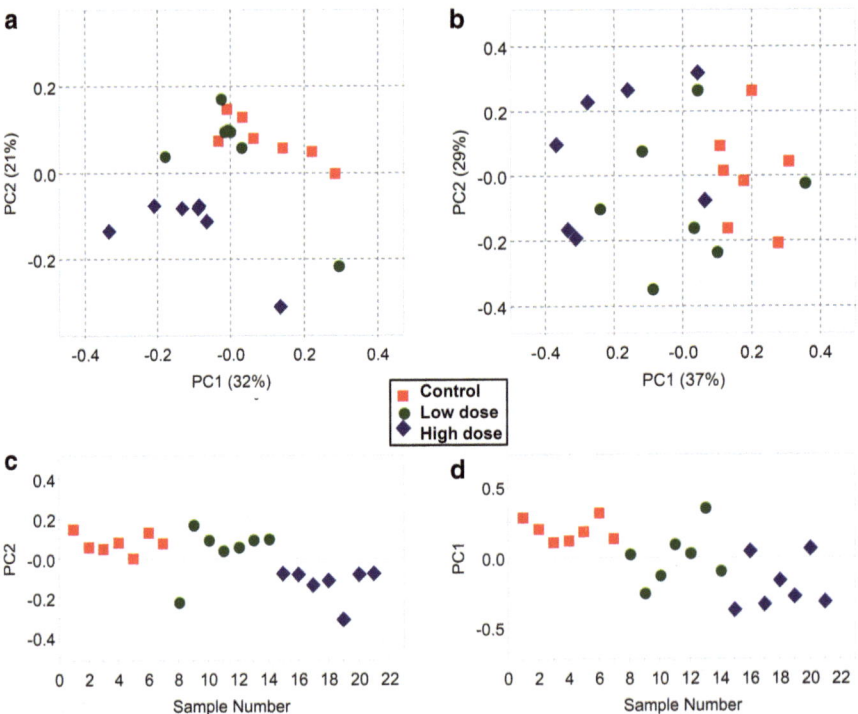

Fig. 9.16 Scores plots of two PCA models for human brain cancer cells (LN229) treated with different dose levels of paclitaxel (PTX). One- and two-component PCA scores plots are generated for intracellular extracts (**a, c**) and culture media (**b, d**), respectively. Both models clearly show class separations between control and high-dose (10 nM) sample classes, and scores of the low dose (1 nM) samples spread in between the two classes

9.5 Dada Analysis and Model Interpretation

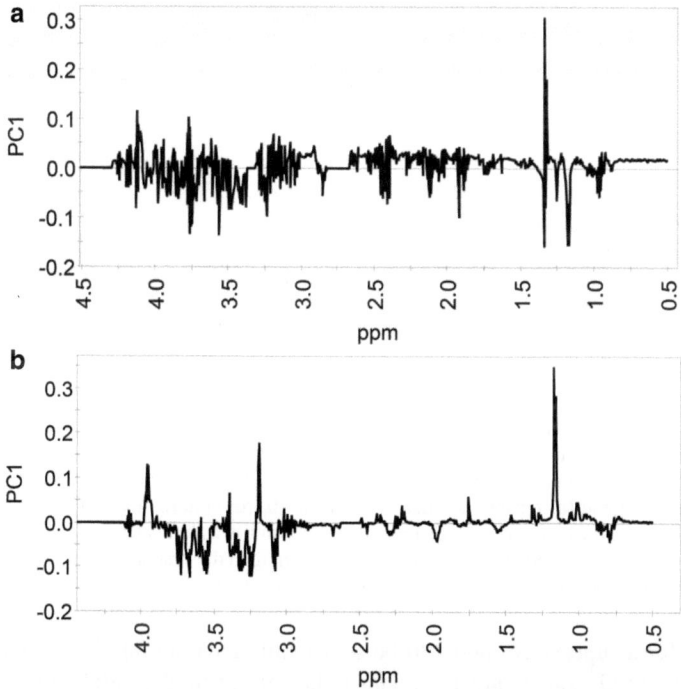

Fig. 9.17 Loadings plots for the two PCA models (summarized in Fig. 9.15). PC2 is used for the loadings plot of cell extracts in (**a**), while PC1 is used for that of culture media in (**b**), because the class separation is primarily along the respective component (see Fig. 9.16a, b). For both plots, the positive loadings indicate that the NMR spectral intensities are higher in controls than high-dose samples, or vice versa

classes of medium samples is larger than those of intracellular extracts since PC1 captures the most variation in the dataset.

Now, the question in one's mind is what metabolites contribute to the class discrimination or which variables vary most between the control and dosed classes. Loadings plots are used to address these questions. Shown in Fig. 9.17 are the 1D loadings plots of the two models. For cellular extracts, the positive PC2 loadings indicate that the NMR spectral intensities are higher in controls than high-dose samples, or vice versa. For medium samples, the positive PC1 loadings represent that the NMR spectral intensities are higher in control samples since most controls are located in the positive PC1 region of the scores plots (Fig. 9.16). It is important to note that, generally, a loadings plot cannot be used to determine the metabolite concentration changes among three classes. However, since in this case, the low-dose treated class is located in the middle region between controls and high-dose treated samples in the PCA scores plots and near the origin of PC1 or PC2, the changes in the loadings plots primarily reflect those between control and high-dose groups.

In the next steps, supervised multivariate analyses (PLS-DA and OPLS-DA) are used to maximize the class separations by taking into account the class information.

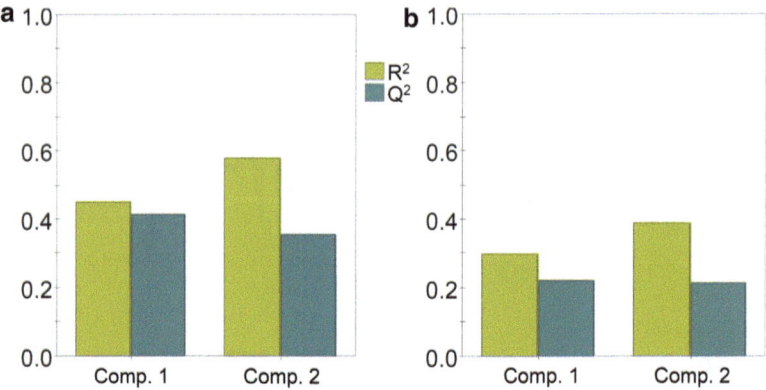

Fig. 9.18 Overview of the quality and predictability of PLS-DA models built using the ^{1}H NMR spectra obtained for human brain cancer cells (LN229) treated with different dose levels of PTX. R^2 and Q^2 measure the quality and predictability of a PLS-DA model, respectively. The PLS-DA models are generated (**a**) for intracellular extracts and (**b**) for culture media. The PLS-DA model for intracellular extracts has a better quality and predictability as indicated by much higher values of R^2 and Q^2. Lower R^2 and Q^2 of the second component than the first one in both models indicate that the first component captures the most class discrimination

In order for a supervised model to be statistically meaningful, the model must be validated and Q^2 values are important indicators in model validation. Using the same set of data preprocessing parameters, PLS-DA models are generated for both datasets (Fig. 9.18). Lower values of R^2 and Q^2 of culture medium model than those of cell extract model indicate that the PLS-DA model for medium is weaker than that for cellular extracts. This is understandable because the culture medium contains nutrients some of which are also cellular metabolites. The high contractions of the nutrients produce large background noise for metabolomic analysis, which is contained in the original dataset. In order for a supervised model to be statistically significant, the model must be validated by means of cross validation and Q^2 values are important indicators in model validation. During a validation, a fraction (usually 10–15%) of the class labels (*Y* variables) are randomly permuted among classes and the model is rebuilt using the permuted class labels. The process is repeated for a few 100 times. The original model is a valid one, if all rebuilt models have Q^2 less than the original model and a majority of rebuilt models have R^2 less than the original model.

The results of validation for the PLS-DA models are shown in Fig. 9.19, which include fitted regression lines for R^2 (triangles) and Q^2 (squares) values of the rebuilt models. Empirically, for a good model, the intercepts of the regression lines should be less than 0.4 for R^2 and less than 0.05 for Q^2 (Eriksson et al. 2006), meaning that rebuilt models with permuted class labels are much worse than the original one. Therefore, the validation results indicate that both models are validated PLS-DA, although the one for medium has smaller R^2 and Q^2 compared to those for cell extracts.

Because the class discrimination in the PLS-DA scores plots (Fig. 9.20) are primarily along PLS1, the 1D PLS1 loadings plots (Fig. 9.21) are used to explore

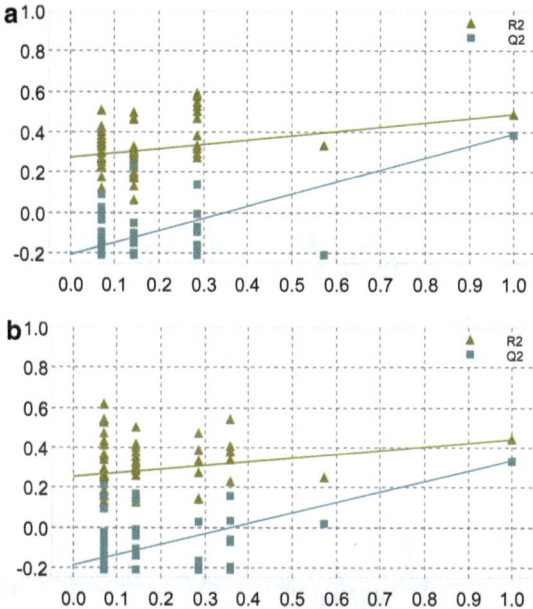

Fig. 9.19 Validation plots of the PLS-DA models (summarized in Fig. 9.18) for (**a**) intracellular extracts and (**b**) culture media. The scale of the y axis represents values of R^2 or Q^2. The x axis reflects the extent of the permutations with the scale of 1.0 representing the case that no class label is permuted and 0.0 meaning that all class labels are permuted. Both models were validated with 50 permutations. The results indicated that either model is valid because none of the rebuilt models had Q^2 values higher than the original model and a few rebuilt models had higher R^2 values with smaller Q^2, which indicated that these rebuilt models were over-fitted

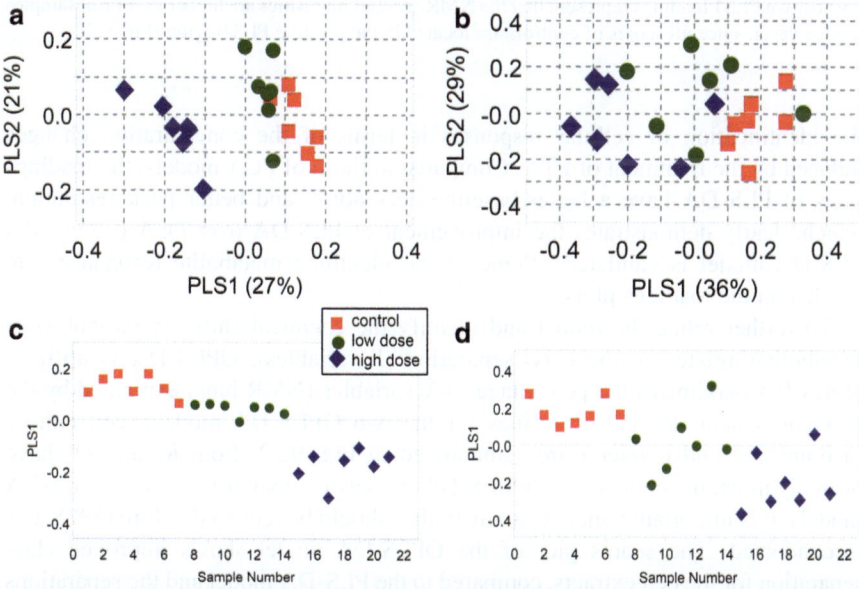

Fig. 9.20 Scores plots of two PLS-DA models for human brain cancer cells (LN229) treated with different doses of PTX. Two- and one-component PLS-DA scores plots are generated for intracellular extracts (**a, c**) and culture media (**b, d**), respectively. Both models clearly show improvement in class separations compared to the PCA models (Fig. 9.16)

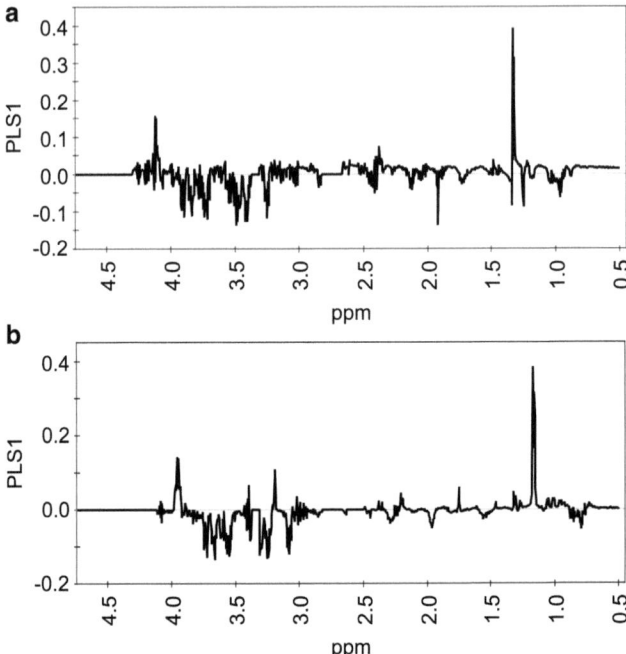

Fig. 9.21 Loadings plots for the two PLS-DA models (summarized in Fig. 9.18) for (**a**) intracellular extracts and (**b**) culture media. The loadings plots show better baseline, less noise, and better peak resolution, which clearly demonstrates the improvement of PLS-DA over PCA. For both plots, the positive PLS1 loadings represent that the NMR spectral intensities are higher in control samples or vice versa, since all scores of controls are located in the positive PLS1region (Fig. 9.20)

the identification of cellular responses in terms of the concentration changes induced by the treatment of PTX. Compared to those of PCA models, the loadings plots of PLS-DA have a better baseline, less noise, and better peak resolution, which clearly demonstrates the improvement of PLS-DA over PCA (provided a PLS-DA model is validated). Some of the identified metabolite resonances are labeled on the loadings plots.

To further refine the model and identify the chemical shifts (X variables) of metabolites related to the class separation (Y variables), OPLS-DA is utilized. OPLS-DA maximizes the percentage of X variables (NMR bins) explained by the first component. R^2 and Q^2 values for the two OPLS-DA models (cell culture medium and cell extracts) are summarized in Fig. 9.22. Both R^2 and Q^2 have been significantly improved in OPLS-DA models compared to those in PLS-DA models. It is important to note that any outlier should be removed before OPLS-DA is conducted. The scores plot of the OPLS-DA model shows improved class separation for the cell extracts, compared to the PLS-DA model and the separations among the three classes are along the first component of the OPLS-DA model (Fig. 9.23). The loadings plots generated from the models are shown in Fig. 9.24,

9.5 Dada Analysis and Model Interpretation

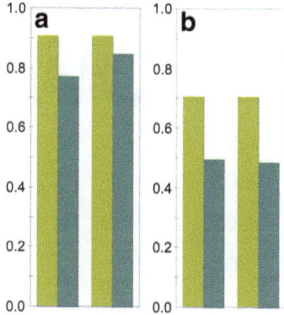

Fig. 9.22 Overview plots of two OPLS-DA models built using ^{1}H NMR data derived from human brain cancer cells (LN229) treated with different concentrations of PTX. R^2 and Q^2 measure the quality and predictability of an OPLS-DA model, respectively. The relatively high values of R^2 and Q^2 of (**a**) intracellular extracts compared to (**b**) the model for culture media indicated that the OPLS-DA model of intracellular extracts had much better quality and predictability than the model for media, provided the model is a valid one. Nonetheless, both models were notably improved over PCA and PLS-DA models

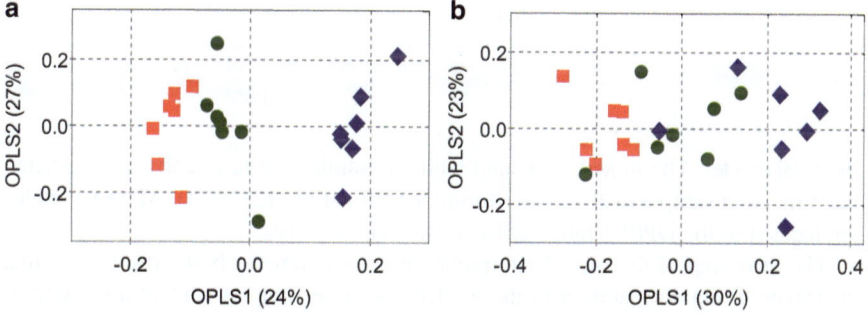

Fig. 9.23 Two-component scores plots of OPLS-DA models for human brain cancer cells (LN229) treated with different dose levels of PTX. (**a**) The separations among the classes of cell extracts are mainly along OPLS1 and the variations within a class spread along OPLS2. (**b**) The class discrimination between control and high-dose groups is largely on OPLS1, except for one outlier of the high-dose group, while the scores of the low dose group distribute between the control and high-dose groups

which are almost identical to those obtained by PLS-DA for cell extracts and medium (Fig. 9.21). OPLS-DA models are validated using the following sevenfold validation procedure slightly different than the procedure used for PLS-DA. First, the dataset is randomly partitioned into 7 subsets of samples, each of which contains 1 sample per class. Then, 6 subsets of samples (approximately 86% of the original data) are used as a training set (6 per class) to build an OPS-DA model and the remaining subset is used as a test set to validate the model. The model is generated and validated seven times (sevenfold) with a different test set (each of the 7 subset is used once as the test set). The results (scores and loadings) from the 7 models are then averaged to obtain

Fig. 9.24 Loadings plots of OPLS-DA models for (**a**) cell extracts and (**b**) culture media of human brain cancer cells (LN229) treated with different dose levels of PTX. For both plots, the negative OPLS1 loadings represent that the NMR spectral intensities are higher in control samples or vice versa, since all scores of controls are located in the negative OPLS1 region

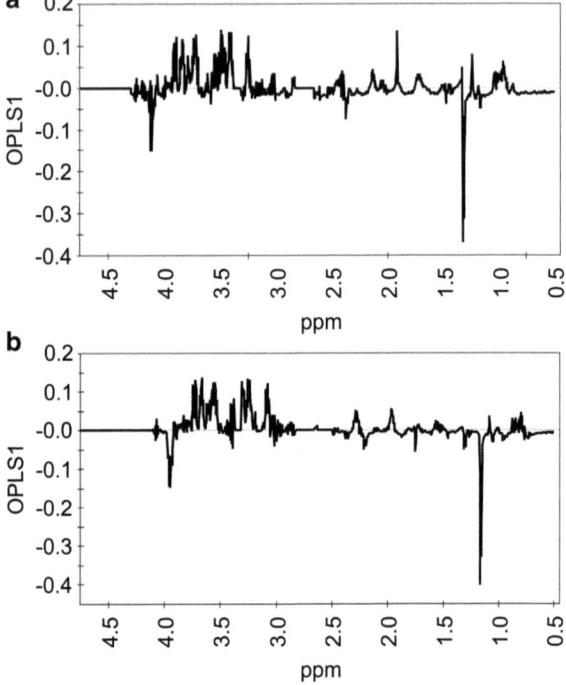

the final model. The models obtained with the training set have correctly predicted the remaining sample in all classes during the validation (Fig. 9.25). All test samples are located in the OPLS1 range of their own sample classes.

The next step of the model interpretation is to assign the NMR resonances that contribute to class separation in the models. Because samples contain a mixture of metabolites, one should pay extra attention to resonance degeneracy. The ^{1}H resonances of metabolites can be assigned by analyzing their ^{1}H–^{1}H spin system coupling patterns and comparing their chemical shifts with reported values (Fan 1996; Lane and Fan 2007; Lindon et al. 1999; Nicholson and Foxall 1995; Teng et al. 2009; Yang et al. 2007a, b), standard compounds (Ulrich et al. 2008; Wishart et al. 2007), and Chenomx database entries (Edmonton, Alberta, Canada). Some resonances observed in the aliphatic region (lower frequency than that of water) can readily be assigned to metabolites such as alanine (Ala), valine (Val), leucine (Leu), isoleucine (Ile), lactate (Lac), glycine (Gly), and aspartate (Asp) based on their characteristic chemical shifts. Assignments of tryptophan (Trp), tyrosine (Tyr), phenylalanine (Phe), and histidine (His) can be done by starting at the aromatic resonances and then extended to side chain protons. NADP$^+$ is usually assigned by the spin system formed near 8.41, 8.81, 9.12, and 9.32 ppm in the TOCSY spectrum. ATP/ADP/AMP (AXP) and GTP/GDP/GMP (GXP) can be assigned based on their readily distinguishable ^{1}H chemical shifts, ^{1}H–^{1}H and ^{1}H–^{13}C coupling patterns. Singlet resonances such as choline (Cho), Phosphatidylcholine (PCh), creatine (Cr), trimethylamine N-oxide (TMAO), Gly, formate (For), Formarate (Frt), pyruvate

Fig. 9.25 Y-predicted scatter plots of OPLS-DA models validated with 86 % of the original data used as a training set (6 per class) and the remaining 14 % (1 per class) used as a test set to validate the model. All test samples are correctly predicted with each locating within the region of its corresponding classes. The models were generated for (**a**) cell extracts and (**b**) culture media of human brain cancer cells (LN229) treated with different dose levels of PTX

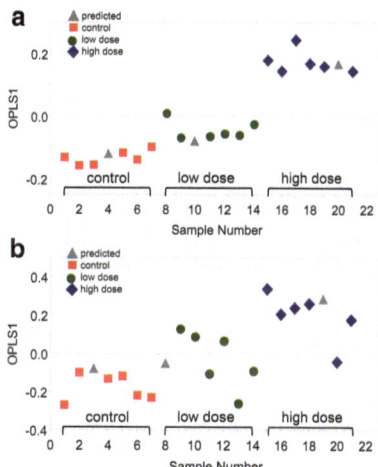

(Pyr), and malonate (Mlo) are often assigned by comparing the resonances with published values. An HSQC spectrum is needed to verify the assigned resonances. Figure 9.26 shows typical PRESAT and HSQC spectra of intracellular extracts. An example protocol to identify metabolites using Chenomx is provided in Table 9.9.

9.6 Metabolomics of Biofluids

Biofluids are mostly used in metabolomics applications (see Fig. 9.1). Since it is not feasible to cover all aspects of the applications of biofluid metabolomics, this section intends to provide a brief introduction on how the metabolomic research is conducted using a metabolomics study of cerebral infarction as a case study (Jung et al. 2011). The key questions to be addressed in this section include the following:

1. What are the metabolic profiles of human urine and plasma observed by NMR spectroscopy?
2. How is multivariate statistical analysis applied to study the changes of urine and plasma profiles?
3. How are the results of the metabolomics study interpreted?
4. Can the results of the biofluid metabolomics be linked to specific metabolic pathways?

Cerebral infarction is an ischemic type of stroke, which occurs when blood flow to a part of the brain stops due to complex health factors such as cardiovascular disease, high cholesterol, and/or diabetes. Because it is very difficult to diagnose and prognose a stroke using traditional methods, NMR-based metabolomics was applied to study metabolic profiles of urine and plasma samples from healthy individuals and patients with cerebral infarction. By applying MVA to the NMR data, metabolic indicators were identified for early diagnosis of cerebral infarction.

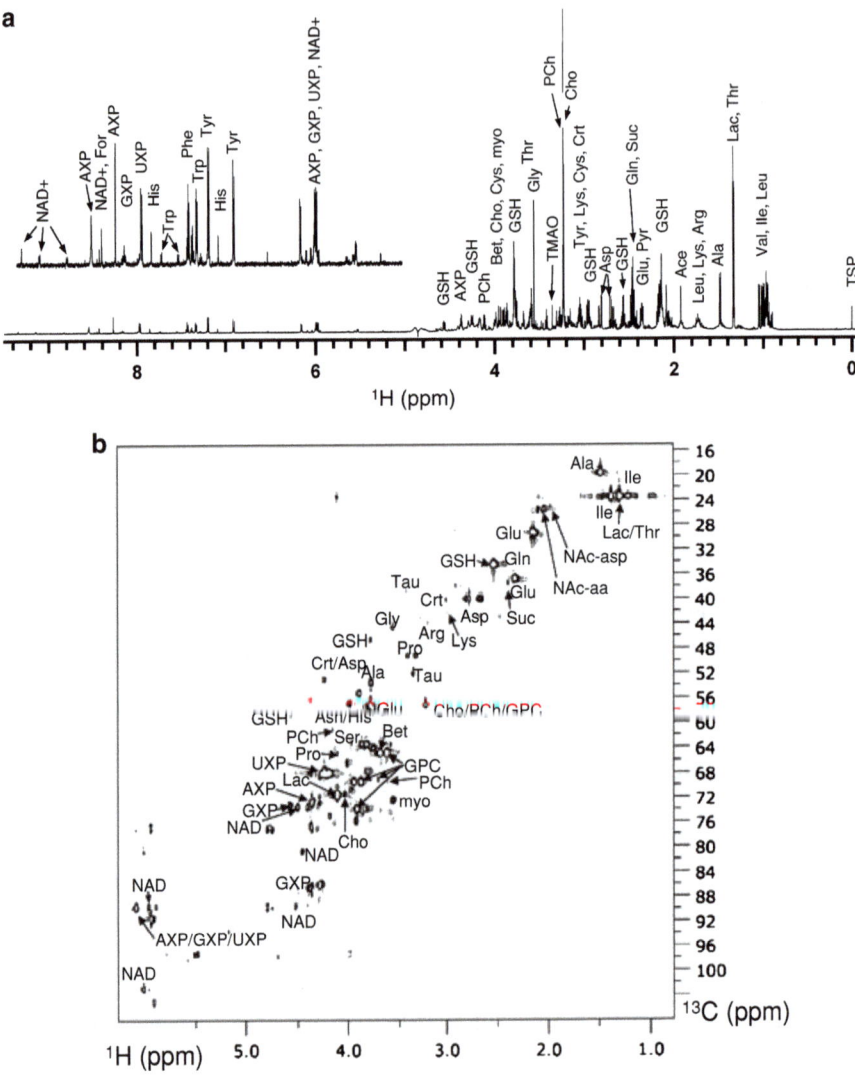

Fig. 9.26 Representative 600-MHz ^{1}H-NMR spectra of a polar cellular extract. (**a**) 1D PRESAT spectrum. (**b**) ^{1}H-^{13}C HSQC. Some of the assigned resonances are labeled in both spectra (reproduced with permission from Teng et al. (2009), Copyright © Springer, 2009)

A total of 101 plasma (47 healthy individuals and 54 patients) and 54 urine (27 for each group) samples were used in the study. For 1D NOESY experiments, 200 μL of plasma and 400 μL of urine were used with a phosphate buffer. After spectral alignment, each spectrum was segmented into 0.005 ppm for plasma samples and 0.003 ppm bins for urine samples and normalized to the total spectral area. Representative ^{1}H NMR spectra of plasma and urine samples are shown in Fig. 9.27. By comparing the spectra of healthy individuals to those of patients,

9.6 Metabolomics of Biofluids

Table 9.9 Protocol for profiling metabolites using Chenomx profiler (Chenomx, Inc.)

1. Start Chenomx Profiler by clicking on Chenomx Profiler shortcut icon

Load data

2. Load a raw 1D NMR data file: Select "File" → Open, and select a .fid file from Open Spectrum window and click Open
3. Input parameters in the Data import window for chemical shift (CS) reference, concentration (mM), pH, etc. Then, click OK

Process data

4. If the spectrum needs to reprocess, send the data to Chenomx Processor: click File → Send to Processor
   (a) Zoom in the reference range by dragging the LMB on the spectral range. To zoom out, drag the LMB on the small spectral overview window (lower right window)
   (b) Move the red triangle to adjust CS reference ppm
   (c) To redo the phase correction, click "Clear last" twice. Click toolbar Processing → Phasing. Sliding the bars for manual phase correction, or select auto with the fine adjustment. Click Accept
   (d) Baseline correction
      (i) Clicking Processing → Baseline Correction
      (ii) Click on Auto Spline (or Auto Linear)
      (iii) Move blue points and/or add more points (click on the spectrum) if necessary to achieve a flat baseline
      (iv) Click on Accept
   (e) Switch to Profiler by clicking "Send to Profiler" icon

Profile metabolites (identify compounds and determine concentrations)

5. Click on a metabolite from the library, then click a CS value. Move the arrows to fit the peak(s). Repeat this step all assigned peaks.
6. Zoom in a range and right-click on an unassigned peak → Filter for compounds
7. Select a compound from the list. Move the blue triangle on x-axis to align CS position and on y-axis to align intensity. Alternatively, the adjustment can be done by dragging the top of the calculated peak to overlap the spectral peak
8. Click on the CS value (if any) under the metabolite name to select another resonance of the metabolite and repeat step (6–7)
9. Click save icon and give a name to the profile
10. Perform similar analysis for all peaks
11. Uncheck Chenomx library compound set (only profiled compound set is checked)

Export analysis results

12. Click File → Export → Profiled Data (.txt)
13. Answer all question accordingly, and use Transposed for export format
14. Click Finish to export. The text file is saved in the same folder as the NMR data file

---

significant changes in metabolic concentrations can be visualized, including plasma very low density lipoprotein (VLDL), low density lipoprotein (LDL) lactate and formate, and urine citrate, DMA, creatinine, and hippurate. PCA and OPLS-DA were used to further investigate metabolic changes between healthy individuals and the patients.

Since PCA scores plots of both plasma and urine models showed slight differentiations between the two classes, OPLS-DA was conducted to minimize the

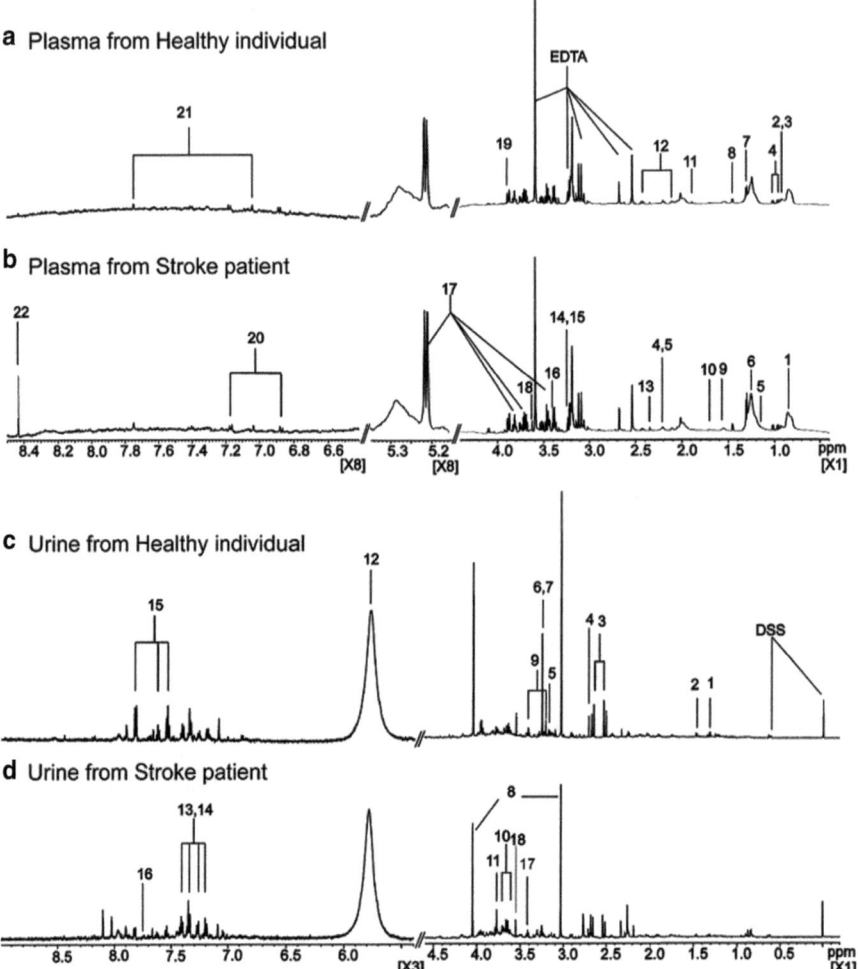

Fig. 9.27 ^{1}H-NMR spectra of plasma (**a, b**) and urine (**c, d**) of healthy individuals (**a, c**) and stroke patients (**b, d**). Labels for plasma samples: 1, very low-density lipoprotein (VLDL)/low-density lipoprotein (LDL) CH$_3$; 2, leucine; 3, isoleucine; 4, valine; 5, 3-hydroxybutylate; 6, VLDL/LDL (CH$_2$)$_n$; 7, lactate; 8, alanine; 9, lipid CH$_2$CH$_2$CO; 10, lipid CH$_2$CH$_2$C_C; 11, acetate; 12, glutamine; 13, pyruvate; 14, trimethylamine-*N*-oxide; 15, betaine; 16, taurine; 17, glucose; 18, methanol; 19, glycolate; 20, 4-hydroxyphenylacetate; 21, τ-methylhistidine; 22, formate. Labels for urine samples: 1, lactate; 2, alanine; 3, citrate; 4, DMA; 5, *O*-acetylcarnitine; 6, trimethylamine-*N*-oxide; 7, betaine; 8, creatinine; 9, carnitine; 10, mannitol; 11, glycolate; 12, urea; 13, phenylacetyglycine; 14, phenylalanine; 15, hippurate; 16, salicylurate; 17, taurine; 18, glycine (reproduced with permission from Jung et al. (2011), Copyright © American Heart Association, Inc., 2011)

9.6 Metabolomics of Biofluids

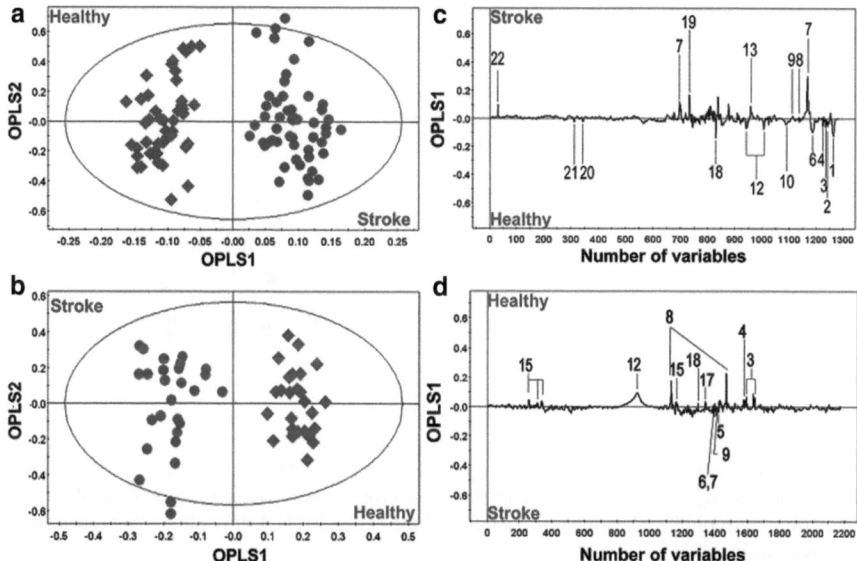

Fig. 9.28 OPLS-DA scores (**a**) and loadings plots (**c**) obtained from the ^{1}H-NMR spectra of plasma samples from healthy individuals ($n = 47$) and stroke patients ($n = 54$). OPLS-DA scores (**b**) and loadings (**d**) plots obtained from the spectra of urine samples from healthy individuals ($n = 27$) and stroke patients ($n = 27$). See Fig. 9.27 for the metabolite labels (reproduced with permission from Jung et al. (2011), Copyright © American Heart Association, Inc., 2011)

contribution of the variability within the classes to the models so as to focus on the variability between the two classes. The scores plots of plasma (Fig. 9.28a) and urine data (Fig. 9.28b) show significant separations between healthy individuals and patients. In addition, the high R^2 and Q^2 values for both plasma and urine ($R^2 > 0.9$, $Q^2 > 0.6$) data indicate that the models have good quality and predictability. The models were validated with 80% samples as a training set and 20% as a test set. The validation results for plasma model are shown in Fig. 9.29. All test data were correctly predicted by the models.

OPLS-DA loadings plots of plasma and urine models were used to identify the metabolites that contribute to the differentiations between the two classes in the scores plots (Fig. 9.28c, d). The loadings plot of plasma model reveals that compared to healthy individuals, plasma of the stroke patients have elevated levels of several metabolites including lactate, alanine, lipid -CH_2CH_2CO, pyruvate, glycolate, and formate, along with decreased levels of VLDL/LDL, leucine, isoleucine, valine, unsaturated lipid, glutamine, methanol, 4-hydroxyphynellactate, and τ-methylhistidine. The urine loadings plot indicated that urine metabolic profiles of the stroke patients were characterized by increased levels of O-acetylcarnitine, TMAO, betaine, and carnitine, and decreased levels of citrate, DMA, creatinine, taurine, glycine, and hippurate. To further investigate the significance of the metabolite changes, Student's t-test was applied to the NMR bins of the metabolites

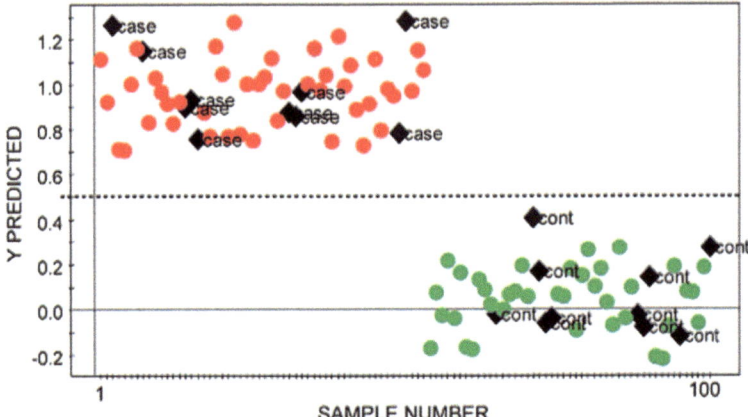

Fig. 9.29 Validation plot of OPLS-DA model validated with 80 % (44 stroke, 37 healthy) of the data as the training set and the remaining 20 % [10 stroke (case) and 10 healthy (control)] as the prediction set (reproduced with permission from Jung et al. (2011), Copyright © American Heart Association, Inc., 2011)

(see Table 9.10 for an example protocol of Student's t-test application). The changes with calculated p value of 0.05 or less is usually considered as statistically significant.

Shown in Fig. 9.30 is the summary of metabolic pathways involving the metabolites of the stroke patients with statistically significant changes compared to healthy individuals. The pathway map is used to explain the pathogenesis of a stroke from the metabolic biology's point of view based on the results of the multivariate statistical analyses. As the map indicates, the metabolites with altered levels in stroke patients are primarily involved in three pathways: anaerobic glycolysis, folic acid pathway, and hyperhomocysteinemia.

Anaerobic glycolysis. It is well known that under hypoxia, cells rely more on glycolysis to generate energy than on the tricarboxylic acid cycle (TCA cycle), which converts glucose into pyruvate to form the high-energy compound ATP (adenosine triphosphate) and reducing agent NADH (reduced nicotinamide adenine dinucleotide). Under the hypoxia condition, a majority of pyruvate does not enter the TCA cycle and instead is oxidized to form lactate. The observed higher levels of lactate and pyruvate in the patient's plasma indicate the enhanced anaerobic glycolysis of plasma glucose to meet the energy consumption because of limited availability of oxygen in the patient's body. Additionally, a decreased citrate level is an indicator that a limited amount of pyruvate entered the TCA cycle.

Folic acid deficiency. In this study, the increased level of formate coupled with the decreased levels of glycine and hippurate was related to folic acid deficiency, which has been considered as a risk factor for stroke (Howard et al. 2002). One of the folic acid metabolic activities is to form tetrahydrofolate (THF) that is further converted

Table 9.10 Protocol for generating *t*-test different spectrum and plotting using Igor (WaveMetrix)

Generate text file of t-test different spectrum

1. Read an excel file for PCA analysis
2. In cell P2, write a formula for average difference = SUM(I2:O2)/7-SUM(B2:H2)/7 (note: 7 controls and 7 exposed samples)
3. In cell Q2, write *t*-test formula: =TTEST(I2:O2,B2:H2,1,1) (one-tail and paired)
4. In cell R2, write IF formula: =IF(Q2 < 0.05,P2,0) [this means that if the value in cell Q2 is less than 0.05, then copy the value of cell P2 to R2. If same or greater, write 0 in R2]
5. Change #DIV/0! to 0 (by copying 0 and then pasting 0 to those cells)
6. Copy column 1 to a new file
7. Copy last column and paste the values to the new file and save the file as *.txt (tab delimited)
8. Repeat steps 2–7 for a different dose class of samples

Plot the text file using Igor

9. Run Igor
10. Click File → Open File → Notebook and brow the directory where the text file is stored
11. Click Data → Load wave → general text (not delimited text)
12. Select the text file and click open
13. Type in name for the first column, e.g., wave1 and check double precision, overwrite, make table
14. Click on the second name field and type in a name, wave2 click Load
15. Click Windows → New graph. Select wave2 for Y Wave (left axis), wave1 for X Wave (bottom axis)
16. Click do it
17. Click Graph → Modify trace appearance and change color, then click Do it
18. Right-click on the graph and select axis properties
19. Click Axis Range
20. Click wave2, then select bottom (up left-corner)
21. Click Swap. The spectrum should swap *X*-axis. If not, click Swap a few more times until it swaps
22. Click wave2 and select left for axis
23. Click Manual Min and increase the range a little bit
24. Click Do it
25. Click File → Save graphics and give a name for the graph. Make sure it is in JPEG format and set resolution to higher than 600
26. If there are nonzeros in the cutoff ranges (water and HEPES), copy zeros to the rows in the text file and reload wave file (step 9)
27. To zoom the spectrum, left-click and drag the mouse arrow on the spectrum and adjust the select box. Right-click inside the box and select Expand
28. To reduce the spectrum, select Shrink in the box
29. You can also change the axis range in Axis properties menu
30. To remove labels and axis line(s), click Axis submenu and change color to white
31. To change labels, click Axis label submenu and type text
32. To add ticks, bottom axis, Auto Ticks, Approximately 10, Minor ticks in 5 steps
33. To save the experiment, click File → Save experiment as and give a name
34. To load the graph in an experiment, open the experiment and then click Windows → graph macro and select the macro
35. To save the graph into a pdf, use 0.1 size for line: Graph → Modify Trace Appearance

(continued)

Table 9.10 (continued)

36. For multiple plots in same file
    (a) To add plots onto an existing graph, Click Graph → Append Traces to Graph. Select the data you would like to add, then click Dot It. Select all of the datasets you would like to add. The X Wave should never change; it is always the ppm values
    (b) The plots will appear on top of the other graph
    (c) Right-click on the new graph and click Modify Trace Appearance. All of the datasets will appear in the box under Trace. Highlight the one you would like to modify, and change the color to black
    (d) For the new plots, check the box that says Offset. A dialog box will open. Change the value of Y offset so that the two graphs are not touching. Click Ok to close the dialog box, then click Do It to apply the changes
    (e) The new graph should now be offset from the old one and in black
    (f) Continue to offset each new dataset by the same intervals, making sure that none of them are touching
    (g) To add labels, click Graph → Add Annotation. In the area that says Annotation, type the name of the dataset you would like to label. Click Frame, and in the dropdown box next to Annotation Frame, select None. Then click Do It
    (h) A text box should appear somewhere on the graph. Drag the text to the plot that you would like to label
    (i) Label each plot separately

Notes

1. If there are gaps in the plot
    a. The cut regions of solvents in the original spectra will cause gaps in the t-test filtered difference spectra.
    b. To correct this, go into the text or excel files, and insert zeros
    c. When you reconstruct the plot, there should now be a continuous line
2. If the plot is inverted from what it should be
    a. Check the Mnova file for the dataset
    b. If all of the graphs are upside down, this is the cause
    c. Highlight all of the samples, and click on the phase correction icon. Click the box that shows +180
    d. The graphs should now all be flipped over and upright
    e. Reprocess the data to get new normalized data and t-test
    f. Reconstruct the plotting in Igor to get a corrected difference plot
3. If the plot shows an abnormal amount of dispersion (lines zigzagging above and below the baseline, looking like positive and negative readings for the same points)
    a. Make sure all of the samples have had their phase adjusted properly
    b. The likely cause is miscalibration of the TSP points
    c. Check peak alignment using Mnova
    d. Reprocess the data to get new normalized data and t-test
    e. Reconstruct the plotting in Igor to get a corrected difference plot
4. To change the title of a graph
    a. Press Ctrl+y
    b. A window will open where you can edit the title of the graph

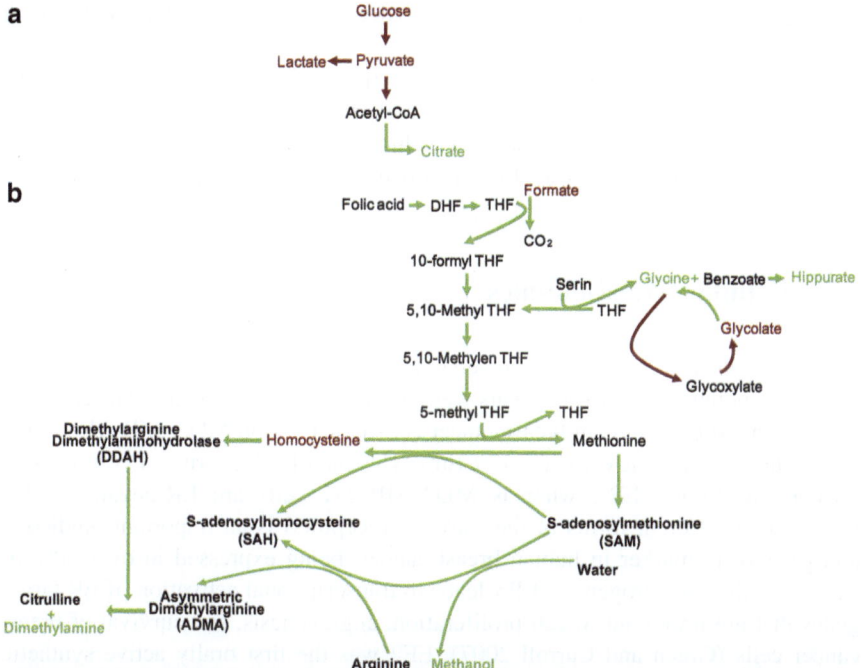

Fig. 9.30 Summary of the metabolic pathways including some of the detected metabolites. *Red* indicates metabolites with increased level due to cerebral infarction. *Black* indicates metabolites not found or not identified. *Green* indicates metabolites with decreased levels. The anaerobic glycolysis pathway was particularly enhanced after cerebral infarction (**a**). Many metabolites were detected because of folic acid deficiency (**b**) (reproduced with permission from Jung et al. (2011), Copyright © American Heart Association, Inc., 2011)

to 10-formyl-THF via the oxidization of formate in the folate pathway. The elevated level of formate may be an indicator of folic acid deficiency. Additionally, glycine is biosynthesized from serine through the biodegradation of folate, which is metabolically converted to hippurate via the conjugation with benzoate in the liver (Gatley and Sherratt 1977). Decreased levels of glycine and hippurate further support the explanation of a folic acid deficiency.

Hyperhomocysteinemia. Hyperhomocysteinemia is a high level of homocysteine in the blood, as a result of folic acid deficiency (Miller et al. 1994) and has been linked to an increased risk of stroke (Homocysteine Studies Collaboration 2002; Howard et al. 2002). A high level of homocysteine causes oxidative injury to vascular endothelial cells and damages the production of nitric oxide in the endothelium, which is a strong vascular relaxing factor (Dardik et al. 2000). In addition, hyperhomocysteinemia increases platelet adhesion to endothelial cells, which may obstruct blood vessels. The observed increased level of homocysteine in the patients is consistent with the previous finding. *S*-adenosylmethionine is metabolically converted to methanol, which is formed via the homocysteine biodegradation (Axelrod and Daly 1965). Therefore, decreased methanol in stroke patients may also be related to hyperhomocysteinemia that leads to limited availability of methionine and S-adenosylmethionine.

In summary, by applying metabolomics, the study of stroke patient's NMR metabolic profiles has identified a set of metabolites as metabolic indicators for early diagnosis of cerebral infarction, which are involved in three metabolic pathways: anaerobic glycolysis, folic acid pathway, and hyperhomocysteinemia.

9.7 Cellular Metabolomics

In this section, a case study is discussed to illustrate how a cell-based metabolomics study is conducted. The experiments were designed to study the cellular responses of different types of human breast cancer cells (MCF-7 and MDA-MB-231) to the treatment of a synthetic estrogen, 17α-Ethynylestradiol (EE2). MCF-7 cells possess estrogen receptors (ER), whereas MDA-MB-231 cells are ER-negative. ERα (one of two different forms of the estrogen receptors) is an important predictive and prognostic marker in human breast cancer, being expressed in over 60% of cases. Binding of estrogens to ERα leads to transcriptional activation of ER target genes that are important in cell proliferation, angiogenesis, and survival of breast cancer cells (Green and Carroll 2007). EE2 was the first orally active synthetic steroidal estrogen in contraceptive pills, and is now commonly used in the estrogen–progestin combination preparations. EE2 is hormonally effective by activating the estrogen receptor.

9.7.1 Experiments

9.7.1.1 Cell Treatment and Quench

In this section, examples will be described on:

1. How are adherent human cells cultured for metabolomics?
2. How are cells seeded and treated with a stressor?
3. How are the treated cells quenched for metabolomics studies?
4. What is a suitable cell quench method for metabolomic studies?

Human breast cancer cell lines MCF7 and MDA-MB-231 were cultured in a humidified atmosphere of 5% CO_2 using low glucose DMEM supplemented with 10% (v/v) heat-inactivated FBS, 100 units/mL penicillin, 100 µg/mL streptomycin, and 25 mM HEPES buffer at 37°C. It is worth noting that high glucose medium is not suitable for culturing cancer cells because low glucose medium has a glucose concentration to mimic the physiological concentration of glucose in solid tumor. In addition, a high glucose level can cause interference during metabolomic data analysis because the intense glucose peaks will dominate the contribution to the statistical modeling. The cells need to be thawed on ice, if they are stored in a liquid

9.7 Cellular Metabolomics

nitrogen dewar or a −80°C freezer. The medium for cell storage should be removed immediately after cells are thawed because 5% DMSO in the solution is toxic to the cells, which is used as a cryoprotectant to protect the cells from damage caused by ice formation when frozen. The cells were initially seeded in a T-25 tissue culture flask with 5 mL DMEM described above and cultured at 37°C and 5% CO_2 for 2 weeks before they were expanded to four T-75 tissue culture flasks with 12 mL DMEM in each flask. The medium was changed twice weekly.

When they reached 80% confluence, the cells were harvested and seeded onto 21 dishes (6 cm) for 24 h treatment with approximately 3×10^5 cells in 3 mL DMEM per dish without HEPES buffer and 21 dishes for 48 h treatment with approximately 2×10^5 cells per dish. For each treatment, there were 3 dose levels, including 0 μg/L (controls), 5 μg/L (low dose), and 10 μg/L (high dose) and 7 replicates per class (per dose level). After seeded, the cells were allowed to be incubated for 2 h to be able to attach on the culture dish surface before the treatment, which increases the reproducibility of the experiments. For each dish, a 3 μL dose solution was added, which were 0, 0.5, or 1.0 μg/L EE2 for control, low dose, or high-dose samples, respectively.

When the treatment was done (at the end of 24 or 48 h), the cells were quenched immediately to instantly stop all enzymatic reactions using a direct quench method (Sect. 9.3.6, Teng et al. 2009). Briefly, after the culture medium was removed from the culture dish, cells were quickly washed twice with ice cold PBS (pH 7.4). The residual PBS was removed by vacuum (Fig. 9.6a). Cells were then quenched using 0.8 mL HPLC grade methanol (Fig. 9.6b). Next, the cells were gently detached from the culture dish using a cell lifter (Fig. 9.6c). The methanol solution containing the quenched cells was pipetted into a 2 mL Safe-lock® centrifuge tube containing a 3.2 mm stainless steel bead for extraction.

9.7.1.2 Extraction of Intracellular Metabolites

Intracellular metabolites were extracted using a dual phase extraction procedure (Sect. 9.3.4, Ekman et al. 2008; Viant 2007). First, 127.5 μL ice cold DI water was added to each tube. The tubes were placed in a tissuelyser (e.g., a Qiagen tissuelyser) and shaken for 10 min at 15 Hz. The second step is to add 300 μL ice cold chloroform, then shake cell samples again for 20 min at 15 Hz. Next, 300 μL ice cold chloroform and 300 μL ice cold DI water were added followed by shaking the samples for another 10 min at 15 Hz. The samples were centrifuged at $1,000 \times g$ at 4°C for 10 min. The solutions were separated into an upper methanol/water phase containing polar metabolites and a lower chloroform phase containing lipophilic metabolites. Proteins and other biological macromolecules are precipitated by the addition of methanol and chloroform, and entrapped in the middle layer between the aqueous and organic phase. The upper and lower phases were collected as described in Sect. 9.3.6.

9.7.1.3 NMR Sample Preparation

The solvents were removed under vacuum at room temperature using a vacuum concentrator (or speedVac; 4 h for chloroform samples and 6 h for aqueous samples). The samples were reconstituted in 230 μL 0.1 M phosphate buffer (pH 7.4) with 20 μM TSP, vortexed for 30 s and centrifuged at 10,000 × g for 10 min. The last step of the sample preparation was to transfer the sample into 3 mm NMR tubes using a gel loading tip for each sample. Careful inspection of each NMR tube should be done for air bubbles. If air bubbles are formed inside the tube, storing the tube at 4°C will help remove the air bubbles. If air bubbles still exist inside the tube after a few hours in a refrigerator, vacuum can be used to remove the air bubbles by placing the tubes in a 1 L bottle with a screw cap through which vacuum is applied.

9.7.1.4 NMR Experiments

The NMR data for all samples were acquired using 1D PRESAT experiment. For experiment setup, ^{1}H channel of the probe was tuned by minimizing the power reflection of the transmitter. The shimming for the first sample was done by gradient shimming of Z_1–Z_4. Since the diameter of a 3 mm tube is relatively small, changes in higher order of Z shims (Z_5 and Z_6) will not influence spectral line width noticeably. Next, transverse shims were optimized by adjusting x, zx, y, zy, xy, and x^2–y^2. After shimming was done, ^{2}H lock frequency was set on ^{2}H$_2$O resonance, followed by adjusting lock power, gain and phase. RF calibration includes the calibrations of 90° pulse width, transmitter frequency, and saturation frequency that were set on water resonance.

The dead delay (the delay between last RF pulse and start of acquisition) were adjusted so as to eliminate error in the first-order phase correction—the first-order phase correction equals to zero. This step makes it easier to phase the whole set of the 1D spectra during data processing (see Sect. 9.7.2). The last step was to set experimental parameters, including spectral width of 12 ppm, acquisition time of 2 s (or longer), saturation duration of 2 s, dummy scans of 8, number of transients, and receiver gain. Before the acquisition of each sample, autoshimming must be performed by autogradient shimming of Z_1–Z_4. Autolock should not be used during any stage of the experiment because it will change lock frequency, resulting in poor water suppression. The data was stored every 16 transients. As a measure of spectral quality, the line width of DSS (or TSP) in a 1D ^{1}H NMR spectrum acquired on a 600 MHz NMR spectrometer is usually less than 1.0 Hz.

9.7.2 NMR Data Processing

After the experiments are done, the data can be transferred to an off-line computer for data processing. There are a variety of NMR software packages that can be used for metabolomics, including Mnova, Spinworks, ACD 1D NMR, Nuts, MestreC, etc. Because of its useful features such as peak alignment, spectral normalization and spectral superimposement, Mnova is used for all data processing discussed here. All data for the treatment study are loaded into Mnova. Line broadening of 0.3 Hz is applied to all spectra. The first-order phase constant is set to zero, followed by adjusting zero-order phase correction. The phases of all data are usually identical if the data is collected in automation for the whole set of samples. The baseline of the spectra is corrected using Bernstein polynomial fit with third polynomial order. All spectra are then superimposed together, after chemical shifts are referenced to internal standard DSS (or TSP). The residual water peak is then removed from the superimposed spectrum and chemical shift reference is inspected again.

To reduce the size of the data points that are variables for the statistical modeling, the spectra are binned into fewer data points before submitted for statistical analysis. Spectra with a range of 0.50–10.00 ppm are segmented into 0.005 ppm wide bins. In order to remove the unwanted systematic errors among the samples, spectral normalization is frequently used before the NMR data is used for statistical modeling. Therefore, the total area of the binned spectra is normalized to 1. The binned normalized superimposed spectrum is then saved as a text file and imported into Microsoft® Excel®. The sample labels are inserted into the first row of the text file. The resulted text file contains chemical shift values in the first column of the data file, which are variables, while the rest of the columns are the data for samples that are observations for PCA or other statistical analyses.

9.7.3 Principal Component Analysis

As discussed in Sect. 9.2, PCA is used to reduce dimensionality of the data for finding the variance among the observations (or samples). In this case, PCA decreases dimensionality from >2,000 variables to two principal components (PC1 and PC2). There are several statistical analysis software packages available such as SIMCA-P (Umetrics), PLS-Toolbox (or Solo, a stand-alone version of PLS-Toolbox; Eigenvector Research), MetoboAnalyst (web-based freeware; http://www.metaboanalyst.ca), and R (a freeware for statistical computing and graphics; http://www.r-project.org/).

Regardless of the statistical software used for metabolomics studies, it is necessary to preprocess the binned NMR data before the analysis is preformed. After all bins are mean-centered and Parato-scaled, principal components analysis (PCA) was conducted using all binned spectra. All samples of MCF-7 ($n = 6$) and MDA-MB-231 cells ($n = 7$) were observed to fall inside of the Hotelling's T2 ellipse at

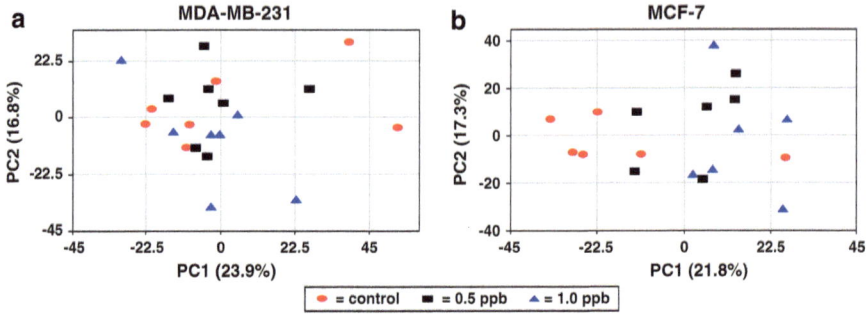

Fig. 9.31 Two-component scores plots from PCA models built with ^{1}H NMR spectra of the extracts of the aqueous extracts of (**a**) estrogen receptor (ER)-negative human breast cancer cells (MDA-MB-231), and (**b**) ER-positive MCF-7 cells following the treatment of 17a-ethynylestradiol (EE2) at the level of 0 (control), 0.5 ppb and 1.0 ppb for 48 h (reproduced with permission from Teng et al. (2009), Copyright © Springer, 2009)

the 95% confidence interval in scores plots of the first two components, meaning that there were no outliers. Loadings plots were used to identify the metabolites responsible for the separation in scores plots.

Shown in Fig. 9.31a is a two-component (PC1/PC2) scores plot of a PCA model built for the MDA-MB-231 cell line treated with EE2, which shows that there is no discernable separation in score values among the three groups, where group (or class) is defined by treatment level (control cells, cells treated with 0.5 ppb EE2, and those treated with 1.0 ppb EE2). The results indicate that MDA-MB-231 cells do not have significant response to the treatment of EE2. This can be explained by the fact that MDA-MB-231 cells are ER negative, and thus are not influenced much by the presence of estrogen EE2. The same type of plot is displayed for the MCF-7 cells treated with EE2 (Fig. 9.31b). In the MCF-7 scores plot, the classes are well separated primarily along the PC1 that is the component describing the maximum variation in the dataset. Scores for the 1.0 ppb treated class are clustered to the right region of the origin, while those for the control class are clustered to the left, with the exception of one outlier in the control class. Scores for the low-dose treated class (0.5 ppb) distribute mostly between the control and the high-dose treated classes. Thus, PCA (a widely used and unsupervised multivariate method) is able to capture the dose-dependent changes in metabolite profiles associated with the treatment for the case of ER-positive cells (MCF-7), which is a widely used and unsupervised multivariate method. In contrast, the treatment with EE2 does not induce significant metabolic changes in ER-negative cells (MDA-MB-231). These results demonstrate that cell-based metabolomics is a valuable tool to study the cellular responses to the presence of external stressors.

To further investigate the effect of EE2 on metabolic profiles in these two cell lines, univariate analysis can be conducted using the normalized and binned spectral spreadsheet (but not scaled) (Ekman et al. 2008). First, an "average class spectrum" is obtained by averaging the binned spectra across all class members ($n = 6$ for MCF-7 or $n = 7$ for MDA-MB-231). Next, a difference spectrum is generated by

9.7 Cellular Metabolomics

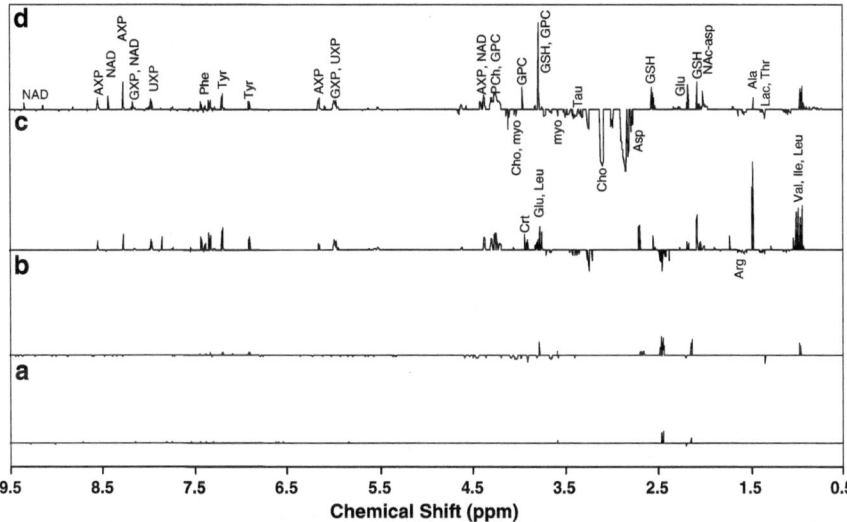

Fig. 9.32 The "*t*-test filtered difference spectra" generated from ^{1}H NMR spectra of cell extracts after the treatment with 17α-ethynylestradiol (EE2): (**a**) MDA-MB-231, 0.5 ppb, (**b**) MDA-MB-231, 1.0 ppb, (**c**) MCF-7, 0.5 ppb, and (**d**) MCF-7, 1.0 ppb (ppb = μg/L). Positive peaks correspond to metabolites that increase upon treatment, whereas negative peaks are from metabolites that decrease upon treatment (reproduced with permission from Teng et al. (2009), Copyright © Springer, 2009)

subtracting the average class spectrum of the control class from that of each treated class. Then, a Student's *t*-test is applied to each bin to test the statistical significance of each averaged bin using a *P*-value <0.05 where *P* is the higher probability of the Student's paired *t*-test. If the *P*-value is less than 0.05, the average for the treated class is considered to differ significantly from that of the relevant control class. If not, the bin value for the difference spectrum is replaced with a zero. This results in a "*t*-test filtered difference spectrum" for each treatment class. Positive peaks represent metabolites that increase in concentration (with statistical significance) upon treatment, whereas negative peaks are from metabolites that decrease in concentration.

Shown in Fig. 9.32 are four *t*-test filtered difference spectra of MDA-MB-231 and MCF-7 cells treated with EE2. Because all four difference spectra are displayed with the same intensity scale, the overall changes of the metabolic profile for each treatment class can be evaluated by comparing the number and intensity of peaks in each spectrum. Being consistent with PCA results, these spectra confirm that the treatment of EE2 induces much larger changes in overall metabolite profiles of MCF-7 cells (Fig. 9.32c, d) than those of MDA-MB-231 cells (Fig. 9.32a, b). Additionally, it is evident that the effect of EE2 on MCF-7 cells is dose-dependent, by comparing the difference spectrum from the low-dose treatment (Fig. 9.32c) with that of the high-dose treatment (Fig. 9.32d). It is worth noting that many metabolites with low-frequency resonances (i.e., lower ppm values) have an increase in concentration in the EE2-treated MCF-7 cells (Fig. 9.32c, d) compared to the controls, including valine (Val), leucine (Leu), isoleucine (Ile), alanine (Ala), threonine (Thr), glutamate (Glu), N-acetyl aspartate (NAc-asp), Glutathione (GSH), and PCh, while decreases were observed for myoinositol (myo), choline (Cho), taurine (Tau), and glutamine (Gln). In the high-frequency region of the water resonance (i.e., higher ppm values), the treated MCF-7 cells show increases in concentrations of nicotinamide adenine dinucleotide (NAD^+), phenylalanine (Phe), Tryptophan (Trp), tyrosine (Tyr), histidine (His), and the nucleotides Guanosine-phosphate (GXP, where X is M, D, or T) and adenosine phosphate (AXP). There is very minor change in the metabolic profiles of the treated MDA-MB-231 cells compared to the treated MCF-7 cells.

9.7.4 Partial Least Squares for Discriminant Analysis

Partial Least Squares for Discriminant Analysis (PLS-DA) is a regression extension of PCA, which builds a linear regression model using class information to discriminate the classes in a new set of components. Because it uses the class information to achieve the utmost separation, PLS-DA is a supervised statistical analysis, and hence models generated by PLS-DA must be undergone a validation process.

Shown in Fig. 9.33a, b are the PLS-DA scores plots for the first two components of MDA-MB-231 and MCF-7 cells, respectively. Similar to its PCA scores plot, MDA-MB-231 cells do not show any clear separations between the classes. Because PCA determined one of MCF-7 control samples as an outlier, the outlier was excluded in the PLS-DA. The class separation is improved slightly in PLS-DA, compared to that in PCA excluding C5. The validation results with 50 permutations are shown in Fig. 9.34, with R^2 and Q^2 intercepts of 0.19 and -0.27, respectively. All 50 permutations have generated models with smaller Q^2 than those of the original model, indicating that overall PLS-DA of MCF-7 samples yields a valid model.

In the 1920s, Warburg et al. discovered that one of the hallmarks of cancer is the increase of glucose uptake by cancer cells to meet the energy demand for the rapid cell growth (Warburg et al. 1924; Warburg 1956), which can cause the increases in intracellular concentrations of energy-containing molecules such as ATP, GTP, and NDA^+. Because of the lack of estrogen receptors, the treatment of EE2 should not cause any impact on the intracellular metabolic profile of the MDA-MB-231 cells. This is clearly shown by the t-test filtered difference spectra of treated MDA-MB-231 cells (Fig. 9.32a, b). The changes in the metabolic profiles of treated MDA-MB-231 cells are negligible, although there were few changes in the high-dose treatment (Fig. 9.32b). In the ER-positive MCF-7 cells, however, the treatment of EE2 significantly increases the levels of ATP, GTP, and NAD^+ (Fig. 9.32c, d), indicating that the EE2 treatment may promote cell proliferation in the MCF-7 cell line. Furthermore, the increase in concentration of several key amino acids in the treated MCF-7 cells indicates that the treatment of EE2 may also increase the availability of the amino acids for protein synthesis in MCF-7 cells. The rapid growth of cancer cells requires not only energy but also proteins in addition to lipids and fatty acids.

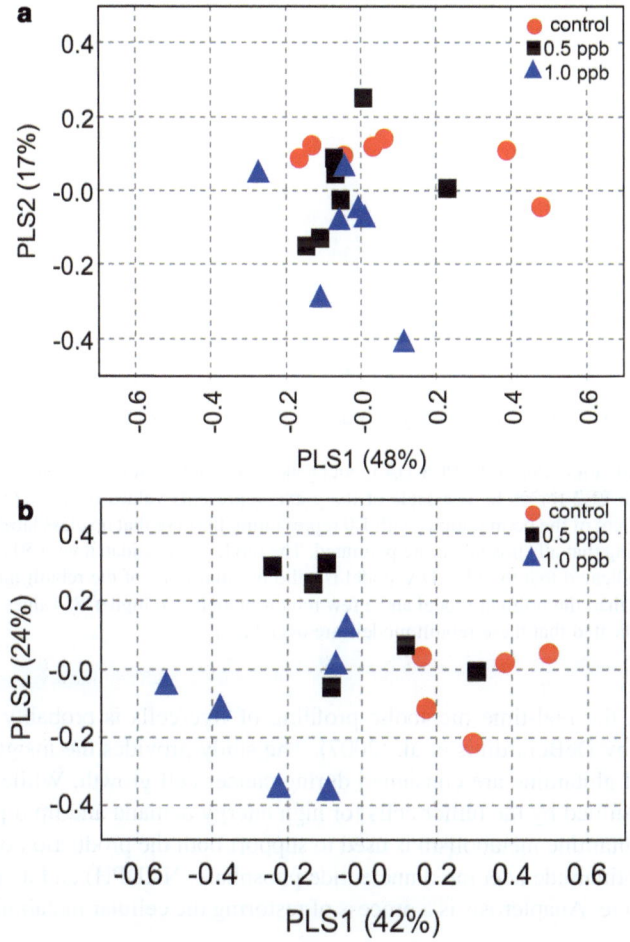

Fig. 9.33 Two-component scores plot from a PLS-DA model built with the same data as the PCA model (Fig. 9.31) from ^{1}H NMR spectra of the aqueous extracts of (**a**) MDA-MD-231 and (**b**) MCF-7, excluding the control outlier. The distribution of the scores is very similar to that of PCA model

9.8 Metabolomics of Live Cell

In mammalian cells, glucose and glutamine, as major nutrients in plasma, account for most carbon and nitrogen metabolism. According to the Warburg effect (Warburg et al. 1924; Warburg 1956), the metabolic hallmark of tumor cells is the rapid glucose consumption and lactate production to meet the requirements of rapid synthesis of biomacromolecules including nucleotides, proteins, and lipids. Dynamic ^{13}C NMR has been used in the real-time observation of metabolic fluxes in glucose and glutamine metabolism of live cells. The most comprehensive

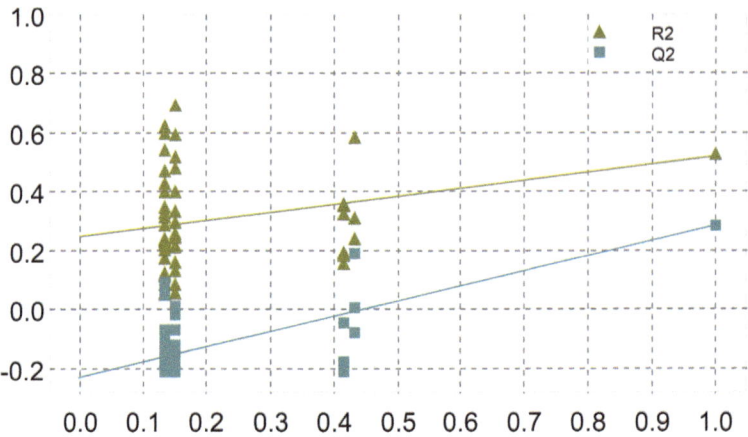

Fig. 9.34 Validation plot of the PLS-DA model generated from polar extracts of MCF-7 after the treatment with EE2 for 48 h. The scale of the y axis represents values of R^2 or Q^2. The x axis reflects the extent of the permutations with 1.0 representing the case that no class label is permuted and 0.0 meaning that all class labels are permuted. The model was validated with 50 permutations. The results indicated that the PLS-DA model is valid because none of the rebuilt models had Q^2 values higher than the original model and a few rebuilt models had higher R^2 values with smaller Q^2, which indicated that these rebuilt models are over-fitted

research on the real-time metabolic profiling of live cells is probably the recent work done by DeBerardinis et al. (2007). The study provides the insights on how glucose and glutamine are consumed during cancer cell growth. While glucose is rapidly consumed by the tumor cells for high energy demand and lipid production, the rapid glutamine metabolism is used to support both the production of reductive power (nicotinamide adenine dinucleotide phosphate, NADPH) and anaplerosis of the TCA cycle. Anaplerosis is a process of restoring the cellular metabolic intermediate pools.

As much of the study is not easily carried out, here the attempt is only to outline some of the results of the live cell study. Shown in Fig. 9.35 is a typical instrument setup for dynamic NMR experiments of live cells. Cells grown in microcarriers are loaded in an NMR tube and maintained by circulating the culture medium using a peristaltic pump. After standard setup and calibration is done (tuning, shimming, and 90° pulse calibration), culture medium is switched to the medium containing ^{13}C-labeled substances (^{13}C glucose and/or ^{13}C glutamine). To study the glucose metabolism, the live SF188 glioblastoma cells were maintained inside the NMR magnet by feeding culture medium with constant oxygenation, while a series of 1D NOE-enhanced ^{13}C NMR experiments were acquired at the rate of 18 min/experiment, the cells were then given a bolus of medium with 10 mM 1,6-^{13}C$_2$-glucose. After 2.5 h, 4 mM 1,6-^{13}C$_2$-glucose medium was continuously flowed through the NMR tube.

The real-time ^{13}C NMR spectra (Fig. 9.36) observed in the 1,6-^{13}C$_2$-glucose perfusion experiment have shown that 3-^{13}C lactate is the most abundant metabolite

9.8 Metabolomics of Live Cell

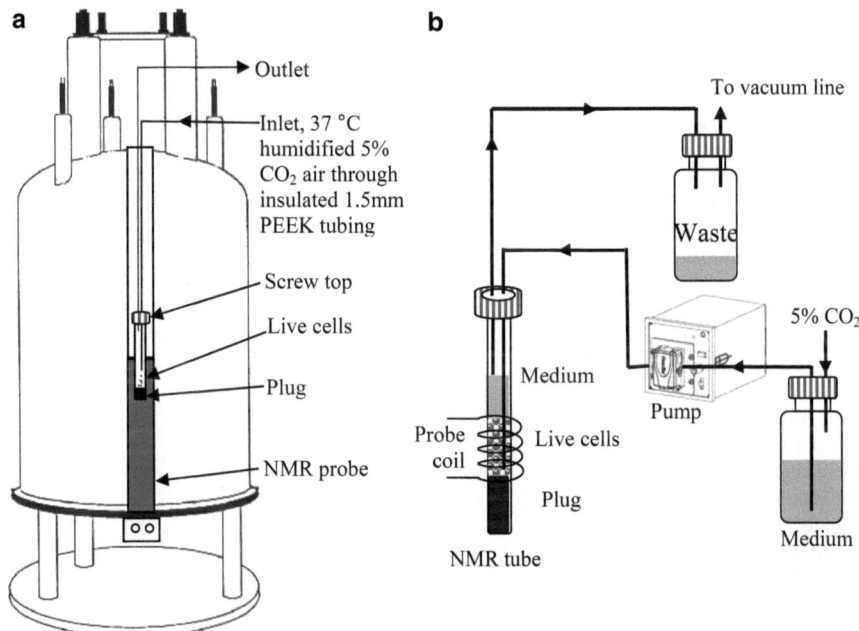

Fig. 9.35 Schematic drawing of NMR setup for live cell experiments. (**a**) The culture medium containing ^{13}C labeled nutrients is constantly flowing through the NMR tube inside the magnet. (**b**) The medium is pumped to the NMR tube containing live cells from the medium reservoir where 5 % CO_2 is maintained. The 5 % CO_2 gas can be adjusted for hypoxia or nomoxia condition

produced from the labeled glucose, which was predominantly extracellular. This suggests that the cancer cells consume glucose to meet the energy demands predominantly through rapid glycolysis producing lactate as the end product, which is consistent with the Warburg Effect. The activities of the TCA cycle were also observed by the real-time experiments, as evidenced by the increased intensity of 4-^{13}C-glutamate (Fig. 9.36b) as a function of perfusion time. The formation of 4-^{13}C-glutamate from 1,6-$^{13}C_2$-glucose is an indicative of TCA cataplerotic reactions. As shown in Fig. 9.36a, 4-^{13}C-glutamate is formed by TCA cycle intermediate α-ketoglutarate (α-KG). The cells also consumed glucose for lipid synthesis through citrate efflux of the TCA cycle (Fig. 9.36a). The graduate increase in intensity of fatty acid ^{13}C signals $(CH_2)_n$ suggests that acetyl-CoA formed by the TCA cycle was used for lipid biosynthesis by the sequential addition of acetyl-CoA. The study determined that approximately 60% of the lipogenic acetyl-CoA pool comes from glucose.

Progressive growth of tumor cells requires rapid lipid synthesis for cell proliferation. To fulfill this need, it is necessary for cells to possess two supporting pathways for maintaining sufficient reductive power (NADPH as an electron donor for lipid synthesis) and to restore oxaloacetate for the continuation of the TCA cycle during citrate export (for the production of lipogenic precursor acetyl-CoA).

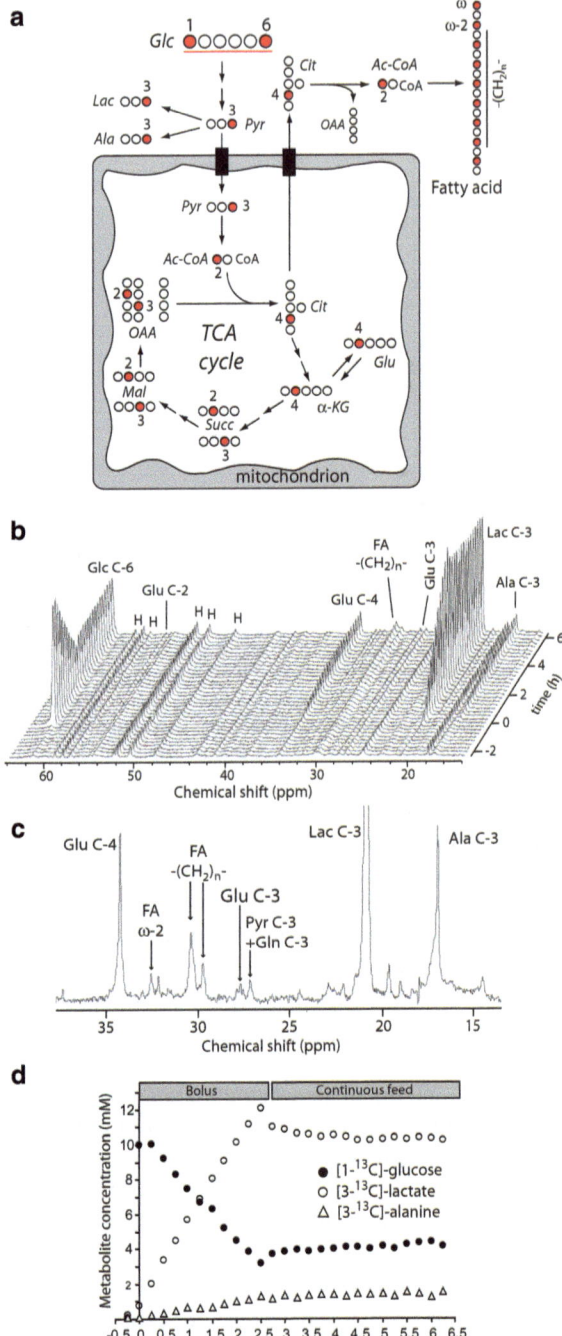

Fig. 9.36 Glucose metabolism in proliferating glioblastoma cells (SF188). (**a**) The ^{13}C carbons (filled circles) of [1,6-^{13}C$_2$]glucose distribute into various metabolites in several major pathways, including glycolysis, anaerobic pyruvate metabolism, TCA cycle, and fatty acid synthesis. (**b**) Stacked spectra from a real-time [1,6-^{13}C$_2$] glucose perfusion experiment in SF188 cells. (**c**) An extended spectrum from time 6 h shows the enrichment of metabolic intermediates, after the subtraction of background signals from the time 0-h spectrum (**d**) Concentration plots of ^{13}C labeled glucose, lactate, and alanine as the function of the experimental time. (reproduced with permission from DeBerardinis et al. (2007), Copyright © The National Academy of Sciences of the USA. 2007)

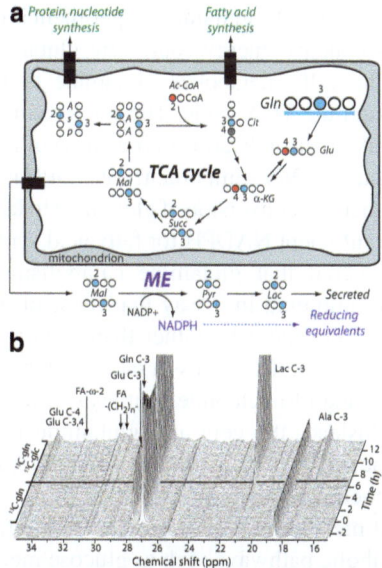

Fig. 9.37 Glutamine metabolism in proliferating glioblastoma cells (SF188). (**a**) The labeling pathway of ^{13}C from [3-^{13}C]glutamine (*blue*; at the *top right* of the mitochondrion) highlights glutaminolysis that produces NADPH by converting glutamine-derived carbon to lactate using malic enzyme (ME), [2-^{13}C]Ac-CoA (*red*) comes from [1,6-$^{13}C_2$]glucose. (**b**) Stacked spectra acquired during the two-stage perfusion experiment (reproduced with permission from DeBerardinis et al. (2007), Copyright © The National Academy of Sciences of the USA. 2007)

Similar to glucose, glutamine is metabolized by proliferating cells using a variety of pathways to support bioenergetics and biosynthesis. It was proposed that the high rate of glutamine metabolism can meet the both needs to support the lipid synthesis. A two-stage perfusion was used for investigating the activation of pathways involved in the glutamine metabolism. In the first stage, cells received medium containing 4 mM 3-^{13}C glutamine and 10 mM unlabeled glucose, while in the second stage, the medium contained both 4 mM 3-^{13}C glutamine and 10 mM 1,6-$^{13}C_2$ glucose (Fig. 9.37b). Glutamine provided a secondary source of carbon for lipid synthesis. It was determined by the real-time measurement of the $(^{13}CH_2)_n$ signal that the rate of ^{13}C labeling in $-(CH_2)_n-$ during stage 2 was triple the rate in stage 1, consistent with glucose's role as the major lipogenic precursor. The study showed that glutamine contributed to approximately 25% of total lipid carbon, while glucose accounted for 50%.

During stage 1, labeled 3-^{13}C glutamate was the most abundant ^{13}C metabolite produced rapidly from 3-^{13}C glutamine (Fig. 9.37b). The immediate appearance of ^{13}C labeled aspartate in C-2 and C-3 after 3-^{13}C glutamate was formed indicated that anaplerotic oxaloacetate was derived from the glutamine pool, which suppressed pyruvate carboxylation despite substantial mitochondrial pyruvate metabolism. This depicts that some of the glutamine pool entered the TCA cycle for protein production (through Asp synthesis) and lipid synthesis (formation of Acetyl-CoA).

The other important finding of the glutamine experiment is that the cancer cells also utilize glutamine as a secondary energy source by glutaminolysis. Similar to the utilization of glucose, the cells metabolize glutamine to lactate, rather than the complete oxidation through the TCA cycle. The observed formation of ^{13}C labeled lactate at C-2 and C-3 is likely the indication of oxidation of labeled malate at C-2 and C-3 derived from the glutamine. More importantly, NADPH is produced by the malate oxidation. It was proven that this conversion of glutamine to lactate (glutaminolysis) is rapid enough to produce sufficient NADPH for fatty acid synthesis.

The study also uncovered that glutamine catabolism was accompanied by secretion of alanine and ammonia in a way that most of the amino groups from glutamine were released from the cells rather than used by other molecules. The results demonstrate that cancer cells possess a rapid glutamine consumption, which exceeds the nitrogen demand of nucleotide synthesis or maintenance of nonessential amino acid pools. Instead, the cells use glutamine metabolism as a carbon source in the same manner as to use glucose-derived carbon and TCA cycle intermediates as biosynthetic precursors.

In summary, live cell metabolomic profiling is a powerful tool to study, in real time, the important metabolic pathways such as glucose metabolism and glutamine metabolism. It can bridge the gap between high-resolution metabolite profiling of cell or tissue extracts and the low-resolution MRS studies in animal models and patients. The results of real-time metabolic profiling can provide insights on the metabolic fluxes of cancer cells.

9.9 Metabolomics Applied to Cancer Research

In recent years, metabolomics has been successfully applied in cancer research to study cancer biochemical networks and metabolic pathways. For example, metabolomics has been used to differentiate different cancer cell lines (Donato et al. 2008; Ramanathan et al. 2005), to monitor metabolic processes in cancer cells during apoptosis, to detect and prognose cancers, including liver (Yang et al. 2007a, b), leukemia (MacIntyre et al. 2010), colon (Piotto et al. 2009; Ritchie et al. 2010), breast (Oakman et al. 2011), ovarian (Odunsi et al. 2005), and prostate cancer (Raina et al. 2009). Metabolic profiles of cancerous and noncancerous tissues have been studied to characterize cell growth and death, specific tumor types, and pathologic states of tumors (Griffin and Kauppinen 2007; Griffin and Shockcor 2004; Serkova et al. 2007; Serkova and Niemann 2006). In this section, two in vivo case studies are used as examples to describe how NMR-based metabolomics can be applied in cancer research. The first example involves a quantitative characterization of the metabolic profile of a silibinin-treated mouse tumor model to assess the chemopreventive efficacy of silibinin. The second example describes the application of metabolomics to predict overall survival time of patients with metastatic colorectal cancer (CRC).

9.9.1 Silibinin Anticancer Efficacy

Silibinin (also known as silybin) has long been recognized to have inhibitory effects on tumor growth, progression, and migration (Bhatia et al. 1999; Raina et al. 2007; Hogan et al. 2007). Although the mechanism of the tumor inhibition was studied (Raina et al. 2008), the lack of information on silibinin-induced metabolic perturbations of the tumor tissue makes it difficult to assess its chemopreventive efficacy. To quantitatively characterize the metabolic profile of a silibinin-treated mouse tumor model, an NMR-based metabolomics study was conducted (Raina et al. 2009). The study aimed to specifically identify a set of biomarkers that can sensitively detect metabolic changes in the tumor tissue upon treatment with silibinin.

Four prostate tumor tissue samples were used for a control or a silibinin-treated group. The silibinin treatment was started at week 4 for 20 weeks. Approximately 0.1 g of frozen tissues were extracted and both aqueous (polar) and lipophilic (nonpolar) fractions were dried at room temperature using a speedVac. For NMR analysis, aqueous samples were reconstituted in 450 μL ^2H_2O and lipophilic samples in 600 μL deuterated chloroform/methanol mixture (2:1, v/v) and PRESAT or one-pulse experiments were conducted for aqueous or lipophilic samples, respectively. For each tissue, a total of 38 metabolites were assigned and their concentrations were calculated based on the concentration of an external TSP reference standard (0.5 mM for polar and 1.2 mM for nonpolar extracts) and subsequently normalized to the unit (wet) weight of individual tissues. It was the concentrations of the metabolites (plus 4 concentration ratios) that were used as variables for MVA, not NMR spectral bins.

A PCA model was built using the 42 concentration variables. The scores plot of the PCA model (Fig. 9.38a) clearly shows that the treated group is well separated from the control group. The shattered loadings plot of PCA model (Fig. 9.38b) shows the contributions of individual metabolites to the group discrimination described by the model. By comparing the average concentrations of treated and control group, a total of 14 biomarkers contributed to the class discrimination, including silibinin-induced elevated levels of citrate, glucose, hydrophilic choline, glycerophosphocholine (GPC), polyols, and glutathione (reduced form; GSH), as well as significant decrease in levels of lactate, cholesterol, alanine, lipophilic choline (Cho), and phosphatidylcholine (PtdCho). One of the hallmarks of prostate cancer is a decrease in citrate levels due to the increased consumption of citrate in the TCA cycle along with abnormal secretion functions and zinc metabolism (Costello and Franklin 2000, 2005). In normal prostate cells, citrate is not a functional intermediate of the TCA cycle, but an end product due to the high concentration of zinc that is involved in inhibition of the conversion of citrate into isocitrate. On the other hand, the decreased amount of zinc in prostate tumors leads to the formation of isocitrate from citrate, causing increased activities of the TCA cycle and decreased levels of citrate to meet the increased energy demand and fatty acid synthesis (Costello and Franklin 2000). Additionally, the decrease in cholesterol indicated that the increased citrate was not utilized by cholesterogenesis. Its ability to restore

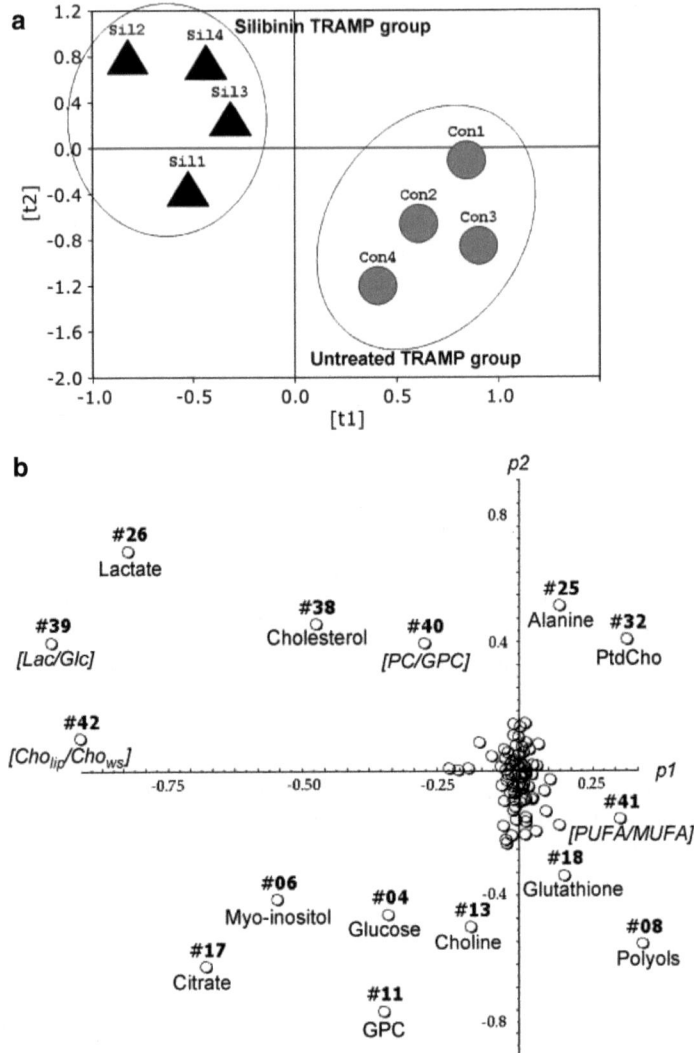

Fig. 9.38 Two-component scores (**a**) and loadings (**b**) plots of a PCA model generated from concentrations of metabolites. The concentrations of water-soluble and lipid metabolites were calculated from ^{1}H-NMR spectra of prostate tissue extracts. The scores plot clearly showed class separations between control and silibinin-treated groups. Axis labels t_1 and t_2 in (**a**) are PC1 and PC2, respectively (reproduced with permission from Raina et al. (2009), Copyright © American Association for Cancer Research, 2009)

the citrate accumulation and secretion of the prostate tissue indicates that silibinin possesses a remarkable chemopreventive efficacy.

Silibinin induced a significant decrease in PtdCho level (a major membrane phospholipid) and lipophilic fraction of choline (a precursor of PtdCho) compared to the untreated controls, which are intermediates of choline phospholipid metabolism. The decrease in levels of both metabolites has been observed in the malignant tissues (Glunde and Serkova 2006; Glunde et al. 2006). Other bioindicators of silibinin chemopreventive efficacy includes the significant increase in concentrations of polyols (decreased during prostate tumor progression) and glutathione (an antioxidant), meaning that silibinin exhibited inhibitory effect on progression of prostate cancer.

In summary, the concentrations of metabolites in tissue extracts of a mouse prostate tumor model were used to generate a PCA model that showed the silibinin-induced metabolic changes of the treated tumors compared to the untreated controls. The metabolic changes related to silibinin chemopreventive efficacy against prostate cancer include the restoration of normal citrate concentrations, decrease in membrane phospholipid synthesis, and increase in polyols and antioxidants.

9.9.2 Metabolomic Profiling of Colon Cancer

Colorectal cancer (CRC, or colon cancer) is the third leading cause of cancer-related deaths in the United States. Most of CRC symptoms may not be noticeable until the progression stage in which many patients have metastases and the tumor has spread to other organs. Therefore, early detection of CRC, especially for patients with metastatic CRC (mCRC), can significantly improve the treatment and survival outcomes. An NMR-based metabolomics study (Bertini et al. 2012) was conducted to determine whether there is a significant difference in the metabolomic profiles, which can be used to predict overall survival (OS) time.

Serum samples from 139 healthy individuals and 153 patients with mCRC (resistant to 5-FU, oxaliplatin, and irinotecan) were used to determine the metabolic signature of mCRC. The samples were collected and allowed to clot at room temperature for 30–120 min. For NMR analysis, serum was mixed with an equal volume of 70 mM PB containing 6.15 mM NaN_3 and 6.64 mM TSP at pH 7.4. CMPG spectra were acquired for metabolite profiling and J-resolved (JRES) spectra were used to identify the metabolites. The residual water resonance of 4.2–5.0 ppm was excluded and metabolite resonances were aligned. The spectral region of 0.5–8.5 ppm was then segmented into 0.02 ppm bins and normalized over the total spectral intensity of the region.

The whole dataset (all binned spectra) was split into a training set with $n = 141$ (96 healthy individuals and 45 patients) for generating a PLS model, and a validation set with $n = 151$ (43 healthy individuals and 108 patients) for assessing the prediction accuracy of the model. The PLS model was generated and subsequently validated by a tenfold cross validation. First, all data of the training set were partitioned into 10 subsets of samples, of which 9 subsets were used to produce a PLS model and the remaining subset was used as a test set to validate the model. The model generation and cross validation were repeated nine more times (tenfold) using each of the 10 subsets as the test set. The random partition was repeated 100 times, resulting in a total of 1,000 random subsamplings (tenfold cross validation is repeated 100 times). The results (scores and loadings)

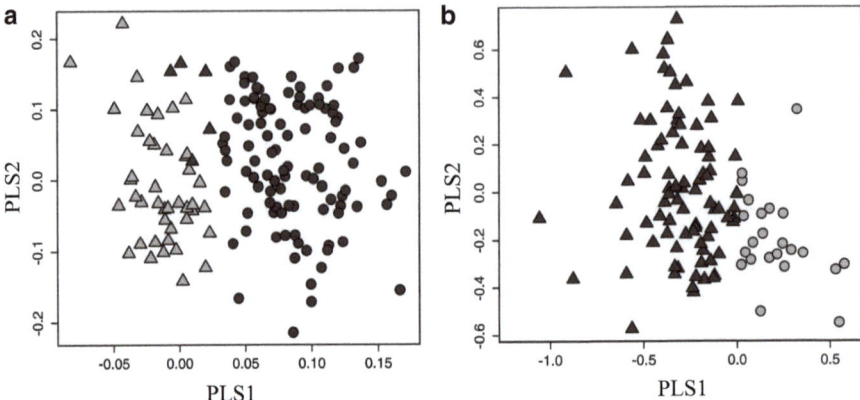

Fig. 9.39 Two-component PLS scores plots. (**a**) The score plot for the healthy subjects (*triangles*) and patients with mCRC (*circles*) of the validation set based on the PLS model built on the NMR spectra of the training set (see text for definition of training and validation set). *Triangles* are predicted as healthy subjects and circles as mCRC. (**b**) The score plot for the long OS group (*triangles*) and short OS group (*circles*) of the validation set based on the model built on the NMR spectra of the training set (reproduced with permission Bertini et al. (2012), Copyright © American Association for Cancer Research, 2012)

for the training set were then averaged to obtain the final model. A validated PLS model was then used to predict classifications of samples using the validation set. The results are shown in Fig. 9.39a. The model clearly shows a good separation along PLS1 (the first component) between the healthy and patient group of the validation set, meaning that the spectral patterns (or metabolite profiles) of serum from the patients with mCRC have characteristic features compared to those of healthy individuals. Note that the validation set in this study has a different meaning than the commonly defined test set, which was not used to generate or cross validate the PLS models.

One of the main goals of the study was to determine whether NMR metabolic profiles of serum from mCRC patients can be related to OS time. To test the prognostic classification using serum metabolic profiles, a model was generated using 20 samples from mCRC patients of the training set, of which 10 patients had the shortest OS and the remaining 10 had the longest OS. After the model was validated with cross validation (1,000 times of permutations), the model was used to predict OS group classification of mCRC patients in the validation set. The results (the scores plot, Fig. 9.39b) showed two slightly overlapped clusters of long and short OS group. The model assigned 85 of 108 mCRC patients in the validation set to the long OS class and 23 patients to the short OS class, with mean OS values of 12.3 and 5.2 months, respectively. This suggested that the model has reasonable accuracy for predicting OS time of mCRC patients.

Characterization of metabolites related to the mCRC serum profiles was conducted using the loadings plot generated with all sets of healthy individuals and patients (training set and validation set). After the NMR resonances in the

9.9 Metabolomics Applied to Cancer Research

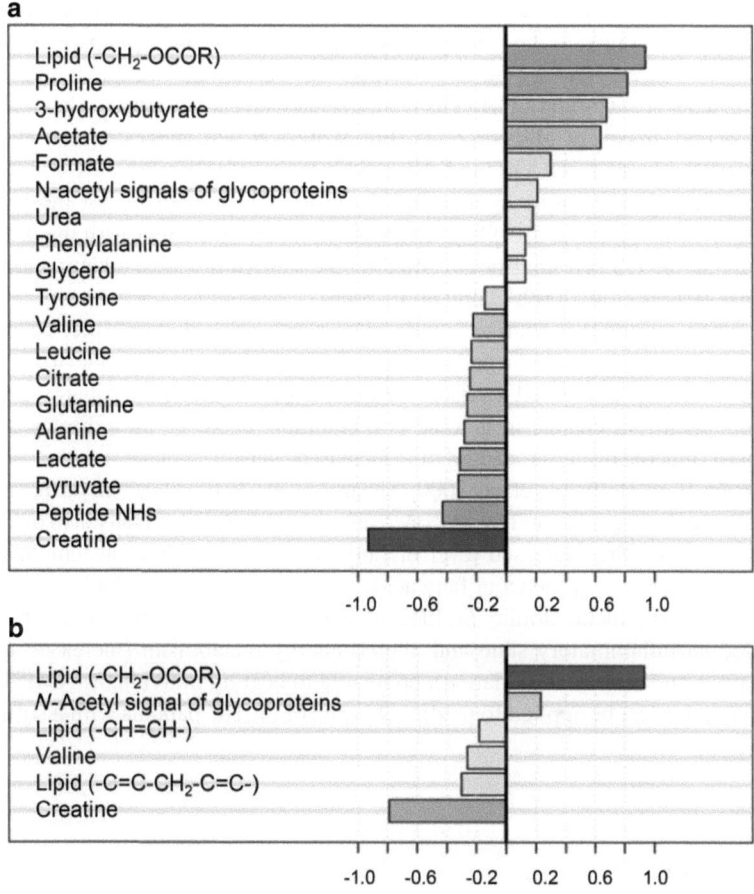

Fig. 9.40 Values of ln(FC) obtained (**a**) by comparing healthy subjects to patients with mCRC, and (**b**) by comparing patients with mCRC and long OS (>24 months) to short OS (<3 months). FC represents fold change. (**a**) ln(FC) with a positive value indicates an increased serum concentration in the patients with mCRC, whereas a negative value means a decreased serum concentration in the patients with mCRC. (**b**) ln(FC) with a positive value indicates an increased serum concentration present in the patients with mCRC and short survival (<3 months), whereas a negative value means a decreased serum concentration in the patients with mCRC and short survival (<3 months) (reproduced with permission Bertini et al. (2012), Copyright © American Association for Cancer Research, 2012)

loadings plot were identified, the concentration changes of the metabolites were estimated by comparing the average spectral integrates of the metabolites in serum samples of patients with those of healthy individuals. The analysis showed that mCRC patients are characterized by lower serum levels of creatine, peptide NHs, pyruvate, lactate, alanine, glutamine, citrate, leucine, valine, and tyrosine, and higher levels of lipid ($-CH_2-OCOR$), proline, 3-hydroxybutyrate, acetate, formate, N-acetyl groups of glycoproteins, phenylalanine, and glycerol (Fig. 9.40a).

Comparison of the average spectra of serum samples from patients with an OS shorter than 3 months to those longer than 24 months revealed that patients with a short OS have lower serum levels of creatine, valine, and lipid (−C=C−C$\underline{H}_2$−C=C− and −C$\underline{H}$=C$\underline{H}$−) and higher levels of lipid (−CH$_2$−OCOR) and N-acetyl groups of glycoproteins (Fig. 9.40b). It was suggested that elevated levels of lipid −CH$_2$−OCOR and N-acetyl groups of glycoproteins reflected a nonspecific inflammatory response and the effect is more severe for the patients with shorter OS (Torri et al. 1999). These results were consistent with the previous finding that the OS is primarily governed by the state of the local adaptive immune response (Pages et al. 2009; Galon et al. 2006).

Lactate, pyruvate, alanine, and glutamine are glucose precursors of gluconeogenesis—a metabolic pathway that generates glucose from non-carbohydrate carbon substrates. Decreased levels of the glucose precursors in the patients, coupled with no significant change in glucose concentration, could be an indicator that an increased gluconeogenesis caused an increased hepatic uptake of the glucose precursors (Leij-Halfwerk et al. 2000; Shearer et al. 2010; Holroyde et al. 1975). The increased energy demand for gluconeogenesis was fulfilled via fatty acid oxidation as evidenced by the decreased level of fatty acids along with an accumulation of 3-hydroxybutyrate—a product of fatty acid oxidation (Tiziani et al. 2009).

In summary, metabolomic profiles of serum samples from mCRC patients revealed an inflammatory state and altered energy metabolism (increased gluconeogenesis and fatty acid oxidation), compared to healthy individuals. NMR-based metabolomics can be applied to predict OS time of mCRC patients with a sufficient accuracy.

References

Axelrod J, Daly J (1965) Pituitary gland: enzymic formation of methanol from S-adenosyl-methionine. Science 150:892–893

Baxter MJ (1995) Standardization and transformation in principal component analysis, with applications to archaeometry. Appl Statist 44:513–527

Beloueche-Babari M, Chung YL, Al-Saffar NM, Falck-Miniotis M, Leach MO (2010) Metabolic assessment of the action of targeted cancer therapeutics using magnetic resonance spectroscopy. BrJ Canc 102:1–7

Bernini P, Bertini I, Luchinat C, Nincheri P, Staderini S, Turano P (2011) Standard operating procedures for pre-analytical handling of blood and urine for metabolomic studies and biobanks. J Biomol NMR 49:231–243

Bertini I, Cacciatore S, Jensen BV, Schou JV, Johansen JS, Kruhøffer M, Luchinat C, Nielsen D, Turano P (2012) Metabolomic NMR fingerprinting to identify and predict survival of patients with metastatic colorectal cancer. Cancer Res 72:356–364

Bhatia N, Zhao J, Wolf DM, Agarwal R (1999) Inhibition of human carcinoma cell growth and DNA synthesis by silibinin, an active constituent of milk thistle: comparison with silymarin. Cancer Lett 147:77–84

Britten RJ, McClure FT (1962) The amino acid pool in Escherichia coli. Bacteriol Rev 26:292–335

Costello LC, Franklin RB (2005) 'Why do tumour cells glycolyse?': from glycolysis through citrate to lipogenesis. Mol Cell Biochem 280:1–8

Costello LC, Franklin RB (2000) The intermediary metabolism of the prostate: a key to understanding the pathogenesis and progression of prostate malignancy. Oncology 59:269–282

Dardik R, Varon D, Tamarin I, Zivelin A, Salomon O, Shenkman B, Savion N (2000) Homocysteine and oxidized low density lipoprotein enhance platelet adhesion to endothelial cells under flow conditions: distinct mechanisms of thrombogenic modulation. Thromb Haemost 83:338–344

Davis JM (ed) (2002) Basic cell culture, practical approach series, 2nd edn. Oxford University Press, USA

DeBerardinis RJ, Mancuso A, Daikhin E, Nissim I, Yudkoff M, Wehrli S, Thompson CB (2007) Beyond aerobic glycolysis: transformed cells can engage in glutamine metabolism that exceeds the requirement for protein and nucleotide synthesis. PNAS 104:19345–19350

de Koning W, van Dam K (1992) A method for the determination of changes of glycolytic metabolites in yeast on a subsecond time scale using extraction at neutral pH. Anal Biochem 204:118–123

Deprez S, Sweatman BC, Connor SC, Haselden JN, Waterfield CJ (2002) Optimisation of collection, storage and preparation of rat plasma for 1H NMR spectroscopic analysis in toxicology studies to determine inherent variation in biochemical profiles. J Pharm Biomed Anal 30:1297–1310

Donato MT, Lahoz A, Castell JV, Gómez-Lechón MJ (2008) Cell lines: a tool for in vitro drug metabolism studies. Curr Drug Metab 9:1–11

Ekman DR, Teng Q, Villeneuve DL et al (2008) Investigating compensation and recovery of fathead minnow (Pimephales promelas) exposed to 17alpha-ethynylestradiol with metabolite profiling. Environ Sci Technol 42:4188–4194

Eriksson L, Johansson E, Kettaneh-Wold N, Wold S (2006) Multi- and megavariate data analysis. Umetrics Academy, Sweden

Fan TWM (1996) Metabolite profiling by one- and two-dimensional NMR analysis of complex mixtures. Prog Nucl Magn Reson Spectr 28:161–219

Fiehn O (2001) Combining genomics, metabolome analysis, and biochemical modelling to understand metabolic networks. Comp Funct Genom 2:155–168

Freshney RI (2005) Culture of animal cells: a manual of basic technique, 5th edn. Wiley-Liss, Hoboken, NJ

Freshney RI (2010) Culture of animal cells: a manual of basic technique and specialized applications, 6th edn. Wiley-Blackwell, Hoboken, NJ

Galon J, Costes A, Sanchez-Cabo F, Kirilovsky A, Mlecnik B, Lagorce-Pages C et al (2006) Type, density, and location of immune cells within human colorectal tumors predict clinical outcome. Science 313:1960–1964

Gatley SJ, Sherratt HS (1977) The synthesis of hippurate from benzoate and glycine by rat liver mitochondria. submitochondrial localization and kinetics. Biochem J 166:39–47

Gates SC, Sweeley CC (1978) Quantitative metabolic profiling based on gas chromatography. Clin Chem 24:1663–1673

Glunde K, Serkova NJ (2006) Therapeutic targets and biomarkers identified in cancer choline phospholipid metabolism. Pharmacogenomics 7:1109–1123

Glunde K, Jacobs MA, Bhujwalla ZM (2006) Choline metabolism in cancer: implications for diagnosis and therapy. Expert Rev Mol Diagn 6:821–829

Goodacre R (2005) Metabolomics – the way forward. Metabolomics 1:1–2

Goodacre R, Vaidyanathan S, Dunn WB, Harrigan GG, Kell DB (2004) Metabolomics by numbers: acquiring and understanding global metabolite data. Trends Biotechnol 22:245–252

Green KA, Carroll JS (2007) Oestrogen-receptor-mediated transcription and the influence of cofactors and chromatin state. Nat Rev Cancer 7:713–722

Griffin JL, Shockcor JP (2004) Metabolic profiles of cancer cells. Nat Rev Cancer 4:551–561

Griffin JL, Kauppinen RA (2007) Tumour metabolomics in animal models of human cancer. J Proteome Res 6:498–505

Hogan FS, Krishnegowda NK, Mikhailova M, Kahlenberg MS (2007) Flavonoid, silibinin, inhibits proliferation and promotes cell-cycle arrest of human colon cancer. J Surgical Res 143:58–65

Holroyde CP, Gabuzda TG, Putnam RC, Paul P, Reichard GA (1975) Altered glucose metabolism in metastatic carcinoma. Cancer Res 35:3710–3714

Homocysteine Studies Collaboration (2002) Homocysteine and risk of ischemic heart disease and stroke: a meta-analysis. JAMA 288:2015–2022

Horning EC, Horning MG (1971) Metabolic profiles: gas-phase methods for analysis of metabolites. Clin Chem 17:802–809

Hotelling H (1931) The generalization of student's ratio. Ann Math Stat 2:360–378

Howard VJ, Sides EG, Newman GC, Cohen SN, Howard G, Malinow MR, Toole JF et al (2002) Changes in plasma homocyst(e)ine in the acute phase after stroke. Stroke 33:473–478

Jackson JE (1991) A user's guide to principal components. Wiley, New York

Jung JY, Lee HS, Kang DG, Kim NS, Cha MH, Bang OS, Ryu DH, Hwang GS (2011) 1H-NMR-based metabolomics study of cerebral infarction. Stroke 42:1282–1288

Keenan MR, Kotula PG (2004) Accounting for poisson noise in the multivariate analysis of ToF-SIMS spectrum images. Surf Int Anal 36:203–212

Khoo SHG, Al-Rubeai M (2007) Metabolomics. Al-Rubeai M, Fussenegger M (eds) Cell engineering vol. 5. system biology. Springer

Kurien BT, Everds NE, Scofield RH (2004) Experimental animal urine collection: a review. Lab Anim. 38:333–361

Lane AN, Fan TWM (2007) Quantification and identification of isotopomer distributions of metabolites in crude cell extracts using 1H TOCSY. Metabolomics 3:79–86

Leij-Halfwerk S, Dagnelie PC, van Den Berg JWO, Wattimena JDL, Hordijk-Luijk CH, Wilson JHP (2000) Weight loss and elevated gluconeogenesis from alanine in lung cancer patients. Am J Clin Nutr 71:583–589

Lindon JC, Nicholson JK, Everett JR (1999) NMR spectroscopy of biofluids. Ann Rev NMR Spectr 38:1–88

MacIntyre DA, Jimenez B, Lewintre EJ, Martin CR, Schafer H, Ballesteros CG et al (2010) Serum metabolome analysis by 1H-NMR reveals differences between chronic lymphocytic leukaemia molecular subgroups. Leukemia 24:788–797

Miller JW, Nadeau MR, Smith D, Selhub J (1994) Vitamin B-6 deficiency vs folate deficiency: comparison of responses to methionine loading in rats. Am J Clin Nutr 59:1033–1039

Moestue S, Sitter B, Bathen TF, Tessem MB, Gribbestad IS (2011) HR MAS MR spectroscopy in metabolic characterization of cancer. Curr Top Med Chem 11:2–26

Nicholson JK, Foxall PJ, Spraul M, Farrant RD, Lindon JC (1995) 750 MHz 1H and 1H-13C NMR spectroscopy of human blood plasma. Anal Chem 67:793–811

Nicholson JK, Lindon JC, Holmes E (1999) 'Metabonomics': understanding the metabolic responses of living systems to pathophysiological stimuli via multivariate statistical analysis of biological NMR spectroscopic data. Xenobiotica 29:1181–1189

Oakman C, Tenori L, Biganzoli L, Santarpia L, Cappadona S, Luchinat C, Di Leo A (2011) Uncovering the metabolomic fingerprint of breast cancer. Int J Biochem Cell Biol 43:1010–1020

Odunsi K, Wollman RM, Ambrosone CB, Hutson A, Mccann SE, Tammela J et al (2005) Detection of epithelial ovarian cancer using 1H-NMR-based metabonomics. Int J Cancer 113:782–788

Oliver SG, Winson MK, Kell DB, Baganz B (1998) Systematic functional analysis of the yeast genome. Trends Biotechnol 16:373–378

Pages F, Kirilovsky A, Mlecnik B, Asslaber M, Tosolini M, Bindea G et al (2009) In situ cytotoxic and memory T cells predict outcome in patients with early-stage colorectal cancer. J Clin Oncol 27:5944–5951

Patt SL, Sykes BD (1972) Water eliminated Fourier transform NMR spectroscopy. Chem Phys 56:3182

Pearson K (1901) (1901) On lines and planes of closest fit to systems of points in space. Phil Mag 2:559–572

Piotto M, Moussallieh FM, Dillmann B, Imperiale A, Neuville A, Brigand C et al (2009) Metabolic characterization of primary human colorectal cancers using high resolution magic angle spinning 1H magnetic resonance spectroscopy. Metabolomics 5:292–301

Raina K, Blouin MJ, Singh RP et al (2007) Dietary feeding of silibinin inhibits prostate tumor growth and progression in transgenic adenocarcinoma of the mouse prostate model. Cancer Res 67:11083–11091

Raina K, Serkova NJ, Agarwal R (2009) Silibinin feeding alters the metabolic profile in TRAMP prostatic tumors: 1H-NMRS–based metabolomics study. Cancer Res 69:3731–3735

Raina K, Rajamanickam S, Singh RP et al (2008) Stage-speacific inhibitory effects and associated mechanisms of silibinin on tumor progression and metastasis in transgenic adenocarcinoma of the mouse prostate model. Cancer Res 68:6822–6830

Ramanathan A, Wang C, Schreiber SL (2005) Perturbational profiling of a cell-line model of tumorigenesis by using metabolic measurements. PNAS 102:5992–5997

Rencher AC (2002) Methods of multivariate analysis. Wiley, New York

Ritchie SA, Ahiahonu PWK, Jayasinghe D, Heath D, Liu J, Lu YS et al (2010) Reduced levels of hydroxylated, polyunsaturated ultra long-chain fatty acids in the serum of colorectal cancer patients: implications for early screening and detection. BMC Med 8:13–32

Serkova NJ, Spratlin JL, Eckhardt SG (2007) NMR-based metabolomics: translational application and treatment of cancer. Curr Opin Mol Ther 9:572–585

Serkova NJ, Niemann CU (2006) Pattern recognition and biomarker validation using quantitative 1H-NMR-based metabolomics. Expert Rev Mol Diagn 6:717–731

Shearer JD, Buzby GP, Mullen JL, Miller E, Caldweell MD (1984) Alteration in pyruvate metabolism in the liver of tumor-bearing rats. Cancer Res 44:4443–4446

Simpson AJ, Brown SA (2005) Purge NMR: effective and easy solvent suppression. J Magn Reson 175:340–346

Smeaton JR, Elliott WH (1967) Selective release of ribonuclease-inhibitor from Bacillus subtilis cells by cold shock treatment. Biochem Biophys Res Commun 26:75–81

Teng Q, Ekman DR, Huang W, Collette TW (2012) Push-through direct injection NMR: an optimized automation method applied to metabolomics. Analyst 137:2226–2232

Teng Q, Huang W, Collette TW, Ekman DR, Tan C (2009) A direct cell quenching method for cell-culture based metabolomics. Metabolomics 5:199–208

Tiziani S, Lopes V, Gunther UL (2009) Early stage diagnosis of oral cancer using 1H NMR-based metabolomics. Neoplasia 11:269–276

Torri GM, Torri J, Gulian JM, Vion-Dury J, Viout P, Cozzone PJ (1999) Magnetic resonance spectroscopy of serum and acute-phase proteins revisited: a multiparametric statistical analysis of metabolite variations in inflammatory, infectious and miscellaneous diseases. Clin Chim Acta 279:77–96

Ulrich EL, Akutsu H, Doreleijers JF et al (2008) BioMagResBank. Nucleic Acids Res 36: D402–408

van den Berg RA, Hoefsloot HCJ, Westerhuis JA, Smilde AK, van der Werf MJ (2006) Centering, scaling, and transformations: improving the biological information content of metabolomics data. Genomics 7:142–156

Van der Greef J, Smilde AK (2005) Symbiosis of chemometrics and metabolomics: past, present, and future. J Chemomet 19:376–386

Viant M (2003) Improved methods for the acquisition and interpretation of NMR metabolomic data. Biochem Biophys Res Commun 310:943–948

Viant MR (2007) Revealing the metabolome of animal tissues using 1H nuclear magnetic resonance spectroscopy. In: Methods in molecular biology, Weckwerth W (ed). Clifton, NJ, Humana Press, 358:229–246

Villas-Boas SG, Nielsen J, Smedsgaard J et al (2007) Metabolome Analysis: An introduction. Wiley, Hoboken, NJ

Warburg O, Posener K, Negelei E (1924/1930) Ueber den Stoffwechsel der Tumoren. Biochem Z (German) 152:319–344 (Reprinted in Warburg O. On metabolism of tumors. Constable, London)

Warburg O (1956) On respiratory impairment in cancer cells. Science 124:269–270

Williams R (1956) Biochemical Individuality, the basics for the genetotrophic concept. Univ of Texas Press, Austin

Wishart DS, Tzur D, Knox C et al (2007) HMDB: the human metabolome database. Nucleic Acids Res 35:D521–D526

Wold S, Ruhe A, Wold H, Dunn WJ III (1984) The collinearity problem in linear regression. the partial least squares (PLS) approach to generalized inverses. SIAM J Sci Stat Comput 5:735–743

Wold S, Antti H, Lindgren F, Ohman J (1998) Orthogonal signal correction of near-infrared spectra. Chemometr Intell Lab Syst 44:175–185

Wold S, Esbensen K, Geladi P (1987) Principal component analysis. Chemometr Intell Lab Syst 2:37–52

Yang C, Richardson AD, Smith JW, Osterman A (2007a) Comparative metabolomics of breast cancer. Pac Symp Biocomput 12:181–192

Yang YX, Li CL, Nie X, Feng X, Chen W, Yue Y et al (2007b) Metabonomic studies of human hepatocellular carcinoma using high-resolution magic-angle spinning 1H NMR spectroscopy in conjunction with multivariate data analysis. J Proteome Res 6:2605–2614

Kurien BT, Everds NE, Scofield RH (2004) Experimental animal urine collection: a review. Lab Anim 38:333–361

Appendix I: Multiple Choice Questions

1. Assuming that no ^{1}H signal can be observed for an aqueous sample, which of the following is most likely not a cause of the problem?

    a. The cable is not connected to the probe after probe tuning
    b. There is a loose cable connection around the probe
    c. The sample is not shimmed well
    d. The probe has a problem

2. Which of the following is most likely not a cause of a VT problem?

    a. Heater is not on
    b. VT air is disconnected.
    c. Sample is not in the magnet
    d. The set temperature exceeds the maximum set temperature

3. In which of the following situations does it use a ¼ wavelength cable?

    a. ^{15}N decoupler channel
    b. ^{2}H observation
    c. Lock channel
    d. ^{13}C decoupler channel

4. Which of the following delays should be used in a Jump-return experiment on a 500 MHz instrument to have maximum intensity at 9 ppm? Assume that the water resonance is at 4.8 ppm.

    a. 467 µs
    b. 119 µs
    c. 1.9 ms
    d. 238 µs

5. What coil configuration is most likely used for a triple-resonance probe?

    a. The inner coil is double-tuned to ^{1}H and ^{13}C
    b. The ^{1}H and lock channel use the inner coil, ^{13}C and ^{15}N use the outer coil

c. The probe can be used to observe the correlation of ^{1}H, ^{13}C, ^{15}N, and ^{31}P simultaneously
d. One of the two probe coils is used for ^{1}H, ^{13}C, and ^{15}N, the other is used for ^{2}H

6. Which of the following is not a property of an RF amplifier?

   a. It has a linear dependence of output power on attenuation
   b. Its output is gated by a transmitter controller
   c. It amplifies the signal from the probe
   d. The output of the amplifier for a heteronuclear channel is higher than that of the ^{1}H channel in a high resolution NMR instrument

7. A mixer is used to

   a. Subtract the frequencies of two input signals
   b. Add the frequencies of two input signals
   c. Multiply the frequencies of two input signals
   d. Produce IF frequency

8. Which the following frequencies cannot be an intermediate frequency (IF)?

   a. 20 MHz
   b. 10 MHz
   c. 30 MHz
   d. 200 MHz

9. Which of the following has the lowest frequency value?

   a. Carrier
   b. LO
   c. IF
   d. Lock frequency

10. What is the purpose of using LO?

    a. To combine with carrier frequency at transmitter
    b. To make a frequency higher than the spectrometer base frequency
    c. To use a fixed-frequency receiver for all nuclei
    d. To use a fixed-frequency preamplifier for all nuclei

11. Which of the following does not describe IF?

    a. It is the fixed frequency of a receiver
    b. Its frequency value is much lower than that of carrier or LO
    c. Its frequency changes for different carrier frequencies
    d. Either the carrier or LO frequency is made from IF

12. Which of the following statements about the effect of salt concentration on a probe is correct?

   a. High salt concentration affects a conventional probe more severely than a cryogenic probe because it is less sensitive than a cryogenic probe
   b. High salt concentration affects a conventional probe less severely than a cryogenic probe because it is operated at room temperature
   c. High salt concentration affects a cryogenic probe is more severely than a conventional probe because the salt of sample may precipitate in the cryogenic probe
   d. High salt concentration affects a cryogenic probe more severely than a conventional probe because the high Q value of a cryogenic probe is dramatically decreased due to the dielectric influence of the salt concentration

13. The sensitivity of a cryogenic probe on a 500 MHz spectrometer is close to that of a conventional probe on a spectrometer of

   a. 600 MHz
   b. 750 MHz
   c. 900 MHz
   d. 1,050 MHz

14. Which of the following pulses should be tried first for water suppression by presaturation for a 90%H_2O/10%^2H_2O sample? Assume a 50 W ^1H amplifier and a carrier on the water resonance.

   a. 3 s pulse with power attenuation of −55 dB from the maximum power
   b. 3 s pulse with power attenuation of −35 dB from the maximum power
   c. 3 s pulse with power attenuation of −45 dB from the maximum power
   d. 5 s pulse with power attenuation of −60 dB from the maximum power

15. Assuming that ^1H data are acquired on a 600 MHz spectrometer with an acquisition time of 64 ms and the data are Fourier transformed without zero-filling and linear prediction, what is the digital resolution of the ^1H spectrum?

   a. 7.8 Hz/pt
   b. 0.013 ppm/pt
   c. 3.9 Hz/pt
   d. 0.026 ppm/pt

16. If the size of the above ^1H data is doubled by zero-filling, what is the digital resolution of the spectrum?

   a. 15.6 Hz/pt
   b. 7.8 Hz/pt
   c. 3.9 Hz/pt
   d. 0.026 ppm/pt

17. Which parameter can saturate the lock signal if it is set too high?

    a. Lock gain
    b. Lock phase
    c. Lock power
    d. Lock field (or z_0)

18. Which of the following is not true?

    a. Magnets (200 MHz–900 MHz) are made of superconducting wires
    b. The magnet solenoid is in a liquid helium vessel
    c. Liquid helium and liquid nitrogen are needed to maintain the magnetic field
    d. Room temperature shims are in a liquid nitrogen vessel

19. By using cryogenic shims, field homogeneity can be as good as

    a. 1 ppm
    b. 10 ppm
    c. 1 ppb
    d. 0.01 ppm

20. The water-flip-back sequence provides superior water suppression. How is the result achieved?

    a. The selective pulse on water saturates some portion of the water magnetization
    b. The selective pulse on water keeps the water magnetization on the xy plane so that the water magnetization is suppressed by the watergate sequence
    c. The selective pulse on water brings the water magnetization to the z axis
    d. The selective pulse on water keeps the water magnetization in the xy plane so that the water magnetization is destroyed by the gradient pulse

21. Which of the following gives a wider decoupling bandwidth for the same amount of RF power?

    a. CW
    b. Waltz16
    c. GARP
    d. BB

22. Which of the following is a better way to setup a water-flip-back experiment after probe tuning, shimming, and locking?

    a. Calibrate VT, ^{1}H 90° pulse, transmitter offset and set the offset at the center of the spectrum
    b. Calibrate ^{1}H 90° pulse, transmitter offset, ^{1}H 90° selective pulse and set the offset at the center of the spectrum
    c. Calibrate ^{1}H 90° pulse, transmitter and decoupler offsets, ^{1}H 90° selective pulse and set the transmitter offset on water
    d. Calibrate ^{1}H 90° pulse, transmitter offset, ^{1}H 90° selective pulse and set the offset on water

23. Which of the following data are most likely processed with doubling the size by forward linear prediction, 90°-shifted squared sine bell function, zero-filling once, and Fourier transformation?

    a. 1D watergate data
    b. ^{1}H dimension of 3D data
    c. ^{15}N dimension of 3D data
    d. 2D COSY

24. What is the correct way to tune a probe for a triple-resonance experiment?

    a. Tune ^{1}H channel first, then ^{13}C and ^{15}N last without filters
    b. Tune ^{1}H channel first, then ^{13}C and ^{15}N last with filters
    c. Tune ^{15}N channel first, then ^{13}C and ^{1}H last without filters
    d. Tune ^{15}N channel first, then ^{13}C and ^{1}H last with filters

25. An NMR transmitter consists of

    a. Frequency synthesizer, RF signal generator, transmitter controller, and receiver
    b. CPU, RF signal generator, transmitter controller, and RF amplifier
    c. Frequency synthesizer, RF signal generator, and transmitter controller
    d. Frequency synthesizer, RF signal generator, transmitter controller, and RF amplifier

26. Which of the following product operators describe the coherence of a two weakly coupled two spin (I and S) system from an initial coherence of $-I_y$ after an INEPT $\tau \to \pi(I_x + S_x) \to \tau$, when $\tau = 1/(4J_{IS})$?

    a. $-I_z S_x$
    b. $I_x S_x$
    c. $-I_x S_z$
    d. I_x

27. Assuming that on a 600 MHz NMR spectrometer the ^{13}C 90° pulse length is 15 μs at 60 dB and a higher dB value means more power for a pulse, what is most likely the power setting for ^{13}C GARP decoupling over a 50 ppm band width?

    a. 45 dB
    b. 47 dB
    c. 49 dB
    d. 51 dB

28. Which of the following is most likely a Gly NH cross peak?

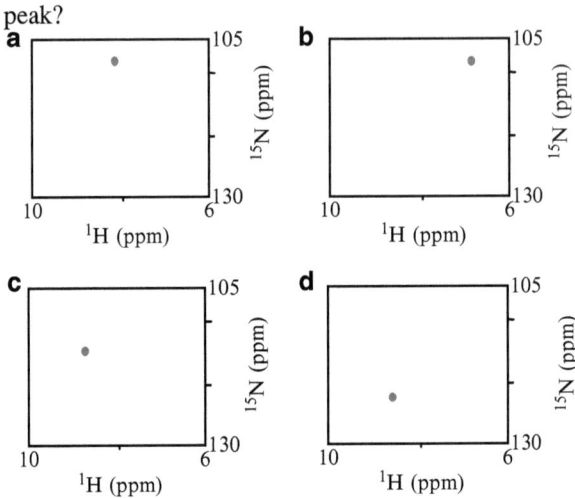

29. Assuming that on a 500 MHz NMR spectrometer the ^{15}N 90° pulse length is 35 μs at 60 dB and a higher dB value means more power for a pulse, what is most likely the power setting for ^{15}N WALTZ-16 decoupling over a 30 ppm band width?

   a. 40 dB
   b. 42 dB
   c. 45 dB
   d. 49 dB

30. Assuming that on a 500 MHz NMR spectrometer the ^{15}N 90° pulse length is 35 μs at 60 dB and a higher dB value means more power for a pulse, what is most likely the power setting for ^{15}N GARP decoupling over a 30 ppm band width?

   a. 35 dB
   b. 40 dB
   c. 45 dB
   d. 49 dB

31. If all four buffer solutions work ok for a protein sample, which one should be used to make the NMR sample?

   a. 100 mM Tris–HCl, pH 7, 100 mM KCl
   b. 50 mM phosphate, pH 7, 200 mM KCl
   c. 100 mM Tris–HCl, pH 7, 20 mM KCl
   d. 50 mM phosphate, pH 7, 50 mM KCl

32. On a 600 MHz instrument, if the resonance frequency of DSS is 599.7685438, what is the reference frequency for ^{15}N using liquid NH$_3$ as reference?

a. 60.2563985
b. 60.7740175
c. 59.9856329
d. 60.995357

33. On the same spectrometer as in question 32, what is the frequency at 177 ppm of ^{13}C using DSS as ^{13}C reference?

    a. 150.8644358
    b. 150.8458732
    c. 150.8329679
    d. 150.8382120

34. On the same spectrometer as in question 32, what is the frequency at 118 ppm of ^{15}N?

    a. 60.6088655
    b. 60.7811888
    c. 59.9771635
    d. 60.7902741

35. ^{13}C chemical shift has a much wider range (~300 ppm) than ^{1}H (~10 ppm) because

    a. The contribution of the diamagnetic shielding of ^{13}C is much larger than ^{1}H due to the small ^{13}C energy gap
    b. The contribution of the paramagnetic shielding of ^{13}C is much larger than ^{1}H due to the small ^{13}C energy gap
    c. The electron density of ^{1}H is almost always spherically symmetrical
    d. The reason is unknown

36. If a $^3J_{H^N H^\alpha}$ coupling constant of a residue has a value of ~10 Hz, what could the torsion angle ϕ be?

    a. Approximately 0°
    b. Approximately −120°
    c. Approximately −180°
    d. Approximately 120°

37. What is a better criterion to measure the accuracy of the calculated structure?

    a. RMSD of backbone atoms
    b. The total number of distance violations
    c. Quality factor of residual dipolar couplings
    d. Quality factor of NOE intensity

38. Why is DSS used instead of TMS as ^{1}H chemical shift reference for biological sample?

    a. TMS can denature proteins
    b. The chemical shift of TMS is dependent on temperature
    c. DSS has a higher solubility in aqueous solution
    d. DSS is widely used in protein sample preparation

39. At equilibrium state

    a. There is no $-z$ component of nuclear magnetization
    b. There is slightly larger $+z$ component than $-z$ component of nuclear magnetization
    c. There are equal $+z$ and $-z$ components of nuclear magnetization
    d. There is slightly larger $-z$ component than $+z$ component of nuclear magnetization

40. In a magnetic field, nuclear dipoles (nuclear spins with a spin quantum number of ½)

    a. Precess around the magnetic field direction randomly
    b. Are motionless along the direction of the magnetic field
    c. Do not exist
    d. Precess around the magnetic field direction at the Larmor frequency

41. A B_1 field used to interact with nuclear dipoles in order to generate an NMR signal has the following property:

    a. The orientation of the B_1 field is fixed in the rotating frame
    b. Components of the B_1 field rotate in laboratory frame with Larmor frequency
    c. The B_1 field is a linear alternating magnetic field in the laboratory frame
    d. **All of the above**

42. If your NMR spectrum has a distorted baseline, the problem is most likely because

    a. Spectral window (SW) is too low
    b. Receiver gain is too high
    c. Receiver gain is too low
    d. y-Axis scale of display is too high

43. Which of the following quantities is *not* changed at different magnetic field strength?

    a. Chemical shift (in Hz)
    b. Nuclear spin population in an energy state
    c. J-coupling constant
    d. Energy difference between two energy states of nuclei with nonzero spin quantum number

44. Chemical shifts originate from

    a. Magnetic momentum
    b. Electron shielding
    c. Free induction decay
    d. Scalar coupling (J-coupling)

Appendix I

45. Chemical shifts of protons have a frequency range of about

    a. Megahertz
    b. 250 MHz
    c. Kilohertz
    d. 10 Hz

46. DSS has a chemical shift value of 0.0 ppm or 0.0 Hz because

    a. Its absolute chemical shift value is 0.0
    b. The chemical shift value of DSS is chosen as the chemical shift reference
    c. H_2O has ^1H chemical shift of 4.76 ppm and DSS chemical shift is 4.76 ppm lower than H_2O
    d. None of the above

47. What is the ^{13}C resonance frequency on a 600 MHz NMR spectrometer?

    a. 600 MHz
    b. 92 MHz
    c. 60 MHz
    d. 150 MHz

48. The pulse angle is dependent on

    a. Transmitter power (pulse power)
    b. Pulse length
    c. Receiver gain
    d. Both (a) and (b)

49. If number of time domain points equals 4k and dwell time equals 100 μsec, then acquisition time equals

    a. 100 μs × 4,000
    b. 100 μs × 4,096
    c. 100 μs × 4,000 × 2
    d. 100 μs × 4,096 × 2

50. The signal-to-noise ratio (S/N) of an NMR spectrum can be increased by the accumulation of acquisitions. Compared to the one recorded with 2 scans, a spectrum with 32 scans has an S/N ratio

    a. 16 times higher
    b. 4 times higher
    c. 8 times higher
    d. 32 times higher

51. The wider frequency range covered by an RF pulse (pulse bandwidth) is achieved by the pulse with

    a. Lower power and longer pulse length
    b. Higher power and longer pulse length
    c. Higher power and shorter pulse length
    d. None of the above

52. Integration of ^{1}H signal intensities (or peak area) gives information about

   a. The absolute number of protons corresponding to the resonance frequencies
   b. The ratio of the number of protons corresponding to the resonance frequencies
   c. The types of protons corresponding to the resonance frequencies
   d. Intensities of protons relative to the solvent peak

53. ^{13}C spectra without decoupling show multiplicity of ^{13}C peaks due to the coupling of ^{1}H to ^{13}C. In a 1D ^{1}H spectrum of an unlabeled sample (natural abundance ^{13}C), the coupling of ^{13}C to ^{1}H is neglected because

   a. The NMR spectrometer decouples ^{13}C from ^{1}H automatically
   b. A large portion of protons are bound to ^{12}C which is NMR inactive
   c. The J_{CH} coupling constant is small compared to the linewidths of ^{1}H peaks
   d. Both b and c

54. Improper shimming

   a. Can be eliminated by spinning the sample
   b. Can broaden the line shape of the NMR signal
   c. Can shorten T_1 relaxation
   d. Does not have any effect on NMR spectra

55. The dwell time is defined as

   a. The total time needed to acquire an FID
   b. The time difference between two adjacent time domain data points
   c. The time delay between the last pulse and acquisition
   d. The time delay before the first pulse

56. The reason that an exponential function (EM) with line broadening (LB) of 3–5 Hz is used for a ^{13}C FID before Fourier transformation is

   a. To increase the resolution of the ^{13}C spectrum
   b. To increase sensitivity of the spectrum
   c. To improve the baseline of the spectrum
   d. To make the FID look nicer

57. The relative sensitivity of ^{15}N to ^{1}H for a 100% ^{15}N enriched sample is

   a. 0.1
   b. 1.0×10^{-3}
   c. 3.7×10^{-5}
   d. 1.0×10^{-2}

58. Choose the correct statement(s) about the steady state transients (ss or dummy scans):

   a. ss are executed before parameter nt (number of transients or scans)
   b. During ss, the experiment is performed except without data acquisition
   c. ss are used to ensure a steady state before data acquisition
   d. All of the above

Appendix I

59. The natural abundance of ^{13}C is about

    a. Four times less than ^1H
    b. 0.11% of total carbon
    c. 1.1% of total carbon
    d. 99% of total carbon

60. Deuterated solvent in an NMR sample is used to

    a. Stabilize the magnetic field
    b. Set the chemical shift reference
    c. Obtain good field homogeneity across the sample
    d. Both (a) and (c)

61. Digital resolution (Hz/point) can be improved by

    a. Decreasing the number of time domain points
    b. Zero-filling the FID
    c. Decreasing the spectral width
    d. Both (a) and (c)

62. Quadrature detection uses

    a. Single detector on the x axis
    b. Two detectors which are opposite each other
    c. Two detectors which are perpendicular to each other
    d. Four detectors which are on the $x, y, -x$ and $-y$ axes

63. Tuning the probe is to

    a. Match the probe impedance to the 50 Ω cable impedance
    b. Tune the probe frequency to the carrier frequency
    c. Make shimming easier
    d. (a) and (b)

64. $^3J_{HH}$ coupling constants may have a value of

    a. 140 Hz
    b. 35 Hz
    c. 8 Hz
    d. 70 Hz

65. $^1J_{CH}$ coupling constants may have a value of

    a. 140 Hz
    b. 35 Hz
    c. 8 Hz
    d. 70 Hz

66. $^3J_{HH}$ coupling constants are dependent on

   a. Magnetic field strength
   b. Relative orientation of the coupled protons
   c. Sample concentration
   d. 90° pulse width

67. Assume that a proton is scalar coupled to proton(s) with different chemical environments. If this proton shows a triplet signal, how many proton(s) does it is scalar coupled to?

   a. One
   b. Two
   c. Three
   d. Four

68. Which of the following molecules has the largest $^3J_{HH}$ coupling constant between H_a and H_b?

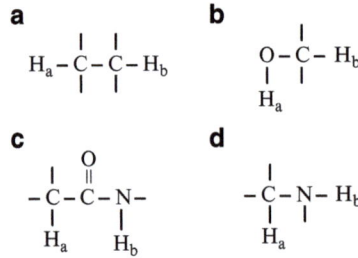

69. In which of the following cases is the J_{HH} coupling constant between H_a and H_b most likely less than 1 Hz?

a.
$$H_a-\overset{|}{\underset{|}{C}}-\overset{|}{\underset{|}{C}}-H_b$$

b.
$$O-\overset{|}{\underset{|}{C}}-H_b$$
$$\quad\ H_a$$

c.
$$-\overset{|}{\underset{|}{C}}-\overset{O}{\overset{\|}{C}}-N-$$
$$\ \ H_a\ \ \ \ H_b$$

d.
$$-\overset{|}{\underset{|}{C}}-N-H_b$$
$$\ \ H_a$$

70. If the last ^{13}C and ^{1}H 90° pulses are omitted in the INEPT experiment (Fig. 1.24), can the ^{13}C NMR signal be observed?

   a. Yes, because there exists transverse magnetization

b. No, because the two components of the transverse magnetization will be canceled each other out
c. No, because there is no ^{13}C transverse magnetization.
d. None of the above

71. How is quandrature detection in the indirect dimension achieved by the States–TPPI method?

   a. For each t_1 increment, one FID is acquired. The phase of the pulse prior to the evolution time is shifted by 90° for each FID. The interferogram is transformed with real Fourier transformation
   b. For each t_1 increment, two FIDs are acquired. The phase of the pulse prior to the evolution time is shifted by 90° for each t_1 increment. The interferogram is transformed with complex Fourier transformation
   c. For each t_1 increment, two FIDs are acquired. The phase of the pulse prior to the evolution time is shifted by 90° for each FID. The interferogram is transformed with complex Fourier transformation
   d. For each t_1 increment, two FIDs are acquired. The phase of the pulse prior to the evolution time is shifted by 90° for the second FID of each t_1 increment. The interferogram is transformed with complex Fourier transformation

72. How is quandrature detection in the indirect dimension achieved by the TPPI method?

   a. For each t_1 increment, one FID is acquired. The phase of the pulse prior to the evolution time is shifted by 90° for each FID. The interferogram is transformed with real Fourier transformation
   b. For each t_1 increment, two FIDs are acquired. The phase of the pulse prior to the evolution time is shifted by 90° for each t_1 increment. The interferogram is transformed with complex Fourier transformation
   c. For each t_1 increment, two FIDs are acquired. The phase of the pulse prior to the evolution time is shifted by 90° for each FID. The interferogram is transformed with complex Fourier transformation
   d. For each t_1 increment, two FIDs are acquired. The phase of the pulse prior to the evolution time is shifted by 90° for the second FID of each t_1 increment. The interferogram is transformed with complex Fourier transformation

73. How is quandrature detection in the indirect dimension achieved by the States method?

   a. For each t_1 increment, one FID is acquired. The phase of the pulse prior to the evolution time is shifted by 90° for each FID. The interferogram is transformed with real Fourier transformation
   b. For each t_1 increment, two FIDs are acquired. The phase of the pulse prior to the evolution time is shifted by 90° for each t_1 increment. The interferogram is transformed with complex Fourier transformation

c. For each t_1 increment, two FIDs are acquired. The phase of the pulse prior to the evolution time is shifted by 90° for each FID. The interferogram is transformed with complex Fourier transformation

d. For each t_1 increment, two FIDs are acquired. The phase of the pulse prior to the evolution time is shifted by 90° for the second FID of each t_1 increment. The interferogram is transformed with complex Fourier transformation

74. Assuming that a HMQC experiment is collected on a 600 MHz spectrometer with a spectral window of 60 ppm, $pw_{90(^1H)} = 7\,\mu s$, $pw_{90(^{13}C)} = 15\,\mu s$, and the evolution element $90°(^{13}C)-½t_1 - 180°(^1H)-½t_1 - 90°(^{13}C)$ using the States–TPPI method, what is the value of $t_1(0)$ chosen for the phase correction of 0° (zero order) and −180° (first order)? $\left[t_1(0) = \frac{1}{2SW} - pw_{180(^1H)} - \frac{4 \times pw_{90(^{13}C)}}{\pi}\right]$

a. 55.5 μs
b. 111.1 μs
c. 78 μs
d. 22.5 μs

75. Assuming that a constant time HMQC experiment is collected on a 600 MHz spectrometer with a spectral window of 60 ppm, $pw_{90(^1H)} = 7\mu s$, $pw_{90(^{13}C)} = 15\,\mu s$, and the evolution element $90°-½t_1 - T - 180° - (T-½t_1) -90°(^{13}C)$ using the States–TPPI method, what is the value of $t_1(0)$ chosen for the phase correction of 0° (zero order) and −180° (first order)?

a. 55.5 μs
b. 111.1 μs
c. 78 μs
d. 22.5 μs

76. A larger value of the squared generalized order parameter S^2 describes a internal motion:

a. The bond vector is more flexible in the molecular frame
b. The motion is faster
c. The bond vector is more rigid in the molecular frame
d. The motion is slower

77. The squared generalized order parameter S^2 is highly sensitive to the angle θ between the equilibrium orientation of the bond vector and the diffusion tensor of the bond vector.

a. The value of S^2 becomes smaller as the angle θ decreases
b. The value of S^2 becomes smaller as the angle θ increases
c. The value of S^2 is constant in the region of $\theta = 0°–20°$ and then rapidly decreases as θ increases
d. The value of S^2 is large in the region of $\theta = 70°–90°$

78. Smaller S^2 with a smaller τ_e characterizes the molecular internal dynamics of
    a. Rigid and slow
    b. Flexible and fast
    c. Rigid and fast
    d. Flexible and slow

79. Certain NMR parameters are measured for the study of protein dynamics. Which of the following parameter is *not* measured for the protein dynamics?
    a. ^{15}N T_1 relaxation rate
    b. $^1H-^1H$ NOE
    c. $^1H-^{15}N$ NOE
    d. ^{15}N T_2 relaxation rate

80. NOE connectivities are assigned for structure calculation. How are they usually assigned?
    a. A majority of the NOESY cross peaks are assigned based on the sequence-specific assignment of chemical shift resonances
    b. A small fraction of the NOESY cross peaks are assigned for initial structure calculation and more connectivities are added by the iterative NOE analysis
    c. A majority of the NOESY cross peaks are assigned and the iterative NOE analysis is used to refine the calculated structures
    d. Almost all of the NOE connectivities can be assigned by setting up chemical shift tolerance within values less than 0.2 ppm

81. The nuclear relaxation characterized by T_1 relaxation is *not*:
    a. Spin–lattice relaxation
    b. Longitudinal relaxation
    c. Spin–spin relaxation
    d. Relaxation along the z axis

82. The purpose of shimming is to
    a. Stabilize the static magnetic field
    b. Obtain homogeneity of the B_1 field
    c. Find the lock frequency
    d. Obtain homogeneity of the static magnetic field

83. The TROSY experiment is based on the property of
    a. Cross relaxation caused by dipolar coupling (DD)
    b. Cross-correlated relaxation caused by the interference between DD and CSA (chemical shift anisotropy)
    c. Cross-correlated relaxation caused by the interference between DD and CSA
    d. Auto-correlated relaxation caused by the interference between DD and CSA

84. After a B_1 field is applied along the x axis, the transverse magnetization when relaxing back to the equilibrium state during acquisition,

   a. Rotates about the x axis of the rotating frame
   b. Rotates about the direction of the B_0 field in the laboratory frame
   c. Is stationary along the $-y$-axis in the laboratory frame
   d. Relaxes only along the z axis

85. During an INEPT subsequence, the magnetization transfers for all three types of CH groups (methine, methylene, and methyl) are optimized by setting the delay τ (half of the INEPT period) to

   a. 11 ms
   b. 2.75 ms
   c. 2.2 ms
   d. 3.6 ms

86. The C^α cross peaks of CBCANH have opposite sign relative to C^β cross peaks because

   a. C^β cross peaks are folded in the spectrum
   b. The magnetization transfer from the C^α has opposite sign to that from the C^β by setting the delay for the $C^{\alpha,\beta} \rightarrow N$ transfer to 11 ms
   c. The INEPT delay for $H \rightarrow C^{\alpha,\beta}$ transfer is setting to 2.2 ms
   d. None of the above

87. Which of the following sequences can be used for a ROESY spin lock?

   a. 5 kHz off resonance DIPSI-3
   b. 5 kHz on resonance MLEV17
   c. 2 kHz off resonance CW
   d. 6 kHz off resonance CW

88. Compared to the random coil values, ^{13}C chemical shifts of a α-helix secondary structure usually

   a. Smaller
   b. Larger
   c. No significant change
   d. Unpredictable

89. Relative to a 2D, a 3D experiment has a better

   a. S/N ratio
   b. Resolution
   c. Baseline
   d. Line shape

Appendix I 409

90. Which of the following mixing times is most likely used in a homonuclear TOCSY to mainly observe the correlations of H^N to H^α and to all aliphatic H^{aliph} ?

    a. 30 ms
    b. 5 ms
    c. 60 ms
    d. 100 ms

91. Which of the following mixing times is most likely used in a homonuclear TOCSY to mainly observe the correlation between H^N and H^α?

    a. 30 ms
    b. 5 ms
    c. 60 ms
    d. 100 ms

92. Which of the following mixing times is most likely used in a NOESY experiment for a 20 kDa protein sample?

    a. 30 ms
    b. 100 ms
    c. 300 ms
    d. 500 ms

93. Which of the following experiments has been used for measuring ^1H–^{15}N residual dipolar coupling?

    a. ^1H–^{15}N HSQC
    b. IPAP ^1H–^{15}N HSQC
    c. ^1H–^{15}N NOE HSQC
    d. PEP ^1H–^{15}N HSQC

94. What is a constant time evolution period used for?

    a. Increase the sensitivity of the experiment
    b. Increase the resolution of the experiment
    c. Decouple J-coupling
    d. Suppress artifacts

95. Which of the following experiments should be used for studying a complex formed by ^{13}C–^{15}N labeled protein and unlabeled peptide?

    a. Isotope-edited experiment
    b. Isotope-filtered experiment
    c. 3D H(C)CH-TOCSY
    d. Saturation transfer experiment

96. It is necessary to perform shimming before an experiment can be run. The shimming is done by:

    a. Adjusting the lock power to obtain highest lock level
    b. Optimizing lock gain to obtain highest lock level
    c. Adjusting the current to room temperature shim coils when monitoring the locklevel
    d. Adjusting the current to cryogenic shim coils when monitoring the lock level

97. Assuming that a cryogenic probe has a sensitivity 4 times higher than a conventional room temperature probe and both probes give same line widths, the signal intensity of the spectrum obtained using the cryogenic probe compared to the room temperature probe with the same amount of experimental time is

    a. 16 times higher
    b. 8 times higher
    c. 4 times higher
    d. 2 times higher

98. For a 200 μL ^{15}N labeled protein sample, which of the following probe is the best for a ^{1}H–^{15}N HSQC experiment?

    a. Triple-resonance HCX probe (X: ^{15}N–^{31}P)
    b. Dual broadband ^{1}H–^{19}F/^{15}N–^{31}P
    c. High resolution MAS (magic-angle spinning) probe
    d. Triple-resonance HCN probe

99. For 150 μL ^{15}N/^{13}C labeled protein containing 100 mM salt, which of the following probes is the best for triple-resonance experiments?

    a. 5 mm room temperature HCN probe
    b. 3 mm room temperature HCN probe
    c. 5 mm cryogenic HCN probe using a 5 mm NMR microtube
    d. 5 mm cryogenic HCN probe using a 3 mm NMR microtube

100. Which of the following statements about preamplifier, IF amplifier, and RF amplifier is wrong?

    a. The outputs of IF and RF amplifiers are adjustable
    b. A preamplifier is located near or inside a probe
    c. They are all frequency tunable amplifiers
    d. RF amplifier has a linear dependence of attenuation

Answers to Multiple Choice Questions

1. c.	2. c.	3. b.	4. b.	5. b.	6. c.	7. c.	8. d.	9. c.	10. c.
11. c.	12. d.	13. d.	14. a.	15. d.	16. b.	17. c.	18. d.	19. a.	20. c.
21. c.	22. d.	23. c.	24. c.	25. d.	26. c.	27. b.	28. a.	29. c.	30. b.
31. d.	32. b.	33. d.	34. b.	35. b.	36. b.	37. c.	38. c.	39. b.	40. d.
41. d.	42. b.	43. c.	44. b.	45. c.	46. b.	47. d.	48. d.	49. b.	50. b.
51. c.	52. b.	53. b.	54. b.	55. b.	56. b.	57. b.	58. d.	59. c.	60. d.
61. b.	62. c.	63. d.	64. c.	65. a.	66. b.	67. b.	68. c.	69. c.	70. c.
71. c.	72. a.	73. d.	74. d.	75. a.	76. c.	77. b.	78. b.	79. b.	80. b.
81. c.	82. d.	83. b.	84. b.	85. c.	86. b.	87. c.	88. b.	89. b.	90. d.
91. a.	92. b.	93. b.	94. b.	95. b.	96. c.	97. c.	98. d.	99. d.	100. c.

Appendix J: Nomenclature and Symbols

ADC	Analog-to-digital converter (conversion)
B_0	Magnetic field strength
B_1	Oscillating RF magnetic field strength
B_{eff}	Effective B_1 field strength
C	Capacitance
COS	Coherence order selection
COSY	Correlation spectroscopy
CP	Cross polarization
CPMG	Carr–Purcell–Meiboom–Gill
CSA	Chemical shift tensor
CT	Constant time
CTAB	Cetyl (hexadecyl) trimethyl ammonium bromide
DAC	Digital-to-analog converter (conversion)
dB	Decibel
dB_m	dB relative to 1 mW
DD	Dipole–dipole
DEPT	Distortionless enhancement by polarization transfer
DHPC	Dihexanoylphosphatidylcholine
DI NMR	Direct injection NMR
DMEM	Dulbecco's Modified Eagle Medium
DMPC	Dimyristoylphosphatidylcholine
DQF	Double-quantum filter
DSBSC	Double sideband band suppression carrier
DSS	2,2-Dimethyl-2-silapentane-5-sulfonic acid
ER	Estrogen receptor
F	Noise figure
F_1, F_2	Frequency domain of multidimensional experiments
FBS	Fetal bovine serum
f_d	Decoupling efficiency
Ft, FT	Fourier transformation

g	Gradient
G	Pulse field gradient strength
GND	Ground
HCN	Proton/carbon/nitrogen
HCX	Proton/carbon/heteronuclei
HMQC	Heteronuclear multiquantum correlation
HR MAS	High resolution magic-angle spinning
HSQC	Heteronuclear single quantum coherence
i	Current, imaginary unit
I	Spin quantum number, spin I, current
IF	Intermediate frequency
INEPT	Insensitive nuclei enhanced by polarization transfer
IPAP	In-phase anti-phase
j	Imaginary unit
J	Scalar (indirect, spin–spin) coupling constant (Hz)
$J(\omega)$	Spectral density function at frequency ω
JRES	J-resolved COSY
K_A	Association constant
K_D	Dissociation constant
L	Inductance
LC	Liquid crystal, inductor-capacitor
LED	Longitudinal eddy-current delay
LO	Local oscillator
LSB	Least significant bit
M	Magnetization
M_0	Equilibrium bulk (macroscopic) magnetization
MAS	Magic-angle spinning
MVA	Multivariate analysis
MSB	Most significant bit
MVDA	Multivariate data analysis
NMR	Nuclear magnetic resonance
NOE	Nuclear Overhauser enhancement
NOESY	Nuclear Overhauser spectroscopy
OPLS-DA	Orthogonal PLS-DA
p	Coherence order
P	Nuclear angular momentum, power levels of the signals (W)
PAS	Principal axis system
PB	Phosphate buffer
PBS	Phosphate-buffered saline
PCA	Principal component analysis
PEP	Preservation of equivalent pathway
PLS-DA	Partial least squares discriminant analysis
PSD	Phase sensitive detector
PT DI NMR	Push-through direct injection NMR

PURGE	Presaturation utilizing relaxation gradients and echoes
pw_{90}	90° Pulse length
Q	Nuclear quadrupole moment, quality factor, adiabatic factor, probe quality factor
Q^2	Predictability measurement of a statistical model
R	Resistance
R_1	Spin–lattice (longitudinal) relaxation rate
R^2	Quality measurement of a statistical model
R_2	Spin–spin (transverse) relaxation rate
RDC	Residual dipolar coupling
RF	Radio frequency
rMD	Restrained molecular dynamics
rmsd	Root-mean-square deviation
ROESY	Rotating frame Overhauser spectroscopy
RT	Real time
S	Saupe order matrix, spin S, generalized order parameter
S^2	Squared generalized order parameter
SAR	Structure–activity relationship
SC	Superconducting
SDS	Sodium dodecyl sulfate
se	Sensitivity enhancement
SL	Spin lock
SQ	Single quantum
SSB	Single sideband
ST	Sweep-tune
T/R	Transmitter/receiver
T_1	Spin–lattice (longitudinal) relaxation time
t_1, t_2	Time domain of multidimensional experiments
$T_{1\rho}$	T_1 of spin locked magnetization in rotating frame
T_2	Spin–spin (transverse) relaxation time
T_2^*	Effective T_2
$T_{2\rho}$	T_2 of spin locked magnetization in rotating frame
TCA	Tricarboxylic acid
TMS	Tetramethylsilane
TOCSY	Total correlation spectroscopy
TROSY	Transverse relaxation optimized spectroscopy
TSP	Sodium 3-(trimethylsilyl) propionate-2,2,3,3-d_4
V	Voltage
V_{pp}	Peak-to-peak voltage
V_{rms}	Root-mean-square amplitude of a signal (V)
VT	Variable temperature
w_0, w_1, w_2	Transition probabilities for zero-, single-, and double-quantum transitions
X	Heteronucleus

Z	Impedance, generalized resistor
Ω	Frequency offset, free precession frequency
Ξ	Frequency ratio of chemical shift reference
δ	Chemical shift (ppm)
γ_I	Gyromagnetic ratio of nucleus I
η	Nuclear Overhauser enhancement, filling factor of probe coil, asymmetric parameter of principal axis system, viscosity
λ	Wavelength, decoupling scaling factor
μ	Magnetic dipole moment, nuclear angular moment
μ_0	Permeability of vacuum
ν_D	Dipolar coupling constant
ρ	Density operator
σ	Chemical shift tensor, conductivity of sample, cross relaxation rate
σ_{dia}	Diamagnetic shielding
σ_{para}	Paramagnetic shielding
τ_c	Correlation time
τ_e	Effective correlation time
τ_m	Global correlation time
τ_s	Internal correlation time
ω	Angular frequency
ω_1	Frequency of B_1 field (rad/s), Larmor frequency in the rotating frame
ω_L	Larmor frequency (rad/s), carrier frequency
ω_R	Intermediate frequency (rad/s)

Index

2D (two-dimensional) NMR, 158–159
2QF-COSY, 164–166
3D (three-dimensional) NMR, 189–190
90°C pulse, 128, 130
   excitation null, 128, 132
   resonance offset, 129
96-well plate, 323, 324, 333, 334
180°C pulse, 128, 130
   catenation, 209
   composite, 136
   decoupling (*see* Decoupling, 180° pulse)
   excitation null, 128, 132
   imperfect, 136, 145
   nonresonant phase shift, 161
   pulsed field gradients, 145
   refocusing, 176
   resonance offset, 129

A

Absorptive line shape, 164
AB spin system, 29
Acquisition period, direct, 158–159
   indirect, 158–159
   initial delay, 155, 157, 176, 179, 194
      adjusting, 155, 157 (*see also* Evolution period)
ADC. *See* Analog-to-digital converter (ADC)
Adiabatic pulse, 139
Adiabatic relaxation contributions, 45
Adjoint operator, 56
Aggregation, 110, 111
Air bubble, 342, 372
Alanine
   chemical shifts, 24
Aliasing and folding, 18
Alignment media, 111–113

Alignment, spectral, 348, 350, 352, 362
Allowed transition, 4
Amide. *See also* Side chain
   chemical shifts, 24
Amide acid chemical shifts
   proteins, 24
   random coil, 253
Amide proton chemical shift
   dispersion, 217
Amide proton exchange, 110, 147, 148, 187, 191
Amino acid chemical shifts, 209
   13C-13C correlation (HCCH), 209–211
Amplifier rf, 69
Amplitude modulated pulse, 134
Amplitude rf, 69
Analog signal, 87
Analog-to-digital converter (ADC), 65, 87–90
Analysis, data, 150, 274, 275, 289, 306–308, 313, 342, 370
Angle phase, 96
Angle polar, 34
Angular momentum, 2, 3
Angular momentum quantum number, 2, 3, 11. *See also* Quantum number
Anisotropic chemical shift. *See* Chemical shift anisotropy (CSA)
Anisotropic diffusion. *See* Diffusion rotational anisotropic
Annealing simulated (SA); rMD, 249–251, 260
Antiphase lineshape, 164
Apodization, 151–154
   heteronuclear correlation, 177, 190
   multidimensional NMR, 154
   resolution enhancement, 151–152
Apodization function
   cosine bell, 153

Apodization function (*cont.*)
  exponential, 152
  Lorentzian-to-Gaussian, 153
  sine bell phase shifted transformation, 153
Aromatic resonance, 342, 360
Aromatic rings, 24
  local magnetic fields, 24
Aromatic spin systems
  chemical shifts, 24
Artifact suppression, 145
  phase cycling, 161
  pulsed field gradient, 161
Assignments resonance
  $13C^\beta$-$13C^\alpha$ correlations (CBCA), 188, 203
  ^{13}C-^{13}C correlations (HCCH), 188, 209–211
  heteronuclear-edited NMR, 188, 216
  homonuclear NMR, 217
  sequential, 187
  stereospecific, 259
  triple-resonance NMR, 217–218
Asymmetric peak, 336
Atomic coordinates, 251–253
Attenuation rf power, 70–71, 128
Audiofrequency filter. *See* Filter, audiofrequency
Audiofrequency signal, 1, 74, 75
Autocorrelation function, 40, 41
Autocorrelation or diagonal peak, 159, 164
Autolock, 372
Automation, 332–336, 373
Autoshimming, 121, 372
Average chemical shift, 25
Average dipolar coupling, 31, 40
Average scalar coupling, 26
Axial symmetry, 34
AX_n spin system, 30
AX spin system, 29

B

Backbone
  conformation, 264, 265
  fingerprint region, 216, 217
  spin system, 217
Backbone resonance assignment, 268
Backbone to side-chain correlations, 217, 219
Back-calculation, 264, 267
Balance mixer; BM, 71
Bandwidth, frequency, 16, 74, 97
Baseline correction, 150
Baseline distortions, 149, 151
Basis functions, 54

B_1 field. *See* rf magnetic field
Bicelles, 32, 111, 112
Binning, 341, 348, 350, 352, 353
Bins, 314, 317, 344, 348–350, 358, 362, 365, 373–376, 383, 385
Biofluid, 321, 324, 333, 361–370
Bloch equations
  derivation, 11
  free-procession, 11
  laboratory frame, 12
  limitations, 12
  relaxation, 12
  rotating frame, 11
Boltzmann distribution
  spin system, 5
Bras and kets, 55
Breast cancer cell, 370, 374
Broadband decoupling. *See* Decoupling, phase-modulated
Broadband isotropic mixing. *See* Isotropic mixing
Bulk angular momentum, 5

C

^{13}C–^{13}C correlations (HCCH), 209–211
^{13}Cα–^{13}Cβ decoupling, 207
^{13}Cβ–^{13}Cα correlations (CBCA), 207, 218, 264. *See also* CBCA(CO)NH
^{13}Cα and ^{13}CO
  decoupling, 128
  nonresonant phase shift, 161
  off resonance excitation, 132–133
  resonance frequency difference, 132
  rf field strength excitation, 132
^{13}C-glucose, 107, 108, 114, 187, 242, 377–381
^{13}C isotropic labeling, 104–108
Calibration, NOESY, 256
Cancer, 327, 352, 354, 356, 357, 359, 360, 370, 374, 376–379, 382–388
Capacitance, 78–80
Carrier frequency, 9, 69, 71–75
Carr-Purcell-Meiboom-Gill (CPMG), 293, 297, 300, 302–304, 336, 338–340, 342, 343
Cascades, pulse
CBCA(CO)NH, 192, 206–209
CBCANH, 188, 190, 203–206, 218, 219, 268
^{13}C chemical shift. *See* Chemical shifts, ^{13}C
Cell culture; tissue culture, 106, 107, 326–329, 333, 353, 358, 371

Cell lifter, 329, 330, 332, 371
Cell line, 106–108, 326–328, 330, 352, 370, 374, 382
Chemical exchange, 147, 229, 297
Chemical shift
  average, 25
  conformation-dependent, 253–254
  degeneracy, 263
    3D heteronuclear-edited NMR, 217, 218
    dispersion, 217
    2D NMR, 247
  evolution, product operators, 59
  isotropic, 23
  reference, 20–22
  secondary, 257
  structure, dependence, 253–254
  tensor, 23
  water, 21
Chemical shift anisotropy (CSA), 23, 24, 42, 46
  axial symmetry, 185
  ^{13}C, 186
  heteronuclear, 185
  ^{15}N, 186
  relaxation, 41–42
Chemical shift frequency range, 6–7
Chemical shift overlap. *See* Chemical shift, degeneracy
Chemical shifts, ^{13}C, 22, 24
  assignments, 187
  ^{13}Cα and ^{13}Cβ distinguishing, 205
  ^{13}Cα and ^{13}CO resonance difference between proteins, 128, 132
Chemical shifts, ^{1}H, 22, 24
  proteins, 24
  random coil, 24
  secondary structure dependence, 253–254
    resonance frequency difference, 24, 128, 132
Chenomx, 360, 361, 363
Cholesterol, 361, 383
CH$_n$, 203, 205–207, 211
Circuits
  open, 87
  parallel, 77–80
  parallel-series, 80, 81
  series, 77–79
  series-parallel, 80, 81
  short, 80, 81
Citrate, 323, 363–366, 379, 383–385, 387
Class, 256, 270, 330, 344, 345, 347–348, 352–361, 365, 367, 371, 374–376, 383–386

Class group, 348
Class label, 344, 356, 357, 378
Class separation; class discrimination, 355–361, 376, 383
Class spectrum, average, 375
Coherence
  antiphase, 52
  double quantum (DQ), 47
  multiple quantum, 178, 181, 183
  in phase, 51, 52
  single quantum, 173, 181, 185, 186
  zero quantum (ZQ), 37
Coherence, dephasing, 134, 142, 161, 162
  pulsed field gradient, 160–166
  relaxation, 45
  static magnetic field inhomogeneity, 45
Coherence, phase shift, 161, 162
  frequency discrimination, 159
  rf pulse, 161
Coherence level. *See* Coherence order
Coherence level diagram, 48, 161–163
Coherence order, 161–163
  change in, 161–163
  herteronuclear, 163
Coherence selection, 161–162
Coherence transfer, 162, 163, 165
  antiphase, 164
  continuous (CW) rf field, 136
  COSY-type, 163–164
  heteronuclear, 181–183
  HMQC-type, 181
  INEPT, 48
  in-phase, 47
    COSY-type, 164
    TOCSY-type, 166–167
  isotropic mixing, 137, 166
  magnetization transfer, 136–137, 166, 174, 175
  rf pulse, 47
  single quantum, 47
  through-bond, 167
  through-space, 167
  TOCSY-type, 166
  triple-resonance NMR, 187–187 (*see also* Triple resonance)
Coherence transfer pathway, 48
  selection, 161–162
    phase cycle, 161–162
    pulsed field gradient, 161–162
Coil, rf, 77, 78, 80–85
Coil, SC shim, 65. *See also* Cryogenic shims
  cryoshims, 67
Coil, shim, 69
Colon cancer, 385–388

Complete assignment, 189, 216–218
Complex Fourier transformation, 13–18
Composite pulse, 136–138
Composite pulse decoupling, 48, 136–138.
    See also Decoupling, phase-
    modulated
Computer-aided assignments, 263, 264
Concentration, sample, 104
Conformation, protein, 253, 254. *See also*
    Protein structure
Connectivity, 188, 198, 206, 216
Constant ranges, 27
Constant-time (CT) evolution period,
    179, 181
Constant-time (CT) HMQC, 181
Constraints, accuracy, 264
Constraints, experimental, 253–259
    chemical shifts, 253–254
    dihedral angle, J coupling, 254
    distance, NOE, 255–256
    hydrogen-bond, 251, 260, 265
    orientational RDC, 256–258
Constraints, short range, 251
Constraints geometry, 253, 262, 264
Constraints per residue, 263
Constraint violations, 251, 253, 263
Continuous (CW) rf field, 37
Continuous signal, 87, 121
Convolution, 15
Convolution difference low pass
    filter, 150
Convolution theorem, 15
Coordinate frame, 7
    laboratory, 7, 9
    rotating, 6–9
Coordinates, atomic, 251, 252
Correlated spectroscopy. *See* COSY
Correlation, heteronuclear. *See* Heteronuclear
    correlation
Correlations, side-chain to backbone
    heteronuclear, 173–176
Correlation time, 40, 291, 292, 295
    effective, 291, 307
    rotational, 291
Cosine bell apodization, 90º shifted sine
    apodization, 153
Cosine-modulated signal, 164
COSY, 163–164
    antiphase, 164
    lineshape, 164
    N type, 164
    phase twisted, 164
    P type, 164

scalar coupling constraints, measuring
    from, 164
self-cancellation of resonances in, 164
CPMG. *See* Carr-Purcell-Meiboom-Gill
    (CPMG)
Cross-correlation, 46, 47, 293,
    296–298, 304
Cross-peak, 47, 158, 159, 164, 187
    antiphase, 164
    line shape, 164
    overlap, 160, 166, 189, 209, 214
    phase twisted, 164
Cross-polarization, 136, 137, 166, 167.
    See also Isotropic mixing, TOCSY
Cross-relaxation, 36–39, 167–169, 185.
    See also Dipolar relaxation; NOE;
    NOESY; ROE; ROESY
    dipolar, 304
    laboratory frame, 38, 166
    NOE and ROE, compensating for, 147
    rotating frame, 38, 166
    through-space correlation by, 166
Cryogenic probe, 83, 92, 104, 110
Cryogenic shims, 67, 68
CSA. *See* Chemical shift anisotropy (CSA)
CTAB, 112
Culture medium, 327–329, 352, 353, 356, 358,
    371, 378, 379
Cutoff, filter, 20
CW rf field. *See* continuous (CW) rf field

D

Dataset, 19, 164, 175, 178, 179, 182–184, 196,
    201, 214, 232, 301, 302, 314,
    317–321, 344, 348, 349, 351, 355,
    356, 359, 368, 374, 385
dB. *See* Decibel (dB)
dBm, 72, 73
DC offset, 150
Dead delay, 177, 372
Decay constant, 14. *See also* Relaxation rate
    constants
Decibel (dB), 70, 71
Decoupling, constant-time evolution, 158, 174,
    176, 178
Decoupling, 180º pulse, 137, 173. *See also*
    Isotropic mixing
    HSQC, 174
    off resonance, 128
    phase-modulated, 120, 121
    proton, 53
    supercycle, 137

Decoupling, phase-modulated, 134, 137, 139
Degenerate resonances, 36, 187, 203, 206, 217, 218, 247, 264
Denaturation, 114
Density matrix, 54–58
 diagonal elements, 56
 off-diagonal elements, 50
Density matrix formalism, 54–58
Dephasing, coherence, 162
Deshielding, 25
Detection, phase-sensitive, 159, 164
Detection, proton, 144–146
Detection, quadrature, 76, 159
Detector, phase-sensitive, 71, 75
Deuterium, 23, 69, 71, 108, 123
Diagnosis, 361, 370
Diagonal peaks, 159, 164, 169
Diffusion, rotational, 297
 anisotropic, 23, 24, 26, 30, 111, 297
 correlation time, 291, 297
 isotropic, 298
 relaxation, 298
Diffusion, spin, 39, 147, 167, 168, 255, 256, 264
Digital resolution, 145, 154, 158, 164, 190
Digital signal processing, 91
Digitizer, 91
Dihedral angle, 28, 29, 251, 253–255, 260
 constraints, 255, 260, 262
 Karplus, curve, 28, 254, 255
2,2-Dimethyl-2-silapentane-5-sulfonic acid (DSS), 21–25, 124, 125
Dimethyl sulfoxide (DMSO), 21, 125, 129, 130, 138, 371
DI NMR. *See* Direct injection NMR (DI NMR)
DI NMR; flow NMR, 322, 333–335
Dipolar coupling, 30, 31
 constraints, 30
 distance dependence, 30
 heteronuclear, 34, 35
 orientational dependence, 30
Dipolar relaxation, 39
DIPSI-2, 137, 168
DIPSI-3, 128, 137
Dirac model, 26, 27
Dirac notation, 54
Direct injection NMR (DI NMR), 335
Dispersion, chemical shift, 217
Dispersive phase, 121
Dispersive signal, 126, 127, 164
Distance, lower bound, 250, 251, 256, 263

Distance, upper bound, 250, 251, 263
Distance constraints, 250, 255–256
Distance geometry, 249, 250
DMEM. *See* Dulbecco's Modified Eagle Medium (DMEM)
DMPC, 111–113
DMPC/DHPC bicelles, 111–113
DMSO. *See* Dimethyl sulfoxide (DMSO)
Dose-dependent, 374
Double quantum filter (DQF) coherence, 164–166
Double quantum transition, 37
Downfield, 25, 155, 198, 296, 342
DSS, 21–25, 124, 125
Dulbecco's Modified Eagle Medium (DMEM), 326, 327, 370
Duplexer, 92
Dynamical simulated annealing (SA), 250, 260
Dynamic range, 88, 90
Dynamics, internal, 291, 292

E
eBURP, 146, 148, 197
EDTA, 107–109, 323, 327
Effective correlation time, 291, 307, 308
Effective magnetic field, 12, 302
Eigenstate, 55
Eigenvalue, 54
Electromotive force (EMF), 84
Electron, 20, 23, 24, 26
Electronic shielding, 20, 26
Energy, 1–5
Energy levels, 29, 37, 38
Enhancement, NOE, 38–39
Ensemble of structures, 249, 250, 253, 267
Equation of motion, 252
Equilibrium, thermal, 5
 magnetization, 5, 12
Estrogen receptor (ER), 370, 374, 376
17α-ethynylestradiol (EE2), 370, 371, 374–377
Evolution, 158
Evolution period, 158–160
 constant time, 180, 181
 initial delay, 156, 157, 180, 194
Exchangeable protons, 147, 148. *See also* Amide proton exchange
Excitation null, 128, 130–133, 197
Excited state, 5, 40, 42
EXORCYCLE, 144
Expansion, 5, 57

Expectation value, 56
Exponential apodization, 151, 152
Expression system, 104–105
Expression vectors, 104–106
Extract, cell, tissue, 328–331
Extraction, 321, 324–326, 329, 331, 371
Extreme narrowing, 39, 45, 169

F

F_1, 158–160, 163
F_2, 158–160, 163
FBS. *See* Fetal bovine serum (FBS)
Fetal bovine serum (FBS), 326, 327, 340, 370
FFT (fast Fourier transformation), 13
FID (free-induction decay), 12–14, 19, 41, 45, 46, 52
Field gradient. *See* Pulse field gradient (PFG)
Filter, audiofrequency, 20, 75
Filling factor, 82, 84
Filter, 18, 19, 74, 75, 94, 96, 97
  analog, 18
  bandpass, 74
  cutoff, 20
  digital, 18–19
  isotope (*see* Isotope filter)
  solvent, 150
  transition band, 15, 16
  tunable, 75
Filter function. *See* Apodization function
Filtering, 98, 107, 150
Fingerprint region, 216, 217
First-order phase correction, 156–158
Flip angle, pulse, 57, 70, 71, 126, 127, 133, 137
Flip-back, 146, 148–149, 298
Flow cell, 322, 333–336
Flow NMR; DI NMR, 322, 333–335
Folding, 18, 19
Folic acid; forlate, 366, 369, 370
Forbidden transition, 37
Force constant, 252, 253, 261, 262
Fourier transform, 13–16, 41, 134
Fourier transform algorithms
  complex, 159–160
  fast (FFT), 14
  inverse, 44
  real, 160
  scaling first point, 157
Fourier transform theorems, 13–15
Free-induction decay, 11–14, 20, 41, 45, 52

Free-procession, 50, 57, 59
  product operator formalism, 59
Frequency-dependent phase error; First order phase correction, 154–157
Frequency dimensions, 158
Frequency discrimination, 164, 173. *See also* Quadrature detection
  hypercomplex (States), 160–161
  N/P selection for, 162–163
  PEP (preservation of equivalent pathways), 175, 177, 178
  pulsed field gradient, 141–145
  Redfield's method, 76
  time-proportional phase-incrementation (TPPI) method, 76, 159–161
  TPPI-States, 159–161
Frequency-independent phase error; Zero order phase correction, 154–157
Frequency labeling, 163, 174, 176
Full-width-at-half-height, 45

G

GARP, 128, 130, 137, 138
Gaussian lineshape, 153
Gaussian pulse, 16, 17, 134–135
Germinal protons, 27, 28
  scalar coupling constant, 27, 28
Global energy minimum, 251, 253
Global fold, 253, 261–263
Glucose, 107, 108, 114, 238, 242, 243, 311, 326, 327, 364, 366, 370, 376–379, 381–383, 388
Gradient echo, 148
Gradient-enhanced heteronuclear single quantum coherence (HSQC), 173–180
Gradient-enhanced HNCA, 191–198
Gradient, pulse field, 121, 141–145
Group; class, 348, 386
Gyromagnetic ratio, 3

H

Hamiltonian, 54, 56
HCCH-TOCSY, 192, 209–212
^{1}H chemical shift, 20–22, 24
Healthy individual, 361–366, 385, 386, 388
Heat transfer processes, 66
α-helix. *See* Secondary structure
Heteronuclear coherence order, 47, 144
Heteronuclear correlation, 175, 184

Heteronuclear-edited NMR, 233–235
Heteronuclear-edited NOESY, 215, 233, 234
Heteronuclear-edited TOCSY, 216–217
Heteronuclear multiple quantum coherence (HMQC), 173, 174, 180–182
Heteronuclear scalar coupling constants, 28, 181, 187
Heteronuclear single quantum coherence (HSQC), 173–180
High-resolution 3D structure, 254
High resolution MAS (HR MAS), 341–342, 410
His-tag, 104, 105, 115
HMQC (heteronuclear multiple quantum coherence), 173, 174, 180–182
HMQC-type coherence transfer, 181
HNCA, 188, 189, 191–196
HN(CO)CA, 188, 192, 198–201
HNCO, 189, 191–196
HN(CA)CO, 188, 189, 192, 201–206
$^1H^N$-$^1H^\alpha$ fingerprint region, 216, 217
HOHAHA; TOCSY, 166–167
Homogeneity, magnetic field, 65, 67, 69, 91, 118, 120, 121, 147
Homonuclear chemical shifts, 25
Homonuclear Hartman-Hahn, 169
Homonuclear NOE enhancement, 39
Homonuclear scalar coupling, 26, 27
Homonuclear spin, 51, 59
Hotelling's T2 ellipse, 314, 373
HR MAS. *See* High resolution MAS (HR MAS)
HSQC (Heteronuclear single quantum coherence), 173–180
HSQC-NOESY, 214–216
HSQC-TOCSY, 217
Hydrogen bond, 251, 253, 260
Hypercomplex (States) frequency discrimination, 159–161

I
IF; Intermediate frequency, 67, 69, 71–73, 75, 96, 97
Impedance, 72, 73, 77–83, 85, 86, 92, 99, 118, 119
Imperfect 180° pulse, 144, 145, 212, 234
Indirect detection, 156, 159–160
Indirect dimension, 145, 151, 156, 159–160, 164, 190, 191, 302
Indirect evolution period, 159
Inductance, 77–79, 82
Induction, IPTG, 107

Inductor-capacitor (LC), 77–81
INEPT (insensitive nuclei enhanced by polarization transfer), 48, 53, 174–178, 183, 186, 191
Inhomogeneity, rf pulse; magnetic field homogeneity, rf, 166
Injection, 334–336
In-phase coherence transfer, 40, 47
In-phase lineshape, 164, 175
Instability, instrument, 141
Interleaved acquisition, 232
Intermediate frequency; IF, 71–73, 75, 97
Intermolecular interaction, 41
Internuclear distance vector, 31–36
Internuclear interaction, 40
Interresidue connectivity, 187
Intracellular, 226, 328, 331, 348, 353—359, 371, 376
Inversion, magnetization, 36
Inversion recovery, 58
IPTG, 106
Isolated spin, 50, 58
Isotope filter, 235–238
Isotopic labeling, 103, 110
Isotopic mixing sequences, 136–137
Isotropic mixing, 136–138, 166, 169, 209
Isotropic mixing, TOCSY, 136–137

J
J coupling, 26–36, 166, 176
J-resolved (JRES), 336, 338, 341, 351, 385
Jump-return, 146, 149

K
Karplus curve, 29, 255
Kronecker delta function, 33

L
Labels, isotopic Labeling, isotopic, 103, 110
Laboratory frame, 3, 6, 7, 9, 12
Lactate; lactic acid, 327, 360, 363–366, 377, 378, 381–383, 387
Larmor frequency, 1, 4, 6, 7
LB media, 108
Linear prediction, 151
    initial data points, 151
    mirror image, 151
Linear regression, 376
Line-broadening, 99, 152

Lineshape, 1, 14, 15, 47, 84, 92, 109, 112, 120
  absorptive, 160, 177
  dispersive, 156
  Gaussian, 134
  Lorentzian, 14, 153
Linewidth, 23, 45, 69, 85, 92, 111, 152, 184, 185
  Gaussian, 153
  Lorentzian, 153
Liouville-von Neumann equation, 56
Lipari-Szabo formalism, 291, 295
Lipid, 32, 112, 339, 340, 342, 364, 365, 376, 379, 384, 385, 387
Lipophilic, 321, 324–326, 336, 343, 371, 383, 384
Lipoprotein, 363, 364
Liquid crystal, 31, 32, 34
Liquid crystalline media, 111, 112
Liquid helium vessel, 66, 67
Liquid NH_3, 21, 22, 24
Liquid nitrogen vessel, 67
Live cell, 377–382
Loadings, 314–317, 319, 320, 324–326, 343–348, 355–363, 365, 372, 374, 383, 385, 386
Local error function, 295, 307
Local magnetic fields, 20, 24, 26
Local minimum, 252, 253
Local motion, 253, 307
Local oscillator; LO, 71, 97
Lock, field frequency, 21, 22, 43, 69, 123, 125
Log transformation, 319, 320
Longitudinal eddy-current delay; LED, 228
Longitudinal magnetization, 84, 85, 144, 145, 305, 306
Longitudinal relaxation, 40, 298, 302
Long range NOE, 216, 255
Lorentzian lineshape, 14
Lowering operator, 52
Low pass filter, 97, 150
Lysis buffer, 108, 115

M

Magic T, 92
Magnet, superconducting, 66, 67, 396
Magnetic field
  homogeneity, 67, 68, 91, 117, 120, 147
  inhomogeneity, 141
Magnetic field mapping, 68
Magnetic field, shimming, 91, 120
Magnetic moment, 2
Magnetic quantum number, 2, 3
Magnetization, 5–7, 9–11
Magnetization transfer, 136, 166, 167
Magnitude mode, 341
Mapping, magnetic field, 68
Matching, 77, 80–82, 118, 119
Matrix, 33, 34, 48, 50, 54–58, 185, 249, 250
Medium, 31, 36, 38, 107, 110, 112–114, 169, 170, 256–258, 326–330, 352–358, 370, 378, 381
MetaboAnalyst, 344, 349, 351, 373
Metabolic flux, 375, 380
Metabolic pathway, 223, 238–243, 366, 369, 382, 388
Metabolite, 238, 242, 311–313, 317–319, 321–326, 328, 331, 336, 337, 339, 341–343, 346, 349, 353–355, 358–361, 365–366, 369–371, 374–376, 378–381, 383–388
Methanol, temperature calibration, 122, 124
Methylene group, 203, 205, 207, 408
Microcarrier, 326–328, 378
Micro NMR tube, 115
Mixer, 71, 72, 97
Mixing period, 137, 159, 166, 209
Mixing time, 137, 166–169, 211, 256
MLEV-17, 136, 137, 166
Mnova, 344, 350–352, 368, 373
Model; statistical, 313, 317, 370, 373
Molecular dynamics, restrained; rMD, 187, 249, 251–253, 261, 262
Molecular frame, 32–34, 291, 295
Multidimensional NMR, 53, 84, 107, 137, 138, 161
  2D (two-dimensional), 158–169
  3D (three-dimensional), 118, 189–190
  4D (four-dimensional), 154
Multipilcity. *See* CH_n
Multiple quantum coherence, 48, 178, 181, 183, 194
Multiplet, 46, 181, 184
Multivariate analysis (MVA), 313, 344, 345, 349, 351, 352, 361, 383
Multivariate data analysis (MVDA), 313
MVA. *See* Multivariate analysis (MVA)
MVDA. *See* Multivariate data analysis (MVDA)

N

^{15}N reference, 125
^{15}N relaxation, 46
Natural abundance, 103, 174

Index

Net magnetization transfer, 194
NMR spectrometer, 1, 12, 14, 65, 66, 71, 72, 87
NOE, 36–39. *See also* ROE
  build up curve, 168
  difference, 232, 263, 265
  distance restraint, 251, 260
  enhancement, 38, 39, 169
  heteronuclear, 48, 304–306
  rate constant, 305
  steady-state, 36, 119, 127, 169, 298, 305
  transferred, 229–231
  transient, 304
NOE ROESY, 168
NOESY, 167–169, 187, 216
NOESY-HSQC, 215
Noise figure, 98, 99
Normalization, 265, 344–352, 373
N/P selection, 164
N-type coherence, 165
Nuclear Overhauser effect. *See* NOE; ROE
Nuclear Overhauser effect spectroscopy. *See* NOESY; ROESY
Nuclei, active, 2, 257
Nuclei, spin, 2, 7, 20
  coupled, 26, 28, 30
Nyquist frequency, 18
Nyquis theorem, 18

O

Observable coherence, 47, 181, 203, 206, 237
Observable magnetization, 5, 58, 178, 183, 194, 195
Observations, 39, 121, 254, 298, 314–317, 345, 354, 373, 377, 393
Off-resonance pulse, 117, 132–133, 197
Offset, 117, 121, 123, 126–130
  dc, 150
  resonance, 304
Omics, 311, 312, 326
One-dimensional (1D), 158, 190, 197
One-pulse experiment, 13, 48, 57, 126, 129, 135, 142, 147
On-resonance, 8, 9, 22, 57, 82, 121, 125, 129, 132–134, 137, 149, 150
Operator, 48–54
OPLS-DA. *See* Orthogonal PLS-DA (OPLS-DA)
Order parameter, 258, 291–295
Orientation of alignment tensor, 258
Orientation of angular moment, 2

Orientation of CSA tensor, 296
Orientation of internuclear vector, 31, 33, 35, 291, 298
Orthogonal PLS-DA (OPLS-DA), 313, 344–346, 351, 355, 357, 360–366
OS. *See* Overall survival (OS)
Oscillating fields, 8, 9, 65, 84, 87, 92–96, 118
Oscilloscope, 72, 87, 92–96
Out-and-back, 188, 191, 199, 201
Outlier, statistical, 314
Overall survival (OS), 382, 385–388
Oversampling, 19

P

Paramagnetic shielding, 24
Pareto scaling, 318–320, 349, 353
Partial least squares discriminant analysis (PLS-DA), 313, 344, 345, 349, 351, 355–361, 376–377
Parts per million (ppm), 21, 22, 24
Pascal triangle, 30
Patient, 242, 243, 311, 361–363, 365–366, 369, 370, 382, 385–388
PB. *See* Phosphate buffer (PB)
PCA. *See* Principal component analysis (PCA)
PCA; principal component analysis
Peak-to-peak voltage, 94–96
PEP (preservation of equivalent pathways), 175, 177–179, 191
Peptide, 105, 110, 111, 166, 198, 233, 234, 260, 268
Permutation, 357, 376, 378, 386
Perturbation, 36, 133
PFG. *See* Pulse field gradient (PFG)
Phase, NMR spectrum, 19
Phase correction, 154–157
  first-order, 155–156
  zero-order, 155–156
Phase cycle, 179
  EXORCYCLE, 144
  frequency discrimination, 164, 175
Phase error, 134, 149
Phase-modulated pulse, 134, 139
Phase modulated sequence, 137
Phase sensitive, 159, 160, 164, 177, 179
Phase sensitive detector; PSD, 71, 75
Phasing,. *See* Phase correction
Phosphate buffer (PB), 109, 115, 321–324, 326, 327, 342, 362, 372, 385
Plasma, 323–324, 339, 361–366, 377

Plot
    loadings, 343–348, 355–358, 360–362, 365, 374, 383, 386
    scores, 313, 314, 316, 341, 343, 345, 347–349, 353, 355–361, 374, 376, 377, 383–386
    Y-predicted, 359
PLS-DA. *See* Partial least squares discriminant analysis (PLS-DA)
PLS Toolbox, 344, 347, 351, 373
Poisson scaling, 318
Polynomial, 32, 150, 151, 297, 350, 373
Population, 1, 3–5, 36–38, 41, 45, 49, 54, 69, 147
    Boltzmann distribution, 5, 41, 69
    inversion, 36, 137, 139
    saturation, 36–38
Post-acquisition solvent suppression. *See* Solvent suppression filter
Power, pulse, 6, 10, 69, 70, 78, 92, 125–128
Preacquisition delay; predelay, 146, 148
Preamplifier, 74, 75, 83, 87, 99, 100
Precession, 2–4, 7, 11, 12, 20, 24, 50, 57, 59, 61, 141, 143
Precision, three dimensional structures, 233, 247, 249
Predictability measurement of a statistical model (Q^2), 313, 319–321, 344, 353–359, 365, 367, 376, 378
Prediction, 14, 150, 151, 167, 169, 177, 190, 253, 254, 302, 366, 385, 395, 397
Prediction accuracy, 383
Preparation period, 158, 215, 298
Presaturation, 127, 134, 146–147
Presaturation utilizing relaxation gradients and echoes (PURGE), 336–340
Preservation of equivalent pathway. *See* PEP (preservation of equivalent pathways)
Primary sequence. *See* Protein sequence
Principal axis, 34, 36, 296
Principal component, 35, 314, 315, 317, 373
Principal component analysis (PCA), 313–317, 320, 341, 343–345, 347, 349, 351, 353–357, 363, 367, 373–377, 383, 384
Probability, 6, 34, 375
Probe, 22, 77–85
    impedance, 72, 73, 77–87, 93, 99, 118
    probe coil, 77–87
    quality factor, 77–85
    resonance frequency, 77–85
    tuning and matching, 77, 81, 118, 119

Probe volume, 334–336
Product operator formalism, 48–54
Profile, metabolic, 311, 329, 334, 344, 361, 365, 370, 374–376, 382, 383, 385
Prostate cancer, 383, 385
Protein
    chemical shift, 25
    denatured, 111, 115
    expression, 106–107
    isotopic labeling, 103, 110
    sequence, 256, 260
    structure, 247–268
Proton detection. *See* Indirect detection
PT DI NMR. *See* Push-through direct injection NMR (PT DI NMR)
P-type signal, 163, 164
Pulse
    coherence order, 161–163
    composite, 136–138
    flip angle; pulse angle, 57, 70, 125–127, 133
    length; pulse width, 10, 69, 372, 404
    off-resonance, 59, 117, 132–135
    selective (*see* Selective pulses)
    shaped (*see* Selective pulses)
    spin lock (*see* Spin lock pulses)
    strength; pulse power, 6, 10, 69, 70, 92, 129–135, 147, 149
Pulsed field gradient profile, 117
Pulse field gradient (PFG), 120, 141–145
    coherence selection, 161–162
    dephasing coherence, 162
    frequency discrimination, 164
    shape factor, 162
    solvent suppression using, 141, 145–146, 175
    strength, 142, 144, 228, 229, 243
    z-axis, 92, 194
Pulse sequences
    heteronuclear
        HCCH-TOCSY, 189, 209–214
        HMQC, 180–182
        HSQC, 180–182
        HSQC-TOCSY, 189
        NOESY-HSQC, 214–216
    homonuclear
        COSY, 163–164
        DQF-COSY, 164–166
        flip-back, 146, 148–149, 177
        jump-return, 146, 149
        NOESY, 167–169, 187, 216
        ROESY, 167–169, 187, 216
        TOCSY, 166–167
        WATERGATE, 146–148, 177

Index 427

triple-resonance
  CBCANH, 188, 192, 203–206
  CBCA(CO)NH, 188, 192, 206–209
  HNCA, 188, 191–198
  HN(CO)CA, 188, 198–201
  HNCO, 188, 191–198
  HN(CA)CO, 188, 201–202
Pure absorption, 166, 179
Pure phase, 214, 301, 303
PURGE. *See* Presaturation utilizing relaxation gradients and echoes (PURGE)
Purity, 109, 115, 322
Push-through direct injection NMR (PT DI NMR), 335
Push volume, 335, 336
Pyruvate; pyruvic acid, 326, 360, 364–366, 380, 383, 387

Q

Quadrature detection, 76, 159–161
Quadrature image, 92
Quadrupolar coupling, 296
Quadrupolar relaxation, 42, 296
Quadrupole interaction, 290, 296
Quadrupole moment, 296
Quality factor, 77, 79, 264, 267
Quality measurement of a statistical model (R^2), 79, 313, 319–321, 344, 353–357, 359, 365, 376, 378
Quantum number, 2, 3
Quarter-wave length cable; $1/4$ wave length cable, 85–87
Quench (molecular biology), 324–326, 328–332, 370–371

R

R_1. *See* Spin-lattice relaxation
$R_{1\rho}$; R1 in rotating frame, 303, 304
R_2. *See* Spin-lattice relaxation; Spin-spin relaxation
Radiation damping, 77, 84, 85, 146–148
Radiofrequency magnetic field. *See* rf magnetic field
Raising operator, 52
Random coil chemical shifts, 24, 253
Rate constants, 290, 292
Real Fourier transformation, 76, 160
Real time (RT), 19, 20, 85, 97, 121, 198, 377–382,
Receiver, 10, 12, 65, 74–77
Receiver gating time, 155

Rectangle function, 338
Rectangular selective pulse, 146–148, 301
Recycle delay, 299, 301
Redfield's method for frequency discrimination, 76
Reference, chemical shift, 20–22
Reference frame
  laboratory, 7, 9
  molecular, 32–34, 291, 295
  rotating, 6–10
Reference frequency, 21–23, 69
Refinement, 3D structures, 248, 253, 260, 261, 265, 267
Reflection bridge, 92–93
Refocused INEPT, 53, 298
Refocusing, gradient, 142, 162, 164, 174, 197
Refocusing, 180° pulses, 144, 157, 193, 194, 212
Refocusing period, IPAP, 183
Relaxation
  cross-correlation, 46, 184, 293, 296, 297, 301, 304
  cross-relaxation, 38, 39, 46
  Lipari-Szabo formalism (*see* Lipari-Szabo formalism)
  longitudinal (*see* Spin-lattice relaxation)
  rate constants, tables of, 290, 292
  rotating frame, 169, 303
  spin-lattice, 37, 39–44, 290, 292, 294, 296, 297, 303, 306
  spin-spin, 40, 44–47, 290, 292, 294, 296, 297, 303, 306
  transverse (*see* Spin-spin relaxation)
Relaxation measurements
  inversion recovery, 44, 298–302
  T2 and T1ρ, 302–304
Relaxation mechanisms, 39–47
Relaxation rate constants, 290, 291
Relaxation theory, 39–47, 290–298
Reorientation, molecular. *See* Diffusion, rotational
Replicates, 371
Resistance, 70, 77, 80–82, 84
Resolution
  digital, 134, 145, 154, 158, 164, 190
  enhancement, 151, 152
Resonance degeneracy, 187, 203, 206, 217, 218
Resonance dispersion, 217
Resonance frequency, 4, 6, 17, 20, 22, 25, 26, 37, 77, 79–82, 84
Resonance offset, 304
Restrained molecular dynamics; rMD, 249–251, 260–262, 264

Reverse refocused INEPT, 298
R factor, 267
rf coil, 77–85
rf magnetic field, 6–13
ROE, 167. *See also* Cross-relaxation; NOE ROESY
ROESY, 167
Root-mean-square-deviation (rmsd), 264, 266, 275
Root-mean-square noise, 99
Root-mean-square voltage, 95
Rotating frame transformation, 6–8
Rotating reference frame, 6–8
Rotational correlation time. *See* Diffusion, rotational correlation time
Rotational diffusion, 40, 297
RT. *See* Real time (RT)

S
Sample preparation, 103–115
Samples, temperature, 5
Sampling delay, 156, 157
Sampling interval, 18
Sampling rate, 18, 76
Sampling theorem; Nyguist theorem, 18–20
SAR; Structure activity relationship, 224–227
Saturation, 36, 37, 39, 121, 127, 146, 147
Saturation transfer, 147, 148, 175, 177, 179, 186, 232–233
Scalar coupling, 26–29, 181, 187
Scalar coupling constants
 heteronuclear, 28, 181, 187
 homonuclear, 27
Scaling, 15, 35, 56, 137, 184, 313, 317–320, 343, 345, 349, 351, 353
Scaling, interferogram, first point, 157
Scaling factor, 56, 137, 184, 317
Scaling factor, decoupling, 137
Scores, 267, 313–317, 319, 320, 341, 343–349, 353–360, 363, 365, 374, 376, 383–386
Secondary chemical shifts, 257
Secondary structure, 217, 248, 253, 255, 258–259
 chemical shifts, 253–254
 determination, 254, 256, 258
 $^1H^N$-$^1H^\alpha$ scalar coupling for, 255
 ^{15}N-$^{13}C^\alpha$ scalar coupling for, 255
 NOE, 255, 257
SEDUCE, 128, 130, 137, 197
Segment, 285, 338, 341, 348
Selective decoupling, 137

Selective pulses, 133–135, 146
Sensitivity, 5, 53, 69, 72, 75, 82, 83, 89, 92, 97
 cryogenic probe, 77, 83, 84, 92
 heteronuclear NMR, 53, 174
 high field, 5, 69, 110
 noise; signal-to-noise (S/N) ratio, 19, 69, 74, 83, 92, 98–100, 147, 152, 154, 174
 probe, 82, 83, 92
Sensitivity-enhancement. *See* PEP (preservation of equivalent pathways)
Sequence specific assignment, 187, 217–218
Sequential assignment strategy, 187, 188, 217–218
Sequential NOE, 258
Serum, 323–324, 326, 339, 340, 385–388
Shaped pulses, 133–135
β-Sheet. *See also* Secondary structure
 shielding, chemical, 42
 shielding, diamagnetic, 24
 shielding, magnetic, 290
 shielding, paramagnetic, 24
Shielding constant, 20, 23, 24
Shielding electronic, 20, 26
Shielding tensor, 23
Shim assembly; shim coil, 21, 67, 69, 87, 91
Shimming, 21, 69, 91, 120–123
Side-chain, 113, 166, 167, 173, 187–189, 203, 209–214, 216, 218, 268, 307, 360
Side-chain spin systems, 218
Signal
 amplitude, 45, 70, 73, 89, 152, 300
 analog, 18, 87
 antiphase, 53
 audiofrequency, 1, 74
 background, 232, 242
 complex, 48, 163
 cosine-modulated, 164
 degeneracy, 247, 259
 deuterium lock, 123, 129
 dispersive, 126, 127, 135
 distortion, 75, 85
 double sideband, 71
 IF, 71, 75
 intensity, 4, 44, 126, 228
 N-type, 163–165
 overlap, 184, 214, 240, 302
 P-type, 163–165
 quadrature frequency, 72
 reference, 88
 resolved, 240, 259
 RF, 69, 74, 75, 95

sine, 74
sinusoidal, 73, 94, 96
solvent/solute, 85, 126, 127, 146, 147, 150
time domain, 13, 17, 19, 43, 155
waveform, 74
Signal-to-noise (S/N) ratio, 19, 69, 92, 98, 147
   field gradient, 141, 179
   heteronuclear NMR, 53, 173, 174
   PEP, 175, 177, 178
SIMCA-P, 344, 345, 351, 353, 373
Similarity theorem; scaling theorem, 15, 134, 152
Simulated annealing (SA); rMD, 250–253, 260–262
Sinc pulse, 134, 142, 144, 146, 149
Sine bell apodization, 153, 154
Single quantum coherence. See Coherence, single quantum
Single quantum relaxation. See Spin-spin relaxation
Single sideband (SSB), 71
Sinusoidal signal, 73, 94, 96
Slow tumbling, 181, 185
Sodium 3-(trimethylsilyl) propionate-2,2,3,3-d_4 (TSP), 322–324, 326, 337, 342, 368, 372, 373, 383, 385
Solvent exchange. See Amide proton exchange; Chemical exchange
Solvent suppression, 85, 134, 136, 141
   flip-back, 146, 148–149, 298
   jump-return, 146, 149
   postacquisition, 150–151
   presaturation, 127, 134, 146–148
   pulsed field gradient, 146, 148
   sample concentration, 145
   saturation transfer, 147, 148, 175, 177, 179, 232–233
   selective pulses, 133–135
Solvent suppression filter, 150–151, 190
Spectral alignment, 348, 350, 352, 362
Spectral density function, 40–41, 185, 290–294, 296, 297
   autocorrelation, for, 40
   cross-correlation, for, 46, 184, 293, 296–298, 304
   Lipari-Szabo, 291, 295
Spectrometer, 65–85
Spectrum analyzer, 96–98
Spectrum, frequency-domain, 15–18, 41, 45, 96, 134, 154, 159
Spin diffusion, 39, 147, 167, 169, 256, 264
Spin diffusion limit, 39, 169
Spin echo, 45, 48, 50–52

Spin echo period, 338, 339
Spin-lattice relaxation, 37–42, 290–297, 303, 306
   Bloch formulation, 42
   CSA, 42
   dipolar, 42
   ^{1}H, 42
   heteronuclear, 290–297
   inversion recovery, 42
   quadrupolar, 42
   scalar relaxation, 42
Spin lock filter, 231, 233
Spin lock pulses, 127, 128, 136, 166, 231, 232
   ROESY, 167–169
   saturation transfer, 232–233
   T_2 measurement, 302–304
   transferred NOE, 229–231
Spin lock pulses, TOCSY, 127
Spinning rate, 342
Spin-1/2 nuclei, 2–5
Spin quantum number, 2, 3
Spin-spin coupling. See Scalar coupling
Spin-spin relaxation, 40, 44–45, 47, 290, 292, 294, 296, 297, 303, 306, 407
   Bloch formulation, 44
   CPMG, 300, 302, 304
   CSA, 46, 47, 290, 291
   dipolar, 46
   heteronuclear, 291–297
   linewidth, 45
   magnetic field inhomogeneity, 45
   spin echo, 45
Spin system assignments, 185–189, 216–218
Spin systems
   AB, 29
   AX, 29
   AX_n, 30
   two-spin (see Two-spin system)
Spontaneous emission, 40
Stability, 21, 69, 104, 108, 112, 125, 141, 268
States (hypercomplex) frequency discrimination, 159–161
States-TPPI, 159–161
Static magnetic field, 2–6, 9, 11, 25, 39, 43, 65, 69, 120, 293, 307
Statistical significance, 356, 366, 375
Steady-state NOE, 36, 119, 127, 129, 169
Stereospecific assignments, 259
Straight-through pulse sequence. See CBCA (CO)NH, CBCANH
Stroke, 361, 364–366, 369, 370

Structural constraints, 253–259
Structure determination, 247–268
   accuracy, 256, 264, 267
   back-calculation, 264, 267
   distance geometry, 249–251
   global fold, 253, 261–263
   precision, 249, 256, 266
   restrained molecular dynamics, 249, 251–253, 261–263
   secondary structure, 253, 255–259
   simulated annealing (SA), 249, 251–253, 261–263
Student's t-test, 365, 375
Subset, 189, 359, 385
Supercycle, 137
Superimposement, spectral, 373
Suppression, solvent. *See* Solvent suppression

T
T_1, 40–42, 291–294, 303
t_1, 158–160
t_2, 158–160, 163
T^*_2, 45
T_2 delay, 342
T_2; Spin-spin relaxation, 40, 44–47, 290, 292, 294, 296, 297, 303, 306
TCA cycle, 366, 378–383
Temperature, 5, 21, 68, 82, 99, 106, 112, 119, 122
   calibration, 123–124
Tensor, chemical shift, 23
Tertiary structure, 217, 248, 253, 255, 259–267
Test set, 359, 361, 365, 385
Tetramethyl silane (TMS), 21, 22, 124
Three-dimensional structure, protein, 247–268
Through-bond correlation, 167
Through-space correlation, 167
Time-proportional phase-incrementation (TPPI) frequency discrimination, 76, 159–161
Tissue, 311, 321, 324–326, 341–343, 382–385
Tissue culture; cell culture, 106, 107, 326–329, 333, 353, 358, 371
TMS. *See* Tetramethyl silane (TMS)
TOCSY, heteronuclear edited, 217, 218
TOCSY (total correlation spectroscopy); HOHAHA, 166–167
   coherence transfer, 166
   DIPSI-2, 137, 168
   DIPSI-3, 128, 137
   Hartmann-Hahn matching, 169
   isotropic mixing, 136–137
   mixing time, 166
   MLEV17, 136, 137, 166
   spin lock, 166
   trim pulses, 167
TPPI. *See* Time-proportional phase-incrementation (TPPI) frequency discrimination
Trace element solution, 107
Training set, 359–361, 365, 366, 385, 386
Transferred NOE, 229–231
Transient NOE, 304
Transmitter, RF, 69–74
Transverse magnetization, 10, 12, 39, 43, 47, 51, 84, 134, 141, 144, 145, 207
Transverse relaxation. *See* Spin-spin relaxation
Treatment, 48, 82, 114, 161, 181, 184, 256, 352, 358, 370–371, 373–377, 383, 385
Trim pulse. *See* Spin lock pulses, TOCSY
Triple resonance, 71, 84, 92, 101, 118, 132, 177, 184, 187–190
Triple-resonance, NMR, 187–189
Tris-HCl, Tris buffer, 108, 114
Trypsin, 109, 327–329
Tuning and matching, 77, 80, 81, 118, 119
Two-spin system
   AB, 29
   AX, 29
   coupled, 51–52, 59–61
   energy diagram, 37
   energy state, 1–5, 7, 27, 37
   product operator, 50, 59, 60
   relaxation matrix, 185, 264

U
Unfolding, 109, 248
Uniformly labeled, 107, 190, 235
Unit variance; UV scaling, 318–320
Unsupervised multivariate, 374
Upfield, 124, 155, 166, 177, 232, 296, 342
Urea, ^{15}N, 21, 125, 129, 130, 138
Urine, 311, 322–324, 337, 361–366

V
Vacuum, 66, 67, 83, 290, 322, 324–326, 329, 330, 332, 371, 372
Validation, 267, 356, 357, 359, 360, 365, 366, 376, 378, 385
Validation set, 385, 386

Valine
  chemical shift, 24
  diastereotopic protons, 259
Variables, 13, 32, 33, 83, 93, 94, 122–124, 135, 158, 187, 228, 238, 249, 256, 258, 314, 316–320, 331, 344–345, 348, 349, 355, 356, 358, 373, 383
Variance, 313, 314, 317, 318, 320, 343, 371
Variation, 157, 314, 333, 344, 349, 353, 355, 359, 374
Vicinal protons, 28
Viscosity, 42, 227
Voltage, 70, 72–74, 78, 90, 93, 96, 119

W
WALTZ-16, 128, 130, 136–138, 166, 306
Warburg effect, 376, 377
Wash solvent, 334, 335
WATERGATE, 146
Water radiation damping; radiation damping, 77, 84, 85, 146
Water suppression. *See* Solvent Suppression

Wavefunction; eigenstate, 54
Wave length, 85–87
Weak coupling, 26, 29, 201
Window function. *See* Apodization Function
WURST, 140, 238

X
X variable, 358

Y
Y variable, 344, 345, 356, 358

Z
Zeeman
  energy levels, 4–5, 37
  transitions, 5
Zero filling, 154
Zero-order phase correction, 372
Zero quantum (ZQ) coherence, 47, 236
Zero quantum transition, 37
z gradient, 120, 121, 142–144

If you have any concerns about our products,
you can contact us at
ProductSafety@springernature.com

In case Publisher is established outside the EU,
the EU authorized representative is:
Springer Nature Customer Service GmbH
Europaplatz 3, 69115 Heidelberg, Germany

Printed by LION Printer GmbH
in Hamburg, Germany

MIX
Papier aus verantwortungsvollen Quellen
Paper from responsible sources
FSC® C105338

If you have any concerns about our products,
you can contact us on
ProductSafety@springernature.com

In case Publisher is established outside the EU,
the EU authorized representative is:
**Springer Nature Customer Service Center GmbH
Europaplatz 3, 69115 Heidelberg, Germany**

Printed by Libri Plureos GmbH
in Hamburg, Germany